Nonlinear Analysis and Optimization

CONTEMPORARY MATHEMATICS

659

Nonlinear Analysis and Optimization

IMU/AMS Special Session on
Nonlinear Analysis and Optimization
June 16–19, 2014
Bar-Ilan University and Tel Aviv University, Tel Aviv, Israel
Workshop on
Nonlinear Analysis and Optimization
June 12, 2014
Technion–Israel Institute of Technology, Haifa, Israel

Boris S. Mordukhovich
Simeon Reich
Alexander J. Zaslavski
Editors

American Mathematical Society
Providence, Rhode Island

2010 *Mathematics Subject Classification.* Primary 14-xx, 20-xx, 26-xx, 28-xx, 30-xx, 31-xx, 32-xx, 37-xx, 42-xx, 46-xx, 47-xx, 49-xx, 54-xx, 90-xx, 91-xx, 93-xx.

Library of Congress Cataloging-in-Publication Data

Names: Mordukhovich, B. Sh. (Boris Sholimovich), editor. | Reich, Simeon, editor. | Zaslavski, Alexander J., editor.
Title: Nonlinear analysis and optimization : Workshop on Nonlinear Analysis and Optimization, June 12, 2014, Technion–Israel Institute of Technology, Haifa, Israel : IMU/AMS Special Session on Nonlinear Analysis and Optimization, June 16-19, 2014, Bar-Ilan University and Tel Aviv Universities, Ramat-Gan and Tel-Aviv, Israel / Boris S. Mordukhovich, Simeon Reich, Alexander J. Zaslavski, editors.
Description: Providence, Rhode Island : American Mathematical Society, [2016] | Series: Contemporary mathematics ; volume 659 | Includes bibliographical references.
Identifiers: LCCN 2015037183 | ISBN 9781470417369 (alk. paper)
Subjects: LCSH: Nonlinear functional analysis – Congresses. | Mathematical optimization – Congresses. | AMS: Algebraic geometry. msc | Group theory and generalizations. msc | Functions of a complex variable. msc | Partial differential equations. msc | Functional analysis. msc | Operator theory. msc | Calculus of variations and optimal control; optimization. msc | General topology. msc | Game theory, economics, social and behavioral sciences. msc | Systems theory; control. msc
Classification: LCC QA321.5 .N65155 2016 | DDC 515/.7248–dc23
LC record available at http://lccn.loc.gov/2015037183

DOI: http://dx.doi.org/10.1090/conm/659

∞ The paper used in this book is acid-free and falls within the guidelines
established to ensure permanence and durability.
Visit the AMS home page at http://www.ams.org/

10 9 8 7 6 5 4 3 2 1 21 20 19 18 17 16

Contents

Preface

A Special Session on Nonlinear Analysis and Optimization was part of the Second Joint International Meeting of the Israel Mathematical Union (IMU) and the American Mathematical Society (AMS), which took place at Bar-Ilan and Tel-Aviv Universities, Israel, during June 16-19, 2014. Boris S. Mordukhovich, Simeon Reich and Alexander J. Zaslavski were the organizers of this special session. The special session brought together a select group of over twenty experts from Israel and the USA. It was devoted to various issues, results and open problems regarding inverse dynamic programming, Denjoy-Wolff theorems, nonlinear input-output equilibrium, analyticity and nonanalyticity of solutions of delay-differential equations, weak convergence in metric spaces, gradient flows of entropy, turnpike properties of solutions to optimal control problems, optimal control of the sweeping process over moving sets, differential variational systems, and proximal alternating linearized minimization for nonconvex and nonsmooth problems.

A one-day Workshop on Nonlinear Analysis and Optimization organized by Simeon Reich and Alexander J. Zaslavski was held at the Technion - Israel Institute of Technology on June 12, 2014. It was a satellite event of the Second Joint International Meeting of the Israel Mathematical Union and the American Mathematical Society.

This volume is the tangible record of both the Special Session on Nonlinear Analysis and Optimization, and the Technion Workshop on Nonlinear Analysis and Optimization. Most of the papers assembled here have been contributed by participants in the session and in the workshop. In some cases, they have chosen to submit manuscripts which depart from the texts of their lectures. Several invited speakers who were unable to attend the meeting also contributed papers to this volume. All submissions have been carefully refereed and revised whenever necessary. The papers collected in this volume cover many different topics in Nonlinear Analysis and Optimization: Taylor domination property for analytic functions in the complex disk, mappings with upper integral bounds for p-moduli, multiple Fourier transforms and trigonometric series in line with Hardy's variation, finite-parameter feedback control for stabilizing damped nonlinear wave equations, implicit Euler approximation and optimization of one-sided Lipschitz differential inclusions, Bolza variational problems with extended-valued integrands on large intervals, first order singular variational problem with nonconvex cost, gradient and extragradient methods for the elasticity imaging inverse problem, discrete approximations of the entropy functional for probability measures on the plane, optimal irrigation scheduling for wheat production, existence of a fixed point of nonexpansive mappings in uniformly convex Banach spaces, strong convergence properties of m-accretive bounded operators, the Reich-Simons convex analytic inequality, nonlinear input-output equilibrium, differential linear-quadratic Nash games with mixed state-control constraints, and excessive revenue models of competitive markets.

We feel that this volume will be very valuable for many mathematicians and practitioners, who are interested in recent developments in Nonlinear Analysis and Optimization, as well as in their numerous and diverse applications.

The Editors

Contemporary Mathematics
Volume **659**, 2016
http://dx.doi.org/10.1090/conm/659/13162

Taylor domination, Turán lemma, and Poincaré-Perron sequences

Dmitry Batenkov and Yosef Yomdin

ABSTRACT. We consider "Taylor domination" property for an analytic function $f(z) = \sum_{k=0}^{\infty} a_k z^k$, in the complex disk D_R, which is an inequality of the form

$$|a_k|R^k \leq C \max_{i=0,\ldots,N} |a_i|R^i, \ k \geq N+1.$$

This property is closely related to the classical notion of "valency" of f in D_R. For f - rational function we show that Taylor domination is essentially equivalent to a well-known and widely used Turán's inequality on the sums of powers.

Next we consider linear recurrence relations of the Poincaré type

$$a_k = \sum_{j=1}^{d} [c_j + \psi_j(k)]a_{k-j}, \quad k = d, d+1, \ldots, \qquad \text{with} \lim_{k\to\infty} \psi_j(k) = 0.$$

We show that the generating functions of their solutions possess Taylor domination with explicitly specified parameters. As the main example we consider moment generating functions, i.e. the Stieltjes transforms

$$S_g(z) = \int \frac{g(x)\,dx}{1-zx}.$$

We show Taylor domination property for such S_g when g is a piecewise D-finite function, satisfying on each continuity segment a linear ODE with polynomial coefficients.

1. Introduction

"Taylor domination" for an analytic function $f(z) = \sum_{k=0}^{\infty} a_k z^k$ is an explicit bound of all its Taylor coefficients a_k through the first few of them. This property was classically studied, in particular, in relation with the Bieberbach conjecture, finally proved in [**12**]: for univalent f always $|a_k| \leq k|a_1|$. See [**4, 5, 10, 15**] and references therein. To give an accurate definition, let us assume the radius of convergence of the Taylor series for f is $\hat{R}$, $0 < \hat{R} \leqslant +\infty$.

DEFINITION 1.1. Let a positive *finite* $R \leq \hat{R}$, a natural N, and a positive sequence $S(k)$ of a subexponential growth be fixed. The function f is said to

The first author was supported by the Adams Fellowship Program of the Israel Academy of Sciences and Humanities.

The second author is supported by ISF, Grants No. 639/09 and 779/13, and by the Minerva foundation.

possess an $(N, R, S(k))$ - Taylor domination property if for each $k \geq N+1$ we have

$$|a_k|R^k \leqslant S(k) \max_{i=0,\dots,N} |a_i|R^i.$$

For $S(k) \equiv C$ a constant we shall call this property (N, R, C)-Taylor domination.

The parameters $(N, R, S(k))$ of Taylor domination are not defined uniquely. In fact, each nonzero analytic function f possesses this property, with N being the index of its first nonzero Taylor coefficient a_k:

PROPOSITION 1.2. *If $0 < \hat{R} \leqslant +\infty$ is the radius of convergence of $f(z) = \sum_{k=0}^{\infty} a_k z^k$, with $f \not\equiv 0$, then for each finite and positive $0 < R \leqslant \hat{R}$, f satisfies the $(N, R, S(k))$-Taylor domination property with N being the index of its first nonzero Taylor coefficient, and $S(k) = R^k|a_k|(|a_N|R^N)^{-1}$, for $k > N$.*

Conversely, let f possess $(N, R, S(k))$-domination. Then the series $\sum_{k=0}^{\infty} a_k z^k$ converges in a disk of radius R^* satisfying $R^* \geqslant R$. For each $R' < R$ the function f possesses (N, R', C)-domination, with the constant C depending on R'/R and on the sequence $S(k)$.

PROOF. It follows just by noticing that $\hat{R}^{-1} = \limsup_{k\to\infty} |a_k|^{\frac{1}{k}}$. This implies that for each $R \leq \hat{R}$ the sequence $S(k) = R^k|a_k|(|a_N|R^N)^{-1}$ has a subexponential growth. □

Consequently, the Taylor domination becomes really interesting only for those *families* of analytic functions f where we can specify the parameters N, R, $S(k)$ in an explicit and uniform way.

In this paper we study Taylor domination for a_k generated by linear non-stationary homogeneous recurrence relations of a fixed length:

$$a_k = \sum_{j=1}^{d} c_j(k) \cdot a_{k-j}, \quad k = d, d+1, \dots, \tag{1.1}$$

If for $j = 1, \dots, d$ the coefficients $c_j(k)$ have a form $c_j(k) = c_j + \psi_j(k)$, with fixed c_j and with $\lim_{k\to\infty} \psi_j(k) = 0$, then recurrence relation (1.1) is said to be a *linear recurrence relation of Poincaré type* (see [**19**, **21**]). Such recurrences appear in numerous problems of Analysis, Differential Equations, Combinatorics. We attempt to produce accurate bounds in Taylor domination for a_k generated by Poincaré type recurrence relations, and pose some open questions in this direction.

The paper is organized as follows: after presenting in Section 2 some classical connections between Taylor domination for f and its zeroes distribution, we start in Section 3 with recurrence relations (1.1) with constant coefficients. Those produce Taylor coefficients of rational functions $f(z)$. Surprisingly, Taylor domination in this case turns out to be a nontrivial fact, essentially equivalent to an important and widely used inequality on the sums of powers - Turán's lemma ([**17**, **18**, **25**, **26**]). In Section 3 we prove this equivalence, obtaining, as a byproduct, a very short proof of Turán's lemma from the classical Biernacki theorem ([**5**]) on multivalent functions, and from Bezout bound on zeroes of rational functions.

In Section 4 we obtain an explicit Taylor domination for solutions of general recurrence relations (1.1) with uniformly bounded coefficients. No additional information, besides the size of $c_j(k)$, is required. So we get this result, and hence (via Section 2 below) also explicit bounds on the number of zeroes, for a fairly wide class of analytic functions. On the other hand, in contrast with Turán's lemma, the disk where we get Taylor domination is usually significantly smaller than the expected disk of convergence.

In Section 5 we obtain some more accurate results for recurrence relations of Poincaré type. This includes (only partially explicit) Taylor domination in the full disk of convergence, and an explicit and uniform (depending only the size of perturbations $\psi_j(k)$) one in a smaller disk.

Finally, in Section 6, we consider moments of "D-finite" functions, produce a linear recurrence relation for them, describe the cases when it is of Poincaré type, and provide some conclusions concerning Taylor domination.

The authors would like to thank O. Friedland for a careful reading of the first version of this paper, and for numerous important remarks and suggestions.

2. Taylor domination and counting zeroes

Taylor domination allows us to compare the behavior of $f(z)$ with the behavior of the polynomial $P_N(z) = \sum_{k=0}^N a_k z^k$. In particular, the number of zeroes of f can be easily bounded in this way. In one direction the bound is provided by the classical results of [**5**, **10**]. To formulate them, we need the following definition (see [**15**] and references therein):

DEFINITION 2.1. A function f regular in a domain $\Omega \subset \mathbb{C}$ is called p-valent there, if for any $c \in \mathbb{C}$ the number of solutions in Ω of the equation $f(z) = c$ does not exceed p.

THEOREM 2.2 (Biernacki, 1936, [**5**]). *If f is p-valent in the disk D_R of radius R centered at $0 \in \mathbb{C}$ then for each $k \geq p+1$*

$$|a_k|R^k \leq (A(p)k/p)^{2p} \max_{i=1,\dots,p} |a_i|R^i,$$

where $A(p)$ is a constant depending only on p.

In our notations, Theorem 2.2 claims that a function f which is p-valent in D_R, possesses a $(p, R, (Ak/p)^{2p})$ - Taylor domination property.

For univalent functions, i.e. for $p = 1$, $R = 1$, Theorem 2.2 gives $|a_k| \leq A(1)^2 k^2 |a_1|$ for each k, while the sharp bound of the Bieberbach conjecture is $|a_k| \leq k|a_1|$.

A closely related result (obtained somewhat earlier) is the following:

THEOREM 2.3 (Cartwright, 1930, [**10**]). *If f is p-valent in the disk D_1 of radius 1 centered at $0 \in \mathbb{C}$ then for $|z| < 1$*

$$|f(z)| < B(p) \max_{i=0,\dots,p} |a_i|(1-|z|)^{-2p},$$

where $B(p)$ is a constant depending only on p.

Various forms of inverse results to Theorem 2.2, Theorem 2.3 are known. In particular, an explicit bound for the number of zeroes of f possessing Taylor domination can be obtained by combining Proposition 1.2 above and Lemma 2.2 from [**24**]:

THEOREM 2.4. *Let the function f possess an $(N, R, S(k))$ - Taylor domination property. Then for each $R' < R$, f has at most $M(N, \frac{R'}{R}, S(k))$ zeros in $D_{R'}$, where*

$M(N, \frac{R'}{R}, S(k))$ is a function depending only on N, $\frac{R'}{R}$ and on the sequence $S(k)$, satisfying $\lim_{\frac{R'}{R}\to 1} M = \infty$ and $M(N, \frac{R'}{R}, S) = N$ for $\frac{R'}{R}$ sufficiently small.

We can replace the bound on the number of zeroes of f by the bound on its valency, if we exclude a_0 in the definition of Taylor domination (or, alternatively, if we consider the derivative f' instead of f).

3. Taylor domination for rational functions and Turán's lemma

We shortly recall some basic facts concerning Taylor coefficients of rational functions. Consider a rational function $R(z) = \frac{P(z)}{Q(z)}$ with $Q(z) = 1 - \sum_{j=1}^{d} c_j z^j$ and $\deg P(z) \le d-1$. To simplify the presentation we shall assume that all the roots $s_1, \dots, s_d$ of Q are pairwise different (and they are clearly nonzero since $Q(0) = 1$). All the results below remain valid in the general case of multiple roots. Now $R(z)$ can be represented as a sum of elementary fractions:

$$R(z) = \sum_{j=1}^{d} \frac{\alpha_j}{s_j - z} = \sum_{j=1}^{d} \frac{\beta_j}{1 - \sigma_j z}, \quad \text{with } \beta_j = \frac{\alpha_j}{s_j},\ \sigma_j = \frac{1}{s_j}. \tag{3.1}$$

Developing into geometric progressions, we obtain

$$R(z) = \sum_{k=0}^{\infty} a_k z^k, \quad \text{with } a_k = \sum_{j=1}^{d} \beta_j \sigma_j^k. \tag{3.2}$$

The radius of convergence of this series is $R \stackrel{\text{def}}{=} \min_{i=1,\dots d} \left|\sigma_i^{-1}\right|$, which is the distance from the origin to the nearest pole of $R(z)$.

It is well known that the Taylor coefficients a_k of $R(z)$ satisfy a linear recurrence relation with constant coefficients

$$a_k = \sum_{j=1}^{d} c_j a_{k-j}, \quad k = d, d+1, \dots, \tag{3.3}$$

where c_j are the coefficients of the denominator $Q(z)$ of $R(z)$. Conversely, for any initial terms $a_0, \dots, a_{d-1}$ the solution sequence of (3.3) forms a sequence of the Taylor coefficients a_k of a rational function $R(z)$ as above. The equation $\sigma^d - \sum_{j=1}^{d} c_j \sigma^{d-j}$ is called the characteristic equation, and its roots $\sigma_1, \dots, \sigma_d$ defined above are called the characteristic roots of (3.3).

Taylor domination property for rational functions is provided by the following theorem, which is, essentially, equivalent to the "first Turán lemma" ([**17**, **25**, **26**], see also [**18**]):

THEOREM 3.1. *Let $\{a_j\}_{j=1}^{\infty}$ satisfy recurrence relation (3.3) and let $\sigma_1, \dots, \sigma_d$ be its characteristic roots. Put $R \stackrel{def}{=} \min_{i=1,\dots d} \left|\sigma_i^{-1}\right|$. Then for each $k \ge d$*

$$|a_k|\, R^k \le\ Q(k,d) \max_{i=0,\dots,d-1} |a_i| R^i, \tag{3.4}$$

where $Q(k,d) = (\frac{2ek}{d})^d$.

PROOF. The original Turán's result is as follows: let $b_j, z_j,\ j = 1, \dots, d$, be given complex numbers, with $\min_j |z_j| = 1$. Denoting $B \stackrel{\text{def}}{=} (b_1, \dots, b_d)$ and $Z \stackrel{\text{def}}{=} (z_1, \dots, z_d)$, define g_ν as the power sum $g_\nu(B, Z) = \sum_{j=1}^{d} b_j z_j^\nu$.

THEOREM 3.2 (Turán, 1953, [**25**]). *For each natural m we have*

$$|b_1 + \ldots + b_d| \leq Q(m,d) \max_{\nu=m+1,\ldots,m+d} |g_\nu|. \tag{3.5}$$

where $Q(m,d) = [2e(\frac{m}{d}+1)]^d$.

Put $k = m + d$. We immediately obtain that for any $F = (f_1, \ldots, f_d)$ and $W = (w_1, \ldots, w_d)$ with $\max |w_j| = 1$ we have

$$|g_k(F,W)| \leq Q(m,d) \max_{i=0,\ldots,d-1} |g_i(F,W)|, \tag{3.6}$$

by applying Theorem 3.2 with $z_j = w_j^{-1}$ and $b_j = f_j w_j^k$. Notice that $w_j \neq 0$.

Now we return to the sequence a_k satisfying recurrence relation (3.3). Put $S = (\sigma_1 R, \ldots, \sigma_d R)$ and $D = (\beta_1, \ldots, \beta_d)$. Then, according to (3.2), we get $|g_k(D,S)| = |a_k| R^k$. Clearly, $\max |\sigma_i R| = 1$ and so the inequality (3.6) can be applied, giving the required inequality (3.4). □

Theorem 3.1 provides an interpretation of Turán's lemma as a statement about Taylor domination for rational functions. This fact allows one to give a very short proof of Turán's lemma, albeit with a less sharp bound. Indeed, by Bezout theorem, rational functions of degree d are globally d-valent. In particular, $R(z)$ as above is d-valent in its maximal disk of convergence. Applying Biernacki's Theorem 2.2, we get an inequality

$$|a_k|\, R^k \leq [\frac{A(d)k}{d}]^{2d} \max_{i=1,\ldots,d} |a_i| R^i. \tag{3.7}$$

The transformations as above (shifted by 1) show that this is equivalent to Turán's lemma, with $Q(k,d) = [\frac{A(d)k}{d}]^{2d}$. This is less sharp than Turán's expression. Notice, that the best possible constant is given in [**9**]:

$$Q(k,d) = \sum_{j=0}^{d-1} \binom{k+j}{j} 2^j.$$

Theorem 3.1 provides a uniform Taylor domination for rational functions in their maximal disk of convergence D_R, in the strongest possible sense. Indeed, after rescaling to the unit disk D_1 the parameters of (3.4) depend only on the degree of the function, but not on its specific coefficients.

Turán's lemma can be considered as a result on exponential polynomials, and in this form it was a starting point for many deep investigations in Harmonic Analysis, Uncertainty Principle, Analytic continuation, Number Theory (see [**17, 18, 25, 26**] and references therein). Recently some applications in Algebraic Sampling were obtained, in particular, estimates of robustness of non-uniform sampling of "spike-train" signals ([**3, 14**]). One can hope that apparently new connections of Turán's lemma with Taylor domination, presented in this paper, can be further developed.

A natural open problem, motivated by Theorem 3.1, is a possibility to extend uniform Taylor domination in the maximal disk of convergence D_R, as provided by Theorem 3.1 for rational functions, to wider classes of generating functions of Poincaré type recurrence relations. Some initial examples in this direction were provided in [**30**], via techniques of Bautin ideals. An inequality closely related to Turán's lemma was obtained in [**28**], via techniques of finite differences. Presumably, the latter approach can be extended to Stieltjes transforms of certain natural classes of functions, providing uniform Taylor domination in the maximal disk of convergence.

4. Recurrence relations with uniformly bounded coefficients

We will denote by $\mathcal{R}$ a non-stationary linear recurrence relation of the form (1.1). Let $\mathcal{S}$ be the class of such $\mathcal{R}$ with uniformly bounded coefficients $c_j(k)$. We would like to write the bounds on $c_j(k)$ in a form

$$|c_j(k)| \leqslant K\rho^j, \qquad j = 1, \dots, d, \ k = d, d+1, \dots ,$$

for certain positive constants K, ρ. So for each $\mathcal{R} \in \mathcal{S}$ we define $K(\mathcal{R})$ and $\rho(\mathcal{R})$ to be the pair of constants providing the required bounds on $c_j(k)$, for which the product $\nu(\mathcal{R}) = (2K(\mathcal{R}) + 2) \cdot \rho(\mathcal{R})$ is minimal possible. We put $R(\mathcal{R}) \stackrel{\text{def}}{=} \nu(\mathcal{R})^{-1}$.

THEOREM 4.1. *Let $\{a_k\}_{k=0}^{\infty}$ be a solution of the recurrence relation $\mathcal{R} \in \mathcal{S}$. Put $K = K(\mathcal{R})$, $\rho = \rho(\mathcal{R})$, $R = R(\mathcal{R})$. Then the series $f(z) = \sum_{k=0}^{\infty} a_k z^k$ converges in the open disk D_R and possesses there $(d-1, R, (2K+2)^{d-1})$ Taylor domination.*

PROOF. Put $\hat{R} \stackrel{\text{def}}{=} \rho^{-1}$. First we show that for each $k \geqslant d$ we have

$$|a_k| \hat{R}^k \leqslant (2K+2)^k \max_{i=0,\dots,d-1} |a_i| \hat{R}^i. \tag{4.1}$$

The proof is by induction on k. Denote

$$M \stackrel{\text{def}}{=} \max_{i=0,\dots,d-1} |a_i| \hat{R}^i,$$

$$\eta \stackrel{\text{def}}{=} 2K+2,$$

and assume that $|a_\ell| \hat{R}^\ell \leqslant \eta^\ell M, \quad \ell \leqslant k-1$. We have

$$|a_k|\hat{R}^k = \hat{R}^k |\sum_{j=1}^{d} c_j(k) a_{k-j}| \leq K\hat{R}^k \sum_{j=1}^{d} |a_{k-j}|\rho^j = K \sum_{j=1}^{d} |a_{k-j}|\hat{R}^{k-j}.$$

By the inductive assumption $|a_{k-j}| \hat{R}^{k-j} \leqslant \eta^{k-j} M$, therefore we conclude that

$$|a_k| \hat{R}^k \leqslant KM \sum_{j=1}^{d} \eta^{k-j} = KM\eta^{k-1} \sum_{s=0}^{d-1} \eta^{-s} \leqslant 2KM\eta^{k-1} < \eta^k M.$$

This completes the induction step and proves inequality (4.1). Since by definition $R = \frac{\hat{R}}{\eta}$, dividing (4.1) by η^k gives

$$|a_k| R^k \leqslant \max_{i=0,\dots,d-1} |a_i| R^i \eta^i \leqslant \eta^{d-1} \max_{i=0,\dots,d-1} |a_i| R^i.$$

This is precisely the definition of $\left(d-1, R, \eta^{d-1}\right)$-Taylor domination, which, in turn, implies convergence in the open disk D_R. □

By a proper rescaling, Theorem 4.1 can be easily extended to non-stationary linear recurrences with a subexponential (or exponential) growth of the coefficients $c_j(k)$. Consequently, generating functions of such recurrences allow for explicit bounds on their valency. On the other hand, a drawback of this approach is that in the case of linear recurrences with constant coefficients (and for Poincaré-type recurrences - see below) the disk D_R where the uniform Taylor domination is guaranteed, is much smaller than the true disk of convergence.

Example 1. Consider a recurrence $\mathcal{R}$ with constant coefficients of form (3.3): $a_k = \sum_{j=1}^{d} c_j a_{k-j}$, $k = d, d+1, \ldots$. Let $\sigma_1, \ldots, \sigma_d$ be its characteristic roots, i.e. the roots of its characteristic equation $\sigma^d - \sum_{j=1}^{d} c_j \sigma^{d-j} = 0$. Put $\rho = \max_j |\sigma_j|$. Then for each $j = 1, \ldots, d$ we have

$$|c_j| = |e_j(\sigma_1, \ldots, \sigma_d)| \leqslant \binom{d}{j} \rho^j \leqslant 2^d \rho^j,$$

where $e_j(\cdot)$ is the elementary symmetric polynomial of degree j in d variables. We conclude that $K(\mathcal{R}) \le 2^d$, $\rho(\mathcal{R}) \le \rho$, $R(\mathcal{R}) \ge 2^{-d}\rho^{-1}$. It is easy to see that these bounds are sharp. So the radius of convergence is ρ^{-1}, while Theorem 4.1 provides Taylor domination only in the concentric disk of a $(2^{d+1}+2)$ times smaller radius. In the next section we discuss in some detail the problem of extending uniform Taylor domination to the full disk of convergence for Poicaré-type recurrences.

5. Taylor domination for Poincaré-type recurrences

Now we consider recurrence relations $\mathcal{R}$ of Poincaré-type, i.e. of the form

$$a_k = \sum_{j=1}^{d} [c_j + \psi_j(k)] \cdot a_{k-j}, \quad k = d, d+1, \ldots, \quad \lim_{k \to \infty} \psi_j(k) = 0. \tag{5.1}$$

We denote this class by $\mathcal{S}_P$. The characteristic polynomial $\sigma^d - \sum_{j=1}^{d} c_j \sigma^{d-j} = 0$ and the characteristic roots $\sigma_1, \ldots, \sigma_d$ of $\mathcal{R} \in \mathcal{S}_P$ are those of its constant part. We put $\rho(\mathcal{R}) = \max_j |\sigma_j|$.

The asymptotic behaviour of the solutions of such recurrences has been extensively studied, starting from Poincaré's own paper [**21**]. A comprehensive overview of the subject can be found in e.g. [**13**, Chapter 8]. The general idea permeating these studies is to compare the solutions of (5.1) to the solutions of the corresponding unperturbed recurrence relation (3.3). In this last case, as we have seen above, for some σ_j

$$\limsup_{k \to \infty} \sqrt[k]{|a_k|} = |\sigma_j|.$$

O.Perron proved in [**19**] that this relation holds for a general recurrence of Poincaré type, but with an additional condition that $c_d + \psi_d(k) \neq 0$ for all $k \in \mathbb{N}$. In [**20**] M.Pituk removed this restriction, and proved the following result.

THEOREM 5.1 (Pituk's extension of Perron's Second Theorem, [**20**]). *Let $\{a_k\}_{k=0}^{\infty}$ be any solution to a recurrence relation $\mathcal{R}$ of Poincaré class $\mathcal{S}_P$. Then either $a_k = 0$ for $k \gg 1$ or*

$$\limsup_{k \to \infty} \sqrt[k]{|a_k|} = |\sigma_j|,$$

where σ_j is one of the characteristic roots of $\mathcal{R}$.

This result, together with Proposition 1.2, implies the following:

THEOREM 5.2. *Let $\{a_k\}_{k=0}^{\infty}$ be any nonzero solution to a recurrence relation $\mathcal{R}$ of Poincaré class $\mathcal{S}_P$ with initial data $\bar{a}$, and let R be the radius of convergence of the generating function $f(z)$. Then necessarily $R > 0$, and in fact $R = |\sigma|^{-1}$ where σ is some (depending on $\bar{a}$) characteristic root of $\mathcal{R}$. Consequently, f satisfies $(d-1, R, S(k))$-Taylor domination with $S(k)$ as defined in Proposition 1.2.*

PROOF. The only thing left to show is that $a_m \neq 0$ for some $m = 0, 1, \ldots, d-1$. Assume on the contrary that

$$a_0 = a_1 = \cdots = a_{d-1} = 0.$$

Plugging this initial data into the recurrence (5.1), we immediately conclude that $a_k = 0$ for all $k \in \mathbb{N}$, a contradiction. □

Taylor domination in the maximal disk of convergence provided by Theorem 5.2, is not effective. Indeed, Proposition 1.2 guarantees that the sequence $S(k) = R^k |a_k| \cdot (\max_{i=0}^{d-1} R^i |a_i|)^{-1}$ is of subexponential growth but gives no further information on it. We can pose a natural question in this direction. For a sequence $\Delta = \{\delta_k\}$ of positive numbers tending to zero, consider a subclass $\mathcal{S}_{P,\Delta}$ of $\mathcal{S}_P$, consisting of $\mathcal{R} \in \mathcal{S}_P$ with $|\psi_j(k)| \leq \delta_k \cdot \rho(\mathcal{R})^j,\ j = 1, \ldots, d,\ k = d, d+1, \ldots$

Problem 1. *Do solutions of recurrence relations $\mathcal{R} \in \mathcal{S}_{P,\Delta}$ possess $(N, R, S(k))$-Taylor domination in the maximal disk of convergence D_R, with $S(k)$ depending only on d and Δ? Is this true for specific Δ, in particular, for $\Delta = \{1, \frac{1}{2}, \frac{1}{3}, \ldots, \}$, as it occurs in most of examples (solutions of linear ODE's, etc.)?*

Taking into account well known difficulties in the analysis of Poincaré-type recurrences, this question may be tricky. Presumably, it can be easier for Δ with $\sum_{k=1}^{\infty} \delta_k < \infty$.

In the remainder of this section we prove a version of Theorem 4.1 for Poincaré-type recurrences $\mathcal{R}$. It provides Taylor domination in a smaller disk, but with explicit parameters, expressed in a transparent way through the constant part of $\mathcal{R}$, and through the size of the perturbations.

THEOREM 5.3. *Let $\{a_k\}_{k=0}^{\infty}$ satisfy a fixed recurrence $\mathcal{R} \in \mathcal{S}_P$. Put $\rho \stackrel{def}{=} \rho(\mathcal{R})$, $R = 2^{-(d+3)}\rho^{-1}$. Let $\hat{N}$ be the minimal of the numbers n such that for all $k > n$ we have $|\psi_j(k)| \leq 2^d \rho^j,\ j = 1, \ldots, d$. We put $N = \hat{N} + d$, and $C = 2^{(d+3)N}$.*

Then $\{a_k\}_{k=0}^{\infty}$ possesses (N, R, C)-Taylor domination property.

PROOF. By the calculations in Example 1 above, we have for the constant part of $\mathcal{R}$ the bounds $|c_j| \leq 2^d \rho^j$. Since for $k > N$ we have $|\psi_j(k)| \leq 2^d \rho^j$, we obtain for such k that $|c_j(k)| \leq 2^{d+1} \rho^j$. Now repeating the proof of Theorem 4.1 (with the starting point of the recurrence shifted by $\hat{N}$) we complete the proof. □

Theorem 5.3 provides a partial answer to Problem 1:

COROLLARY 5.4. *Let $\Delta = \{\delta_k\}$ be a sequence of positive numbers tending to zero. Define $\hat{N}(\Delta)$ as a minimal number n such that for $k > n$ we have $\delta_k \leq 2^d$. Then for each $\mathcal{R} \in \mathcal{S}_{P,\Delta}$ solution sequences of $\mathcal{R}$ possess (N, R, C)-Taylor domination, where $N = \hat{N}(\Delta) + d,\ R = 2^{-(d+3)}\rho(\mathcal{R})^{-1},\ C = 2^{(d+3)N}$.*

PROOF. By definition of the class $\mathcal{S}_{P,\Delta}$ the number $\hat{N}(\Delta)$ coincides with $\hat{N}$ in Theorem 5.3. Application of this theorem completes the proof. □

One can consider at least two possible approaches to the extension of these results to the full disk of convergence D_R. First, asymptotic expressions in [**6, 20**] may be accurate enough to provide an inequality of the desired form. If this is a case, it remains to get explicit bounds in these asymptotic expressions.

Second, one can use a "dynamical approach" to recurrence relation (1.1) (see [**7, 11, 16, 22, 27**] and references therein). We consider (1.1) as a non-autonomous linear dynamical system T. A "non-autonomous diagonalization" of T is a sequence $\mathcal{L}$ of linear changes of variables, bringing this system to its "constant model" T_0, provided by the limit recurrence relation (3.3).

If we could obtain a non-autonomous diagonalization $\mathcal{L}$ of T with an explicit bound on the size of the linear changes of variables in it, we could get the desired inequality as a pull-back, via $\mathcal{L}$, of the Turán inequality for T_0. There are indications that the second approach may work in the classes $\mathcal{S}_{P,\Delta}$, for Δ with a finite sum.

6. Piecewise D-finite functions

In this section we investigate a certain class of power series, defined by the Stieltjes integral transforms

$$f(z) = S_g(z) = \int_a^b \frac{g(x)\,dx}{1-zx}, \tag{6.1}$$

where $g(x)$ belongs to the class of the so-called *piecewise D-finite functions* [**1**], which are solutions of linear ODEs with polynomial coefficients, possessing a finite number of discontinuities of the first kind.

Using the expansion $(1-zx)^{-1} = \sum_{k=0}^{\infty}(zx)^k$ for $|z| < \frac{1}{|x|}$, we obtain the following useful representation of $S_g(z)$:

Proposition 6.1. *Let $g : [a,b] \to \mathbb{R}$ be bounded and integrable on $[a,b]$. Then the Stieltjes transform (6.1) of g is regular at the origin, and it is given by the moment-generaing function*

$$S_g(z) = \sum_{k=0}^{\infty} m_k z^k, \quad \text{where } m_k \stackrel{def}{=} \int_a^b x^k g(x)\,dx.$$

Definition 6.2. A real-valued bounded integrable function $g : [a,b] \to \mathbb{R}$ is said to belong to the class $\mathcal{PD}(\mathfrak{D}, p)$ if it has $0 \leqslant p < \infty$ discontinuities (not including the endpoints a, b) of the first kind, and between the discontinuities it satisfies a linear homogeneous ODE with polynomial coefficients $\mathfrak{D}\, g = 0$, where

$$\mathfrak{D} = \sum_{j=0}^{n} p_j(x)\left(\frac{d}{dx}\right)^j, \quad p_j(x) = \sum_{i=0}^{d_j} a_{i,j} x^i.$$

Obtaining uniform Taylor domination for S_g, where g belongs to particular subclasses of $\mathcal{PD}$ (in particular, g being piecewise algebraic), is an important problem with direct applications in Qualitative Theory of ODEs (see [**8**] and references therein). In this paper we provide initial results in this direction.

Let $g \in \mathcal{PD}(\mathfrak{D}, p)$, with $\mathfrak{D}$ as above. Denote the discontinuities of g by $a = x_0 < x_1 < \cdots < x_p < x_{p+1} = b$. In what follows, we shall use some additional notation. Denote for each $j = 0, \ldots, n$, $\alpha_j \stackrel{\text{def}}{=} d_j - j$. Let $\alpha \stackrel{\text{def}}{=} \max_j \alpha_j$, and denote for each $\ell = -n, \ldots, \alpha$

$$q_\ell(k) \stackrel{\text{def}}{=} \sum_{j=0}^{n} (-1)^j a_{\ell+j,j} (k+\ell+j)_j. \tag{6.2}$$

where $(a)_b = a(a-1)\times\cdots\times(a-b+1) = \frac{\Gamma(a)}{\Gamma(b)}$ is the Pochhammer symbol for the falling factorial.

Our approach is based on the following result:

THEOREM 6.3 ([1]). *Let $g \in \mathcal{PD}(\mathfrak{D}, p)$. Then the moments $m_k = \int_a^b x^k g(x)\,dx$ satisfy the recurrence relation*

$$\sum_{\ell=-n}^{\alpha} q_\ell(k)\, m_{k+\ell} = \varepsilon_k, \quad k = 0, 1, \dots, \tag{6.3}$$

where

$$\varepsilon_k = \sum_{\ell=0}^{p+1} \sum_{j=0}^{n-1} x_\ell^{k-j} (k)_j\, c_{\ell,j},$$

with each $c_{\ell,j}$ being a certain homogeneous bilinear form in the two sets of variables

$$\{p_m(x_\ell), p_m'(x_\ell), \dots, p_m^{(n-1)}(x_\ell)\}_{m=0}^{n},$$
$$\{g(x_\ell^+) - g(x_\ell^-),\ g'(x_\ell^+) - g'(x_\ell^-), \dots, g^{(n-1)}(x_\ell^+) - g^{(n-1)}(x_\ell^-)\}.$$

The recurrence (6.3) is inhomogeneous, and the coefficient of the highest moment is different from one. Accordingly, we first transform (6.3) into a homogeneous matrix recurrence. Next, we give conditions for this last recurrence to be of Poincaré type. Finally, we apply an appropriate version of the results in Section 5 to get Taylor domination.

It is well-known that the sequence $\{\varepsilon_k\}_{k=0}^{\infty}$ satisfies a recurrence relation $\mathcal{S}$ of the form (3.3) with constant coefficients, whose characteristic roots are precisely $\{x_0, \dots, x_{p+1}\}$, each with multiplicity n. Let the characteristic polynomial $\Theta_{\mathcal{S}}(z)$ of degree $\tau \stackrel{\text{def}}{=} n(p+2)$ be of the form

$$\Theta_{\mathcal{S}}(\sigma) = \prod_{j=0}^{p+1} (\sigma - x_j)^n = \sigma^\tau - \sum_{i=1}^{\tau} b_i \sigma^{\tau-i}, \tag{6.4}$$

then

$$\varepsilon_k = \sum_{j=1}^{\tau} b_j \varepsilon_{k-j}, \quad k = \tau, \tau+1, \dots.$$

Rewrite this last recurrence as

$$\varepsilon_{k+\tau} = \sum_{j=0}^{\tau-1} b_{\tau-j} \varepsilon_{k+j}, \quad k = 0, 1, \dots. \tag{6.5}$$

Now denote the vector function $\vec{w}(k) : \mathbb{N} \to \mathbb{C}^{\alpha+\varkappa+\tau}$ as

$$\vec{w}(k) \stackrel{\text{def}}{=} \begin{bmatrix} m_{k-n} \\ \vdots \\ m_{k+\alpha-1} \\ \varepsilon_k \\ \vdots \\ \varepsilon_{k+\tau-1} \end{bmatrix}.$$

Then by (6.3) and (6.5) we see that $\vec{w}(k)$ satisfies the linear system

(6.6)

$$\vec{w}(k+1)=$$

$$\left[\begin{array}{cccccc|ccccc}
0 & 1 & 0 & \dots & 0 & & & & & & \\
0 & 0 & 1 & \dots & 0 & & & \vec{0}^{(\alpha+n-1)\times\tau} & & & \\
\dots & & & & & & & & & & \\
-\frac{q_{-n}(k)}{q_\alpha(k)} & -\frac{q_{-n+1}(k)}{q_\alpha(k)} & \dots & & -\frac{q_{\alpha-1}(k)}{q_\alpha(k)} & & \frac{1}{q_\alpha(k)} & 0 & \dots & & 0 \\
\hline
 & & & & & & 0 & 1 & 0 & \dots & 0 \\
 & & & & & & 0 & 0 & 1 & \dots & 0 \\
 & & \vec{0}^{\tau\times(\alpha+n)} & & & & \dots & & & & \\
 & & & & & & b_\tau & b_{\tau-1} & \dots & & b_1
\end{array}\right]\vec{w}(k).$$

Now we can show Taylor domination for the Stieltjes transform $S_g(z)$, utilizing the system version of Theorem 5.1.

Definition 6.4. The vector function $\vec{y}(k):\mathbb{N}\to\mathbb{C}^{\varkappa}$ is said to satisfy a linear system of Poincaré type, if

$$\vec{y}(k+1)=(A+B(k))\,\vec{y}(k), \tag{6.7}$$

where A is a constant $n\times n$ matrix and $B(k):\mathbb{N}\to\mathbb{C}^{\varkappa\times\varkappa}$ is a matrix function satisfying $\lim_{k\to\infty}\|B(k)\|=0$.

Theorem 6.5 ([**20**]). *Let the vector $\vec{y}(k)$ satisfy the perturbed linear system of Poin-*
caré type (6.7). *Then either $\vec{y}(k)=\vec{0}\in\mathbb{C}^{\varkappa}$ for $k\gg1$ or*

$$\lim_{k\to\infty}\sqrt[k]{\|\vec{y}(k)\|}$$

exists and is equal to the modulus of one of the eigenvalues of the matrix A.

Lemma 6.6. *The recurrence system* (6.6) *is of Poincaré type if and only if*

$$\alpha_n\geqslant\alpha_j\qquad j=0,1,\dots,n. \tag{6.8}$$

Proof. Clearly, a necessary and sufficient condition for (6.6) being of Poincaré type is that

$$\deg q_\ell(k)\leqslant\deg q_\alpha(k),\quad \ell=-n\dots,\alpha-1.$$

We will show that this condition is equivalent to (6.8).

Recall the definition (6.2). The highest power of k in any $q_\ell(k)$ is determined by the maximal index $j=0,\dots,n$ for which $a_{i,j}\neq0$ and $i-j=\ell$. Consider $\ell=\alpha_n=d_n-n$, then obviously since $a_{d_n,n}\neq0$ we have $\deg q_{\alpha_n}(k)=n$.

(1) Now let's assume that for some $s<n$ we have $\alpha_s>\alpha_n$, i.e. $d_s-s>d_n-n$, and consider the polynomial $q_{\alpha_s}(k)$:

$$q_{\alpha_s}(k)=\sum_{j=0}^{n}(-1)^j a_{j+\alpha_s,j}\,(k+j+\alpha_s)_j\,.$$

By assumption, $\alpha_s+n>d_n$, and therefore in this case $\deg q_{\alpha_s}(k)<n$. Thus if $\alpha_s>\alpha_n$ for some $s<n$, we have $\deg q_{\alpha_n}>\deg q_{\alpha_s}$. In particular, $\alpha\geqslant\alpha_s>\alpha_n$ and therefore $\deg q_\alpha<\deg q_{\alpha_n}$. This proves one direction.

(2) In the other direction, assume that $\alpha=\max_j\alpha_j=\alpha_n$. Clearly $\deg q_\alpha=\deg q_{\alpha_n}=n$, but on the other hand it is always true that $\deg q_\ell\leqslant n$.

This concludes the proof. □

REMARK 6.7. The condition (6.8) in fact means that the point $z = \infty$ is at most a regular singularity of the operator $\mathfrak{D}$.

So in the remainder of the section we assume that (6.8) is satisfied and $n > 0$. The constant part of the system (6.6) is the matrix

$$A = \begin{bmatrix} 0 & 1 & 0 & \dots & 0 & & & & & \\ 0 & 0 & 1 & \dots & 0 & & & \vec{0}^{d_n \times \tau} & & \\ \dots & & & & & & & & & \\ -\beta_{-n} & -\beta_{-n+1} & \dots & & -\beta_{-n+d_n-1} & & & & & \\ & & & & & 0 & 1 & 0 & \dots & 0 \\ & & & & & 0 & 0 & 1 & \dots & 0 \\ & & \vec{0}^{\tau \times d_n} & & & \dots & & & & \\ & & & & & b_\tau & b_{\tau-1} & \dots & & b_1 \end{bmatrix},$$

where

$$\beta_{-n+s} \stackrel{\text{def}}{=} \lim_{k\to\infty} \frac{q_{-n+s}(k)}{q_{\alpha_n}(k)} = \frac{a_{s,n}}{a_{d_n,n}}.$$

PROPOSITION 6.8. *The set Z_A of the eigenvalues of the matrix A is precisely the union of the roots of $p_n(x)$ (i.e. the singular points of the operator $\mathfrak{D}$) and the jump points $\{x_i\}_{i=0}^{p+1}$.*

PROOF. This is immediate, since $A = \begin{bmatrix} C & \vec{0} \\ \vec{0} & D \end{bmatrix}$, where C is the companion matrix of $p_n(x)$ and D is the companion matrix of the characteristic polynomial $\Theta_{\mathcal{S}}(z)$ as defined in (6.4). □

In [2] we study the following question: *how many first moments $\{m_k\}$ can vanish for a nonzero $g \in \mathcal{PD}(\mathfrak{D}, p)$?* In particular, we prove the following result.

THEOREM 6.9 ([2]). *Let the operator $\mathfrak{D}$ be of Fuchsian type (i.e. having only regular singular points, possibly including ∞). In particular, $\mathfrak{D}$ satisfies the condition (6.8). Let $g \in \mathcal{PD}(\mathfrak{D}, p)$.*

1. *If there is at least one discontinuity point $\xi \in [a, b]$ of g at which the operator $\mathfrak{D}$ is nonsingular, i.e. $p_n(\xi) \neq 0$, then vanishing of the first $\tau + d_n - n$ moments $\{m_k\}_{k=0}^{\tau+d_n-n-1}$ of g implies $g \equiv 0$.*
2. *Otherwise, let $\Lambda(\mathfrak{D})$ denote the largest positive integer characteristic exponent of $\mathfrak{D}$ at the point ∞. In fact, the indicial equation of $\mathfrak{D}$ at ∞ is $q_\alpha(k) = 0$. Then the vanishing of the first $\Lambda(\mathfrak{D}) + 1 + d_n - n$ moments of g implies $g \equiv 0$.*

Everything is now in place in order to obtain the following result.

THEOREM 6.10. *Let $g \in \mathcal{PD}(\mathfrak{D}, p)$ be a not identically zero function, with $\mathfrak{D}$ of Fuchsian type. Then the Stieltjes transform $S_g(z)$ is analytic at the origin, and the series*

$$S_g(z) = \sum_{k=0}^{\infty} m_k z^k$$

converges in a disk of radius R which satisfies

$$R \geqslant R^* \stackrel{def}{=} \min\left\{\xi^{-1} : \xi \in Z_A\right\},$$

where Z_A is given by Proposition 6.8. *Furthermore, for every*

$$N \geqslant \max\{\tau - 1, \Lambda(\mathfrak{D})\} + d_n - n,$$

S_g satisfies $(N, R, S(k))$ Taylor domination, where $S(k)$ is given by Proposition 1.2.

PROOF. By Lemma 6.6 and Theorem 6.5 it is clear that

$$\limsup_{k\to\infty} \sqrt[k]{|m_k|} \leqslant \frac{1}{R^*}.$$

By Theorem 6.9, $m_k \neq 0$ for at least some $k = 0, \ldots, N$. Then we apply Proposition 1.2. □

In order to bound the number of zeros of S_g, we would need to estimate the growth of the rational functions $\frac{q_{-n+s}(k)}{q_{\alpha_n}(k)}$, and this can hopefully be done using some general properties of the operator $\mathfrak{D}$. Then we would apply the results of Section 5. We expect that in this way we can single out subclasses of $\mathcal{PD}$ for which uniform Taylor domination takes place. We plan to carry out this program in a future work.

REMARK 6.11. It is possible to obtain Taylor domination for the Stieltjes transforms $S_g(z)$ by another method, based on Remez-type inequalities [**23**, **29**]. We plan to present these results separately.

We complete this paper with one example and one question.

Example 2. Consider a very special case of D-finite functions: polynomial $g(x)$. In this case a uniform Taylor domination (depending only on the degree) in the maximal disk of convergence for $S_g(z)$ was obtained in [**30**] via the study of the "Bautin ideals", generated by the moments $m_k(g)$ (considered as polynomials in the parameters of the problem).

Problem 2. *Moments of the form $m_k(p, Q, a, b) = \int_a^b P^k(x)q(x)dx$ play an especially important role in Qualitative Theory of ODEs (compare* [**8**]*). Here $P(x)$ and $q(x)$ are univariate complex polynomials. Via change of integration variable $w = P(x)$ these moments can be reduced to the usual moments of a piecewise-algebraic g, but along a certain complex curve. Does the generating function of $m_k(p, Q, a, b)$ satisfy uniform Taylor domination (depending only on the degrees of P and q) in its disk of convergence?*

References

[1] Dmitry Batenkov, *Moment inversion problem for piecewise D-finite functions*, Inverse Problems **25** (2009), no. 10, 105001, 24, DOI 10.1088/0266-5611/25/10/105001. MR2545970 (2011a:44011)

[2] D. Batenkov and G. Binyamini. Moment vanishing of piecewise solutions of linear ODE's. *arXiv:1302.0991.*

[3] Dmitry Batenkov, Niv Sarig, and Yosef Yomdin, *Accuracy of algebraic Fourier reconstruction for shifts of several signals*, Sampl. Theory Signal Image Process. **13** (2014), no. 2, 151–173. MR3315358

[4] Ludwig Bieberbach, *Analytische Fortsetzung* (German), Ergebnisse der Mathematik und ihrer Grenzgebiete (N.F.), Heft 3, Springer-Verlag, Berlin-Göttingen-Heidelberg, 1955. MR0068621 (16,913a)

[5] M. Biernacki. Sur les fonctions multivalentes d'ordre p. *CR Acad. Sci. Paris*, 203:449–451, 1936.

[6] Sigrun Bodine and D. A. Lutz, *Asymptotic solutions and error estimates for linear systems of difference and differential equations*, J. Math. Anal. Appl. **290** (2004), no. 1, 343–362, DOI 10.1016/j.jmaa.2003.09.068. MR2032246 (2004m:39017)

[7] Julius Borcea, Shmuel Friedland, and Boris Shapiro, *Parametric Poincaré-Perron theorem with applications*, J. Anal. Math. **113** (2011), 197–225, DOI 10.1007/s11854-011-0004-0. MR2788356 (2012e:39004)

[8] Miriam Briskin, Nina Roytvarf, and Yosef Yomdin, *Center conditions at infinity for Abel differential equations*, Ann. of Math. (2) **172** (2010), no. 1, 437–483, DOI 10.4007/annals.2010.172.437. MR2680423 (2011j:34093)

[9] N. G. de Bruijn. On Turán's first main theorem. *Acta Math. Hung.*, 11:213–2016, 1960.

[10] M. L. Cartwright, *Some inequalities in the theory of functions*, Math. Ann. **111** (1935), no. 1, 98–118, DOI 10.1007/BF01472208. MR1512983

[11] W. Coppel. Dichotomies and stability theory. In *Proceedings of the Symposium on Differential Equations and Dynamical Systems*, pages 160–162. Springer, 1971.

[12] Louis de Branges, *A proof of the Bieberbach conjecture*, Acta Math. **154** (1985), no. 1-2, 137–152, DOI 10.1007/BF02392821. MR772434 (86h:30026)

[13] Saber Elaydi, *An introduction to difference equations*, 3rd ed., Undergraduate Texts in Mathematics, Springer, New York, 2005. MR2128146 (2005j:39001)

[14] Omer Friedland and Yosef Yomdin, *An observation on the Turán-Nazarov inequality*, Studia Math. **218** (2013), no. 1, 27–39, DOI 10.4064/sm218-1-2. MR3125122

[15] W. K. Hayman, *Multivalent functions*, 2nd ed., Cambridge Tracts in Mathematics, vol. 110, Cambridge University Press, Cambridge, 1994. MR1310776 (96f:30003)

[16] Peter E. Kloeden and C. Pötzsche, *Guest editorial: Non-autonomous difference equations and discrete dynamical systems*, J. Difference Equ. Appl. **17** (2011), no. 2, 129–130, DOI 10.1080/10236198.2010.549001. MR2783339

[17] D. S. Mitrinović, J. E. Pečarić, and A. M. Fink, *Classical and new inequalities in analysis*, Mathematics and its Applications (East European Series), vol. 61, Kluwer Academic Publishers Group, Dordrecht, 1993. MR1220224 (94c:00004)

[18] F. L. Nazarov, *Local estimates for exponential polynomials and their applications to inequalities of the uncertainty principle type* (Russian, with Russian summary), Algebra i Analiz **5** (1993), no. 4, 3–66; English transl., St. Petersburg Math. J. **5** (1994), no. 4, 663–717. MR1246419 (94k:42004)

[19] O. Perron. Über summengleichungen und Poincarésche differenzengleichungen. *Mathematische Annalen*, 84(1):1–15, 1921.

[20] Mihály Pituk, *More on Poincaré's and Perron's theorems for difference equations*, J. Difference Equ. Appl. **8** (2002), no. 3, 201–216, DOI 10.1080/10236190211954. MR1887903 (2003g:39009)

[21] H. Poincare, *Sur les Equations Lineaires aux Differentielles Ordinaires et aux Differences Finies* (French), Amer. J. Math. **7** (1885), no. 3, 203–258, DOI 10.2307/2369270. MR1505385

[22] Christian Pötzsche, *Geometric theory of discrete nonautonomous dynamical systems*, Lecture Notes in Mathematics, vol. 2002, Springer-Verlag, Berlin, 2010. MR2680867 (2012a:37003)

[23] E.J. Remez. Sur une propriété des polynômes de Tchebyscheff, Comm. *Inst. Sci. Kharkow*, 13:93–95, 1936.

[24] N. Roytwarf and Y. Yomdin, *Bernstein classes* (English, with English and French summaries), Ann. Inst. Fourier (Grenoble) **47** (1997), no. 3, 825–858. MR1465788 (98h:34009a)

[25] Paul Turán, *Eine neue Methode in der Analysis und deren Anwendungen* (German), Akadémiai Kiadó, Budapest, 1953. MR0060548 (15,688b)

[26] Paul Turán, *On a new method of analysis and its applications*, Pure and Applied Mathematics (New York), John Wiley & Sons, Inc., New York, 1984. With the assistance of G. Halász and J. Pintz; With a foreword by Vera T. Sós; A Wiley-Interscience Publication. MR749389 (86b:11059)

[27] Y. Yomdin, *Nonautonomous linearization*, Dynamical systems (College Park, MD, 1986), Lecture Notes in Math., vol. 1342, Springer, Berlin, 1988, pp. 718–726, DOI 10.1007/BFb0082857. MR970581 (91h:58093)

[28] Y. Yomdin, *Singularities in algebraic data acquisition*, Real and complex singularities, London Math. Soc. Lecture Note Ser., vol. 380, Cambridge Univ. Press, Cambridge, 2010, pp. 378–396. MR2759050 (2012g:94045)

[29] Y. Yomdin, *Remez-type inequality for discrete sets*, Israel J. Math. **186** (2011), 45–60, DOI 10.1007/s11856-011-0131-4. MR2852315 (2012i:30009)

[30] Y. Yomdin, *Bautin ideals and Taylor domination*, Publ. Mat. **58** (2014), no. suppl., 529–541. MR3211848

Department of Mathematics, Weizmann Institute of Science, Rehovot 76100, Israel

Current address: Department of Computer Science, Technion - Israel Institute of Technology, Haifa 32000, Israel

E-mail address: batenkov@cs.technion.ac.il

Department of Mathematics, Weizmann Institute of Science, Rehovot 76100, Israel

E-mail address: yosef.yomdin@weizmann.ac.il

Contemporary Mathematics
Volume **659**, 2016
http://dx.doi.org/10.1090/conm/659/13158

On the value of first order singular optimization problems

Ido Bright

ABSTRACT. First order singular optimization problems arise in material science in the study of phase transition, and in infinite-horizon optimization. We show that the value of first order problems, which are space-independent, is attained by a sequence of functions converging to a constant function. The proof we present employs occupational measures.

1. Introduction

In this paper we study first order singular variational problem with non convex cost, namely, the minimization of:

$$v^* = \lim_{\epsilon \to 0} \inf_u \frac{1}{\mathcal{L}^n(\Omega)} \int_\Omega L(x, u(x), \epsilon \nabla u(x))\, dx, \tag{1.1}$$

where $\Omega \subset \mathbb{R}^n$ is open and $\mathcal{L}^n$ is the n-dimensional Lebesgue measure. Our motivation for studying this problems comes from energy functionals arising in the study of material science, e.g. the phase-transitions models (see Cahn and Hilliard [**5**]), and from infinite-horizon optimization problems in n-dimensions. Indeed, applying a change of variable to (1.1), setting $y = \epsilon^{-1}x$ one obtains the equivalent infinite-horizon minimization problem

$$v^* = \lim_{\epsilon \to 0} \inf_u \frac{1}{\mathcal{L}^n(\epsilon^{-1}\Omega)} \int_{\epsilon^{-1}\Omega} L(\epsilon y, u(y), \nabla u(y))\, dy, \tag{1.2}$$

where $\epsilon^{-1}\Omega = \{y \in \mathbb{R}^n | \epsilon y \in \Omega\}$. The optimization is with respect to the average cost in domains that are stretched by a factor of ϵ^{-1}. Setting $L(x, u, v) = (u^2 - 1)^2 + v^2$ in the original equation (1.1), we obtain a reparameterization of the Van der-Waals energy functional (Van der-Waals [**13**], [**5**]), which arises in the study of phase transitions. There, the minimization criterion is

$$v^* = \lim_{\epsilon \to 0} \inf_u \frac{1}{\epsilon \mathcal{L}^n(\Omega)} \int_\Omega \left(\left[(u(x))^2 - 1 \right]^2 + [\epsilon \nabla u(x)]^2 \right) dx, \tag{1.3}$$

with an additional integral constraint $\int_\Omega u(x)\, dx = C$. This problem is generally studied with Γ-convergence (see Sternberg [**11**]). It was also applied to space-dependent costs by Bouchitt [**4**] for two-phase transition models, minimizing a cost of the form $\lim_{\epsilon \to 0} \inf_u \epsilon^{-1} \int_\Omega L(x, u(x), \epsilon \nabla u(x))\, dx$.

Key words and phrases. Singular optimization problems, infinite-horizon problems, occupational measures.

Our main result is in extending a previous result for non-convex costs by Artstein and Leizarowitz [**2**] for the case $n = 1$, establishing that the optimal value is attained by an optimal sequence converging to a constant function inside the open set Ω. This result allows an easy computation of the minimal value. The tool we use is probability measures, specifically, occupational measures that provide the statistical properties of a solution in space. This representation allows us to take meaningful limits of the functions in a compact space of measures.

Previous results in the field provide the existence of a periodic optimal solution for scalar ($n = 1$) second order Lagrangians in Leizarowitz and Mizel [**10**]. This result was extended by Artstein and Bright [**1**] to first order control systems in the plane (i.e., to functions $u : \mathbb{R} \to \mathbb{R}^2$). Infinite horizon optimization problems in the plane ($n = 2$) where studied by Leizarowitz. Overtaking solutions for first order variational problems where obtained in [**8**], and second order variational problems with a radial symmetry where reduced to second order scalar problems in [**9**], this reduction implies the existence of a periodic radially symmetric optimal solution by [**10**].

Higher order Lagrangians appear in the study of copolymers via a non-local Cahn-Hilliard equation. Empirical and numerical results provide evidence of the existence of periodic solutions for the case $n > 1$ (See Choksi [**6**] and references within.). However there is no analytic basis for these results. The main tool used to study these problems is Γ-convergence, which is applied to a cost similar to (1.3) with a different renormalization. Although this renormalization provides finer details on the solutions and cost, this paper shows that a qualitative result on the structure of the solution can still be obtained by the normalization used in this paper, using occupational measures.

The structure of the paper is as follows: In the following section we present the notation used throughout the paper. In Section 3 we present some lemmas regarding limits of functions, which do not converge in the standard sense. In Section 4 we present our results for space-invariant costs, and in the final section we present our results for space-dependent costs.

2. Notations

The following notations are used throughout the paper. We denote the n-dimensional Euclidean space by $\mathbb{R}^n$ and the standard Euclidean norm by $|\cdot|$. The closure of a set $X \subset \mathbb{R}^n$ is denoted by $\bar{X}$ and its boundary by ∂X. We denote by $\mathcal{L}^n$ the n-dimensional Lebesgue measure, and denote the L^∞ norm of $u(\cdot)$ in a set Ω by $\|u(\cdot)\|_{\Omega,\infty}$. We shall use the following Sobolev spaces:

$$W^{1,1}(\Omega) = \left\{u(\cdot) \in L^1(\Omega) \,|\, |\nabla u(\cdot)| \in L^1(\Omega)\right\},$$

where $\nabla u(\cdot)$ corresponds to the weak derivative, and

$$W_0^{1,1}(\Omega) = \left\{u(\cdot) \in W^{1,1}(\Omega) \,|Tr(u(\cdot)) \equiv 0\right\},$$

where $Tr : W^{1,1}(\Omega) \to L^1(\partial\Omega)$ denotes the trace operator. We also define the set of functions satisfying a Lipschitz condition in Ω by $Lip(\Omega)$.

Given a set Y in a Euclidean space, we denote its probability space by $P(Y)$, and by $P_0(Y) \subset P(Y)$ all the measures with first moment being $\mathbf{0}$, namely, measures satisfying $\int_Y v d\mu(v) = \mathbf{0} \in \mathbb{R}^n$. We endow the space of probability measures $P(Y)$ by the weak topology, namely, a sequence $\mu_i \in P(Y)$ converges to the measure $\mu_0 \in P(Y)$ if for every bounded continuous function $g(\cdot)$ we have that $\lim_{i\to\infty} \int_Y g(y)\, d\mu_i(y) = \int_Y g(y)\, d\mu_0(y)$. Within this topology it is easy to see that the the set $P_0(Y)$ is closed.

Another tool we use is disintegration of measures. Given two Euclidean spaces Y_1 and Y_2 and a probability measure $\mu \in P(Y_1 \times Y_2)$, we denote its disintegration by $\mu = p \circledast \mu^x$. The marginal measure $p \in P(Y_1)$ is defined by $p(A) = \mu(A \times Y_2)$ for every measurable set $A \subset Y_1$. The measure valued function $\mu^x \in P(Y_2)$ is the disintegration of μ with respect to p, for p-almost every x. With this notation, $\mu(A_1 \times A_2) = \int_{A_2} \mu^x(A_1) dp(x)$ for every pair of measurable sets $A_1 \subset Y_1$ and $A_2 \subset Y_2$.

The domain we consider is a bounded connected open set $\Omega \subset \mathbb{R}^n$, having a C^1 boundary.

REMARK 2.1. To simplify the presentation of our results we consider domains which are connected sets with C^1 boundary. Indeed, our results can be extended to test to non-smooth domains Ω, such as domains with a Lipschitz boundary, which satisfy the divergence theorem used in Lemma 3.1.

3. General Lemmas

To verify our results in optimization, we need to take meaningful limits of $W^{1,1}$-functions, which do not necessarily converge in the appropriate Sobolev norm. To this end we shall use probability measures to represent the statistical information of these functions, and take the limit in the measure space. Specifically, two types of representations by probability measures are considered. The first type of measure provides general statistical information in the domain Ω, and a function $u(\cdot) \in W^{1,1}(\Omega)$ is represented by the probability measure $\mu \in P(\mathbb{R} \times \mathbb{R}^n)$, also known as an occupational measure, defined by

$$\mu(I \times V) = \frac{1}{\mathcal{L}^n(\Omega)} \mathcal{L}^n(x \in \Omega | u(x) \in I, \nabla u(x) \in V), \tag{3.1}$$

for every open sets $I \subset \mathbb{R}$ and $V \subset \mathbb{R}^n$. The second type of probability measures provides also local information in the phase space. It is an element of $P(\Omega \times \mathbb{R} \times \mathbb{R}^n)$, and is defined by

$$\rho(O \times I \times V) = \mathcal{L}^n(x \in O | u(x) \in I, \epsilon \nabla u \in V)$$

for every open set $O \subset \Omega$, $I \subset \mathbb{R}$ and $V \subset \mathbb{R}^n$. Note that the the gradient here is multiplied by ϵ.

In this section we provide several results on measures corresponding to functions, which rely on the following lemma.

LEMMA 3.1. Suppose $u(\cdot) \in W^{1,1}(\Omega)$ and $\Omega \subset \mathbb{R}^n$ is open, bounded and has a C^1-boundary. The following holds:

(1) If the trace of $u(\cdot)$ is in $L^\infty(\partial\Omega)$ then

$$\left| \int_\Omega \nabla u(x)\, dx \right| \le \|Tr(u(\cdot))\|_{\partial\Omega,\infty}.$$

(2) When the trace of $u(\cdot)$ is constant almost everywhere (a.e.) in $\partial\Omega$ then

$$\int_\Omega \nabla u(x)\, dx = \mathbf{0}.$$

(3) For every interval (a,b) we have that

$$\left|\int_{\{x\in\Omega|u(x)\in(a,b)\}}\nabla u(x)\,dx\right|\leq(b-a)\,\mathcal{L}^{n-1}(\partial\Omega)\,.$$

PROOF. The first result follows from Stokes theorem, since

$$\left|\int_{\Omega}\nabla u(x)\,dx\right|=\left|\int_{\partial\Omega}Tr\,(u(\cdot))\,(x)\,n_{\Omega}(x)\,dx\right|\leq\|Tr\,(u(\cdot))\|_{\partial\Omega,\infty}\,\mathcal{L}^{n-1}(\partial\Omega)\,,$$

where $n_{\Omega}(\cdot)$ denotes the outer unit normal vector. The second result follows from the first by replacing $u(\cdot)$ by $u(\cdot)-u(x_0)$ for some $x_0\in\partial\Omega$.

Setting

$$\bar{u}(x)=\begin{cases}0 & u(x)\leq a\\ u(x)-a & a<u(x)<b\\ b-a & b\leq u(x)\end{cases},$$

we obtain

$$\begin{aligned}\left|\int_{\{x\in\Omega|u(x)\in[a,b]\}}\nabla u(x)\,dx\right| &= \left|\int_{\Omega}\nabla\bar{u}(x)\,dx\right|\\ &\leq \|Tr\,(\bar{u}(\cdot))\|_{\partial\Omega,\infty}\,\mathcal{L}^{n-1}(\partial\Omega)\\ &\leq (b-a)\,\mathcal{L}^{n-1}(\partial\Omega)\,.\end{aligned}$$

□

We now apply the latter lemma to limits of $W^{1,1}$ functions, obtaining local properties on the limit of their corresponding measures. When considering functions which are constant on $\partial\Omega$, the following result holds.

COROLLARY 3.2. Suppose $\Omega\subset\mathbb{R}^n$ is open, bounded, and has a C^1 boundary, and $u(\cdot)\in W^{1,1}(\Omega)$ has constant trace a.e.. Then

$$\int_{\{x\in\Omega|u(x)\in[a,b]\}}\nabla u(x)\,dx=\mathbf{0}\in\mathbb{R}^n,$$

for every $a<b$. Furthermore, let $\mu\in P(\mathbb{R}\times\mathbb{R}^n)$ be an occupational measure defined by (3.1) for every open sets $I\subset\mathbb{R}$ and $V\subset\mathbb{R}^n$. Then the disintegration of the measure $\mu=\mu^u\circledast p$ satisfies $\mu^u\in P_0(\mathbb{R}^n)$ for p-almost every $u\in\mathbb{R}$.

PROOF. Suppose $u(\cdot)\equiv c$ in $\partial\Omega$. We set

$$\bar{u}(x)=\begin{cases}a & u_i(x)\leq a\\ u(x) & a<u_i(x)<b\\ b & b\leq u_i(x)\end{cases}.\tag{3.2}$$

The function $\bar{u}(\cdot)$ is constant on $\partial\Omega$, where it is equal to c if $c\in[a,b]$, and otherwise either to a or to b. Thus, by Lemma 3.1

$$\int_{\{x\in\Omega|u(x)\in[a,b]\}}\nabla u(x)\,dx=\int_{\Omega}\nabla\bar{u}(x)\,dx=\mathbf{0}.$$

The second result follows from the fact that for every $a < b$,

$$\begin{aligned}\int_{[a,b]}\left(\int_{\mathbb{R}^n} v\mu^x(v)\right)p(dx) &= \int_{[a,b]\times\mathbb{R}^n} vd\mu(v)\\ &= \frac{1}{\mathcal{L}^n(\Omega)}\int_{\{x\in\Omega|u(x)\in[a,b]\}} \nabla u(x)\,dx = 0.\end{aligned}$$

Since the latter equality holds for every interval $[a,b]$, the Lebesgue-Besichovich theorem (see, Evans and Gariepy [**7**, page 43]) implies the desired result. □

The latter corollary implies that when the function $u(\cdot)$ is constant on the boundary of $\partial\Omega$, then the total gradient $\nabla u(\cdot)$ in every vertical slice of the function $u(\cdot)$ in Ω is zero.

This result is now extended to non-homogenous boundary conditions. Here we use a different occupational measure, that takes into account the position as well as the value of $u(\cdot)$ and $\nabla u(\cdot)$ (in a similar manner to Young measures). Notice that now our measures are in $P(\Omega\times\mathbb{R}\times\mathbb{R}^n)$ rather than $P(\mathbb{R}\times\mathbb{R}^n)$, and that $\nabla u(\cdot)$ is multiplied by ϵ.

In this case a weaker result holds.

COROLLARY 3.3. Let $u_i(\cdot)\in W^{1,1}(\Omega)$ and $\epsilon_i\to 0$. For every i we define the probability measure $\rho_i\in P(\Omega\times\mathbb{R}\times\mathbb{R}^n)$ by

$$\rho_i(O\times I\times V) = \mathcal{L}^n(x\in O|u_i(x)\in I,\epsilon_i\nabla u_i\in V), \tag{3.3}$$

for every open sets $O\subset\Omega$, $I\subset\mathbb{R}$ and $V\subset\mathbb{R}^n$. If the sequence of measures ρ_i converge to the measure ρ_0. Then for every interval $[a,b]$ and every open set $O\subset\Omega$ with a C^1 boundary

$$\int_{O\times[a,b]\times\mathbb{R}^n} vd\rho_0(x,u,v) = \mathbf{0}\in\mathbb{R}^n.$$

Furthermore, the disintegration of the measure, $\rho_0 = \rho_0^{x,u}\circledast p_0$, satisfies

$$\int_{\mathbb{R}^n} vd\rho_0^{u,x}(v) = \mathbf{0}\in\mathbb{R}^n,$$

for p_0-almost every $x\in\Omega, u\in\mathbb{R}$.

PROOF. By the weak convergence, for every ball $B\subset\mathbb{R}^n$ and interval $[a,b]$ we have that

$$\begin{aligned}\left|\int_{B\times[a,b]\times\mathbb{R}^n} vd\rho_0(x,u,v)\right| &= \lim_{i\to\infty}\left|\int_{B\times[a,b]\times\mathbb{R}^n} vd\rho_i(x,u,v)\right|\\ &= \lim_{i\to\infty}\left|\int_{\{x\in B|u_i(x)\in[a,b]\}}\epsilon_i\nabla u_i(x)\,dx\right|\\ &= \lim_{i\to\infty}\left|\int_B \epsilon_i\nabla\bar{u}_i(x)\,dx\right|,\end{aligned}$$

where $\bar{u}_i(\cdot)$ is defined by (3.2). Lemma 3.1 bounds the latter by

$$\lim_{i\to\infty}\left(\epsilon_i\mathcal{L}^{n-1}(\partial B)(b-a)\right) = 0,$$

and the Lebesgue-Besichovich theorem completes the proof. □

Our main result for space independent costs of the form $L(u,v)$, establishes that the minimum value for such costs is $v^* = \inf_{\nu\in P_0(\mathbb{R}^n)} \int_{\mathbb{R}^n} L(\alpha,v)\,d\nu(v)$, for some $\alpha \in \mathbb{R}$. According to the lemma below it suffices to consider measures with a finite support.

LEMMA 3.4. Suppose $G : \mathbb{R}^n \to \mathbb{R}$ is continuos. Then the set

$$A = \left\{ \int_{\mathbb{R}^n} G(v)\,d\nu(v)\,|\nu \in P_0(\mathbb{R}^n) \right\} \subset \mathbb{R}$$

is equal to the set

$$\left\{ \sum_{j=1}^{n+2} \lambda_j G(v_j)\,|v_j \in \mathbb{R}^n,\ \sum_{j=1}^{n+2} \lambda_j v_j = 0,\quad \sum_{j=1}^{n+2} \lambda_j = 1,\ \lambda_j \in [0,1] \right\}. \tag{3.4}$$

PROOF. The set

$$\hat{B} = \left\{ \int_{\mathbb{R}^n} [G(v), v] d\mu(v)\,|\mu \in P(\mathbb{R}^n) \right\} \subset \mathbb{R}^{n+1},$$

is convex as a linear map of the convex set $P(\mathbb{R}^n)$. The extreme points of $P(\mathbb{R}^n)$ are Dirac measures, hence, the extreme points of $\hat{B}$ correspond to Dirac measures, and by Caratheodory's theorem we conclude that

$$\hat{B} = \left\{ \sum_{j=1}^{n+2} \lambda_j [G(v_j), v_j]\,|v_j \in \mathbb{R}^n,\quad \sum_{j=1}^{n+2} \lambda_j = 1,\ \lambda_j \in [0,1] \right\} \subset \mathbb{R}^{n+1}.$$

With this definition we obtain that the set

$$\begin{aligned} \hat{A} &= \left\{ \int_{\mathbb{R}^n} [G(v), v]\,d\nu(v)\,|\nu \in P_0(\mathbb{R}^n) \right\} \\ &= \left\{ \left[\int_{\mathbb{R}^n} G(v)\,d\nu(v), 0 \right] |\nu \in P_0(\mathbb{R}^n) \right\} \subset \hat{B}, \end{aligned}$$

equals to

$$\left\{ \left[\sum_{j=1}^{n+2} \lambda_j G(v_j), \mathbf{0} \right] |v_j \in \mathbb{R}^n,\quad \sum_{j=1}^{n+2} \lambda_j = 1,\ \lambda_j \in [0,1],\ \sum_{j=1}^{n+2} \lambda_j v_j = 0 \right\}.$$

Since A is the projection onto the first coordinate of $\hat{A}$ we are done. □

In the construction of the optimal sequence, we employ a lemma found in Sychev [**12**, Lemma 4.2], which implies the following result.

LEMMA 3.5. Suppose that $\alpha \in \mathbb{R}$, and $\nu \in P_0(\mathbb{R}^n)$ is of the form $\nu = \sum_{i=1}^m \lambda_i \delta_{v_i}$, where $\delta_{v_i} \in P(\mathbb{R}^n)$ is the Dirac measure concentrated at $v_i \in \mathbb{R}^n$, $\lambda_i > 0$ and $\sum_{i=1}^m \lambda_i = 1$. There exists a sequence of function $u_i(\cdot) \in W^{1,1}(\Omega)$ satisfying the following conditions:

(1) The sequence $u_i(\cdot)$ converges to the constant function $u_0(\cdot) \equiv \alpha$ in the supremum norm.
(2) The sequence of gradients $\nabla u_i(\cdot) \in L^\infty(\Omega)$ is uniformly bounded in norm by $\max_{i=1,\dots,m} |v_i|$.
(3) The sequence of gradients $\nabla u_i(\cdot)$ converges to the homogenous (constant) Young measure $\nu(\cdot) \equiv \nu$.

4. Space Invariant Costs

In this section we study space independent costs of the form $L(u,v)$. We establish that the optimal value is attained by a sequence of functions that converges point-wise in Ω, to a constant function. This constant may not be unique, and there also may be other optimal solutions containing no converging subsequence.

Our assumptions on the cost function $L(u,v)$ are stated below.

ASSUMPTION 4.1. We assume that the function $L(u,v)$ is continuous and there exists $u_{opt} \in \mathbb{R}$ such that $L_0 = L_0(u_{opt})$ satisfies $L_0(\alpha) \geq L_0 > -\infty$ for every $\alpha \in \mathbb{R}$, where

$$L_0(\alpha) = \inf\left\{\int_{\mathbb{R}^n} L(\alpha, v)\, d\nu(v) \,|\nu \in P_0(\mathbb{R}^n)\right\}.$$

REMARK 4.2. The condition of Assumption 4.1 implies that while $L(u,v)$ can obtain arbitrary small values, the rate it decreases is at most linear, namely, there exists a $C_1, C_2 > 0$ such that $L(\alpha, v) \geq -C_1 |v| - C_2$ for all $\alpha \in \mathbb{R}$ and $v \in \mathbb{R}^n$ sufficiently large. Furthermore, one can see that for every $\alpha \in \mathbb{R}$ the function $L_0(\alpha)$ equals $L_v^{**}(\alpha, \mathbf{0})$ is the biconjugate function of $L(\alpha,\cdot)$, i.e., the largest convex lower semi-continuous function smaller or equal to $L(\alpha,\cdot)$. Our assumption imply that $L_v^{**}(\cdot, \mathbf{0})$ attains its minimum at u_{opt}.

Note that we do not assume that $L(u,v)$ is neither convex in v nor bounded. For example the function

$$L(u,v) = \begin{cases} 2v & v > 0 \\ v & v \leq 0 \end{cases}$$

satisfies our assumptions.

We first prove our result for homogeneous boundary conditions, and then extended it for more general boundary conditions. Note, that all the results we present in this section enforce constraints only on the boundary conditions, and there are no constraints inside Ω.

PROPOSITION 4.3. Suppose that $L(u,v)$ satisfies Assumption 4.1. The minimal value of the optimization problem

$$v^* = \lim_{\epsilon \to 0} \inf_{u \in W_0^{1,1}(\Omega)} \frac{1}{\mathcal{L}^n(\Omega)} \int_\Omega L(u(x), \epsilon \nabla u(x))\, dx$$

is $v^* = L_0$. Moreover, it is attained by a sequence of functions in $W_0^{1,1}(\Omega)$ converging point-wise, in Ω, to a constant function.

PROOF. We start by showing that $v^* \geq L_0$. Let $u(\cdot) \in W_0^{1,1}(\Omega)$ and $\epsilon > 0$ be arbitrary. As in Corollary 3.2, we define the occupational measure corresponding to $u(\cdot)$ by (3.1) for every open set $I \subset \mathbb{R}$ and $V \subset \mathbb{R}^n$, and its disintegration by $\mu = \mu^u \circledast p$. With this definition we have that the value corresponding to $u(\cdot)$ equals

$$\begin{aligned} \frac{1}{\mathcal{L}^n(\Omega)} \int_\Omega L(u(x), \epsilon \nabla u(x))\, dx &= \int_{\mathbb{R}\times\mathbb{R}^n} L(u, \epsilon v)\, d\mu(u,v) \\ &= \int_{\mathbb{R}} \left(\int_{\mathbb{R}^n} L(u, \epsilon v)\, d\mu^u(v) \right) dp(u). \end{aligned}$$

By Corollary 3.2, for p-almost every $u \in \mathbb{R}$ we have that $\mu^u \in P_0(\mathbb{R}^n)$, thus, according to Assumption 4.1

$$\int_{\mathbb{R}} \left(\int_{\mathbb{R}^n} L(u, \epsilon v)\, d\mu^u(v) \right) dp(u) \geq \int_{\mathbb{R}} L_0(u)\, dp(u) \geq L_{opt}, \tag{4.1}$$

and $v^* \geq L_{opt}$.

To show that the value $L_0(u_{opt})$ can be attained we apply Lemma 3.5. We fix a decreasing sequence $\epsilon_i \to 0$. For each i we choose a probability measure $\nu_i \in P_0(\mathbb{R}^n)$, with a finite support (by Lemma 3.4), so that

$$\int_{\mathbb{R}^n} L(u_{opt}, v)\, d\nu_i(v) \leq L_0 + \epsilon_i. \tag{4.2}$$

For each i we assume that ν_i is supported in a ball $B(\mathbf{0}, r_i)$, and we choose $\delta_i < 1$ so that

$$|L(\alpha, v) - L(\alpha + \delta, v)| < \epsilon_i, \tag{4.3}$$

for every $v \in B(\mathbf{0}, r_i)$ and $\delta \in [-\delta_i, \delta_i]$. Applying Lemma 3.5, for every i we construct a function $\tilde{u}_i(\cdot) \in W^{1,1}(\Omega)$ with values in $[\alpha - \delta_i, \alpha + \delta_i]$ and with gradients contained in the ball $B(0, r_i)$, satisfying

$$\left| \frac{1}{\mathcal{L}^n(\Omega)} \int_{\Omega} L(u_{opt}, \epsilon_i \nabla \tilde{u}_i(x))\, dx - \int_{\Omega} L(u_{opt}, v)\, d\nu_i(v) \right| \leq \epsilon_i. \tag{4.4}$$

Inequality (4.3) implies that

$$\frac{1}{\mathcal{L}^n(\Omega)} \left| \int_{\Omega} L(\tilde{u}_i(x), \epsilon_i \nabla \tilde{u}_i(x))\, dx - \int_{\Omega} L(u_{opt}, \epsilon_i \nabla \tilde{u}_i(x))\, dx \right| \leq \epsilon_i. \tag{4.5}$$

Combining (4.2), (4.4) and (4.5) we obtain

$$\left| \frac{1}{\mathcal{L}^n(\Omega)} \int_{\Omega} L(\tilde{u}_i(x), \epsilon_i \nabla \tilde{u}_i(x))\, dx - L_0 \right| \leq 3\epsilon_i,$$

hence the values of the sequence $\tilde{u}_i(\cdot)$ converge to L_0. However $\tilde{u}_i(\cdot)$ may not have trace 0. Modifying each $\tilde{u}_i(\cdot)$ in a small neighborhood of $\partial\Omega$, while controlling their gradients, one obtains a sequence in $W_0^{1,1}(\Omega)$ with values converging to L_0. This implies that $v^* \leq L_0$, and $v^* = L_0$. □

The latter result can be extended to non-homogeneous boundary conditions.

THEOREM 4.4. Suppose $g(\cdot) \in Lip(\partial\Omega)$. The minimal value of the optimization problem

$$v^* = \lim_{\epsilon \to 0} \inf_{u \in \mathcal{S}^g} \frac{1}{\mathcal{L}^n(\Omega)} \int_{\Omega} L(u(x), \epsilon \nabla u(x))\, dx,$$

with respect to functions with a non-homogenous boundary conditions, namely, functions in the set

$$\mathcal{S}^g = \left\{ u(\cdot) \in W^{1,1}(\Omega)\, | Tr(u(\cdot)) = g(\cdot) \right\},$$

is $v^* = L_0$. Moreover, it is attained by a sequence of functions in $\mathcal{S}^g$ converging point-wise, in Ω, to a constant function.

PROOF. Let us fix an optimal sequence $\epsilon_i \to 0$ and $u_i(\cdot) \in \mathcal{S}^g$ satisfying $\lim_{i\to\infty} v_i = v^*$, where

$$v_i = \frac{1}{\mathcal{L}^n(\Omega)} \int_\Omega L(u_i(x), \epsilon_i \nabla u_i(x))\, dx. \tag{4.6}$$

Fix $\eta > 0$ and an open set

$$\Omega \subset \{x | \exists y \in \Omega \quad |x-y| < \eta\} \subset \tilde{\Omega} \subset \{x | \exists y \in \Omega \quad |x-y| < 2\eta\},$$

with a C^1 boundary. We let u_0 be an extension of a function in $\mathcal{S}^g$ to a function in $W_0^{1,1}\left(\tilde{\Omega}\right)$ that satisfies a Lipschitz condition outside Ω.

For every i, we defined the following sequence of functions in $W^{1,1}\left(\tilde{\Omega}\right)$

$$\tilde{u}_i(x) = \begin{cases} u_i(x) & x \in \Omega \\ u_0(x) & x \in \tilde{\Omega}\backslash\Omega \end{cases}.$$

Now we consider the value of the latter sequence in the set $\tilde{\Omega}$, which is

$$\lim_{i\to\infty} \frac{1}{\mathcal{L}^n\left(\tilde{\Omega}\right)} \int_{\tilde{\Omega}} L(\tilde{u}_i(x), \epsilon_i \nabla \tilde{u}_i(x))\, dx. \tag{4.7}$$

Separating the integral to the set Ω and $\tilde{\Omega}\backslash\Omega$ we obtain

$$\lim_{i\to\infty} \frac{1}{\mathcal{L}^n\left(\tilde{\Omega}\right)} \left[\int_\Omega L(u_i(x), \epsilon_i \nabla u_i(x))\, dx + \int_{\tilde{\Omega}\backslash\Omega} L(u_0(x), \epsilon_i \nabla u_0(x))\, dx\right].$$

The limit of the first integral is $\mathcal{L}^n(\Omega)\, v^*$, and since $u_0(\cdot)$ is a fixed function the limit of the second integral converges to

$$\left|\int_{\tilde{\Omega}\backslash\Omega} L(u_0(x), \mathbf{0})\, dx\right| \le \mathcal{L}^n\left(\tilde{\Omega}\backslash\Omega\right) |L_0|.$$

Since $\eta > 0$ can be chosen arbitrary small, we have that the value of (4.7) can be chosen arbitrarily close to v^*, which by Proposition 4.3 is bounded by L_0. Applying a similar construction to the one described in the proof of Proposition 4.3, we conclude that the value L_0 can be attained in $\mathcal{S}^g$, and that $v^* = L_0$. □

The latter result can be extended to consider varying boundary conditions in the following two cases.

COROLLARY 4.5. For every ϵ let $G_\epsilon \subset Lip(\partial\Omega)$ satisfy $\|g(\cdot)\|_{\partial\Omega,\infty} \le M$ and $\|\nabla g(\cdot)\|_{\partial\Omega,\infty} \le K$ for every $g(\cdot) \in G_\epsilon$, and

$$\mathcal{S}^{G_\epsilon} = \left\{u(\cdot) \in W^{1,1}(\Omega) \,|Tr(u(\cdot)) \in G_\epsilon\right\}.$$

The minimal value of the optimization problem

$$v^* = \lim_{\epsilon\to 0} \inf_{u\in\mathcal{S}^{G_\epsilon}} \frac{1}{\mathcal{L}^n(\Omega)} \int_\Omega L(u(x), \epsilon\nabla u(x))\, dx$$

is $v^* = L_0$. Moreover, it is attained by a sequence of functions in $\mathcal{S}^{G_\epsilon}$ converging point-wise, in Ω, to a constant function.

PROOF. We follow the construction in the proof of Theorem 4.4, with the exception that for every $i = 1, 2, \dots$ we extend $u_i(\cdot)$ to an an element of $W_0^{1,1}\left(\tilde{\Omega}\right)$, while keeping it bounded in absolute value by M, and satisfying a Lipschitz condition with constant $\max\left(\epsilon_i^{-1}K, 2M/\eta\right)$ in $\tilde{\Omega}\backslash\Omega$. In this case (4.7) equals

$$(4.8)\quad \lim_{i\to\infty} \frac{1}{\mathcal{L}^n\left(\tilde{\Omega}\right)}\left[\int_\Omega L\left(u_i(x), \epsilon_i \nabla u_i(x)\right) dx + \int_{\tilde{\Omega}\backslash\Omega} L\left(\tilde{u}_i(x), \epsilon_i \nabla \tilde{u}_i(x)\right) dx\right],$$

and since the norm of $\epsilon_i \nabla \tilde{u}_i(\cdot)$ is bounded by K for all large enough i almost everywhere in $\tilde{\Omega}\backslash\Omega$ we bound the second term by

$$\max_{v\in B(\mathbf{0},K)}\left|\int_{\tilde{\Omega}\backslash\Omega} L\left(\tilde{u}_i(x), v\right) dx\right| \le \mathcal{L}^n\left(\tilde{\Omega}\backslash\Omega\right) \max_{|\alpha|\le M, |v|\le K} \left|L(\alpha, v)\right|,$$

and the poof follows. □

When $L(u, v)$ is bounded a stronger result holds.

COROLLARY 4.6. If $L(u, v)$ is bounded in absolute value then the minimal value of the optimization problem

$$v^* = \lim_{\epsilon\to 0} \inf_{u\in W^{1,1}(\Omega)} \frac{1}{\mathcal{L}^n(\Omega)} \int_\Omega L\left(u(x), \epsilon\nabla u(x)\right) dx$$

is $v^* = L_0$. Moreover, it is attained by a sequence of functions in $W^{1,1}$ converging point-wise, in Ω, to a constant function.

PROOF. We follow the proof of Corollary 4.5, with the exception that for every i we extend $u_i(\cdot)$ to an arbitrary function in $W_0^{1,1}\left(\tilde{\Omega}\right)$. In this case, we bound the absolute value of the second term in (4.8) by $\mathcal{L}^n\left(\tilde{\Omega}\backslash\Omega\right) L_M$, where L_M bounds the absolute value of $L(x, v)$, and the proof follows. □

5. Space Dependent Costs

When considering space dependent costs, we cannot expect an optimal sequence of solutions to converge to a constant valued function. However there still is similarity in the nature of the optimal sequence. While the optimal sequence does not converge to a constant function, one can find a sequence where the gradients converge, in the sense of Young measures, to a Young measure with zero mean almost everywhere.

In this case we weaken our assumptions on the cost function $L(x, u, v)$, merely requiring it to be continuous, while strengthening our assumptions on the set of test functions considered. We consider sets of functions, which are uniformly bounded and having a Lipschitz constant proportional to ϵ^{-1}, defined by

$$\mathcal{S}_{M,\epsilon^{-1}K} = \left\{u \in Lip(\Omega) \mid \|u(\cdot)\|_{\Omega,\infty} \le M \text{ and } \|\nabla u(\cdot)\|_{\Omega,\infty} \le \epsilon^{-1}K\right\}.$$

The definition of this set arrises naturally when considering the infinite horizon problem in (1.2) on inflated domains of the form on $\epsilon^{-1}\Omega = \{y | \epsilon y \in \Omega\}$, with respect to uniformly bounded and uniformly Lipschitz functions.

We now provide our main result for space dependent costs, following which, we present two examples that show that our requirements on the set of test functions is necessary.

THEOREM 5.1. If $L(x,u,v)$ is continuous then the minimal value of the optimization problem

$$v^* = \lim_{\epsilon\to 0} \min_{u\in\mathcal{S}_{M,\epsilon^{-1}K}} \frac{1}{\mathcal{L}^n(\Omega)} \int_\Omega L(x,u(x),\epsilon\nabla u(x))\,dx,$$

equals $\frac{1}{\mathcal{L}^n(\Omega)}\int_\Omega L_0(x)\,dx$ where $L_0(x) = \min_{\alpha\in[-M,M]} L_0(x,\alpha)$ and

$$L_0(x,\alpha) = \min\left\{\int_{\mathbb{R}^n} L(x,\alpha,v)\,d\nu(v) \,|\, \nu \in P_0([-K,K]^n)\right\}.$$

Moreover, it is attained by a sequence of functions converging almost everywhere to the function $u_{opt}(x) \equiv \arg\min_{\alpha\in[-M,M]} L_0(x,\alpha)$ in Ω.

PROOF. We fix a minimizing sequence $\epsilon_i \to 0$ and $u_i(\cdot) \in \mathcal{S}_{M,\epsilon_i^{-1}K}$, with corresponding values

$$v_i = \frac{1}{\mathcal{L}^n(\Omega)} \int_\Omega L(x,u_i(x),\epsilon_i\nabla u_i(x))\,dx$$

converging to v^*. For each $u_i(\cdot)$ we define a corresponding probability measure $\rho_i \in P(\Omega\times\mathbb{R}\times\mathbb{R}^n)$ by

$$\rho_i(O\times I\times V) = \mathcal{L}^n(x\in O | u_i(x)\in I, \epsilon_i\nabla u_i \in V),$$

for every open sets $O\subset\Omega$, $I\subset\mathbb{R}$ and $V\subset\mathbb{R}^n$. With this notation

$$v_i = \frac{1}{\mathcal{L}^n(\Omega)} \int_{\Omega\times\mathbb{R}\times\mathbb{R}^n} L(x,u,v)\,d\rho_i(x,u,v)$$

Since all measures are supported on the compact set $\Omega\times[-M,M]\times[-K,K]^n$, by Prohorov theorem (see, Billingsley [**3**, p. 72]), we can extract a subsequence, which we denote as the sequence itself, converging weakly to the measure ρ_0, that is also supported on $\Omega\times[-M,M]\times[-K,K]^n$. The weak convergence implies that

$$v^* = \frac{1}{\mathcal{L}^n(\Omega)} \int_{\Omega\times\mathbb{R}\times\mathbb{R}^n} L(x,u,v)\,d\rho_0(x,u,v).$$

Using the disintegration $\rho_0 = \rho_0^{x,u} \circledast p_0$ we have that

$$v^* = \frac{1}{\mathcal{L}^n(\Omega)} \int_{\Omega\times\mathbb{R}} \left(\int_{\mathbb{R}^n} L(x,u,v)\,d\rho_0^{x,u}(v)\right) dp_0(x,u).$$

By Corollary 3.3 the measure $\rho_0^{x,u} \in P_0(\mathbb{R}^n)$ for p_0 almost every x,u. This implies that

$$v^* \geq \frac{1}{\mathcal{L}^n(\Omega)} \int_\Omega L_0(x)\,dx.$$

To see that the value $\frac{1}{\mathcal{L}^n(\Omega)}\int_\Omega L_0(x)\,dx$ is attained, we partition our domain Ω to cubes with edges of length $1/k$, for a parameter k, and apply a similar approximation procedure seen in the previous section. The function $L(x,u,v)$ is uniformly continuous in $\Omega\times[-M,M]\times[-K,K]^n$. Thus, we can choose piecewise constant approximations $w_k(\cdot)$ of $u_{opt}(x) = \arg\min_{a\in[-M,M]} L_0(x,\alpha)$, and a piecewise constant Young measure $\nu_k^x \in P_0([-K,K]^n)$, with a finite support (by Lemma 3.4), both being constant in every cube of our partition, such that

$$\liminf_{k\to\infty} \frac{1}{\mathcal{L}^n(\Omega)} \int_\Omega\int_{\mathbb{R}^n} L(x,w_k(x),v)\,d\nu_k^x(x)(v)\,dx = \frac{1}{\mathcal{L}^n(\Omega)} \int_\Omega L_0(x)\,dx. \tag{5.1}$$

For each fixed k and each cube of our partition, we apply a construction similar to the one in Proposition 4.3, where we require our function to be 0 on the boundaries of the cubes in the partition. By choosing ϵ_i sufficiently small we can construct a function $\tilde{u}_k(\cdot) \in \mathcal{S}_{M,\epsilon_i^{-1}K}$, approximating $w_k(\cdot)$ in most of Ω, such that its value,

$$\frac{1}{\mathcal{L}^n(\Omega)} \int_\Omega L(x, \tilde{u}_k(x), \epsilon_i \nabla \tilde{u}_k(x))\, dx,$$

is arbitrarily close to

$$\frac{1}{\mathcal{L}^n(\Omega)} \int_\Omega \int_{\mathbb{R}^n} L(x, w_k(x), v)\, d\nu_k^x(x)(v)\, dx.$$

By (5.1) the latter expression approaches v^* as k goes to infinity, which completes the proof. □

When we consider unbounded functions our result does not hold as can be seen in the following example.

EXAMPLE 5.2. Let $n = 1$, $L(x, u, v) = -xv - e^{-100 + \frac{1}{|u|+1}}$ and $\Omega = (-1/2, 1/2)$. We shall consider the minimization of

$$v^* = \lim_{\epsilon \to 0} \min_{u \in \mathcal{S}_{\infty,\epsilon^{-1}}} \frac{1}{\mathcal{L}^n(\Omega)} \int_\Omega L(x, u(x), \epsilon \nabla u(x))\, dx.$$

We choose a sequence $\epsilon_i \to 0$ and the sequence of functions in $\mathcal{S}_{\infty,\epsilon^{-1}}$ defined by

$$u_i(x) = \begin{cases} -\epsilon_i^{-1}(x + 1/2) & x < 0 \\ -\epsilon_i^{-1}(1/2 - x) & x \geq 0 \end{cases}.$$

For every i we denote the value of $u_i(\cdot)$ by

$$v_i = \frac{1}{\mathcal{L}^n(\Omega)} \int_\Omega L(x, u_i(x), \epsilon_i \nabla u_i(x))\, dx.$$

Each v_i is bounded from above by

$$-\int_{-1}^{1} x \epsilon_i \nabla u_i(x)\, dx = \int_{-1/2}^{0} x - \int_0^{1/2} x = -\frac{1}{4}.$$

The values satisfy $v_i < -\frac{1}{4} < L_0(x) \equiv -e^{-99}$, which implies that v^* is strictly smaller than $\frac{1}{\mathcal{L}^n(\Omega)} \int_\Omega L_0(x)\, dx = -e^{-99}$.

When we do not control the growth of the gradient our result does not hold as can be seen in the following example.

EXAMPLE 5.3. Let $n = 1$, $L(x, u, v) = v \sum_{k=1}^{\infty} \left(2^{-k} \cos\left(2^{2k} \pi x\right)\right)$ and $\Omega = (-1/2, 1/2)$. We shall consider the minimization of

$$v^* = \lim_{\epsilon \to 0} \min_{u \in \mathcal{S}_{1,\infty}} \frac{1}{\mathcal{L}^n(\Omega)} \int_\Omega L(x, u(x), \epsilon \nabla u(x))\, dx.$$

The value corresponding to the sequences $\epsilon_i = i^{-1}$ and $u_i(x) = -\sin\left(2^{2i} \pi x\right)$ is

$$\begin{aligned} v_i &= \frac{1}{\mathcal{L}^n(\Omega)} \int_\Omega L(x, u_i(x), \epsilon_i \nabla u_i(x))\, dx \\ &= -2^{2i} i^{-1} \pi \int_{-1/2}^{1/2} \sum_{k=1}^{\infty} \left(2^{-k} \cos\left(2^{2k} \pi x\right)\right) \cos\left(2^{2i} \pi x\right) dx. \end{aligned}$$

Separating the term where $k = i$ from the sum we obtain

$$v_i = -2^{2i}\pi i^{-1}[\sum_{k=1,k\neq i}^{\infty} 2^{-k}\int_{-1/2}^{1/2}\cos\left(2^{2k}\pi x\right)\cos\left(2^{2i}\pi x\right)dx + \int_{-1/2}^{1/2} 2^{-i}\cos^2\left(2^{2i}\pi x\right)dx].$$

Each term in the infinite sum equals to zero, hence, $v_i = -2^{i-1}i^{-1}\pi$. This implies that $\lim_{i\to\infty} v_i \to -\infty$, while $L_0(x) = 0$ for every x.

References

[1] Zvi Artstein and Ido Bright, *Periodic optimization suffices for infinite horizon planar optimal control*, SIAM J. Control Optim. **48** (2010), no. 8, 4963–4986, DOI 10.1137/090772617. MR2735512 (2011k:49008)

[2] Zvi Artstein and Arie Leizarowitz, *Singularly perturbed control systems with one-dimensional fast dynamics*, SIAM J. Control Optim. **41** (2002), no. 2, 641–658 (electronic), DOI 10.1137/S0363012901390889. MR1920515 (2003g:49038)

[3] Patrick Billingsley, *Convergence of probability measures*, John Wiley & Sons, Inc., New York-London-Sydney, 1968. MR0233396 (38 #1718)

[4] Guy Bouchitté, *Singular perturbations of variational problems arising from a two-phase transition model*, Appl. Math. Optim. **21** (1990), no. 3, 289–314, DOI 10.1007/BF01445167. MR1036589 (91d:49016)

[5] John W. Cahn and John E. Hilliard. Free energy of a nonuniform system. i. interfacial free energy. *The Journal of Chemical Physics*, 28:258, 1958.

[6] Rustum Choksi, *Nonlocal Cahn-Hilliard and isoperimetric problems: periodic phase separation induced by competing long- and short-range interactions*, Singularities in PDE and the calculus of variations, CRM Proc. Lecture Notes, vol. 44, Amer. Math. Soc., Providence, RI, 2008, pp. 33–45. MR2528732

[7] Lawrence C. Evans and Ronald F. Gariepy, *Measure theory and fine properties of functions*, Studies in Advanced Mathematics, CRC Press, Boca Raton, FL, 1992. MR1158660 (93f:28001)

[8] Arie Leizarowitz, *Variational problems on unbounded two-dimensional domains*, Math. Models Methods Appl. Sci. **2** (1992), no. 2, 183–201, DOI 10.1142/S0218202592000120. MR1167769 (93e:49005)

[9] Arie Leizarowitz, *Radially symmetric minimizers of variational problems in the plane*, Adv. Differential Equations **10** (2005), no. 3, 241–258. MR2123131 (2005m:49007)

[10] Arie Leizarowitz and Victor J. Mizel, *One-dimensional infinite-horizon variational problems arising in continuum mechanics*, Arch. Rational Mech. Anal. **106** (1989), no. 2, 161–194, DOI 10.1007/BF00251430. MR980757 (90b:49007)

[11] Peter Sternberg, *The effect of a singular perturbation on nonconvex variational problems*, Arch. Rational Mech. Anal. **101** (1988), no. 3, 209–260, DOI 10.1007/BF00253122. MR930124 (89h:49007)

[12] Mikhail A. Sychev, *Characterization of homogeneous gradient Young measures in case of arbitrary integrands*, Ann. Scuola Norm. Sup. Pisa Cl. Sci. (4) **29** (2000), no. 3, 531–548. MR1817708 (2001m:26029)

[13] John S. Rowlinson, *Translation of Johannes D. van der Waals' "The thermodynamic theory of capillarity under the hypothesis of a continuous variation of density"*, J. Statist. Phys. **20** (1979), no. 2, 197–244, DOI 10.1007/BF01011513. MR523642 (82e:01128)

Department of Applied Mathematics, University of Washington
E-mail address: `ibright@uw.edu`

Contemporary Mathematics
Volume **659**, 2016
http://dx.doi.org/10.1090/conm/659/13192

Strong convergence theorems for bounded accretive operators in uniformly smooth Banach spaces

C. E. Chidume

ABSTRACT. Let E be a uniformly smooth real Banach space and $A : E \to 2^E$ be a multi-valued bounded m-accretive mapping. Assume that the inclusion $0 \in Au$ has a solution. An iterative process is constructed which converges *strongly* to a solution of this inclusion. Our theorems complement several important results on *the proximal point algorithm* for approximating a solution of this inclusion (assuming existence). Furthermore, our theorems resolve some important open questions. We achieve this by combining our technique with a celebrated theorem of Reich on the strong convergence of the resolvent of m-accretive operators in uniformly smooth real Banach spaces.

1. Introduction

Let E be a normed space. For a multi-valued map $A : E \to 2^E$, the domain of A, $D(A)$, the image of a subset S of E, $A(S)$, the range of A , $R(A)$ and the graph of A, $G(A)$ are defined as follows:

$$D(A) := \{x \in E \,:\, Ax \neq \emptyset\};\; A(S) := \cup\{Ax \,:\, x \in S\};$$
$$R(A) := A(E);\; G(A) := \{[x,u] :\; x \in D(A),\; u \in Ax\}.$$

A multi-valued map $A : E \to 2^E$ with domain $D(A)$ is called *accretive* if the inequality

$$\|x - y\| \leq \|x - y + s(u - v)\| \tag{1.1}$$

holds for each $x, y \in D(A)$, $u \in Ax$, $v \in Ay$ and for all $s > 0$. It is *maximal accretive* if, in addition, the graph of A is not properly contained in the graph of any other accretive operator $A' : E \to 2^E$. It is called maximal accretive if and only if A is accretive and $R(I + tA) = E$ for all $t > 0$. If E is a real Hilbert space, accretive mappings are called *monotone* and maximal accretive mappings are called *maximal monotone*.

A fundamental problem in the study of monotone operators is that of finding an element $u \in H$ such that $0 \in Au$. This problem has been investigated by many researchers.

Let $f : H \to (-\infty, \infty]$ be a proper, lower semicontinuous and convex function.

2010 *Mathematics Subject Classification.* Primary 47H04, 47H06, 47H15, 47H17, 47J25.
Key words and phrases. Accretive mappings, fixed points, multi-valued mappings.

Then, it is known that the multi-valued map $T := \partial f$, the subdifferential of f, is *maximal monotone* (see, e.g., Minty [**22**], Moreau [**9**]), where for $w \in H$,

$$w \in \partial f(x) \Leftrightarrow f(y) - f(x) \geq \langle y - x, w \rangle \ \forall \ y \in H$$
$$\Leftrightarrow x \in \operatorname{argmin}(f - \langle \cdot, w \rangle).$$

In particular, solving the inclusion $0 \in \partial f(x)$ is equivalent to solving the minimization problem $f(x) = \min_{y \in H} f(y)$.

More generally, consider the following problem:

$$\text{Find } u \in H \text{ such that } 0 \in Au, \tag{1.2}$$

where H is a real Hilbert space and $A : H \to 2^H$ is a *maximal monotone operator* on H. The *proximal point algorithm (PPA)* introduced by Martinet [**21**] and studied extensively by Rockafellar [**23**] and numerous authors is concerned with iterative methods for approximating a solution of $0 \in Au$ where A is a maximal monotone operator on a real Hilbert space. Specifically, given $x_k \in H$, the proximal point algorithm generates the next iterate x_{k+1} by solving the following equation:

$$x_{k+1} = \left(I + \frac{1}{\lambda_k} A\right)^{-1}(x_k) + e_k, \tag{1.3}$$

where $\lambda_k > 0$ is a regularizing parameter.

Rockafellar [**23**] proved that if the sequence $\{\lambda_k\}_{k=1}^{\infty}$ is bounded from above, then the resulting sequence $\{x_k\}_{k=1}^{\infty}$ of proximal point iterates converges *weakly* to a solution of $0 \in Au$, provided that a solution exists. He then posed the following question.

Q1. Does the proximal point algorithm always converge strongly?

This question was resolved in the negative by Güler [**16**] who produced a proper closed convex function g in the infinite dimensional Hilbert space l_2 for which the proximal point algorithm converges *weakly* but *not strongly*, (see also Bauschke *et.al.* [**1**]). This naturally raised the following question.

Q2. Can the proximal point algorithm be modified to guarantee strong convergence?

Before we comment on Q2, we make the following remark.

REMARK 1. The proximal point algorithm (1.3) is not at all convenient to use in some possible applications. This is because *at each step of the iteration process*, one has to compute $\left(I + \frac{1}{\lambda_n} A\right)^{-1}(x_n)$ and this is certainly not convenient.

Consequently, while thinking of modifications of the proximal point algorithm that will guarantee strong convergence, the following question is, perhaps, more important than **Q2**.

Q3. Can an iteration process be developed which will not involve the computation of $\left(I+\frac{1}{\lambda_n} A\right)^{-1}(x_n)$ at each step of the iteration process and which will guarantee *strong convergence* to a solution of $0 \in Au$?

First, we review modifications of the *PPA* obtained by several authors that yield strong convergence.

• Bruck [**2**] considered an iteration process of the *Mann-type* and proved that the sequence of the process converges strongly to a solution of the inclusion $0 \in Au$ in a real Hilbert space where A is a maximal monotone map, *provided the initial vector is chosen in a neighbourhood of a solution of this inclusion.* Chidume [**6**] extended this result to L^p spaces, $p \geq 2$ (see also Reich [**11–13**]). These results of Bruck [**2**] and Chidume [**6**] are not easy to use in any possible application because the neighborhood of a solution in which the initial vector must be chosen is not known precisely.

• Solodov and Svaiter [**17**] proposed a modification of the proximal point algorithm which guarantees strong convergence in a real Hilbert space. Their algorithm is as follows.

Algorithm. Choose any $x^0 \in H$ and $\sigma \in [0,1)$. At iteration k, having x_k, choose $\mu_k > 0$, and find (y_k, v_k), an inexact solution of $0 \in Tx + \mu_k(x - x_k)$, with tolerance σ. Define

$$C_k := \Big\{z \in H : \langle z - y^k, v^k\rangle \leq 0\Big\}.$$

$$Q_k := \Big\{z \in H : \langle z - x^k, x^0 - x^k\rangle \leq 0\Big\}.$$

Take

$$x_{k+1} = P_{C_k \cap Q_k}(x^0), k \geq 1.$$

The authors themselves noted ([**17**], p. 195) that "...at each iteration, there are two subproblems to be solved..." : (i) find an inexact solution of the proximal point algorithm, and (ii) find the projection of x^0 onto $C_k \cap Q_k$. They also acknowledged that these two subproblems constitute a serious drawback in using their algorithm.

• Kamimura and Takahashi [**18**] extended this work of Solodov and Svaiter [**17**] to the framework of Banach spaces that are both uniformly convex and uniformly smooth. Reich and Sabach [**14**] extended this result to refexive Banach spaces.

• **Xu** [**24**] noted that "...Solodov and Svaiter's algorithm, though strongly convergent, does need more computing time due to the projection in the second subproblem ...". He then proposed and studied the following algorithm:

$$(*) \qquad x_{n+1} = \alpha_n x_0 + (1 - \alpha_n)\Big(I + c_n A\Big)^{-1}(x_n) + e_n,\ n \geq 0.$$

He proved that $(*)$ converges strongly to a solution of $0 \in Au$ provided that the sequences $\{\alpha_n\}$ and $\{c_n\}$ of real numbers and the sequence $\{e_n\}$ of errors are chosen appropriately. We note here, however, that the occurence of errors is random and so the sequence $\{e_n\}$ cannot actually be chosen.

• Lehdili and Moudafi [**19**] considered the technique of the proximal map and the Tikhonov regularization to introduce the so-called Prox-Tikhonov method which generates the sequence $\{x_n\}$ by the algorithm:

$$(**) \qquad x_{n+1} = J_{\lambda_n}^{A_n} x_n,\ n \geq 0,$$

where $A_n := \mu_n A + A$, $\mu_n > 0$ and $J_{\lambda_n}^{A_n} := \left(I + \dfrac{1}{\lambda_n} A_n\right)^{-1}$. Using the notion of variational distance, Lehdili and Moudafi [**19**] proved strong convergence theorems for this algorithm and its perturbed version, under appropriate conditions on the sequences $\{\lambda_n\}$ and $\{\mu_n\}$.

• Xu also studied the recurrence relation $(**)$. He used the technique of nonexpansive mappings to get convergence theorems for the perturbed version of the algorithm $(**)$, under much relaxed conditions on the sequences $\{\lambda_n\}$ and $\{\mu_n\}$.

• Another modification of the proximal point algorithm, perhaps the most significant, which yields strong convergence, is implicitly contained in the following theorem of Reich.

THEOREM 1.1 (Reich, [**10**]). *Let E be a uniformly smooth real Banach space, and let $A : E \to 2^E$ be m-accretive. Let $J_t x := (I + tA)^{-1} x$, $t > 0$ be the resolvent of A, and assume that $A^{-1}(0)$ is nonempty. Then, for each $x \in E$, $\lim_{t\to\infty} J_t x$ exists and belongs to $A^{-1}(0)$.*

REMARK 2. We have seen, in response to **Q2**, that **each** *modification* of the classical proximal point algorithm to obtain strong convergence so far studied involved: either the computation of
$\left(I + \frac{1}{\lambda_n} A\right)^{-1}(x_n)$ at each step of the iteration process, or the construction of two convex subsets C_n and Q_n of the space E and projecting the initial vectors onto $C_n \cap Q_n$, at each step of the process. This is certainly not convenient in any possible application of the algorithm.

REMARK 3. The proximal point algorithm, however, can still be very useful in some special cases. For example, the algorithm can be used in signal processing and in image restoration in the cases where the proximal mappings are not very difficult to evaluate.

In response to $Q3$, Chidume and Djitte [**3**] provided an affirmative answer.

To present their result, we need the following definitions.

Let E be a real normed linear space of dimension ≥ 2. The *modulus of smoothness* of E, ρ_E, is defined by:

$$\rho_E(\tau) := \sup\left\{\frac{\|x+y\| + \|x-y\|}{2} - 1 : \|x\| = 1, \|y\| = \tau\right\}; \quad \tau > 0.$$

A normed linear space E is called *uniformly smooth* if

$$\lim_{\tau\to 0} \frac{\rho_E(\tau)}{\tau} = 0.$$

For $q > 1$, let J_q denote the *generalized duality mapping* from E to 2^{E^*} defined by

$$J_q(x) := \left\{f \in E^* : \langle x, f\rangle = \|x\|^q \text{ and } \|f\| = \|x\|^{q-1}\right\},$$

where $\langle .,. \rangle$ denotes the generalized duality pairing. J_2 is called the *normalized duality mapping* and is denoted by J. It is well known that if E is smooth, then J_q is single-valued.

Chidume and Djitte [**3**] studied an iteration process studied earlier by Chidume and Zegeye [**5**], for approximating a solution of the equation $Au = 0$, which does not involve the computation of $\left(I+\frac{1}{\lambda_k}A\right)^{-1} x_k$ at any stage, to approximate zeros of *bounded, m-accretive* maps in 2-*uniformly smooth real Banach spaces.* They proved the following theorem which yields an affirmative answer to Q3 in the setting of the theorem.

Theorem CD (Chidume and Djitte, [**3**]). *Let E be a 2-uniformly smooth real Banach space, and let $A : E \to E$ be a bounded m-accretive operator with $N(A) := \{x \in E \ : \ Ax = 0\} \neq \emptyset$. Let a sequence $\{x_n\}$ be generated from arbitrary $x_1 \in E$ by*

$$x_{n+1} = x_n - \lambda_n A x_n - \lambda_n \theta_n (x_n - x_1), \tag{1.4}$$

where $\{\lambda_n\}$ and $\{\theta_n\}$ are sequences in $(0, 1)$ satisfying the following conditions:

(1) $\lim_{n\to\infty} \theta_n = 0$; $\lambda_n(1 + \theta_n) < 1$, $\sum_{n=1}^{\infty} \lambda_n \theta_n = \infty$, $\lambda_n = o(\theta_n)$;

(2) $\lim_{n\to\infty} \dfrac{\left(\frac{\theta_{n-1}}{\theta_n} - 1\right)}{\lambda_n \theta_n} = 0.$

Then, $\{x_n\}$ converges strongly to a solution of $Au = 0$.

REMARK 4. **We note here that L_p spaces, $p \geq 2$ are 2-uniformly smooth but L_p spaces, $1 < p < 2$ are not**. So, theorem CD does not guarantee strong convergence to a solution of $0 \in Au$ on L_p spaces, for $1 < p < 2$.

REMARK 5. Chidume and Zegeye [**5**] proved that the sequence $\{x_n\}$ defined by (1.4) converges strongly in real Banach spaces having uniformly Gâteaux differentiable norm to a zero of A (assuming $N(A) \neq \emptyset$) when A is a *single-valued Lipschitz accretive* map.

REMARK 6. It is known that if $A : D(A) \subseteq E \to 2^E$ is a multi-valued *continuous accretive* map, then it is always single-valued in the interior of its domain. In fact, this result holds if continuity is replaced by lower semi-continuity (see, *e.g.*, Chidume and Morales [**7**], or Chidume [**4**], chapter 23). Consequently, in the theorem of Chidume and Zegeye [**5**], the operator T cannot be assumed to be multi-valued.

It is our purpose in this paper to modify the recursion formula (1.4) and prove that the sequence of the modified formula converges *strongly* to a zero of A (assuming existence) where A is a *multi-valued* $m-$accretive bounded operator in *arbitrary* uniformly smooth real Banach spaces. We achieve this by combining our technique with a celebrated theorem of Reich [**10**] on the strong convergence of the resolvent of m-accretive operators in uniformly smooth real Banach spaces. Furthermore, in the setting of our theorems, we obtain an affirmative answer to **Q3**.

As has been shown above, our theorems which do not involve the computation of $\left(I + \frac{1}{\lambda_k}A\right)^{-1} x_k$ at any step of the iteration process and which do not require the construction of two convex subsets C_n and Q_n of the space E and projecting the initial vectors onto $C_n \cap Q_n$, at each step of the process, complement several important results obtained by modifying the proximal point algorithm to obtain strong convergence theorems for the approximation of a solution of the inclusion $0 \in Au$. In addition, our main theorem extends theorem CD, first, from single-valued maps to *multi-valued maps* and then, from 2-uniformly smooth real Banach spaces to *arbitrary* uniformly smooth real Banach spaces. In particular, our theorems guarantee *strong* convergence in L_p spaces and in Sobolev spaces $W^{m,p}$, for $1 < p < \infty$, which is not the case with theorem CD.

2. Preliminaries

In the sequel, we shall need the following results.

REMARK 7. As a consequence of Kato [**8**], the definition of accretive maps by inequality (1.1) can be reformulated as follows: A multi-valued map $A : E \to 2^E$ is accretive if and only if for each $x, y \in E$, there exists $j(x-y) \in J(x-y)$ such that

$$\langle u - v, j(x-y)\rangle \geq 0, \ \forall\, u \in Ax, v \in Ay.$$

LEMMA 2.1 (Xu, [**25**]). *Let $\{\rho_n\}$ be a sequence of nonnegative real numbers satisfying the following inequality:*

$$\rho_{n+1} \leq (1-\alpha_n)\rho_n + \alpha_n\sigma_n + \gamma_n, \ \ n \geq 0,$$

where,
(i) $\{\alpha_n\} \subset (0,1)$, $\sum \alpha_n = \infty$; *(ii)* $\limsup\limits_{n\to\infty} \sigma_n \leq 0$;
(iii) $\gamma_n \geq 0$, $\sum \gamma_n < \infty$. *Then,* $\rho_n \to 0$ *as* $n \to \infty$.

LEMMA 2.2. *Let E be a normed real linear space. Then, the following inequality holds:*

$$||x+y||^2 \leq ||x||^2 + 2\langle y, j(x+y)\rangle \ \forall j(x+y) \in J(x+y), \ \forall\, x, y \in E. \tag{2.1}$$

LEMMA 2.3 (Xu and Roach, [**26**]). *Let E be a uniformly smooth real Banach space. Then, there exist constants D and C such that for all $x, y \in E, j(x) \in J(x)$, the following inequality holds:*

$$\begin{aligned} ||x+y||^2 &\leq ||x||^2 + 2\langle y, j(x)\rangle \\ &+ D\max\left\{||x|| + ||y||, \frac{1}{2}C\right\}\rho_E(||y||), \end{aligned}$$

where ρ_E denotes the modulus of smoothness of E.

(See also Reich [**15**], for another inequality).

3. Main Results

In the sequel, D and C are the constants appearing in lemma 2.3. We first prove the following lemma.

LEMMA 3.1. *Let E be a uniformly smooth real Banach space with modulus of smoothness ρ_E, and let $A: E \to 2^E$ be a multi-valued bounded accretive operator with $D(A) = E$ such that the inclusion $0 \in Au$ has a solution. For arbitrary $x_1 \in E$, define a sequence $\{x_n\}$ by,*

$$x_{n+1} = x_n - \lambda_n u_n - \lambda_n \theta_n (x_n - x_1), \ u_n \in Ax_n, \ n \geq 1, \tag{3.1}$$

where $\{\lambda_n\}$ and $\{\theta_n\}$ are sequences in $(0,1)$. Then, there exists a constant $\gamma_0 > 0$ such that if $\frac{\rho_E(M_0\lambda_n)}{M_0\lambda_n} \leq \gamma_0\theta_n$, for some $M_0 > 0$, the sequence $\{x_n\}$ is bounded.

PROOF. Let x^* be a solution of $0 \in Au$ i.e., $0 \in Ax^*$. Then, there exists $r > 0$ such that $x_1 \in \overline{B(x^*, \frac{r}{2})} := \{x \in E : ||x - x^*|| \leq \frac{r}{2}\}$. Define $B = \overline{B(x^*, r)}$. It suffices now to prove that $x_n \in B \ \forall \ n \geq 1$. We prove this by induction. By construction, $x_1 \in B$. Assume $x_n \in B$ for some $n \geq 1$. We prove $x_{n+1} \in B$. Since A is bounded, we have that $A(B)$ is bounded and so we can define the following numbers:

$$M_0 := \sup\{||u + \theta(x - x_1)|| : x \in B, \ u \in Ax; \ 0 < \theta < 1\} + 1.$$

$$M^* := \sup\Big\{D\max\Big\{||x|| + \lambda M_0, \frac{C}{2}\Big\} : x \in B, \ \lambda \in (0,1)\Big\}.$$

$$\gamma_0 := \frac{1}{2}min\Big\{1, \frac{r^2}{4M^*M_0}\Big\}.$$

.

Using the recursion formula (3.1), Lemma 2.3, the fact that $h(\tau) := \frac{\rho_E(\tau)}{\tau}$ is non-decreasing, and denoting $0 \in Ax^*$ by 0^*, we compute as follows:

$$\begin{aligned}
||x_{n+1} - x^*||^2 &= ||x_n - x^* - \lambda_n(u_n + \theta_n(x_n - x_1))||^2 \\
&\leq ||x_n - x^*||^2 - 2\lambda_n\langle u_n + \theta_n(x_n - x_1), j(x_n - x^*)\rangle \\
&+ Dmax\Big\{||x_n - x^*|| + \lambda_n||u_n + \theta_n(x_n - x_1)||, \frac{C}{2}\Big\} \\
&\quad \times \rho_E(\lambda_n\|u_n + \theta_n(x_n - x_1)\|) \\
&\leq ||x_n - x^*||^2 - 2\lambda_n\langle u_n - 0^*, j(x_n - x^*)\rangle \\
&\quad -2\lambda_n\theta_n\langle x_n - x_1, j(x_n - x^*)\rangle \\
&+ M^*\rho_E(\lambda_n\|u_n + \theta_n(x_n - x_1)\|) \\
&\leq ||x_n - x^*||^2 - 2\lambda_n\theta_n||x_n - x^*||^2 \\
&\quad +\lambda_n\theta_n(||x^* - x_1||^2 + ||x_n - x^*||^2) \\
&\quad +M^*\rho_E(\lambda_n\|u_n + \theta_n(x_n - x_1)\|) \\
&\leq (1 - \lambda_n\theta_n)\|x_n - x^*\|^2 + \lambda_n\theta_n||x^* - x_1||^2 \\
&\quad +M^*\frac{\rho_E(\lambda_n M_0)}{\lambda_n M_0}.\lambda_n M_0 \\
&\leq (1 - \lambda_n\theta_n)\|x_n - x^*\|^2 + \lambda_n\theta_n||x^* - x_1||^2 + M^*\gamma_0\theta_n\lambda_n M_0 \\
&\leq \Big(1 - \frac{1}{2}\lambda_n\theta_n\Big)r^2 \leq r^2.
\end{aligned}$$

Hence, $x_{n+1} \in \overline{B(x^*, r)} \ \forall \ n \geq 1$, and so $\{x_n\}$ is bounded. $\square$

We now prove the following theorem.

THEOREM 3.2. *Let E be a uniformly smooth real Banach space with modulus of smoothness ρ_E, and let $A : E \to 2^E$ be a multi-valued bounded $m-$accretive operator with $D(A) = E$ such that the inclusion $0 \in Au$ has a solution. For arbitrary $x_1 \in E$, define a sequence $\{x_n\}$ by,*

$$x_{n+1} = x_n - \lambda_n u_n - \lambda_n \theta_n (x_n - x_1),\ u_n \in Ax_n,\ n \geq 1,$$

where $\{\lambda_n\}$ and $\{\theta_n\}$ are sequences in $(0,1)$ satisfying the following conditions: (i) $\lim_{n\to\infty} \theta_n = 0$, $\{\theta_n\}$ is decreasing; (ii) $\sum \lambda_n\theta_n = \infty$; $\sum \rho_E(\lambda_n M_1) < \infty$, for some constant $M_1 > 0$; (iii) $\lim_{n\to\infty} \frac{\left[\frac{\theta_{n-1}}{\theta_n} - 1\right]}{\lambda_n\theta_n} = 0$. There exists a constant $\gamma_0 > 0$ such that $\frac{\rho_E(\lambda_n)}{\lambda_n} \leq \gamma_0\theta_n$. Then the sequence $\{x_n\}$ converges strongly to a zero of A.

PROOF. By a theorem of Reich [**10**], (theorem 1.1 above), there exists a sequence $\{y_n\}$ defined by $y_n := J_{t_n} x_1$ where x_1 is an arbitrary fixed vector in E; set $t_n = \theta_n^{-1}\ \forall\ n \geq 1$ and observe that with this t_n, the sequence $\{y_n\}$ satisfies the following conditions: there exists $z_n \in Ay_n$ such that

$$\theta_n(y_n - x_1) + z_n = 0,\ n \geq 1, \tag{3.2}$$

$$y_n \to y^* \in A^{-1}(0). \tag{3.3}$$

We now prove that $||x_{n+1} - y_n|| \to 0$ as $n \to \infty$.

By lemma 3.1, $\{x_n\}$ is bounded. Also $\{y_n\}$ is bounded, as a convergent sequence. Hence, using the fact that $h(\tau) := \frac{\rho_E(\tau)}{\tau}$ is nondecreasing, there exist constants $M_1 > 0,\ M_2 > 0$ such that

$$\begin{aligned} ||x_{n+1} - y_n||^2 &= ||x_n - y_n - \lambda_n(u_n + \theta_n(x_n - x_1))||^2 \\ &\leq ||x_n - y_n||^2 - 2\lambda_n\langle u_n + \theta_n(x_n - x_1), j(x_n - y_n)\rangle \\ &+ M_2\rho_E(\lambda_n M_1). \end{aligned}$$

Using the fact that A is accretive, we obtain, for $z_n \in Ay_n$, using equation (3.2), that

$$\begin{aligned} \langle u_n + \theta_n(x_n - x_1), j(x_n - y_n)\rangle &= \langle u_n - z_n, j(x_n - y_n)\rangle + \theta_n||x_n - y_n||^2 \\ &+\langle z_n + \theta_n(y_n - x_1), j(x_n - y_n)\rangle \\ &\geq \frac{\theta_n}{2}\Big|\Big|x_n - y_n\Big|\Big|^2. \end{aligned}$$

Therefore,

$$||x_{n+1} - y_n||^2 \leq (1 - \lambda_n\theta_n)||x_n - y_n||^2 + M_2\rho_E(\lambda_n M_1). \tag{3.4}$$

Using again the fact that A is accretive, we have

$$||y_{n-1} - y_n|| \leq \Big|\Big|y_{n-1} - y_n + \frac{1}{\theta_n}\Big(z_{n-1} - z_n\Big)\Big|\Big|. \tag{3.5}$$

Observing from equation (3.2) that

$$y_{n-1} - y_n + \frac{1}{\theta_n}\Big(z_{n-1} - z_n\Big) = \frac{\theta_n - \theta_{n-1}}{\theta_n}\Big(y_{n-1} - x_1\Big),$$

it follows from inequality (3.5) that

$$\|y_{n-1}-y_n\| \leq \frac{\theta_{n-1}-\theta_n}{\theta_n}\Big\|y_{n-1}-x_1\Big\|. \tag{3.6}$$

By lemma (2.2), we have

$$\begin{aligned} \|x_n-y_n\|^2 &= \|(x_n-y_{n-1})+(y_{n-1}-y_n)\|^2 \\ &\leq \|x_n-y_{n-1}\|^2+2\langle y_{n-1}-y_n, j(x_n-y_n)\rangle \\ &\leq \|x_n-y_{n-1}\|^2+2\|y_{n-1}-y_n\|\|x_n-y_n\|. \end{aligned} \tag{3.7}$$

Using (3.4), (3.5), (3.7) and the fact that $\{x_n\}$ and $\{y_n\}$ are bounded, we have,

$$\begin{aligned} \|x_{n+1}-y_n\|^2 &\leq (1-\lambda_n\theta_n)\|x_n-y_{n-1}\|^2+C^*\Big(\frac{\theta_{n-1}-\theta_n}{\theta_n}\Big)+M_2\rho_E(\lambda_nM_1) \\ &= (1-\lambda_n\theta_n)\|x_n-y_{n-1}\|^2+(\lambda_n\theta_n)\sigma_n+\gamma_n, \end{aligned}$$

for some constant $C^*>0$ where,

$$\sigma_n := \frac{C^*\Big(\frac{\theta_{n-1}-\theta_n}{\theta_n}\Big)}{\lambda_n\theta_n} = C^*\Big(\frac{\frac{\theta_{n-1}}{\theta_n}-1}{\lambda_n\theta_n}\Big), \quad \gamma_n := M_2\rho_E(\lambda_nM_1).$$

Thus, by Lemma 2.1, $x_n-y_{n-1}\to 0$. Using (3.3), it follows that $x_n\to y^*$ and $0\in Ay^*$. This completes the proof. □

COROLLARY 3.3. *Let $E=L_p$, $2\leq p<\infty$, and let $A:E\to 2^E$ be a multi-valued bounded $m-$accretive operator with $D(A)=E$ such that the inclusion $0\in Au$ has a solution. For arbitrary $x_1\in E$, define a sequence $\{x_n\}$ by,*

$$x_{n+1}=x_n-\lambda_nu_n-\lambda_n\theta_n(x_n-x_1),\ u_n\in Ax_n,\ n\geq 1,$$

where $\{\lambda_n\}$ and $\{\theta_n\}$ are sequences in $(0,1)$ satisfying the folliowing conditions:
(i) $\lim_{n\to\infty}\theta_n=0$, $\{\theta_n\}$ is decreasing;
(ii) $\sum\lambda_n\theta_n=\infty$; $\sum\lambda_n^2<\infty$; (iii) $\lim_{n\to\infty}\frac{\Big[\frac{\theta_{n-1}}{\theta_n}-1\Big]}{\lambda_n\theta_n}=0$.
Then, there exists a constant $\gamma_0>0$ such that if $\lambda_n\leq\gamma_0\theta_n$, the sequence $\{x_n\}$ converges strongly to a zero of A.

COROLLARY 3.4. *Let $E=L_p$, $1<p<2$, and let $A:E\to 2^E$ be a multi-valued bounded $m-$accretive operator with $D(A)=E$ such that the inclusion $0\in Au$ has a solution. For arbitrary $x_1\in E$, define a sequence $\{x_n\}$ by,*

$$x_{n+1}=x_n-\lambda_nu_n-\lambda_n\theta_n(x_n-x_1),\ u_n\in Ax_n,\ n\geq 1,$$

where $\{\lambda_n\}$ and $\{\theta_n\}$ are sequences in $(0,1)$ satisfying the folliowing conditions:
(i) $\lim_{n\to\infty}\theta_n=0$, $\{\theta_n\}$ is decreasing;
(ii)) $\sum\lambda_n\theta_n=\infty$; $\sum\lambda_n^p<\infty$; (iii) $\lim_{n\to\infty}\frac{\Big[\frac{\theta_{n-1}}{\theta_n}-1\Big]}{\lambda_n\theta_n}=0$.
Then, there exists a constant $\gamma_0>0$ such that if $\lambda_n^{(p-1)}\leq\gamma_0\theta_n$, the sequence $\{x_n\}$ converges strongly to a zero of A.

REMARK 8. The following estimates have been obtained for $\rho_E(\tau)$ in L_p spaces, $1 < p < \infty$, (Lindenstrauss and Tzafriri, [**20**]).

$$\rho_E(\tau) = \begin{cases} (1+\tau^p)^{\frac{1}{p}} < \frac{1}{p}\tau^p, & \text{if } 1 < p < 2; \\ \frac{p-1}{2}\tau^2 + o(\tau^2) < \frac{p-1}{2}\tau^2, & \text{if } 2 \le p < \infty. \end{cases}$$

With these estimates, we observe that in corollary 3.3, the condition $\sum \lambda_n^2 < \infty$ implies that the condition $\sum \rho_E(\lambda_n M_1) < \infty$ in theorem 3.2 is satisfied; and the condition $\lambda_n \le \gamma_0 \theta_n$ implies that the condition $\frac{\rho_E(\tau_n)}{\tau_n} \le \gamma_0 \theta_n$ in theorem 3.2 is satisfied. Hence, the proof of corollary 3.3 follows from theorem 3.2.

Similarly, the condition $\sum \lambda_n^p < \infty$ in corollary 3.4 implies that the condition $\sum \rho_E(\lambda_n M_1) < \infty$ in theorem 3.2 is satisfied; and the condition $\lambda_n^{(p-1)} \le \gamma_0 \theta_n$ implies that the condition $\frac{\rho_E(\tau_n)}{\tau_n} \le \gamma_0 \theta_n$ in theorem 3.2 is satisfied. Hence, the proof of corollary 3.4 follows from theorem 3.2.

REMARK 9. Real sequences that satisfy conditions of corollary 3.3 are $\lambda_n = (n+1)^{-a}$ and $\theta_n = (n+1)^{-b}$, $n \ge 1$ with $0 < b < a$, $\frac{1}{2} < a < 1$ and $a + b < 1$.

REMARK 10. Real sequences that satisfy conditions of corollary 3.4 are $\lambda_n = (n+1)^{-a}$ and $\theta_n = (n+1)^{-b}$, $n \ge 1$ with $0 < b < (p-1)a$, $p^{-1} < a < 1$ and $a+b < 1$.

In fact, (i) and (ii) are easy to check in both corollaries. For (iii), using the fact that $(1+x)^s \le 1 + sx$, for $x > -1$ and $0 < s < 1$, we have

$$\begin{aligned} 0 &\le \frac{\left(\frac{\theta_{n-1}}{\theta_n} - 1\right)}{\lambda_n \theta_n} = \left[\left(1 + \frac{1}{n}\right)^b - 1\right] \cdot (n+1)^{a+b} \\ &\le b \cdot \frac{(n+1)^{a+b}}{n} = b \cdot \frac{n+1}{n} \cdot \frac{1}{(n+1)^{1-(a+b)}} \to 0 \text{ as } n \to \infty. \end{aligned}$$

References

[1] Heinz H. Bauschke, Eva Matoušková, and Simeon Reich, *Projection and proximal point methods: convergence results and counterexamples*, Nonlinear Anal. **56** (2004), no. 5, 715–738, DOI 10.1016/j.na.2003.10.010. MR2036787 (2004m:47116)

[2] Ronald E. Bruck Jr., *A strongly convergent iterative solution of* $0 \in U(x)$ *for a maximal monotone operator* U *in Hilbert space*, J. Math. Anal. Appl. **48** (1974), 114–126. MR0361941 (50 #14383)

[3] C. E. Chidume and N. Djitté, *Strong convergence theorems for zeros of bounded maximal monotone nonlinear operators*, Abstr. Appl. Anal., posted on 2012, Art. ID 681348, 19, DOI 10.1155/2012/681348. MR2903815

[4] Charles Chidume, *Geometric properties of Banach spaces and nonlinear iterations*, Lecture Notes in Mathematics, vol. 1965, Springer-Verlag London, Ltd., London, 2009. MR2504478 (2010f:47001)

[5] C. E. Chidume and H. Zegeye, *Approximate fixed point sequences and convergence theorems for Lipschitz pseudocontractive maps*, Proc. Amer. Math. Soc. **132** (2004), no. 3, 831–840 (electronic), DOI 10.1090/S0002-9939-03-07101-6. MR2019962 (2004j:47114)

[6] C. E. Chidume, *The iterative solution of the equation* $f \in x + Tx$ *for a monotone operator* T *in* L^p *spaces*, J. Math. Anal. Appl. **116** (1986), no. 2, 531–537, DOI 10.1016/S0022-247X(86)80017-8. MR842819 (87m:47141)

[7] Charles E. Chidume and Claudio H. Morales, *Accretive operators which are always single-valued in normed spaces*, Nonlinear Anal. **67** (2007), no. 12, 3328–3334, DOI 10.1016/j.na.2006.10.015. MR2350889 (2008k:47104)

[8] Tosio Kato, *Nonlinear semigroups and evolution equations*, J. Math. Soc. Japan **19** (1967), 508–520. MR0226230 (37 #1820)

[9] Jean-Jacques Moreau, *Proximité et dualité dans un espace hilbertien* (French), Bull. Soc. Math. France **93** (1965), 273–299. MR0201952 (34 #1829)

[10] Simeon Reich, *Strong convergence theorems for resolvents of accretive operators in Banach spaces*, J. Math. Anal. Appl. **75** (1980), no. 1, 287–292, DOI 10.1016/0022-247X(80)90323-6. MR576291 (82a:47050)

[11] Simeon Reich, *Iterative methods for accretive sets*, Nonlinear equations in abstract spaces (Proc. Internat. Sympos., Univ. Texas, Arlington, Tex., 1977), Academic Press, New York, 1978, pp. 317–326. MR502549 (82c:47065)

[12] Simeon Reich, *Constructive techniques for accretive and monotone operators*, Applied nonlinear analysis (Proc. Third Internat. Conf., Univ. Texas, Arlington, Tex., 1978), Academic Press, New York-London, 1979, pp. 335–345. MR537545 (80g:47059)

[13] Simeon Reich, *Extension problems for accretive sets in Banach spaces*, J. Functional Analysis **26** (1977), no. 4, 378–395. MR0477893 (57 #17393)

[14] Simeon Reich and Shoham Sabach, *Two strong convergence theorems for a proximal method in reflexive Banach spaces*, Numer. Funct. Anal. Optim. **31** (2010), no. 1-3, 22–44, DOI 10.1080/01630560903499852. MR2677243 (2011e:47125)

[15] Simeon Reich, *An iterative procedure for constructing zeros of accretive sets in Banach spaces*, Nonlinear Anal. **2** (1978), no. 1, 85–92, DOI 10.1016/0362-546X(78)90044-5. MR512657 (81b:47065)

[16] Osman Güler, *On the convergence of the proximal point algorithm for convex minimization*, SIAM J. Control Optim. **29** (1991), no. 2, 403–419, DOI 10.1137/0329022. MR1092735 (92c:90086)

[17] M. V. Solodov and B. F. Svaiter, *Forcing strong convergence of proximal point iterations in a Hilbert space*, Math. Program. **87** (2000), no. 1, Ser. A, 189–202. MR1734665 (2000j:90077)

[18] Shoji Kamimura and Wataru Takahashi, *Strong convergence of a proximal-type algorithm in a Banach space*, SIAM J. Optim. **13** (2002), no. 3, 938–945 (electronic) (2003), DOI 10.1137/S105262340139611X. MR1972223 (2004c:90096)

[19] N. Lehdili and A. Moudafi, *Combining the proximal algorithm and Tikhonov regularization*, Optimization **37** (1996), no. 3, 239–252, DOI 10.1080/02331939608844217. MR1396238 (97c:65104)

[20] Joram Lindenstrauss and Lior Tzafriri, *Classical Banach spaces. II*, Ergebnisse der Mathematik und ihrer Grenzgebiete [Results in Mathematics and Related Areas], vol. 97, Springer-Verlag, Berlin-New York, 1979. Function spaces. MR540367 (81c:46001)

[21] B. Martinet, *Régularisation d'inéquations variationnelles par approximations successives* (French), Rev. Française Informat. Recherche Opérationnelle **4** (1970), no. Ser. R-3, 154–158. MR0298899 (45 #7948)

[22] George J. Minty, *Monotone (nonlinear) operators in Hilbert space*, Duke Math. J. **29** (1962), 341–346. MR0169064 (29 #6319)

[23] R. Tyrrell Rockafellar, *Monotone operators and the proximal point algorithm*, SIAM J. Control Optimization **14** (1976), no. 5, 877–898. MR0410483 (53 #14232)

[24] Hong-Kun Xu, *A regularization method for the proximal point algorithm*, J. Global Optim. **36** (2006), no. 1, 115–125, DOI 10.1007/s10898-006-9002-7. MR2256886 (2007e:90108)

[25] Hong-Kun Xu, *Iterative algorithms for nonlinear operators*, J. London Math. Soc. (2) **66** (2002), no. 1, 240–256, DOI 10.1112/S0024610702003332. MR1911872 (2003e:47114)

[26] Zong Ben Xu and G. F. Roach, *Characteristic inequalities of uniformly convex and uniformly smooth Banach spaces*, J. Math. Anal. Appl. **157** (1991), no. 1, 189–210, DOI 10.1016/0022-247X(91)90144-O. MR1109451 (92i:46023)

African University of Sciences and Technology, Abuja, Nigeria
E-mail address: cchidume@aust.edu.ng

Contemporary Mathematics
Volume **659**, 2016
http://dx.doi.org/10.1090/conm/659/13160

On weak convergence in metric spaces

Giuseppe Devillanova, Sergio Solimini, and Cyril Tintarev

Abstract. This note gives an exposition of various extensions of the notion of weak convergence to metric spaces. They are motivated by applications, such as existence of fixed points of non-expansive maps, and analysis of the defect of compactness relative to gauge groups in Banach spaces, where weak convergence is generally less useful than metric notions like asymptotic centers and polar convergence. This note compares notions of convergence of weak type found in literature, in particular the notion of Δ-convergence, polar convergence and the modes of convergence of weak type in Hadamard spaces. Some applications of polar convergence, such as the existence of fixed points for nonexpansive maps and a suitable variant of the Brezis-Lieb Lemma are produced.

Dedicated to the memory of Teck-Cheong Lim

Contents

1. Introduction

In a preliminary version of [**35**] a notion of convergence, similar to weak convergence, but based on the distance rather than on the continuity of linear forms, called polar convergence, has been presented as a natural mode of convergence which has been used, instead of weak convergence, in the analysis of the defect of compactness relative to gauge groups, allowing extension of the results, already

2010 *Mathematics Subject Classification.* Primary 46B10, 46B50, 46N20; Secondary 52A01, 52A07, 56E50, 49N99.

known in the case of Hilbert spaces. Later, the authors became aware that this notion is very close to the notion of Δ-*convergence*, already introduced by T.C. Lim in [**25**] (a similar definition also appears in [**24**]). The two definitions remain meaningful in metric spaces and, while they are different even in general linear spaces, they coincide under some regularity conditions, as for instance, Staples' rotundity (see [**37**] and Definition 3.1 below) which, in the case of linear spaces, is equivalent to uniform convexity. In linear spaces the two definitions give a notion which is distinct from the usual weak convergence. Although they agree in Hilbert spaces ([**18**]), they are generically different even in many uniformly convex Banach spaces including L^p spaces.

Opial, in the pioneering work [**31**], searched for conditions under which iteration sequences of non-expansive maps are weakly convergent, introducing the classical Opial condition (see [**31**, Condition (2)] and Definition 5.1 below). As it soon became clear with the Edelstein's proof ([**14**]) of the Browder fixed point theorem, fixed points are associated with Δ-convergence of iterations (as asymptotic centers of iteration sequences), so that the implicit meaning of Opial condition, in broad terms, is to assert that weak and Δ-convergence coincide. In particular, it is exactly the case in uniformly convex and uniformly smooth Banach spaces that weak and Δ-convergence coincide if and only if the Opial condition holds.

There are also definitions of weak convergence in metric spaces, which are based on the continuity of suitable test maps (essentially distance-minimizing projections onto compact convex sets) that play the role of the continuous linear forms on normed spaces. These notions are studied in detail for Hadamard spaces which are complete geodesic spaces which satisfy a suitable property which we shall call "parallelogram inequality" (see Definition 6.1 below). It is well known that every linear normed Hadamard space is a Hilbert space, so it is not surprising that by analogy with Hilbert spaces, Hadamard spaces satisfy Opial condition, and that the polar convergence and the weak convergence in Hadamard spaces coincide. On the other side, it should be noted that in Banach spaces where Opial condition does not hold, in particular, in L^p spaces, metric weak convergence defined via test maps still coincides with Δ-convergence and not with the usual weak convergence. This indicates that distance-minimizing projections are generally nonlinear maps, and thus they cannot give rise to the usual weak convergence.

In essence, Δ-convergence (with its close variations such as polar convergence) remains the only known mode of convergence of weak type, other than the usual weak convergence in normed vector spaces, and the rich literature on the subject gives a number of definitions of convergence that are equivalent, in the context they are stated, to Δ-convergence, or are its close variants.

Although this paper is focused on convergence of sequences, extension of its notions to nets is mostly trivial. With the exception of Section 4 we do not address the problem of checking if the various modes of convergence we introduce are coming from a topology or not. We have tried, despite of the brevity of the paper, to make a self-contained exposition, including even some proofs of properties readily available in the literature.

The paper is organized as follows. In Section 2 we introduce polar and Δ-convergence, which are directly depending on the distance. In Section 3 we introduce metric spaces which are (uniformly) Staples' rotund. These spaces are a metric counterpart of uniformly convex normed spaces. Similarly to uniformly convex Banach spaces, where weak and weak-star topology agree and some local compactness properties (such as the Banach-Alaoglu compactness theorem) hold, in complete uniformly rotund metric spaces we prove that Δ-convergence and polar convergence agree and enjoy some compactness properties (see Theorem 3.4).

In Section 4 we discuss the question posed by Dhompongsa, Kirk and Panyanak [**12**], whether Δ-convergence (or polar convergence) can be associated with a topology, and we give explicit examples of metric spaces in which polar convergence is not derived from a topology.

In Section 5 we study polar convergence in Banach spaces pointing out, by examples, differences between polar convergence and weak convergence. In particular, when a Banach space is uniformly convex and uniformly smooth, Δ-convergence can be characterized as weak convergence in the dual (see Theorem 5.5).

In Section 6 we recall Opial condition and introduce "parallelogram inequality" on metric spaces which gives an equivalent definition of CAT(0) spaces (see [**15**, Definition 2.9]) or spaces of Alexandrov nonpositive curvature (see [**21**, Definition 2.3.1 and the subsequent remark]), and we deduce their Staples' rotundity. Minimal distance projections are employed in Section 7 where we extend to metric spaces the notion of weak convergence by replacing the linear forms with distance-minimizing projections onto geodesic segments. We point out that on general normed spaces continuous linear forms are not distance-minimizing projections onto straight lines, so while this definition is appropriate for Hadamard spaces, which generalize Hilbert spaces, distance-minimizing projections do not extend the notion of weak convergence to general metric spaces. Moreover in Theorem 7.4 we prove the equivalence of weak and polar convergence. Before ending the section we also recall two topologies $\mathcal{T}_c$ and $\mathcal{T}_w$ defined on Hadamard spaces in [**28**] (while searching for a topology associated with Δ-convergence (or weak convergence)), and we discuss the equivalence of the topology $\mathcal{T}_c$ with the weak topology. Moreover we show that in Hilbert spaces T_w coincides with the topology induced by the norm.

In Section 8 we apply polar (Δ) convergence to the fixed points theory. In particular we introduce a condition (polar asymptotical regularity condition, see Definition 8.3) for a nonexpansive map at a point x which guarantees the polar convergence of the iterations sequence at this point.

Finally we discuss a version of the Brezis-Lieb Lemma [**7**] where, see Theorem 8.5, the assumption of *a.e.* convergence is replaced by the assumption of "double weak" convergence (namely polar convergence and usual weak convergence to the same point, which are both implied by *a.e.* convergence of a bounded sequence in L^p for $1 < p < +\infty$). Moreover, given that there is a stronger version of Brezis-Lieb Lemma in Hilbert spaces, based on weak convergence only, we get an analogous result for Hadamard spaces.

2. Weak convergences of polar type

DEFINITION 2.1. **(Δ-limit)** Let (E, d) be a metric space. A sequence $(x_n)_{n\in\mathbb{N}} \subset E$ is said to Δ-converge to a point $x \in E$ (see [**25**]), and we shall write $x_n \overset{\Delta}{\to} x$, if

$$\limsup_n d(x_{k_n}, x) \le \limsup_n d(x_{k_n}, y) \tag{2.1}$$

for any subsequence $(x_{k_n})_{n\in\mathbb{N}}$ of $(x_n)_{n\in\mathbb{N}}$ and for every $y \in E$.

The following proposition, whose proof is straightforward, gives a characterization of a Δ-limit.

PROPOSITION 2.2. *Let (E, d) be a metric space. A sequence $(x_n)_{n\in\mathbb{N}} \subset E$ Δ-converges to a point $x \in E$ if and only if*

$$\forall y \in E \, : \, d(x_n, x) \leq d(x_n, y) + o(1). \tag{2.2}$$

Here and in what follows the Edmund Landau symbol $o(1)$ denotes a sequence of real numbers convergent to zero.

DEFINITION 2.3. **(Strong Δ-limit)** Let (E, d) be a metric space. A sequence $(x_n)_{n\in\mathbb{N}} \subset E$ is said to strongly-Δ converge to a point $x \in E$ (see [**25**]), and we shall write $x_n \overset{s-\Delta}{\to} x$, if

$$(\exists \lim_n d(x_n, x) \quad \text{and}) \quad \forall y \in E \, : \, \lim_n d(x_n, x) \leq \liminf_n d(x_n, y) \, . \tag{2.3}$$

Note that if a sequence $(x_n)_{n\in\mathbb{N}}$ has no bounded subsequence then it strong$-\Delta$ converges to every point. On the contrary if $(x_n)_{n\in\mathbb{N}}$ is a bounded sequence, obviously, if $x_n \overset{s-\Delta}{\to} x$ then $x_n \overset{\Delta}{\to} x$. The notions of Δ and strong-Δ convergence of a given sequence $(x_n)_{n\in\mathbb{N}}$ can be further clarified by means of the notion of *asymptotic center* (denoted by $\text{cen}_{n\to\infty} x_n$) and *asymptotic radius* (denoted by $\text{rad}_{n\to\infty} x_n$) (definitions of asymptotic radius and asymptotic centers can be found for instance in [**14**] where Edelstein gives a proof of Browder fixed point theorem [**9**] based on these notions. In this connection see page 251 of the paper [**32**] and page 18 in [**18**]). We emphasize that, while the asymptotic radius always exists and is uniquely determined, asymptotic centers may not exist or may be not uniquely determined. Therefore, the symbol $\text{cen}_{n\to\infty} x_n$ must be understood in the same sense as the limit symbol in a topological space which is not assumed to be Hausdorff. Here we use, for the sake of convenience, an elementary reformulation of the original definition of asymptotic centers and asymptotic radius of a sequence $(x_n)_{n\in\mathbb{N}}$ as minimum points and, respectively, infimum value of the following functional on E

$$I_{as}(y) = \limsup_n d(x_n, y). \tag{2.4}$$

REMARK 2.4. Let (E, d) be a metric space. A sequence $(x_n)_{n\in\mathbb{N}} \subset E$ Δ-converges to a point $x \in E$ if and only if x is an asymptotic center of every subsequence. On the other hand the strong-Δ convergence of $(x_n)_{n\in\mathbb{N}}$ to x means that every subsequence has the same asymptotic radius (equal to $I_{as}(x)$). This last property allows to prove that any asymptotic center of the whole sequence $(x_n)_{n\in\mathbb{N}}$ is an asymptotic center of every subsequence and therefore it is a Δ-limit of $(x_n)_{n\in\mathbb{N}}$. (Indeed if $(x_{k_n})_{n\in\mathbb{N}}$ is a subsequence of $(x_n)_{n\in\mathbb{N}}$, since, by strong-Δ convergence, $\text{rad}_{n\to\infty} x_{k_n} = \text{rad}_{n\to\infty} x_n$, the inequality $\limsup_n d(x_{k_n}, x) \leq \limsup_n d(x_n, x) = \text{rad}_{n\to\infty} x_n$ forces x to be an asymptotic center of $(x_{k_n})_{n\in\mathbb{N}}$). Obviously the converse implication is not true.

REMARK 2.5. In other terms, let Ξ be the set of bounded sequences of elements in E, for every ξ, $\zeta \in \Xi$, we shall write $\xi \leq \zeta$ if the sequence ζ is extracted from ξ after a finite number of terms (note that, in spite of the notation, $\leq$ is not an ordering, since it is not antisymmetric). The function f which maps every sequence $\xi = (x_n)_{n\in\mathbb{N}} \in \Xi$ into $f(\xi) = -\text{rad}_{n\to\infty} x_n$ is a "increasing" function (i.e. if $\xi \leq \zeta$ then $f(\xi) \leq f(\zeta)$). We can pass to a coarser relation $\preceq$, which is a true ordering and makes f strictly increasing, by setting, for any $\xi, \zeta \in \Xi$, $\xi \preceq \zeta$ if $\xi = \zeta$ or if $\xi \leq \zeta$ and $f(\xi) < f(\zeta)$. Under this notation, we can reformulate the second part of the previous remark by stating that $x_n \overset{s-\Delta}{\rightarrow} x$ if and only if $(x_n)_{n\in\mathbb{N}}$ is maximal for $\preceq$ and x is an asymptotic center of $(x_n)_{n\in\mathbb{N}}$.

Note that the ordered set $(\Xi, \preceq)$ is countably inductive in the sense specified in [**27**, Appendix A]. Indeed, if $(\xi_n)_{n\in\mathbb{N}} \subset \Xi$ is increasing with respect to $\preceq$, after trowing away a finite number of terms from each sequence ξ_n (in order to make each ξ_{n+1} a subsequence of ξ_n) and passing to a diagonal selection, one obtains an upper bound ξ of the whole sequence.

REMARK 2.6. Note that a (strong) Δ-limit is not necessarily unique, even in the case of a bounded sequence. For instance, if $(A_n)_{n\in\mathbb{N}}$ is a decreasing sequence of measurable sets of $\mathbb{R}$ such that, for any $n \in \mathbb{N}$, $A_n \setminus A_{n+1}$ is of positive measure, by setting, for any $n \in \mathbb{N}$, $x_n = \mathbf{1}_{A_{n+1}} - \mathbf{1}_{A_n \setminus A_{n+1}}$ we get a sequence of bounded functions which (since for any $n \neq m \in \mathbb{N}$ $\|x_n - x_m\|_\infty = 2$) does not admit any subsequence with asymptotic radius strictly smaller than 1. Then, given $\bar{n} \in \mathbb{N}$, any function x such that $\|x\|_\infty \leq 1$ and such that $x_{|A_{\bar{n}}} = 0$ (for instance $x = \mathbf{1}_{\mathbb{R}\setminus A_{\bar{n}}}$) satisfies $\|x_n - x\|_\infty = 1$ for $n > \bar{n}$. Therefore x is an asymptotic center of the sequence $(x_n)_{n\in\mathbb{N}}$ and $\mathrm{rad}_{n\to\infty} x_n = 1$. Since, as already proved, $(x_n)_{n\in\mathbb{N}}$ does not admit any subsequence with asymptotic radius strictly smaller than 1, it follows that $\mathrm{rad}_{n\to\infty} x_{k_n} = 1$ for any subsequence $(x_{k_n})_{n\in\mathbb{N}}$. Therefore, by Remark 2.4, x is a strong-Δ limit of the sequence $(x_n)_{n\in\mathbb{N}}$.

DEFINITION 2.7. **(Polar limit)** Let $(x_n)_{n\in\mathbb{N}}$ be a sequence in a metric space (E, d). One says that $x \in E$ is a *polar limit* of $(x_n)_{n\in\mathbb{N}}$ and we shall write $x_n \rightharpoondown x$, if for every $y \neq x$ there exists $M(y) \in \mathbb{N}$ such that

$$d(x_n, x) < d(x_n, y) \text{ for all } n \geq M(y). \tag{2.5}$$

REMARK 2.8. It is immediate from comparison between (2.2) and (2.5) that polar convergence implies Δ-convergence. Moreover (2.5) guarantees the uniqueness of polar limit. On the other hand, one can deduce from Remark 2.6 that Δ-convergence and polar convergence generally do not coincide.

REMARK 2.9. The following properties for Δ, strong-Δ and polar limits of sequences in a metric space (E, d) are immediate:

(i) If $x_n \overset{s-\Delta}{\rightarrow} x$ or $x_n \rightharpoondown x$, then $x_n \overset{\Delta}{\rightarrow} x$.
(ii) If $x_n \to x$ (in metric), then $x_n \overset{s-\Delta}{\rightarrow} x$ and $x_n \rightharpoondown x$.
(iii) If $(x_n)_{n\in\mathbb{N}}$ is Δ (resp. strong-Δ, resp. polarly)-convergent to a point x, then any subsequence $(x_{k_n})_{n\in\mathbb{N}}$ of $(x_n)_{n\in\mathbb{N}}$ is Δ (resp. strong-Δ, resp. polarly)-convergent to x.
(iv) $(x_n)_{n\in\mathbb{N}}$ is Δ (resp. polarly)-convergent to a point x if and only if any subsequence $(x_{k_n})_{n\in\mathbb{N}}$ of $(x_n)_{n\in\mathbb{N}}$ admits a subsequence which is Δ (resp. polarly)-convergent to x.
(v) If $(x_n)_{n\in\mathbb{N}}$ is Δ-convergent to a point x, then it admits a subsequence which is strong-Δ convergent to x.
(vi) If E is a Banach space, then the sequence $((-1)^n x)_{n\in\mathbb{N}}$, with any $x \neq 0$, has no Δ-limit.

It follows from (iv) and (v) combined that a sequence $(x_n)_{n\in\mathbb{N}}$ Δ-converges to a point x if every subsequence admits a subsequence strong-Δ-convergent to x.

PROPOSITION 2.10. *Let (E, d) be a metric space, $x \in E$ and let $(x_n)_{n\in\mathbb{N}}$ be a precompact sequence (or, in particular, a Cauchy sequence) in E. If $x_n \overset{\Delta}{\to} x$, then $x_n \to x$.*

PROOF. Since the sequence $(x_n)_{n\in\mathbb{N}}$ is precompact, for any $\varepsilon > 0$, there exists a finite ε- net $\mathcal{N}_\varepsilon$ of $(x_n)_{n\in\mathbb{N}}$. Then, by (2.2), for large n, we have $d(x_n, x) \leq \min_{y\in\mathcal{N}_\varepsilon} d(x_n, y) + o(1) < \varepsilon$, i.e. $d(x_n, x) \to 0$. □

DEFINITION 2.11. ([**25**]) A metric space (E, d) is called Δ-complete (or is said to satisfy the Δ-completeness property) if every bounded sequence admits an asymptotic center.

An easy maximality argument, see for instance [**27**, Theorem A.1], yields the following result.

THEOREM 2.12. *([**25**, Theorem 3]) Let (E, d) be a Δ-complete metric space. Then every bounded sequence in E has a strong-Δ-convergent subsequence.*

PROOF. Since the ordered set $(\Xi, \preceq)$ introduced in Remark 2.5 is countably inductive and since the function f which maps every sequence $\xi = (x_n)_{n\in\mathbb{N}} \in \Xi$ into $f(\xi) = -\text{rad}_{n\to\infty} x_n$ is a real strictly increasing function (see Remark 2.5), by using [**27**, Theorem A.1] one obtains that every sequence $(x_n)_{n\in\mathbb{N}} \in \Xi$ has a maximal subsequence for $\preceq$. Since E is Δ-complete, this subsequence has an asymptotic center to which, since it is maximal, it is strongly-Δ convergent (see Remark 2.5). □

This argument has been employed by Lim in [**25**] and, in a very close setting, in the proof of [**17**, Lemma 15.2] (although the lemma is set in Banach spaces, the proof extends to the metric spaces verbatim, as it has been already observed in [**23**]). Note that the existence of a strictly increasing real valued function f and the separability of $\mathbb{R}$ also allow to prove that the countable inductivity leads to inductivity and so one can deduce the existence of a maximal element by Zorn Lemma as in [**25**, Proposition 1]. However, the direct argument in [**27**, Theorem A.1] looks even simpler than the proof of the inductivity.

There is a number of publications where Δ convergence or strong-Δ convergence (not always under that name) is applied to problems related to fixed points in Hadamard spaces - to mention just few, [**3, 13, 23**], but we first consider here a larger class of metric spaces, namely *uniformly rotund spaces* introduced by John Staples [**37**].

3. Rotund metric spaces

DEFINITION 3.1. A metric space (E, d) is a (uniform) SR ("Staples rotund") metric space (or satisfies (uniformly) property SR) if there exists a function $\delta : (\mathbb{R}_+)^2 \to \mathbb{R}_+$ such that for any $r, \overline{d} > 0$, set $\delta = \delta(r, \overline{d})$, for any $x, y \in E$ with $d(x, y) \geq \overline{d}$:

$$\text{(SR)} \qquad \text{rad}(B_{r+\delta}(x) \cap B_{r+\delta}(y)) \leq r - \delta.$$

For the reader's convenience we recall that the Chebyshev radius of a set X in a metric space (E, d) is the infimum of the radii of the balls containing X. In other words $\text{rad}(X) = \inf_{x\in E} \sup_{y\in X} d(x, y)$. Moreover, the Chebyshev radius and the Chebyshev centers of a set $X \subset E$ can also be defined, analogously to the asymptotic radius and asymptotic centers of a sequence, by replacing the functional I_{as} in (2.4) by

$$I_X : x \in E \mapsto \sup_{y \in X} d(x, y). \tag{3.1}$$

REMARK 3.2. If δ is continuous, one can replace, in the above definition, the occurrences of $r + \delta$ by r. Moreover, from Definition 3.1 it is immediate that uniformly convex normed vector spaces (see [**26**, Definition 1.e.1]) are uniformly SR metric spaces. Furthermore, any uniformly SR normed vector space is uniformly convex. To show this, let $\|x\| \leq 1$, $\|y\| \leq 1$ such that $\|x - y\| \geq \varepsilon$. It follows easily that both 0 and $x+y$ belong to $\bar{B}_1(x) \cap \bar{B}_1(y)$. By rotundity, set $\delta = \delta(1, \varepsilon)$, we have $\mathrm{rad}(\bar{B}_1(x) \cap \bar{B}_1(y)) \leq 1 - \delta$. Therefore, $\|x + y - 0\| \leq 2\mathrm{rad}(\bar{B}_1(x) \cap \bar{B}_1(y)) \leq 2 - 2\delta$ and so $\|\frac{x+y}{2}\| \leq 1 - \delta$ follows, thus proving uniform convexity.

REMARK 3.3. Staples' rotundity is close to the uniform ball convexity in Foertsch [**16**] and to the uniform convexity for hyperbolic metric spaces given by Reich and Shafrir [**34**]. The latter properties assume existence of a midpoint map which is not assumed by Staples, although, on the other hand, Staples' definition makes an additional assumption that roughly speaking amounts to the continuity of the modulus of convexity with respect to radius of the considered balls. In Banach spaces the above definitions coincide with the notion of uniform convexity. Particular examples of SR metric spaces are given by Hadamard spaces and, more generally, by CAT(0) spaces which will be discussed in Section 6 below.

For complete Staples rotund metric spaces, [**37**, Theorem 2.5] and [**37**, Theorem 3.3] state, respectively, uniqueness and existence of the asymptotic center of any bounded sequence (giving actually the Δ-completeness of such spaces, see Definition 2.11). We are going to prove the just mentioned properties as one of the following claims which hold true in complete SR metric spaces.

For complete SR metric spaces we have the following results.

- **a**) Δ-convergence and polar convergence coincide.
- **b**) Every bounded sequence has a unique asymptotic center.
- **c**) The space is Δ-complete. Therefore (the sequential compactness property in) Theorem 2.12 also holds for polar convergence as stated in the following theorem.

THEOREM 3.4. *Let (E, d) be a complete SR metric space. Then every bounded sequence in E has a polarly convergent subsequence.*

Actually claim **c**) is just a restatement of claim **b**) and therefore it does not need to be proved.

LEMMA 3.5. *Let (E, d) be a SR metric space. Let $(x_n)_{n\in\mathbb{N}} \subset E$ be a bounded sequence and let $x \in E$ be such that (2.2) holds true. Then, for each element $z \in E$, $z \neq x$, there exist positive constants n_0 and c depending on z such that*

$$d(x_n, x) \leq d(x_n, z) - c \quad \text{for all} \quad n \geq n_0, \tag{3.2}$$

and so $(x_n)_{n\in\mathbb{N}}$ polarly converges to x.

PROOF. If the assertion is false, by (2.2), we can find $z \neq x$ and a subsequence $(x_{k_n})_{n\in\mathbb{N}}$ such that $d(x_{k_n}, x) - d(x_{k_n}, z) \to 0$. Passing again to a subsequence we can also assume that $d(x_{k_n}, x) \to r > 0$. Set $\overline{d} = d(x, z)$ and $\delta = \delta(r, \overline{d})$. Since, for large n, $x_{k_n} \in B_{r+\delta}(x) \cap B_{r+\delta}(z)$, we can deduce from (SR) the existence of $y \in E$ such that $d(x_{k_n}, y) < r - \delta$, in contradiction to (2.2). □

COROLLARY 3.6. *Let (E, d) be a SR metric space. Let $(x_n)_{n\in\mathbb{N}} \subset E$ be a bounded sequence and let $x \in E$ be such that (2.2) holds true. Then, for any compact set $\mathcal{K} \subset E$ such that $x \notin \mathcal{K}$ there exists $\overline{n} = \overline{n}(\mathcal{K})$ such that for any $n \geq \overline{n}$ and for any $z \in \mathcal{K}$ (3.2) holds true.*

By taking into account the characterization of Δ-limit given by Proposition 2.2, Lemma 3.5 guarantees that if $x_n \overset{\Delta}{\rightarrow} x$, then $x_n \rightharpoondown x$. This proves claim **a)**, since, as pointed out in Remark 2.8, polar convergence always implies Δ-convergence to the same point.

LEMMA 3.7. *Let (E, d) be a SR metric space. Let I_{as} be defined by (2.4) relative to a given sequence $(x_n)_{n\in\mathbb{N}} \subset E$. Then, for any $\overline{d} > 0$, there exists $\varepsilon > 0$ such that if $x, y \in E$ satisfy $\max(I_{as}(x), I_{as}(y)) < \inf I_{as} + \varepsilon$, then $d(x, y) < \overline{d}$.*

PROOF. If $\inf I_{as} = 0$ one can take $\varepsilon = \frac{\overline{d}}{2}$, otherwise, since E is a SR metric space (see (SR)), one can take $\varepsilon = \delta(\inf I_{as}, \overline{d})$. □

Two remarkable and immediate consequences of the above lemma trivially follow.

COROLLARY 3.8. *Let (E, d) be a SR metric space and let I_{as} be as above. Then*

(i) the functional I_{as} admits at most one minimum point;
(ii) any minimizing sequence of the functional I_{as} is a Cauchy sequence.

COROLLARY 3.9. *Let (E, d) be a SR complete metric space, then any bounded sequence $(x_n)_{n\in\mathbb{N}} \subset E$, admits a unique asymptotic center. Moreover any minimizing sequence for I_{as} converges to the asymptotic center of the sequence $(x_n)_{n\in\mathbb{N}}$.*

In other words, Corollary 3.9 states in particular that in a SR complete metric space every bounded sequence $(x_n)_{n\in\mathbb{N}}$ has a unique asymptotic center $\mathrm{cen}_{n\to\infty} x_n$, so that claim **b)** is proved.

4. Polar neighborhoods

In this section we discuss the question, posed by Dhompongsa, Kirk and Panyanak in [**12**], if Δ-convergence (or polar convergence) can be associated with a topology.

Let (E, d) be a metric space, let $Y \subset E$, $x \notin Y$, and let

$$\mathcal{N}_Y(x) = \bigcap_{y\in Y} \mathcal{N}_y(x) = \{z \in E \mid d(z, x) < d(z, y), \forall y \in Y\}, \tag{4.1}$$

where

$$\mathcal{N}_y(x) = \{z \in E \mid d(z, x) < d(z, y)\}. \tag{4.2}$$

In other words, the set $\mathcal{N}_Y(x)$ is the set of all points in E which are strictly closer to x than to Y.

REMARK 4.1. A trivial restatement of Definition 2.7 is that, if $x \in E$ and $(x_n)_{n\in\mathbb{N}} \subset E$, then $x_n \rightharpoondown x$ if and only if, for any finite set $Y \not\ni x$, there exists $M(Y) \in \mathbb{N}$ such that $x_n \in \mathcal{N}_Y(x)$ for all $n \geq M(Y)$.

DEFINITION 4.2. Let $x \in E$. We shall call *polar neighborhoods* (briefly p-nbd) of the point x all the subsets $V \subset E$ containing a set $\mathcal{N}_Y(x)$ given by (4.1), where $Y \subset E$ is any finite set such that $x \notin Y$.

REMARK 4.3. Polar convergence can be finally tested, as it follows from Remark 4.1, by using polar neighborhoods. Indeed, if $x \in E$ and $(x_n)_{n\in\mathbb{N}} \subset E$, then $x_n \rightharpoondown x$ if and only if $x_n \in V$ for large n for any polar neighborhood V of x.

One can easily see that the set of polar neighborhoods of a given point x is a filter of parts of E. It is well known (see [**5**, Prop. 1.2.2]) that the union of such filters is a neighborhood base of a (unique) topology $\mathcal{T}$ if and only if the following further property is satisfied

$$\forall x \in E, \forall V \text{ p-nbd. of } x, \exists U \text{ p-nbd of } x \text{ s.t. } \forall z \in U : V \text{ is a p-nbd of } z. \tag{4.3}$$

By Definition 4.2 the above condition can be more explicitly stated as

$$\forall x, y \in E,\ x \neq y, \exists Y \text{ finite set s.t. } \forall z \in \mathcal{N}_Y(x) \exists Z \text{ s.t. } \mathcal{N}_Z(z) \subset \mathcal{N}_y(x). \tag{4.4}$$

When (4.3) (or equivalently (4.4)) is satisfied we shall call $\mathcal{T}$ *polar topology* induced by d on E. Since, for any $x, y \in E$ if $2r = d(x, y)$, $B_r(x) \subset \mathcal{N}_y(x)$, polar topology is in general a coarser topology of that usually induced by d by the classical notion of neighborhood. We shall refer to the latter topology as *strong topology* induced by d when we want to distinguish it from the polar one.

We shall show now that (4.3) is true in some cases and false in others, and even when (4.3) is true, $\mathcal{T}$ not always coincides with the strong topology induced by d.

EXAMPLE 4.4. Let δ be the discrete metric on E. Then, setting $d = \delta$ in (4.2), it is easy to see that $\mathcal{N}_y(x) = \{x\}$ for all $x, y \in E$, $x \neq y$. Therefore, the polar topology induced by δ is the discrete topology as well as the strong topology.

In Hilbert spaces polar topology coincides with weak topology, and thus is different from the strong topology in the infinite-dimensional case.

EXAMPLE 4.5. Let E be a Hilbert space. Then, for any $x, y \in E$, $x \neq y$, we have that $z \in \mathcal{N}_y(x)$ means $\|z-x\|^2 - \|z-y\|^2 < 0$, namely $(2z-(x+y))\cdot(y-x) < 0$. If, with an invertible change of variables, we set $x + y = 2a$, $y - x = v$, we get that $\mathcal{N}_y(x) = \{z \in E \,|\, (z - a)\cdot v < 0\}$. So $\mathcal{N}_y(x)$ gives, for $x, y \in E$, $x \neq y$, a base of the weak topology.

Combining the previous two examples we can discuss the following case.

EXAMPLE 4.6. Let E be a finite dimensional vector space and let d be defined as $d(x, y) = |x - y| + \delta(x, y)$ where δ is as in Example 4.4. Then it is easy to check that the polar convergence induced by d agrees with the natural convergence of vectors while the strong topology induced by d is the discrete topology. In other words the strong topology induced by d is the discrete topology (which, see Example 4.4, coincides with both the strong and the polar topology induced by δ, i.e. by neglecting the contribution of $|x - y|$), while the polar topology induced by d is the natural topology (which, see Example 4.5, coincides with both the strong and the polar topology induced by $|x - y|$, i.e. by neglecting the contribution of δ).

Finally we show an example in which (4.3) is not satisfied and therefore polar topology does not exist.

EXAMPLE 4.7. Let $E = \mathbb{R}$ and let D be the Dirichlet function (i.e. $D(x) = 1$ for $x \in \mathbb{Q}$ and $D(x) = 0$ for $x \notin \mathbb{Q}$). Let d be defined as $d(x, y) = |x - y| + (1 + D(x - y))\delta(x, y)$ with δ as in Example 4.4. It is easy to see that d is a metric (indeed, if $y \notin \{x, z\}$ then $d(x, z) \leq |x - z| + 2 \leq |x - y| + 1 + |y - z| + 1 \leq d(x, y) + d(y, z)$). We shall prove that (4.4) does not hold. Fix any $\overline{x}$, $\overline{y} \in \mathbb{R}$, $\overline{x} \neq \overline{y}$, $|\overline{x} - \overline{y}| \leq 1$ and $\overline{y} - \overline{x} \notin \mathbb{Q}$. Consider $\mathcal{N}_Y(\overline{x})$ with an arbitrary finite set $Y \not\ni \overline{x}$ and let $\varepsilon > 0$ such that $2\varepsilon < |\overline{x} - y|$ for all $y \in Y$. Then fix $\overline{z} \in \mathbb{R}$ such that $|\overline{x} - \overline{z}| < \varepsilon$ and $\overline{x} - \overline{z} \notin \mathbb{Q}$. This choice implies that

$$\forall y \in Y: \quad 2|\overline{x} - \overline{z}| < |\overline{x} - y| \quad \text{and so} \quad |\overline{x} - \overline{z}| < |\overline{z} - y| \tag{4.5}$$

by the triangle inequality. Therefore

$$d(\overline{x}, \overline{z}) = |\overline{x} - \overline{z}| + 1 < |\overline{z} - y| + 1 \leq d(\overline{z}, y) \quad \forall y \in Y, \tag{4.6}$$

proving that $\overline{z} \in \mathcal{N}_Y(\overline{x})$. In the same way, let Z be a given finite set such that $\overline{z} \notin Z$ and fix $\eta > 0$ such that $2\eta < |\overline{z} - z|$ $\forall z \in Z$. Fix $v \in \mathbb{R}$ such that $|\overline{z} - v| < \eta$ and $\overline{x} - v \in \mathbb{Q}$ (therefore $v - \overline{y} \notin \mathbb{Q}$, $|v - \overline{z}| \notin \mathbb{Q}$). With analogous estimates to (4.5) and (4.6) we see that $v \in \mathcal{N}_Z(\overline{z})$. Finally, since $v - \overline{y} \notin \mathbb{Q}$, $|\overline{x} - \overline{y}| \leq 1$ and $\overline{x} - v \in \mathbb{Q}$,

$$d(v, \overline{y}) = |v - \overline{y}| + 1 \leq |v - \overline{x}| + |\overline{x} - \overline{y}| + 1 \leq |v - \overline{x}| + 2 = d(v, \overline{x}).$$

Hence for any $\mathcal{N}_Y(\overline{x})$ we can find $\overline{z} \in \mathcal{N}_Y(\overline{x})$ such that every neighborhood $\mathcal{N}_Z(\overline{z}) \not\subset \mathcal{N}_{\overline{y}}(\overline{x})$ and so (4.4) does not hold.

5. Polar convergence in Banach spaces

In a Hilbert space the polar limit of a sequence is also its weak limit and vice versa, see Example 4.5. The original argument, although brought up to prove a weaker statement, is due to Opial, see [**31**] where the following condition is also formulated, see [**31**, Condition (2)] and [**35**] for more details. We can interpret this condition, in reflexive spaces, as an equivalence between weak and polar convergence.

DEFINITION 5.1 (Opial condition). One says that a Banach space $(E, \| \ \|)$ satisfies Opial condition if for every sequence $(x_n)_{n\in\mathbb{N}} \subset E$, which is weakly convergent to a point $x \in E$,

$$\liminf_n \|x_n - x\| < \liminf_n \|x_n - y\| \text{ for every } y \in E, y \neq x. \tag{5.1}$$

Note that (5.1) with the $<$ sign replaced by $\leq$ remains equivalent in a SR metric space, (the equivalence follows by repeating the argument in the proof of Lemma 3.5).

It is worth mentioning that van Dulst in [**38**] proved that a separable uniformly convex Banach space can be provided with an equivalent norm satisfying the Opial condition.

Opial also gives an example, see [**31**, Section 5], which we can interpret in terms of polar convergence, of a bounded sequence in $L^p((0, 2\pi))$, $p \neq 2$, $1 < p < \infty$, whose polar limit and weak limit do not coincide. We shall come back soon with more powerful tools to the case of L^p but we first present two examples, suggested by Michael Cwikel, which illustrate some differences between polar and weak convergence in a Banach space. They respectively show that a weakly convergent sequence in ℓ^1 does not need to be bounded and that a weakly converging sequence in ℓ^∞ does not need to have a polar limit.

EXAMPLE 5.2. Let E be the sequence space ℓ^1, and let, for any $k \in \mathbb{N}$, $x_k := k(\delta_{k,n})_{n\in\mathbb{N}}$ (we have used the Kronecker delta values, i.e. $\delta_{k,n} = 1$ for $k = n$ and $\delta_{k,n} = 0$ otherwise). Since, for each fixed sequence $\alpha = (\alpha_n)_{n\in\mathbb{N}} \in \ell^1$, we have $\lim_{k\to\infty} \|x_k - \alpha\|_{\ell^1} - \|x_k\|_{\ell^1} = \|\alpha\|_{\ell^1}$ we get that the sequence $(x_k)_{k\in\mathbb{N}}$ is polarly convergent to the zero element of ℓ^1 .

EXAMPLE 5.3. Let E be the sequence space ℓ^∞ and let $x_k := (\delta_{k,n})_{n\in\mathbb{N}}$ for each $k \in \mathbb{N}$. Clearly the sequence $(x_k)_{k\in\mathbb{N}}$ converges weakly to the zero of ℓ^∞. However, zero is not a polar limit of the sequence, since for the sequence $\alpha = (1, 1, 1, \dots)$ we have $1 = d(0, x_k) = d(\alpha, x_k)$ for all k. Let $\beta := (\beta_n)_{n\in\mathbb{N}}$ be an arbitrary nonzero element of ℓ^∞. Then $\beta_{n_0} \neq 0$ for some integer n_0 and we can define the sequence $\gamma := (\gamma_n)_{n\in\mathbb{N}}$ by setting $\gamma_{n_0} = 0$ and $\gamma_n = \beta_n$ for all $n \neq n_0$. We see that $d(\beta, x_k) \geq d(\gamma, x_k)$ for all $k > n_0$ which shows that β cannot be the polar limit of $(x_k)_{k\in\mathbb{N}}$.

Even if most of the assertions in the following statements hold under weaker assumptions as well, we shall work in this section in uniformly convex and uniformly smooth Banach spaces i.e. in a Banach space E such that both E and E' satisfy SR (see [**19**, p. 32]). This implies, in particular, that E is reflexive. A duality relation which holds in this setting is given in [**35**]. We shall use in this section a notion of duality map which to every x belonging to a uniformly convex and uniformly smooth Banach space E associates the unique element x' in the dual space E' such that

$$\langle x, x' \rangle = \|x\|^2 = \|x'\|^2. \tag{5.2}$$

Note that $x' = \|x\|x^*$, where x^* is as in [**11**] or in [**33**]. In other terms, x^* is the Fréchèt derivative of the norm $\|x\|$ while x' is the Fréchèt derivative of $\frac{1}{2}\|x\|^2$, $x' = x$ when E is a Hilbert space. Here and in what follows weak convergence of a sequence $(x_n)_{n\in\mathbb{N}}$ in a Banach space to a point x is denoted as $x_n \rightharpoonup x$. We shall need the following technical lemma.

LEMMA 5.4. *Let $(E, \|\ \|)$ be a Banach space. Let $f : \mathbb{R}_+ \to \mathbb{R}_+$ be a function such that $\lim_{t\to+\infty} f(t) = 0$. Let $(\xi_n)_{n\in\mathbb{N}} \subset E'$ be such that for any $z \in E$, $\|z\| = 1$, and for all $s > 0$ there exists $\overline{n} = \overline{n}(s, z)$ such that for $n \geq \overline{n}$*

$$|\langle \xi_n, z \rangle| \leq f(s\|\xi_n\|)\|\xi_n\|. \tag{5.3}$$

Then ξ_n weakly converges to 0.*

PROOF. It suffices to prove the assertion for a subsequence, therefore we can assume without restrictions that there exists $C_1 > 0$ such that $\|\xi_n\| \geq C_1$ for any n. By taking $s = 1$ in (5.3) we deduce that $\left(f(\|\xi_n\|)\|\xi_n\|\right)^{-1} \xi_n$ is pointwise bounded and by the Uniform Boundedness Principle (see [**6**, Theorem 2.2]) it is bounded. Therefore, by the assumption on f, we deduce that exists $C_2 > 0$ such that $\|\xi_n\| \leq C_2$. Then, by applying (5.3) we deduce $|\langle \xi_n, z \rangle| \leq C_2 f(s\|\xi_n\|)$, and so the assertion follows, since $s\|\xi_n\| \geq sC_1$, by letting $s \to +\infty$. □

The following result, which shows the duality relation between the weak and the polar convergence in Banach spaces, follows from [**35**, Theorem 3.8], which is formulated in an equivalent way in terms of $(x_n^*)_{n\in\mathbb{N}}$, combined with [**35**, Theorem 3.5].

THEOREM 5.5 (Duality principle). *Let $(E, \| \ \|)$ be a uniformly convex and uniformly smooth Banach space. Let $(x_n)_{n\in\mathbb{N}} \subset E$ be any given sequence. Then $x_n \rightharpoondown 0$ in E if and only if $x'_n \rightharpoonup 0$ in E'.*

PROOF. Assume $x_n \rightharpoondown 0$. Note that for any $z \in E$ and for any $n \in \mathbb{N}$ we can find a unitary coefficient $\alpha_n = \alpha_n(z)$ such that

$$|\langle x'_n, z\rangle| = -\alpha_n\langle x'_n, z\rangle, \tag{5.4}$$

then, given $\varepsilon > 0$, since the scalar coefficient $\varepsilon\alpha_n \in \partial B_\varepsilon(0)$, by Corollary 3.6, with $\mathcal{K} = (\partial B_\varepsilon(0))z$, there exists $\overline{n} = \overline{n}(\varepsilon, z)$ such that for any $n \geq \overline{n}$,

$$0 \leq \|x_n + \varepsilon\alpha_n(z)z\|^2 - \|x_n\|^2. \tag{5.5}$$

Since $\frac{1}{2}\|x\|^2$ is uniformly Fréchèt differentiable on the normalized elements of E (see [**26**, pag. 61]), by combining (5.5) with (5.4), we have for large n

$$0 \leq \|x_n\|^2\left(\left\|\frac{x_n}{\|x_n\|} + \frac{\varepsilon\alpha_n}{\|x_n\|}\right\|^2 - 1\right) \leq -2\varepsilon|\langle x'_n, z\rangle| + \|x_n\|^2 r\left(\frac{\varepsilon}{\|x_n\|}\right), \tag{5.6}$$

where r is a function from $\mathbb{R}_+$ into $\mathbb{R}_+$ such that $\frac{r(s)}{s} \to 0$ as $s \to 0$. Then, for large n,

$$|\langle x'_n, z\rangle| \leq \left(\frac{\|x_n\|}{2\varepsilon} r\left(\frac{\varepsilon}{\|x_n\|}\right)\right)\|x_n\|, \tag{5.7}$$

so, since $\|x_n\| = \|x'_n\|$, by applying Lemma 5.4, with $f(t) = 2^{-1}t\,r(t^{-1})$, we deduce $x'_n \rightharpoonup 0$. Conversely, let $x'_n \rightharpoonup 0$, then for any z we get

$$\|x_n\|^2 = \langle x'_n, x_n\rangle = \langle x'_n, x_n - z\rangle + \langle x'_n, z\rangle \leq \|x_n\|\,\|x_n - z\| + o(1), \tag{5.8}$$

so the thesis follows from Lemma 3.5. □

Note that the spaces E and E' in the duality principle above are interchangeable, and therefore weak convergence and polar convergence can be also interchanged in the formulation of Theorem 5.5. Opial's examples in [**31**] of sequences in L^p spaces, $1 < p < \infty$, whose weak and polar limit do not coincide, can be easily explained in the light of Theorem 5.5: one takes normalized sequences of oscillating functions $(f_n)_{n\in\mathbb{N}}$ which weakly converge to zero but have positive and negative oscillations of different heights. Taking into account that the dual elements f'_n in L^p are given by the formula $f' = \|f\|^{2-p}|f|^{p-2}f$, if $p \neq 2$, we easily deduce that the positive and negative oscillations of f'_n are no longer balanced, $f'_n \not\rightharpoonup 0$ and so, by Theorem 5.5 $f_n \not\rightharpoondown 0$.

This situation is completely different if we know that $(f_n)_{n\in\mathbb{N}}$ is bounded and converges pointwise *a.e.* to 0. In such a case, by weak compactness, it also weakly converges to 0 and the dual sequence $(f'_n)_{n\in\mathbb{N}}$ still converges *a.e.* to 0 and therefore weakly. So, by Theorem 5.5, $f_n \rightharpoondown 0$. In other terms we have the following statement.

REMARK 5.6. Bounded *a.e.* converging sequences in L^p are both weakly and polarly convergent to their *a.e.* limits.

This is always the case in ℓ^p spaces, with $1 < p < +\infty$, where weak convergence implies pointwise convergence. So, for $1 < p < +\infty$, the spaces ℓ^p satisfy Opial condition, which is therefore not a characteristic condition of Hilbert spaces. This does not extend to the cases $p = 1$ and $p = +\infty$ discussed in Examples 5.2 and

5.3. Finally, the following result, in contrast to the general situation considered in Example 5.2, was proved in [**35**, Theorem 3.5] and is in some sense implicit in Lemma 5.4.

PROPOSITION 5.7. *If $(E, \| \|)$ is a uniformly convex and uniformly smooth Banach space, then any polarly converging sequence in E is bounded.*

PROOF. If $x_n \rightharpoondown x$, then by Theorem 5.5, $(x_n - x)'_{n\in\mathbb{N}}$ is bounded in E'. By definition, this implies that $(x_n - x)_{n\in\mathbb{N}}$ is bounded in E. □

6. Hadamard spaces

DEFINITION 6.1. Let (E, d) be a metric space, we shall say that (E, d) satisfies the "parallelogram inequality" if given any points $x, y \in E$, there exists a point $m \in E$ such that for every point $z \in E$,

$$d^2(z,m) \leq \frac{1}{2}\left(d^2(z,x) + d^2(z,y)\right) - \frac{1}{4}d^2(x,y). \tag{PI}$$

DEFINITION 6.2. We shall say that a metric space (E, d) is a CAT(0) space (see [**15**, Definition 2.9]) or is a space of Alexandrov nonpositive curvature (see [**21**, Definition 2.3.1 and the subsequent remark]) if it satisfies inequality (PI) and it is geodesic.

REMARK 6.3. Note that a metric space (E, d) which satisfies inequality (PI) is SR.

In order to clarify why we are calling (PI) a "parallelogram inequality" let us first observe that, when x and y have a middle point c (i.e. $\exists c \in E$ such that $d(x, c) = d(y, c) = \frac{1}{2}d(x, y)$) then c is the point m whose existence is asked in (PI).

Given any points $x, y \in E$ if they have a middle point then

$$d(x, y) = 2\text{rad}(\{x, y\}). \tag{6.1}$$

Moreover, spaces which satisfy inequality (PI), have the following property.

PROPOSITION 6.4. *Let (E, d) be a metric space which satisfies inequality (*PI*). Any pair of points $x, y \in E$ admits a (unique) middle point c if and only if (*6.1*) holds. In such cases $c = \text{cen}(\{x, y\})$ and $m = c$ in* (PI).

PROOF. Fix $x, y \in E$ and set $X = \{x, y\}$. Let $(c_n)_{n\in\mathbb{N}}$ be a minimizing sequence of the functional I_X defined in (3.1). Then, by replacing z by c_n in (PI) we deduce that $d(c_n, m) \to 0$ and so $m = \text{cen}(\{x, y\})$. The center is unique since (by Remark 6.3) E is SR. □

Note that on a linear normed space, by setting $z' = 2m - z$, i.e. a reflection of the point z about point m, the points x, y, z and z' are vertices of a parallelogram and so (PI) becomes

$$d^2(z,z') + d^2(x,y) \leq 2\left(d^2(z,x) + d^2(z,y)\right) \tag{6.2}$$

and expresses an inequality involving the squares of the diagonals and of the four sides of the parallelogram and justifies the name of (PI). Moreover, since we can associate to any parallelogram a dual one whose sides are the diagonals of the previous one, equality must hold in (6.2). Therefore the space can be equipped with a scalar product, i.e. linear complete spaces which satisfy (PI) are Hilbert spaces.

REMARK 6.5. Note that if a metric space (E, d) satisfies (6.1) for all x, y and satisfies inequality (PI), then by recursively applying Proposition 6.4, we deduce that for any pair of points $x, y \in E$ and for any dyadic number $\alpha \in [0,1]$ there exists a (unique) point $c_\alpha \in E$ such that

$$d(c_\alpha, y) = \alpha d(x,y) \quad \text{and} \quad d(c_\alpha, x) = (1-\alpha)d(x,y). \tag{6.3}$$

We can also deduce, by recursively applying (PI), that for any $z \in E$

$$d^2(z, c_\alpha) \le (1-\alpha)d^2(z,x) + \alpha d^2(z,y) - d(c_\alpha, x)d(c_\alpha, y) \tag{6.4}$$

or, equivalently,

$$d^2(z, c_\alpha) \le (1-\alpha)d^2(z,x) + \alpha d^2(z,y) - \alpha(1-\alpha)d^2(x,y). \tag{6.5}$$

When (E, d) is in addition a complete metric space, c_α is defined by continuity for any $\alpha \in [0,1]$ and gives rise to a (unique) geodesic, contained in E, which joins x to y. Therefore any complete metric space (E, d) which satisfies inequality (PI) and (6.1) for all x, y is a (uniquely) geodesic space. Finally (6.5) also holds for any $\alpha \in [0,1]$.

DEFINITION 6.6. A metric space (E, d) is an Hadamard space if it is a complete CAT(0) space.

REMARK 6.7. It is easy to conclude, given our observations above about the parallelogram inequality, that linear Hadamard spaces are Hilbert spaces.

As a rule of thumb one may consider Hadamard spaces as a "metric generalization" of Hilbert spaces, and expect that Hadamard spaces have some properties typical of Hilbert spaces but not found in all Banach spaces.

A few general properties of Hadamard spaces follow.

DEFINITION 6.8 (convex sets). Let (E, d) be a geodesic metric space and $C \subset E$. We shall say that C is convex if any two points $x, y \in C$ can be connected by a geodesic contained in C.

Note that, when (E, d) is a geodesic CAT(0) (resp. Hadamard) space and $C \subset E$ is a complete (resp. closed) convex set, then C satisfies (PI). Indeed, the middle point between any two points x, y belongs to the geodesic joining them which is contained in C. Therefore, C satisfies (SR) and so for any $x \in E$, any minimizing sequence of elements in C of the functional $I_{\{x\}}$ (defined by (3.1) for $X = \{x\}$) is a Cauchy sequence (see Corollary 3.8). Therefore, since C is complete, this sequence converges to the unique point $p_C(x)$ of C which minimizes the distance from x. It is easy to prove that $p_C : E \to E$ is a projector (i.e. that $p_C \circ p_C = p_C$). It is called the distance-minimizing projector onto C.

LEMMA 6.9 (Pythagoras Inequality). *Let (E, d) be a* CAT(0) *space. Let $\Gamma \subset E$ be a geodesic, then for all $x \in E$ and for all $y \in \Gamma$*

$$d^2(x,y) \ge d^2(x, p_\Gamma(x)) + d^2(p_\Gamma(x), y). \tag{6.6}$$

PROOF. For any $\alpha \in [0,1]$, by suitably applying (6.4) with c_α as in (6.3), one gets that $d^2(x, c_\alpha) \le (1-\alpha)d^2(x, p_\Gamma(x)) + \alpha d^2(x,y) - \alpha(1-\alpha)d^2(p_\Gamma(x), y)$ and, since $c_\alpha \in \Gamma$ and therefore $d(x, p_\Gamma(x)) \le d(x, c_\alpha)$, we deduce $d^2(x,y) \ge d^2(x, p_\Gamma(x)) + (1-\alpha)d^2(p_\Gamma(x), y)$ the thesis follows by taking α as close to 0 as we want. □

The above property easily extends to any compact convex set.

COROLLARY 6.10. *Let (E,d) be a* CAT(0) *space. Let $C \subset E$ be a compact convex set, then for all $x \in E$ and for all $y \in C$*

$$d^2(x,y) \geq d^2(x,p_C(x)) + d^2(p_C(x),y). \tag{6.7}$$

LEMMA 6.11. *Let (E,d) be a* CAT(0) *space. Let $C \subset E$ be a complete convex set, then the function $p_C : E \to E$ is a nonexpansive map, i.e.*

$$d(p_C(x),p_C(y)) \leq d(x,y) \tag{6.8}$$

holds for all $x,y \in E$.

PROOF. By applying (6.7) we deduce the two symmetric inequalities

$$d^2(x,p_C(y)) \geq d^2(x,p_C(x)) + d^2(p_C(x),p_C(y)) \tag{6.9}$$

$$d^2(y,p_C(x)) \geq d^2(y,p_C(y)) + d^2(p_C(y),p_C(x)). \tag{6.10}$$

Denoting by m the middle point between $p_C(x)$ and y we get from (PI) and (6.10)

$$d^2(x,m) \leq \frac{1}{2}d^2(x,p_C(x)) + \frac{1}{2}d^2(x,y) - \frac{1}{4}d^2(y,p_C(y)) - \frac{1}{4}d^2(p_C(y),p_C(x)) \tag{6.11}$$

$$d^2(p_C(y),m) \leq \frac{1}{4}d^2(y,p_C(y)) + \frac{1}{4}d^2(p_C(y),p_C(x)). \tag{6.12}$$

Then, by Schwartz Inequality and by adding (6.11) to (6.12), we obtain

$$d^2(x,p_C(y)) \leq 2(d^2(x,m) + d^2(m,p_C(y))) \leq d^2(x,p_C(x)) + d^2(x,y). \tag{6.13}$$

From (6.9) combined with (6.13) the assertion follows. □

7. Definitions of convergence via test maps

We consider the following definitions which extend to metric spaces the notion of weak convergence on Hilbert spaces by substituting the linear forms by distance-minimizing projections onto geodesic segments. In the case of normed spaces this does not necessarily define the usual weak convergence, since distance-minimizing projections onto straight lines do not generally coincide with continuous linear forms.

DEFINITION 7.1. Let (E,d) be a Hadamard space. A point $x \in E$ is called the weak limit of a sequence $(x_n)_{n\in\mathbb{N}} \subset E$ if for every geodesic Γ starting at x, $(p_\Gamma(x_n))_{n\in\mathbb{N}}$ converges to x. In this case, we say that $(x_n)_{n\in\mathbb{N}}$ weakly converges to x and we shall write $x_n \rightharpoonup x$.

As far as we know the definition is due to J. Jost (see [**20**, Definition 2.7]) for sequences in geodesic CAT(0) spaces. Subsequently in [**36**] Sosov has introduced essentially the same definition under the name of φ-convergence, and a variant of it called ψ convergence (which utilizes projections onto geodesics trough x rather than geodesics starting from x). As we show below in Theorem 7.4, in Hadamard spaces weak convergence can be equivalently defined as follows.

DEFINITION 7.2. Let (E,d) be a Hadamard space. A point $x \in E$ is called the weak limit of a sequence $(x_n)_{n\in\mathbb{N}} \subset E$ if for every compact convex set $C \subset E$ containing x, the sequence $(p_C(x_n))_{n\in\mathbb{N}}$ converges to x.

Finally we can also drop the requirement of convexity in the definition of weak convergence if we use a multivalued projection, which allows to bring up the following definition which makes sense in every metric space.

DEFINITION 7.3. Let (E,d) be a metric space. A point $x \in E$ is called the weak limit of a sequence $(x_n)_{n\in\mathbb{N}} \subset E$ if for every compact set $C \subset E$ containing x and for all sequences $(x'_n)_{n\in\mathbb{N}} \subset C$ of minimal distance from $(x_n)_{n\in\mathbb{N}}$, the sequence $(x'_n)_{n\in\mathbb{N}}$ converges to x.

By analogy with Hilbert spaces it is natural that Hadamard spaces may satisfy Opial condition (see Definition 5.1). We have the following result (see [**15**], [**2**]).

THEOREM 7.4. *Let (E,d) be a Hadamard space, let $x \in E$ and $(x_n)_{n\in\mathbb{N}} \subset E$ be a sequence. Then the following statements are equivalent:*

a) *$(x_n)_{n\in\mathbb{N}}$ weakly converges to x in the sense of Definition 7.3;*
b) *$(x_n)_{n\in\mathbb{N}}$ weakly converges to x in the sense of Definition 7.2;*
c) *$(x_n)_{n\in\mathbb{N}}$ weakly converges to x in the sense of Definition 7.1;*
d) *$(x_n)_{n\in\mathbb{N}}$ is polarly convergent to x.*

PROOF. a)$\Rightarrow$ b) is trivial and b) $\Rightarrow$ c) follows since any geodesic is a compact convex set.

c)$\Rightarrow$ d) Fix $y \in E$, $y \neq x$ and let Γ be the geodesic connecting x to y, then, by assumption, $p_\Gamma(x_n) \to x$. Then $d(x_n,y) \geq d(x_n,x) + o(1)$ and so the assertion follows from Proposition 2.2 and the equivalence in SR spaces of Δ and polar convergence.

d)$\Rightarrow$ a) Let C be a compact set such that $x \in C$. By compactness, if $(x'_n)_{n\in\mathbb{N}}$ does not converge to x, exists $y \in E$, $y \neq x$ such that (a subsequence of) $(x'_n)_{n\in\mathbb{N}}$ converges to y. Then, we deduce arguing as in the previous part, that $d(x_n,x) \geq d(x_n,x'_n) = d(x_n,y) + o(1)$ in contradiction to Lemma 3.5. □

REMARK 7.5. Note that the equivalence between a) and d), which makes sense in every metric space, holds for every SR space. Indeed, a) $\Rightarrow$ d) easily follows by applying a) for any given $y \in E$ by taking $C = \{x,y\}$.

REMARK 7.6. In [**20**] the following properties of weak convergence are established:

(1) Every bounded sequence has a weakly convergent subsequence ([**20**, Theorem 2.1])
(2) If $(x_n)_{n\in\mathbb{N}}$ is a bounded sequence, it has a renamed subsequence such that the corresponding sequence of "mean values" or "centers of gravity" (see [**20**, Definition 2.3]) converges (see [**20**, Theorem 2.2]).
(3) Consequently, every closed convex set is sequentially weakly closed and every lower semicontinuous convex function is weakly lower semicontinuous.
(4) Every convex coercive function has a minimum (see [**20**, Theorem 2.3]).

Items (1), (3) and (4) also follow from the equivalence in Theorem 7.4 and the following simple result.

PROPOSITION 7.7. *Let (E,d) be a Hadamard space. Then every closed convex set $C \subset E$ is polarly closed, i.e. if $(x_n)_{n\in\mathbb{N}} \subset C$ and $x_n \rightharpoondown x$, then $x \in C$.*

PROOF. Let I_{as} be defined as in (2.4). Then, by Lemma 6.11, $I_{as}(p_C(x)) = \limsup_n d(p_C(x),x_n) = \limsup_n d(p_C(x),p_C(x_n)) \leq \limsup_n d(x,x_n) = I_{as}(x)$, so $I_{as}(p_C(x)) \leq I_{as}(x)$ i.e. $x \in C$. □

The question of topology associated with Δ-convergence (or weak convergence) in Hadamard spaces was discussed by Monod in [**28**] where he proposed another natural definition of weak convergence by defining a topology $\mathcal{T}_C$ as the coarsest topology in which all closed convex sets are closed. We immediately see from (3) or from Proposition 7.7 that weak convergence is finer than $\mathcal{T}_C$ convergence. So, in view of property (1) above the two modes of convergence agree on bounded sequences whenever topology $\mathcal{T}_C$ is separated. On the other side Monod suggests examples of some spaces (see lines between examples 20 and 21 in [**28**]) on which $\mathcal{T}_C$ should be not separated. Therefore, in such a case, since the weak limit is obviously unique, the two modes of convergence would be different at least for nets. In other words, whenever topology $\mathcal{T}_C$ gives a notion of limit different from the weak one, it has bad properties like nonuniqueness of the limit.

Moreover, on Hadamard spaces, Monod introduces another topology $\mathcal{T}_w$ which easily implies Δ-convergence (and thus, weak convergence), but it is unclear how coarse is $\mathcal{T}_w$. $\mathcal{T}_w$ is defined ([**28**, Par. 3.7]) as the weakest topology under which for all $x, y \in E$ the map $z \mapsto d(x, z) - d(y, z)$ is continuous. In Hilbert spaces, if z_k converges to zero in $\mathcal{T}_w$, then $\|x - z_k\| = \|x\| + \|z_k\| + o(1)$ for any x. Calculating squares of this relation one gets, assuming that $\|z_k\|$ is bounded away from zero, that $(x, z_k/\|z_k\|) \to \|x\|$, which, if is true for some x, is false for $-x$. This contradiction implies that $\mathcal{T}_w$-convergence in Hilbert space implies norm convergence, contrary to the statement in [**28**, Example 18].

We refer the reader for further properties of weak convergence in Hadamard spaces, such as properties of Fejer monotone sequences, to the book of Bacak [**3**].

Obviously the study of the notions of convergence introduced so far becomes interesting in metric spaces more general than Hadamard spaces where Opial condition, by analogy with L^p spaces, does not hold. For such spaces we believe that, in order to have a generalization of weak convergence, rather than Δ-convergence, to metric spaces, definitions 7.2 and 7.1 must be modified. Indeed, the proof of Theorem 7.4 still applies in L^p, since the convex sets in L^p spaces are SR (see also Remark 7.5). Therefore, all definitions of weak convergence given in this section are still equivalent, in the case of L^p, to polar convergence. One may perhaps introduce a generalization of weak convergence in metric spaces that will be distinct from polar convergence, by replacing distance-minimizing projections in Definition 7.1 with other maps, but addressing this issue is outside the scope of this note. When two distinct modes of convergence of weak type are defined, one may also consider a notion of "double weak" convergence in which one requires that a point is both a weak and a polar limit of the sequence (as happens, for instance, for *a.e.* sequences in L^p spaces, see Remark 5.6). An application of this notion of double convergence, for which there is no compactness result of Banach-Alaoglu type, in L^p spaces, is given at the end of the next section.

8. Applications

8.1. Convergence of iterations to fixed points. Fixed points of nonexpansive maps (maps with the Lipschitz constant equal to 1) have been obtained as asymptotic centers of iterative sequences (see [**14**, Corollary to Theorem 2] for suitable Banach spaces (see also [**11**] and its review [**33**]) and [**37**, Claim 2.7] for suitable metric spaces). Also in bounded complete SR metric spaces the existence of fixed points for a nonexpansive map T easily follows, since, for any x, the asymptotic center of the iterations sequence $(T^n(x))_{n\in\mathbb{N}}$ is a fixed point of T. However, in general, it is not possible to get a fixed point as a polar limit of an iterations sequence. Indeed, Theorem 2.12 only gives the existence of the polar limit for a subsequence and the latter, as pointed out in the following example, (where the convergence is even in metric) is not necessarily a fixed point.

EXAMPLE 8.1. Consider the map $T(x) = -x$ on a unit ball in any Banach space. Zero is the unique fixed point of T and the iterations sequence $((-1)^n x)_{n\in\mathbb{N}}$ has zero as asymptotic center. On the other hand, if $x \neq 0$, this sequence has no subsequence that Δ-converges to zero and has two subsequences which converge to x and $-x$, which are not fixed points.

On a bounded complete SR metric space, we can give the following characterization of polarly converging iterations sequences of a nonexpansive map.

PROPOSITION 8.2. *Let (E, d) be a bounded complete SR metric space. Let $x \in E$ and let $T : E \to E$ be a nonexpansive map. Then, the following properties are equivalent.*

(a) *The iterations sequence $(T^n(x))_{n\in\mathbb{N}}$ is polarly convergent;*
(b) *The polar limit of any subsequence of $(T^n(x))_{n\in\mathbb{N}}$ (when it exists) is a fixed point of T.*

PROOF. We shall only prove that (b) implies (a) (the converse implication being obvious by (iii) in Remark 2.9). We shall prove, in particular, that if (b) holds true, then the sequence $(T^n(x))_{n\in\mathbb{N}}$ is maximal with respect to the relation $\preceq$ introduced in Remark 2.5. To this purpose let $(T^{k_n}(x))_{n\in\mathbb{N}}$ be a subsequence of $(T^n(x))_{n\in\mathbb{N}}$ and let $r = \operatorname{rad}_{n\to\infty} T^{k_n}(x)$. By Theorem 2.12 we can replace $(T^{k_n}(x))_{n\in\mathbb{N}}$ by a subsequence (which we still denote by $(T^{k_n}(x))_{n\in\mathbb{N}}$, note that the value of r does not increase) which is polarly converging to a point y. Then, if $\varepsilon > 0$ is an arbitrarily fixed real number, there exists $\bar{n} \in \mathbb{N}$ such that $d(T^{k_{\bar{n}}}(x), y) < r + \varepsilon$. Since, by assumption (b), y is a fixed point for T, we get, by induction, $d(T^n(x), y) < r + \varepsilon$, for any $n \geq k_{\bar{n}}$. (Indeed, if $n \geq k_{\bar{n}}$, we have

$$d(T^{n+1}(x), y) = d(T^{n+1}(x), T(y)) \leq d(T^n(x), y) \leq d(T^{k_{\bar{n}}}(x), y) < r + \varepsilon,$$

since T is nonexpansive). Therefore $\operatorname{rad}_{n\to\infty} T^n(x) < r + \varepsilon$. By the arbitrariness of $\varepsilon > 0$ we get $\operatorname{rad}_{n\to\infty} T^n(x) \leq r = \operatorname{rad}_{n\to\infty} T^{k_n}(x)$. $\square$

For spaces where the Opial condition holds, additional conditions, such as asymptotic regularity, are imposed to assure weak convergence of iterative sequences (see [**9**], [**31**, Theorem 2], [**10**] and [**4**]). However, the argument used in [**31**] suggests that the role of Opial condition consists in deducing weak convergence of iterations from their polar convergence.

DEFINITION 8.3. Let (E, d) be a metric space, $T : E \to E$ and $x \in E$. We say that the map T satisfies the PAR-condition (polar asymptotical regularity condition) at x if, for any $y \in E$ polar limit of a subsequence $(T^{k_n}(x))_{n\in\mathbb{N}}$ of the iterations sequence $(T^n(x))_{n\in\mathbb{N}}$ and for all $\varepsilon > 0$, there exists $\bar{n} \in \mathbb{N}$ such that $d(T^{k_n-1}(x), y) < d(T^{k_n}(x), y) + \varepsilon$ for any $n \geq \bar{n}$.

THEOREM 8.4. *Let (E, d) be a bounded complete SR metric space. Let $T : E \to E$ be a nonexpansive map which satisfies the PAR-condition at a point $x \in E$. Then, the sequence $(T^n(x))_{n\in\mathbb{N}}$ polarly converges to a fixed point.*

PROOF. We shall prove (b) in Proposition 8.2. Let $(T^{k_n}(x))_{n\in\mathbb{N}}$ be a polarly converging subsequence of $(T^n(x))_{n\in\mathbb{N}}$ and let y be the polar limit of $(T^{k_n}(x))_{n\in\mathbb{N}}$. Since T is a nonexpansive map which satisfies the PAR-condition in $x \in E$ we have that $d(T^{k_n}(x), T(y)) \le d(T^{k_n-1}(x), y) < d(T^{k_n}(x), y) + \varepsilon$ for any $\varepsilon > 0$ for large n. Therefore also $T(y)$ is an asymptotic center of $(T^{k_n}(x))_{n\in\mathbb{N}}$ and, since in SR spaces the asymptotic center is unique, we get $T(y) = y$. □

8.2. Brezis-Lieb Lemma without *a.e.* convergence. The celebrated Brezis-Lieb Lemma ([**7**]) is stated for (bounded) *a.e.* converging sequences that, see Remark 5.6, are also polarly converging. Remarkably, convergence *a.e.* is not needed when $p = 2$, which suggests that some version of Brezis-Lieb Lemma may hold for other p if *a.e.* convergence in the assumption is replaced by the double weak convergence mentioned in the previous section (weak and polar convergence of L^p). Indeed, in [**35**] the following result is proved.

THEOREM 8.5. *Let (Ω, μ) be a measure space. Assume that $u_n \rightharpoonup u$ and $u_n \rightharpoondown u$ in $L^p(\Omega, \mu)$. If $p \ge 3$ then*

$$\int_\Omega |u_n|^p d\mu \ge \int_\Omega |u|^p d\mu + \int_\Omega |u_n - u|^p d\mu + o(1). \tag{8.1}$$

It is shown in [**1**] that the condition $p \ge 3$ cannot be removed, except when $p = 2$. Moreover, one can see by easy examples with $p = 4$ (see [**35**]) that, in general, the equality does not hold in (8.1).

The elementary relation for Hilbert spaces,

$$u_n \rightharpoonup u \implies \|u_k\|^2 = \|u\|^2 + \|u_k - u\|^2 + o(1),$$

has the following counterpart in CAT(0)-spaces.

THEOREM 8.6. *Let (E, d) be a space which satisfies inequality* (PI) *and property* (6.1) *for all $x, y \in E$. Let $(x_n)_{n\in N} \subset E$, $x \in E$ such that $x_n \rightharpoondown x$. Then, for all $y \in E$*

$$d^2(x_n, y) \ge d^2(x, y) + d^2(x_n, x) + o(1). \tag{8.2}$$

PROOF. Let $y \in E$ and given a dyadic number $\alpha \in]0,1[$ let c_α as in (6.3). For n large

$$d(x_n, x) < d(x_n, c_\alpha). \tag{8.3}$$

So, by taking $z = x_n$ in (6.4) we deduce

$$d^2(x_n, x) \le d^2(x_n, y) - (1-\alpha) d^2(x, y).$$

Since we can take α arbitrarily close to 0, the assertion follows. □

From Theorem 8.6 one can easily deduce some classical properties of weak convergence known for normed spaces. The first one is just another form of item (3) in Remark 7.6.

COROLLARY 8.7. *Let (E, d) be a Hadamard space, then for any $y \in E$ the map $x \in E \mapsto d(x, y) \in \mathbb{R}$ is a polarly lower semicontinuous function.*

COROLLARY 8.8. *Hadamard spaces possess the Kadec-Klee property, i.e. if $(x_n)_{n\in\mathbb{N}} \subset E$ and $x \in E$ are such that $x_n \rightharpoondown x$ and if there exists $y \in E$ such that $d(x_n, y) \to d(x, y)$ then $d(x_n, x) \to 0$.*

Acknowledgment. The authors thank the reviewer of this note for providing a broader perspective to their results, including the crucial references to the work of Lim and Kirk & Panayak cited below. One of the authors (C.T.) thanks the math faculties of the Politecnico and of the University of Bari for their warm hospitality.

References

[1] Adimurthi and Cyril Tintarev, *On the Brezis-Lieb lemma without pointwise convergence*, NoDEA Nonlinear Differential Equations Appl. **22** (2015), no. 5, 1515–1521, DOI 10.1007/s00030-015-0333-2. MR3399189

[2] Miroslav Bačák, *Convex analysis and optimization in Hadamard spaces*, De Gruyter Series in Nonlinear Analysis and Applications, vol. 22, De Gruyter, Berlin, 2014. MR3241330

[3] Miroslav Bačák, *Convergence of nonlinear semigroups under nonpositive curvature*, Trans. Amer. Math. Soc. **367** (2015), no. 6, 3929–3953, DOI 10.1090/S0002-9947-2015-06087-5. MR3324915

[4] J. B. Baillon, R. E. Bruck, and S. Reich, *On the asymptotic behavior of nonexpansive mappings and semigroups in Banach spaces*, Houston J. Math. **4** (1978), no. 1, 1–9. MR0473932 (57 #13590)

[5] Nicolas Bourbaki, *Elements of mathematics. General topology. Part 1*, Hermann, Paris; Addison-Wesley Publishing Co., Reading, Mass.-London-Don Mills, Ont., 1966. MR0205210 (34 #5044a)

[6] Haim Brezis, *Functional analysis, Sobolev spaces and partial differential equations*, Universitext, Springer, New York, 2011. MR2759829 (2012a:35002)

[7] Haïm Brézis and Elliott Lieb, *A relation between pointwise convergence of functions and convergence of functionals*, Proc. Amer. Math. Soc. **88** (1983), no. 3, 486–490, DOI 10.2307/2044999. MR699419 (84e:28003)

[8] Martin R. Bridson and André Haefliger, *Metric spaces of non-positive curvature*, Grundlehren der Mathematischen Wissenschaften [Fundamental Principles of Mathematical Sciences], vol. 319, Springer-Verlag, Berlin, 1999. MR1744486 (2000k:53038)

[9] Felix E. Browder, *Nonexpansive nonlinear operators in a Banach space*, Proc. Nat. Acad. Sci. U.S.A. **54** (1965), 1041–1044. MR0187120 (32 #4574)

[10] Ronald E. Bruck and Simeon Reich, *Nonexpansive projections and resolvents of accretive operators in Banach spaces*, Houston J. Math. **3** (1977), no. 4, 459–470. MR0470761 (57 #10507)

[11] Ioana Cioranescu, *Geometry of Banach spaces, duality mappings and nonlinear problems*, Mathematics and its Applications, vol. 62, Kluwer Academic Publishers Group, Dordrecht, 1990. MR1079061 (91m:46021)

[12] S. Dhompongsa, W. A. Kirk, and B. Panyanak, *Nonexpansive set-valued mappings in metric and Banach spaces*, J. Nonlinear Convex Anal. **8** (2007), no. 1, 35–45. MR2314664 (2008f:47084)

[13] S. Dhompongsa, W. A. Kirk, and Brailey Sims, *Fixed points of uniformly Lipschitzian mappings*, Nonlinear Anal. **65** (2006), no. 4, 762–772, DOI 10.1016/j.na.2005.09.044. MR2232680 (2007b:54063)

[14] Michael Edelstein, *The construction of an asymptotic center with a fixed-point property*, Bull. Amer. Math. Soc. **78** (1972), 206–208. MR0291917 (45 #1005)

[15] Rafa Espínola and Aurora Fernández-León, *CAT(k)-spaces, weak convergence and fixed points*, J. Math. Anal. Appl. **353** (2009), no. 1, 410–427, DOI 10.1016/j.jmaa.2008.12.015. MR2508878 (2010d:47092)

[16] Thomas Foertsch, *Ball versus distance convexity of metric spaces*, Beiträge Algebra Geom. **45** (2004), no. 2, 481–500. MR2093020 (2005f:53133)

[17] Kazimierz Goebel and W. A. Kirk, *Topics in metric fixed point theory*, Cambridge Studies in Advanced Mathematics, vol. 28, Cambridge University Press, Cambridge, 1990. MR1074005 (92c:47070)

[18] Kazimierz Goebel and Simeon Reich, *Uniform convexity, hyperbolic geometry, and nonexpansive mappings*, Monographs and Textbooks in Pure and Applied Mathematics, vol. 83, Marcel Dekker, Inc., New York, 1984. MR744194 (86d:58012)

[19] William B. Johnson and Joram Lindenstrauss, *Basic concepts in the geometry of Banach spaces*, Handbook of the geometry of Banach spaces, Vol. I, North-Holland, Amsterdam, 2001, pp. 1–84, DOI 10.1016/S1874-5849(01)80003-6. MR1863689 (2003f:46013)

[20] Jürgen Jost, *Equilibrium maps between metric spaces*, Calc. Var. Partial Differential Equations **2** (1994), no. 2, 173–204, DOI 10.1007/BF01191341. MR1385525 (98a:58049)

[21] Jürgen Jost, *Nonpositive curvature: geometric and analytic aspects*, Lectures in Mathematics ETH Zürich, Birkhäuser Verlag, Basel, 1997. MR1451625 (98g:53070)

[22] W. A. Kirk, *Fixed point theory for nonexpansive mappings. II*, Fixed points and nonexpansive mappings (Cincinnati, Ohio, 1982), Contemp. Math., vol. 18, Amer. Math. Soc., Providence, RI, 1983, pp. 121–140, DOI 10.1090/conm/018/728596. MR728596 (85a:47062)

[23] W. A. Kirk and B. Panyanak, *A concept of convergence in geodesic spaces*, Nonlinear Anal. **68** (2008), no. 12, 3689–3696, DOI 10.1016/j.na.2007.04.011. MR2416076 (2009m:54061)

[24] Tadeusz Kuczumow, *An almost convergence and its applications* (English, with Russian and Polish summaries), Ann. Univ. Mariae Curie-Skłodowska Sect. A **32** (1978), 79–88 (1980). MR687863 (84f:47066)

[25] Teck Cheong Lim, *Remarks on some fixed point theorems*, Proc. Amer. Math. Soc. **60** (1976), 179–182 (1977). MR0423139 (54 #11120)

[26] Joram Lindenstrauss and Lior Tzafriri, *Classical Banach spaces. II*, Ergebnisse der Mathematik und ihrer Grenzgebiete [Results in Mathematics and Related Areas], vol. 97, Springer-Verlag, Berlin-New York, 1979. Function spaces. MR540367 (81c:46001)

[27] Francesco Maddalena and Sergio Solimini, *Synchronic and asynchronic descriptions of irrigation problems*, Adv. Nonlinear Stud. **13** (2013), no. 3, 583–623. MR3099831

[28] Nicolas Monod, *Superrigidity for irreducible lattices and geometric splitting*, J. Amer. Math. Soc. **19** (2006), no. 4, 781–814, DOI 10.1090/S0894-0347-06-00525-X. MR2219304 (2007b:22025)

[29] Nicolas Monod and Yehuda Shalom, *Cocycle superrigidity and bounded cohomology for negatively curved spaces*, J. Differential Geom. **67** (2004), no. 3, 395–455. MR2153026 (2006g:53051)

[30] Karl Menger, *Untersuchungen über allgemeine Metrik* (German), Math. Ann. **100** (1928), no. 1, 75–163, DOI 10.1007/BF01448840. MR1512479

[31] Zdzisław Opial, *Weak convergence of the sequence of successive approximations for nonexpansive mappings*, Bull. Amer. Math. Soc. **73** (1967), 591–597. MR0211301 (35 #2183)

[32] Simeon Reich, *Remarks on fixed points. II* (English, with Italian summary), Atti Accad. Naz. Lincei Rend. Cl. Sci. Fis. Mat. Natur. (8) **53** (1972), 250–254 (1973). MR0435953 (55 #8904)

[33] Simeon Reich, *Book Review: Geometry of Banach spaces, duality mappings and nonlinear problems*, Bull. Amer. Math. Soc. (N.S.) **26** (1992), no. 2, 367–370, DOI 10.1090/S0273-0979-1992-00287-2. MR1567976

[34] Simeon Reich and Itai Shafrir, *Nonexpansive iterations in hyperbolic spaces*, Nonlinear Anal. **15** (1990), no. 6, 537–558, DOI 10.1016/0362-546X(90)90058-O. MR1072312 (91k:47135)

[35] S. Solimini, C. Tintarev, Concentration analysis in Banach spaces, Comm. Contemp. Math. 2015, available online.

[36] E. N. Sosov, *On analogues of weak convergence in a special metric space* (Russian), Izv. Vyssh. Uchebn. Zaved. Mat. **5** (2004), 84–89; English transl., Russian Math. (Iz. VUZ) **48** (2004), no. 5, 79–83. MR2101682 (2005f:54008)

[37] John Staples, *Fixed point theorems in uniformly rotund metric spaces*, Bull. Austral. Math. Soc. **14** (1976), no. 2, 181–192. MR0415437 (54 #3524)

[38] D. van Dulst, *Equivalent norms and the fixed point property for nonexpansive mappings*, J. London Math. Soc. (2) **25** (1982), no. 1, 139–144, DOI 10.1112/jlms/s2-25.1.139. MR645871 (83e:47040)

Politecnico di Bari, via Amendola, 126/B, 70126 Bari, Italy
E-mail address: `giuseppe.devillanova@poliba.it`

Politecnico di Bari, via Amendola, 126/B, 70126 Bari, Italy
E-mail address: `sergio.solimini@poliba.it`

Uppsala University, box 480, 761 06 Uppsala, Sweden
E-mail address: `tintarev@math.uu.se`

Contemporary Mathematics
Volume **659**, 2016
http://dx.doi.org/10.1090/conm/659/13154

Gradient and extragradient methods for the elasticity imaging inverse problem using an equation error formulation: A comparative numerical study

A. Gibali, B. Jadamba, A. A. Khan, F. Raciti, and B. Winkler

ABSTRACT. This work aims to compare the performance of a variety of gradient-based and extragradient methods for the solution of the elastography inverse problem arising in the identification of cancerous tumors. From a mathematical standpoint, this inverse problem requires the identification of a variable parameter in a system of incompressible elasticity. We use an equation error approach to formulate the inverse problem as a convex optimization problem. The necessary and sufficient optimality condition then leads to a variational inequality which is solved using various extragradient methods, which have received great attention in recent years. Previously, only a few of these methods have been implemented and there is currently no understanding of their relative efficiency and effectiveness. We present a thorough numerical comparison of the projected gradient method, fast projected gradient method (fast iterative shrinkage-thresholding (FISTA)) [**3**], scaled projected gradient method, and several extragradient methods including the Marcotte variants, He-Goldstein-type method, the projection-contraction method proposed by Solodov and Tseng, and a hyperplane method.

1. Introduction

In this paper, we focus on the nonlinear inverse problem of identifying a variable material parameter in a system of incompressible linear elasticity. We pose the inverse problem as an optimization problem using an equation error approach. This inverse problem, which is known as the elastography or the elasticity imaging inverse problem, arises from a relatively new method for detecting tumors inside the human body using the differing elastic properties between healthy and unhealthy tissue. In elastography, a small, external, quasi-static compression force is applied to the body and the tissue's axial displacement field or overall motion are measured. A tumor can then be identified from this measurement by recovering the tissue's underlying elasticity.

The underlying mathematical model for the elastography inverse problem is the following system of partial differential equations which describe an isotropic elastic

2010 *Mathematics Subject Classification.* Primary 47J20, 90C29, 90C30.
Key words and phrases. Inverse problems, elasticity imaging inverse problem, elastography inverse problem, tumor identification, regularization, projected gradient methods, extragradient methods.

object's response to known body forces and traction applied along its boundary:

$$-\nabla \cdot \sigma = f \quad \text{in } \Omega, \tag{1a}$$

$$\sigma = 2\mu\epsilon(u) + \lambda \text{div}\, u\, I, \tag{1b}$$

$$u = f_1 \text{ on } \Gamma_1, \tag{1c}$$

$$\sigma n = f_2 \text{ on } \Gamma_2. \tag{1d}$$

The domain Ω is a subset of $\mathbb{R}^2$ or $\mathbb{R}^3$ and $\partial\Omega = \Gamma_1 \cup \Gamma_2$ is its boundary. In (1), the vector-valued function $u = u(x)$ represents the displacement of the elastic object, f the body force being applied, n the unit outward normal, and $\epsilon(u) = \frac{1}{2}(\nabla u + \nabla u^{\mathrm{T}})$ the linearized strain tensor. The stress tensor σ and the stress-strain law (1b) hold given the assumption that the elastic object is isotropic and the displacement is small enough to maintain a linear relationship. The Lamé parameters, μ and λ, represent the object's variable elastic properties.

The direct problem in the context of (1) is to find the displacement u when the functions f_1, f_2, the coefficients μ and λ, and f are all known. For the elastography inverse problem, we seek to find the parameter μ when a certain measurement z of the displacement u is available. We remark that a related inverse problem for several engineering applications is to find both μ and λ, which typically vary in a small range (see [**14, 16–18**]). However, the elastography inverse problem is modeled under the assumption that the human body is an incompressible material, that is, $\lambda \gg \mu$. Consequently, the elastography inverse problem seeks to recover the parameter μ alone.

To describe the main difficulties associated with near incompressibility, we first introduce some notation. The dot product of two tensors A_1 and A_2 will be denoted by $A_1 \cdot A_2$. Given a sufficiently smooth domain $\Omega \subset \mathbb{R}^2$, the L_2-norm of a tensor-valued function $A = A(x)$ is provided by

$$\|A\|_{L^2}^2 = \|A\|_{L^2(\Omega)}^2 = \int_\Omega A \cdot A = \int_\Omega \left(A_{11}^2 + A_{12}^2 + A_{21}^2 + A_{22}^2\right).$$

On the other hand, for a vector-valued function $u(x) = (u_1(x), u_2(x))^{\mathrm{T}}$, the L_2-norm and the H^1-norm are given by

$$\|u\|_{L_2}^2 = \|u\|_{L_2(\Omega)}^2 = \int_\Omega \left(u_1^2 + u_2^2\right),$$

$$\|u\|_{H^1}^2 = \|u\|_{H^1(\Omega)}^2 = \|u\|_{L_2}^2 + \|\nabla u\|_{L_2}^2.$$

In the following, for the sake of simplicity, we take $f_1 = 0$. The space of test functions, denoted by $\bar{V}$, is then given by

$$\bar{V} = \{\bar{v} \in H^1(\Omega) \times H^1(\Omega) : \ \bar{v} = 0 \text{ on } \Gamma_1\}. \tag{2}$$

By employing Green's identity and applying the boundary conditions from (1c) and (1d), we get the following variational form of (1): Find $\bar{u} \in \bar{V}$ such that

$$\int_\Omega 2\mu\epsilon(\bar{u}) \cdot \epsilon(\bar{v}) + \int_\Omega \lambda(\text{div}\, \bar{u})(\text{div}\, \bar{v}) = \int_\Omega f\bar{v} + \int_{\Gamma_2} \bar{v} f_1, \quad \text{for every } \bar{v} \in \bar{V}. \tag{3}$$

For $T : \bar{V} \times \bar{V} \to \mathbb{R}$ defined by

$$T(\bar{u}, \bar{v}) = \int_\Omega 2\mu\epsilon(\bar{u}) \cdot \epsilon(\bar{v}) + \int_\Omega \lambda(\text{div}\, \bar{u})(\text{div}\, \bar{v}),$$

it can be shown that if μ and $\mu+\lambda$ are bounded away from zero then there are two positive constants $c_1 > 0$ and $c_2 > 0$ with $c_1 \leq \mu$ and $c_2 \geq \lambda + \mu$ such that

$$c_1\|\bar{v}\|_{\bar{V}}^2 \leq T(\bar{v},\bar{v}) \leq c_2\|\bar{v}\|_{\bar{V}}^2, \quad \text{for every } \bar{v} \in \bar{V}.$$

Since $\lambda \gg \mu$, the ratio $c_3 = c_2/c_1$ is large. Because the constant c_3 determines the error estimates (as defined by Céa's lemma), it follows that the actual error could easily outstrip the optimal approximation error. This situation is well-known and has been dubbed the "locking effect".

A wide range of approaches have been given to overcome the locking effect with one of the most popular being the use of mixed finite elements, an approach which we adopt in this work. For this we introduce a "pressure" term $p \in Q = L^2(\Omega)$ by

$$p = \lambda \operatorname{div} \bar{u}, \tag{4}$$

which results in the following variational formulation of (4):

$$\int_\Omega (\operatorname{div} \bar{u})q - \int_\Omega \frac{1}{\lambda} pq = 0, \quad \text{for every } q \in Q. \tag{5}$$

Using (4), the weak form (3) then transforms into the following: Find $\bar{u} \in \bar{V}$ such that

$$\int_\Omega 2\mu\epsilon(\bar{u}) \cdot \epsilon(\bar{v}) + \int_\Omega p(\operatorname{div} \bar{v}) = \int_\Omega f\bar{v} + \int_{\Gamma_2} \bar{v} f_1, \quad \text{for every } \bar{v} \in \bar{V}, \tag{6}$$

where p is also an unknown.

Thus we have moved from finding $\bar{u} \in \bar{V}$ fulfilling (3) to finding $(\bar{u}, p) \in \bar{V} \times Q$ satisfying both (5) and (6). In this work, this saddle point problem constitutes the direct problem.

2. Equation Error Approach

We now proceed to introduce the equation error functional to identify the parameter μ. For this, we define the sets

$$\begin{aligned} \widehat{V} &= \left\{u \in H^1(\Omega) \,:\, u = 0 \text{ on } \Gamma_2\right\}, \\ \overline{V} &= \widehat{V}^2, \\ V &= \overline{V} \times L^2(\Omega), \\ A &= \{\mu \in L^\infty(\Omega) \,:\, \mu \geq \mu_0 \text{ in } \Omega\}, \end{aligned}$$

where $\mu_0 > 0$ is a given constant. For the analysis given below, we will also need the space

$$V^\infty = W^{1,\infty}(\Omega) \times L^\infty(\Omega)$$

with

$$\|u\|_{V^\infty} = \max\left\{\|\overline{u}\|_{W^{1,\infty}}, \|p\|_{L^\infty}\right\},$$

where $u = (\overline{u}, p) \in V^\infty$.

As discussed in the previous section, the boundary valued problem described in (1) is equivalent to the following saddle point problem: Find $u = (\overline{u}, p) \in V$ such that

$$\begin{aligned} &\int_\Omega 2\mu\epsilon(\overline{u}) \cdot \epsilon(\overline{v}) + \int_\Omega p\,(\nabla \cdot \overline{v}) = \int_\Omega f \cdot \overline{v} + \int_{\Gamma_2} f_2 \cdot \overline{v}, \quad \text{for all } \overline{v} \in \overline{V}, \\ &\int_\Omega (\nabla \cdot \overline{u})\, q - \int_\Omega \frac{1}{\lambda} pq = 0, \quad \text{for all } q \in L^2(\Omega). \end{aligned} \tag{7}$$

We define $E_1 : L^\infty(\Omega) \times V \to \overline{V}^*$ and $E_2 : V \to L^2(\Omega)^*$ by

$$\begin{aligned} E_1(\mu, u)v &= \int_\Omega 2\mu\epsilon(\overline{u}) \cdot \epsilon(\overline{v}) + \int_\Omega p\,(\nabla \cdot \overline{v}) \text{ for all } \overline{v} \in \overline{V}, \\ E_2(u)v &= \int_\Omega (\nabla \cdot \overline{u})\, q - \int_\Omega \frac{1}{\lambda} pq \text{ for all } q \in L^2(\Omega), \end{aligned}$$

where $u = (\overline{u}, p)$. We also define $m \in \overline{V}^*$ by

$$m(v) = \int_\Omega f \cdot \overline{v} + \int_{\Gamma_2} f_2 \cdot \overline{v} \text{ for all } \overline{v} \in \overline{V}.$$

Note that although we have defined $E_1(\mu, u)$ for $\mu \in L^\infty(\Omega)$, $u \in V$, this functional is also well-defined for $\mu \in L^2(\Omega)$ and $u \in V^\infty$. That is, we can sacrifice some regularity in μ by imposing more on u. We will take advantage of this below (abusing notation by continuing to write $E_1(\mu, u)$ when μ does not necessarily belong to $L^\infty(\Omega)$).

The equation error method aims to estimate μ^* from a measurement z of u^* by minimizing

$$J(\mu; z, \beta) = \|E_1(\mu, z) - m\|^2_{V^*} + \beta\|\mu\|^2_{H^1}, \tag{8}$$

where it is assumed that $\mu^* \in A$ and $u^* = (\overline{u}^*, p^*) \in V$ satisfy (7).

Clearly it is natural to define J by

$$J(\mu; z, \beta) = \|E_1(\mu, z) - m\|^2_{V^*} + \|E_2(z)\|^2_{V^*} + \beta\|\mu\|^2_{H^1},$$

but since $\|E_2(z)\|^2_{V^*}$ is constant with respect to μ, it makes no difference if this term is included.

The following result, recently proved in [**15**], confirms that $J(\cdot; z, \beta)$ has a unique minimizer in $H^1(\Omega)$ for each $\beta > 0$, provided z belongs to V^∞; See also [**11, 12**].

THEOREM 2.1. *Suppose $z \in V^\infty$. Then, for each $\beta > 0$, there exists a unique μ_β satisfying*

$$J(\mu_\beta; z, \beta) \leq J(\mu; z, \beta) \textit{ for all } \mu \in H^1(\Omega).$$

To give a stability/convergence result, we recall that $\mu^* \in A$ and $u^* = (\overline{u}^*, p^*) \in V$ are assumed to satisfy (7). We do not assume that μ^* is unique in this regard, so let us define $S = \{\mu \in H^1(\Omega) : E_1(\mu, u^*) = m\}$. Since E_1 is affine in μ, S is convex.

The following is a convergence/stability result whose proof can be found in [**15**].

THEOREM 2.2. *Suppose $u^* \in V^\infty$, $\mu^* \in H^1(\Omega)$ satisfy the saddle point problem (7). Let $\{z_n\} \subset V^\infty$ be a sequence of observations of u^* that satisfy, with the sequences $\{\epsilon_n\}$, $\{\beta_n\}$, the conditions*

(1) $\epsilon_n^2 \leq \beta_n \leq \epsilon_n$ *for all* $n \in \mathbb{Z}^+$;
(2) $\epsilon_n^2/\beta_n \to 0$ *as* $n \to \infty$;
(3) $\|z_n - u^*\|_{V^\infty} \leq \epsilon_n$ *for all* $n \in \mathbb{Z}^+$;
(4) $\epsilon_n \to 0$ *as* $n \to \infty$.

For each $n \in \mathbb{Z}^+$, let μ_n be the unique solution of

$$\min_{\mu \in H^1(\Omega)} J(\mu; z_n, \beta_n).$$

Then there exists $\tilde{\mu} \in S$ *such that* $\mu_n \to \tilde{\mu}$ *in* $H^1(\Omega)$. *Moreover,* $\tilde{\mu}$ *satisfies* $\|\tilde{\mu}\|_{H^1} \leq \|\mu\|_{H^1}$ *for all* $\mu \in S$.

2.1. Discrete Formulas for the Equation Error. In this subsection, we collect some basic information concerning the discretization of equation error formulation. Some missing details can be found in [**15**]. As usual, we assume that $\mathcal{T}_h$ is a triangulation on Ω, L_h is the space of all piecewise continuous polynomials of degree d_μ relative to $\mathcal{T}_h$, $\bar{U}_h$ is the space of all piecewise continuous polynomials of degree d_u relative to $\mathcal{T}_h$, and Q_h is the space of all piecewise continuous polynomials of degree d_q relative to $\mathcal{T}_h$.

To represent the discrete saddle point problem in a computable form we proceed as follows. We represent bases for L_h, $\bar{U}_h$ and Q_h by $\{\varphi_1, \varphi_2, \dots, \varphi_m\}$, $\{\psi_1, \psi_2, \dots, \psi_n\}$, and $\{\chi_1, \chi_2, \dots, \chi_k\}$, respectively. The space L_h is then isomorphic to $\mathbb{R}^m$ and for any $\mu \in L_h$, we define $\boldsymbol{\mu} \in \mathbb{R}^m$ by $\boldsymbol{\mu}_i = \mu(x_i), \quad i = 1, 2, \dots, m$, where the nodal basis $\{\varphi_1, \varphi_2, \dots, \varphi_m\}$ corresponds to the nodes $\{x_1, x_2, \dots, x_m\}$. Conversely, each $\boldsymbol{\mu} \in \mathbb{R}^m$ corresponds to $\mu \in L_h$ defined by $\mu = \sum_{i=1}^m \boldsymbol{\mu}_i \varphi_i$. Analogously, $\bar{u} \in \bar{U}_h$ will correspond to $\bar{U} \in \mathbb{R}^n$, where $\bar{U}_i = u(y_i)$, $i = 1, 2, \dots, n$, and $\bar{u} = \sum_{i=1}^n \bar{U}_i \psi_i$, where $y_1, y_2, \dots, y_n$ are the nodes of the mesh defining $\bar{U}_h$. Finally, $q \in Q_h$ will correspond to $Q \in \mathbb{R}^k$, where $Q_i = q(z_i)$, $i = 1, 2, \dots, k$, and $q = \sum_{i=1}^k Q_i \chi_i$, where $z_1, z_2, \dots, z_k$ are the nodes of the mesh defining Q_h.

We next define $S : \mathbb{R}^m \to \mathbb{R}^{n+k}$ to be the finite element solution operator that assigns to each coefficient $\mu_h \in A_h$, the unique approximate solution $u_h = (\bar{u}_h, p_h) \in \bar{U}_h \times Q_h$. Then $S(\boldsymbol{\mu}) = U$, where U is defined by

$$K(\boldsymbol{\mu})U = F, \tag{9}$$

where $K(\boldsymbol{\mu}) \in \mathbb{R}^{(n+k)\times(n+k)}$ is the stiffness matrix and $F \in \mathbb{R}^{n+k}$ is the load vector.

With the above preparation, we easily obtain the discrete version of the equation error functional (8) (without the regularizer), denoted by J_E,

$$J_E(\boldsymbol{\mu}) = \frac{1}{2}\left\langle \widehat{L}(\bar{Z})\boldsymbol{\mu} + B^T P - F, (K+M)^{-1}\left(\widehat{L}(\bar{Z})\boldsymbol{\mu} + B^T P - F\right)\right\rangle,$$

where $\widehat{L}$ is the so-called adjoint stiffness matrix, K is the stiffness matrix, M is the mass matrix, and $\bar{Z}$ is the data for $\bar{u}$ (see [**15**] for a more complete discussion).

Moreover, the first derivative of $J_E(\boldsymbol{\mu})$ is given by:

$$DJ_E(\boldsymbol{\mu})(\delta\boldsymbol{\mu}) = \left\langle \widehat{L}(\bar{Z})\delta\boldsymbol{\mu}, (K+M)^{-1}(\widehat{L}(\bar{Z})\boldsymbol{\mu} + B^T P - F)\right\rangle,$$

implying that the gradient of $J_E(\boldsymbol{\mu})$ is given by

$$\nabla J_E(\boldsymbol{\mu}) = \widehat{L}(\bar{Z})^T (K+M)^{-1}(\widehat{L}(\bar{Z})\boldsymbol{\mu} + B^T P - F). \tag{10}$$

For the second derivative we then have

$$\begin{aligned} D^2 J_E(\boldsymbol{\mu})(\delta\boldsymbol{\mu})(\tilde{\delta}\boldsymbol{\mu}) &= \left\langle \widehat{L}(\bar{Z})\delta\boldsymbol{\mu}, (K+M)^{-1}\widehat{L}(\bar{Z})\tilde{\delta}\boldsymbol{\mu}\right\rangle \\ &= \left\langle \widehat{L}(\bar{Z})^T (K+M)^{-1}\widehat{L}(\bar{Z})\delta\boldsymbol{\mu}, \tilde{\delta}\boldsymbol{\mu}\right\rangle, \end{aligned}$$

which implies that the Hessian of $J_E(\boldsymbol{\mu})$ is

$$\nabla^2 J_E(\boldsymbol{\mu}) = \widehat{L}(\bar{Z})^T (K+M)^{-1}\widehat{L}(\bar{Z}). \tag{11}$$

3. Gradient and Extragradient Methods

Although we perform numerical tests for some gradient-based methods, the main emphasis of this work is on numerical testing of the so-called extragradient methods. We note that extragradient methods, originally proposed to solve minimization and saddle point problems, have received a great deal of attention in recent years, particularly in the context of variational inequalities (see [**1**, **2**, **4**–**10**, **19**–**21**, **23**, **25**, **26**, **28**–**30**, **32**–**35**, **37**, **38**, **40**–**46**]).

Korpelevich [**28**], who introduced the method originally, gave it the name "extragradient" due to the "extra" evaluation of the objective function f being minimized in each iteration and which, when solving a variational inequality, corresponded with the extra evaluation of the gradient of some function.

Earlier developments of these methods were of theoretical nature, however, the recent trend is not only to give convergence analysis but also to test their practical usefulness on numerical examples. Therefore, we anticipate that the novel application of these methods to solve inverse problems for partial differential equations will give a plethora of test problems for examining their efficiency and effectiveness.

In this work, we employ variants of projected gradient methods and extragradient methods to solve the inverse problem of parameter identification by first posing it as a variational inequality. We implement numerous algorithms and present a thorough comparison of the projected gradient method, fast projected gradient method (fast iterative shrinkage-thresholding (FISTA)), scaled projected gradient method and several extragradient methods including the Marcotte variants, He-Goldstein type method, the projection-contraction methods proposed by Solodov and Tseng, and the hyperplane method proposed independently by Konnov and by Iusem and Svaiter. During the last two decades, numerous researches have focused on extragradient-type methods, but to the best of our knowledge, this is the first instance where these methods have been thoroughly compared in the context of an applied problem.

In this section, we implement and test the numerical performance of the following iterative schemes for solving the inverse problem of tumor identification in elastography:

(1) Gradient Projection Using Armijo Line Search
(2) Fast Gradient Projection Using Armijo Line Search (FISTA)
(3) Scaled Gradient Projection Using Barzilai-Borwein Rules
(4) Khobotov Extragradient Method Using Marcotte Rules (3 Variants)
(5) Solodov-Tseng Projection-Contraction Method (2 variants)
(6) Improved He-Goldstein Type Extragradient Method
(7) Two-step Extragradient Method
(8) Hyperplane Extragradient Method

We will solve the elastography inverse problem by formulating the regularized equation error functional, henceforth denoted simply by J, as a variational inequality of finding $\boldsymbol{\mu}^* \in K$ such that

$$\langle \nabla J(\boldsymbol{\mu}^*), \boldsymbol{\mu} - \boldsymbol{\mu}^* \rangle \geq 0, \quad \forall\, \boldsymbol{\mu} \in K,$$

where K is the set of admissible coefficients. In all numerical experiments, we take the set K to be box constrained.

The above variational inequality has a unique solution if ∇J is strongly monotone, that is,

$$\langle \nabla J(\boldsymbol{\mu}_1) - \nabla J(\boldsymbol{\mu}_2), \boldsymbol{\mu}_1 - \boldsymbol{\mu}_2 \rangle \geq c\|\boldsymbol{\mu}_1 - \boldsymbol{\mu}_2\|^2, \quad \forall\, \boldsymbol{\mu}_1, \boldsymbol{\mu}_2 \in K, \quad c > 0,$$

and Lipschitz continuous

$$\|\nabla J(\boldsymbol{\mu}_1) - \nabla J(\boldsymbol{\mu}_2)\| \leq L\|\boldsymbol{\mu}_1 - \boldsymbol{\mu}_2\|, \quad \forall \boldsymbol{\mu}_1, \boldsymbol{\mu}_2 \in K, \quad L > 0.$$

As usual, we convert the above variational inequality into a fixed point problem

$$\boldsymbol{\mu}^* = P_K(\boldsymbol{\mu}^* - \alpha \nabla J(\boldsymbol{\mu}^*)), \quad \alpha > 0,$$

where P_K is the metric projection onto K.

3.1. Gradient Projection Method. The projected gradient algorithm admits the form:

$$\boldsymbol{\mu}^{k+1} = P_K(\boldsymbol{\mu}^k - \alpha \nabla J(\boldsymbol{\mu}^k)).$$

The strong convergence can be established by assuming that

$$\alpha \in \left(0, \frac{2c}{L^2}\right),$$

where c and L are the modulus of strong monotonicity and Lipschitz continuity, respectively.

Note that we do not have information about c and L and hence it is important to use a method to determine the step-length α.

We use Armijo line search to backtrack until the following condition is satisfied:

$$J(\boldsymbol{\mu}^{k+1}) - J(\boldsymbol{\mu}^k) \leq -\alpha\lambda\|\nabla J(\boldsymbol{\mu}^k)\|^2, \quad \text{for } \lambda \in (0,1).$$

3.2. Fast Gradient Projection or Fast Iterative Shrinkage-thresholding Method. Beck and Teboulle [**3**], following the work of Nesterov [**36**], presented a fast version of the projected gradient method which is an optimal first-order method. In the following algorithm, L again represents the Lipschitz constant of ∇J.

Algorithm: FISTA

Choose $\bar{\boldsymbol{\mu}}^1 = \boldsymbol{\mu}^0$, $t_1 = 1$, and N, the maximum number of iterations.
For $k = 0, 1, 2, \ldots, N$, perform the following:
Step 1: $\boldsymbol{\mu}^k = P_K(\bar{\boldsymbol{\mu}}^k - \frac{1}{L}\nabla J(\bar{\boldsymbol{\mu}}^k))$
Step 2: $t_{k+1} = \frac{1+\sqrt{1+4t_k^2}}{2}$
Step 3: $\bar{\boldsymbol{\mu}}^{k+1} = \boldsymbol{\mu}^k + \frac{t_k - 1}{t_{k+1}}(\boldsymbol{\mu}^k - \boldsymbol{\mu}^{k-1})$
End

Convergence is guaranteed under the assumption of Lipschitz continuity. A more practical version of the method (and the one implemented) also uses Armijo line search in order to determine the step-length.

3.3. Scaled Gradient Projection. The scaled gradient projection (SGP) method has the following form

$$\boldsymbol{\mu}^{k+1} = P_K(\boldsymbol{\mu}^k - \alpha_k D_k \nabla J(\boldsymbol{\mu}^k)),$$

where D_k is a scaling matrix. It is a common practice to take the scaling matrix D_k as the main diagonal of the Hessian of $J(\boldsymbol{\mu}^k)$, with all other entries equal to zero.

Algorithm: SGP

Choose $\boldsymbol{\mu}^0 \in \boldsymbol{\mu}_m$, β, $\theta \in (0,1)$, $0 < \alpha_{min} < \alpha_{max}$, $M > 0$
For $k = 0, 1, 2, \ldots,$ perform the following steps:
Step 1: Choose $\alpha_k \in [\alpha_{min}, \alpha_{max}]$ and D_k
Step 2: Projection $Y^k = P_K(\boldsymbol{\mu}^k - \alpha D_k \nabla J(\boldsymbol{\mu}^k))$
If $Y^k = \boldsymbol{\mu}^k$ Stop
Step 3: Descent direction: $d^k = Y^k - \boldsymbol{\mu}^k$
Step 4: Set $\lambda_k = 1$ and $f_{max} = \max\limits_{0 \le j \le \min(k, M-1)} J(\boldsymbol{\mu}^{k-j})$
Step 5: Backtracking loop:
If $J(\boldsymbol{\mu}^k + \lambda_k d^k) \le f_{max} + \beta \lambda_k \nabla J(\boldsymbol{\mu}^k)^T d^k$
Go to Step 6
Else
Set $\lambda_k = \theta \lambda_k$ and go to Step 5
EndIf
Step 6: $\boldsymbol{\mu}^{k+1} = \boldsymbol{\mu}^k + \lambda_k d^k$
End

Here we choose α_k using the Barzilai-Borwein rules. That is, for

$$\begin{aligned} r^{k-1} &= \boldsymbol{\mu}^k - \boldsymbol{\mu}^{k-1} \\ z^{k-1} &= \nabla J(\boldsymbol{\mu}^k) - \nabla J(\boldsymbol{\mu}^{k-1}), \end{aligned}$$

we compute:

$$\begin{aligned} \alpha_k^{(1)} &= \frac{r^{(k-1)^T} D_k^{-1} D_k^{-1} r^{(k-1)}}{r^{(k-1)^T} D_k^{-1} z^(k-1)} \\ \alpha_k^{(2)} &= \frac{r^{(k-1)^T} D_k z^{(k-1)}}{z^{(k-1)^T} D_k^2 z^{(k-1)}}. \end{aligned}$$

Determining α_k: Take a prefixed non-negative integer M_α and $\tau_1 \in (0,1)$:
If $\alpha_k^{(2)}/\alpha_k^{(1)} \le \tau_k$ then
$\alpha_k = \min\left(\alpha_j^{(2)}, j = \max(1, k - M_\alpha), \ldots, k\right)$
$\tau_{k+1} = 0.9\tau_k$
Else
$\alpha_k = \alpha_k^{(1)}$
$\tau_{k+1} = 1.1\tau_k$
EndIf.

Now we explore the extragradient method proposed in [**28**] to relax the conditions on convergence for the projection method:

$$\begin{aligned} \bar{\boldsymbol{\mu}}^k &= P_K(\boldsymbol{\mu}^k - \alpha_k \nabla J(\boldsymbol{\mu}^k)) \\ \boldsymbol{\mu}^{k+1} &= P_K(\boldsymbol{\mu}^k - \alpha_k \nabla J(\bar{\boldsymbol{\mu}}^k)). \end{aligned}$$

where α is constant for all iterations.

Convergence can be proved under the conditions that the solution set is nonempty, ∇J is monotone and Lipshitz (with constant L) and $\alpha \in (0, 1/L)$.

Clearly, when L is unknown, we may have difficulties choosing an appropriate α. As with the gradient projection method, if α is too small, then the algorithm will converge slowly and if α is too big, then it may not converge at all.

3.4. Khobotov Extragradient Method. In the following, we will now consider extragradient methods where α is an adaptive steplength. We are going to implement the adaptive steplength first introduced in [**24**] to remove the constraint that ∇J must be Lipshitz continuous. The adaptive algorithm is of the form:

$$\begin{aligned} \bar{\boldsymbol{\mu}}^k &= P_K(\boldsymbol{\mu}^k - \alpha_k \nabla J(\boldsymbol{\mu}^k)) \\ \boldsymbol{\mu}^{k+1} &= P_K(\boldsymbol{\mu}^k - \alpha_k \nabla J(\bar{\boldsymbol{\mu}}^k)). \end{aligned}$$

Intuitively, we get better convergence when α gets smaller between iterations, however, it is obvious that we must also control how the sequence of $\{\alpha_k\}$ shrinks.

We use the following reduction rule for α_k given in [**24**]:

$$\alpha_k > \beta \frac{\|\boldsymbol{\mu}^k - \bar{\boldsymbol{\mu}}^k\|}{\|\nabla J(\boldsymbol{\mu}^k) - \nabla J(\bar{\boldsymbol{\mu}}^k)\|},$$

where $\beta \in (0, 1)$. Results from [**40**] and [**24**] show that β is usually taken as 0.8 or 0.9, values likewise used successfully in our implementation of the algorithm. The Khobotov extragradient method has the following general form:

Algorithm: Khobotov Extragradient
Choose α_0, $\boldsymbol{\mu}^0$, and $\beta \in (0, 1)$
While $\|\boldsymbol{\mu}^{k+1} - \boldsymbol{\mu}^k\| > \text{TOL}$
Step 1: Compute $\nabla J(\boldsymbol{\mu}^k)$
Step 2: Compute $\bar{\boldsymbol{\mu}}^k = P_K(\boldsymbol{\mu}^k - \alpha_k \nabla J(\boldsymbol{\mu}^k))$
Step 3: Compute $\nabla J(\bar{\boldsymbol{\mu}}^k)$
If $\nabla J(\bar{\boldsymbol{\mu}}^k) = 0$, Stop
Step 4: If $\alpha_k > \beta \frac{\|\boldsymbol{\mu}^k - \bar{\boldsymbol{\mu}}^k\|}{\|\nabla J(\boldsymbol{\mu}^k) - \nabla J(\bar{\boldsymbol{\mu}}^k)\|}$
then reduce α_k and go to Step 5
Step 5: Compute $\boldsymbol{\mu}^{k+1} = P_K(\boldsymbol{\mu}^k - \alpha_k \nabla J(\bar{\boldsymbol{\mu}}^k))$
End.

3.5. Marcotte Choices for Steplength. First Marcotte Variant Khobotov's algorithm gives one workable method for reducing α_k but does not rule out other, perhaps more desirable, methods. Marcotte developed a new rule for reducing α_k along with closely related variants [**31, 40**]. The first Marcotte rule is based on the sequence $a_k = a_{k-1}/2$ and forces α_k to satisfy Step 5 of Khobotov's algorithm by additionally taking:

$$\alpha_k = \min\left\{ \frac{\alpha_{k-1}}{2}, \frac{\|\boldsymbol{\mu}^k - \bar{\boldsymbol{\mu}}^k\|}{\sqrt{2}\|\nabla J(\boldsymbol{\mu}^k) - \nabla J(\bar{\boldsymbol{\mu}}^k)\|} \right\}.$$

Second Marcotte Variant Both the Khobotov and Marcotte reduction rules can still run the risk of choosing an initial α small enough that α_k is never reduced, resulting in slow convergence. Ideally, α_k should then have the ability to increase if α_{k-1} is smaller than some optimal value. This leads to a modified version of Marcotte's rule where an initial α is selected using the rule

$$\alpha = \alpha_{k-1} + \gamma\left(\beta\frac{\|\boldsymbol{\mu}^{k-1} - \bar{\boldsymbol{\mu}}^{k-1}\|}{\|\nabla J(\boldsymbol{\mu}^{k-1}) - \nabla J(\bar{\boldsymbol{\mu}}^{k-1})\|} - \alpha_{k-1}\right)$$

where $\gamma \in (0,1)$.

The reduction rule in Step 5 of Khobotov's algorithm is then replaced with

$$\alpha_k = \max\left\{\widehat{\alpha}, \min\left\{\xi\cdot\alpha, \beta\frac{\|\boldsymbol{\mu}^{k-1} - \bar{\boldsymbol{\mu}}^{k-1}\|}{\|\nabla J(\boldsymbol{\mu}^{k-1}) - \nabla J(\bar{\boldsymbol{\mu}}^{k-1})\|}\right\}\right\}$$

where $\xi \in (0,1)$, and $\widehat{\alpha}$ is some lower limit for α_k (generally taken as no less than 10^{-4}).

3.6. Scaled Extragradient Method. Now we consider a projection-contraction type extragradient method presented by Solodov and Tseng [**39**]. It involves a scaling matrix M to accelerate convergence. The main steps read:

$$\begin{aligned} \bar{\boldsymbol{\mu}}^k &= P_K(\boldsymbol{\mu}^k - \alpha_k\nabla J(\boldsymbol{\mu}^k)) \\ \boldsymbol{\mu}^{k+1} &= \boldsymbol{\mu}^k - \gamma M^{-1}(T_\alpha(\boldsymbol{\mu}^k) - T_\alpha(P_K(\bar{\boldsymbol{\mu}}^k)) \end{aligned}$$

where $\gamma \in \mathbb{R}^+$ and $T_\alpha = (I - \alpha\nabla J)$; here, I is the identity matrix, and α is chosen such that T_α is strongly monotone.

Additional discussion of the scaling matrix is given in [**40**]. However, in both [**39**] and [**40**], test problems take M as the identity matrix. In our numerical experiments, we consider the scaling matrix as both the identity matrix, in Solodov-Tseng (I), and the diagonal of the Hessian of J, in Solodov-Tseng (H).

Algorithm: Solodov-Tseng

Choose $\boldsymbol{\mu}^0, \alpha_{-1}, \theta \in (0,2), \rho \in (0,1), \beta \in (0,1), M \in \mathbb{R}^{m\times m}$
Initialize: $\bar{\boldsymbol{\mu}}^0 = 0, k = 0, rx = \text{ones}(m,1)$
While $\|rx\| > \text{TOL}$
 Step 1: if $\|rx\| < \text{TOL}$ then Stop
  else $\alpha = \alpha_{k-1},\ flag = 0$
 Step 2: if $\nabla J(\boldsymbol{\mu}^k) = 0$ then Stop
 Step 3: While $\alpha(\boldsymbol{\mu}^k - \bar{\boldsymbol{\mu}}^k)^T(\nabla J(\boldsymbol{\mu}^k) - \nabla J(\bar{\boldsymbol{\mu}}^k)) > (1-\rho)\|\boldsymbol{\mu}^k - \bar{\boldsymbol{\mu}}^k\|^2$ or $flag = 0$
  If $flag \neq 0$ Then $\alpha = \alpha_{k-1}\beta$ endif
   update $\bar{\boldsymbol{\mu}}^k = P_K(\boldsymbol{\mu}^k - \alpha\nabla J(\boldsymbol{\mu}^k))$, compute $\nabla J(\bar{\boldsymbol{\mu}}^k)$
   $flag = flag + 1$
  endwhile
 Step 4: update $\alpha_k = \alpha$
 Step 5: compute $\gamma = \theta\rho\|\boldsymbol{\mu}^k - \bar{\boldsymbol{\mu}}^k\|^2/\|M^{1/2}(\boldsymbol{\mu}^k - \bar{\boldsymbol{\mu}}^k - \alpha_k\nabla J(\boldsymbol{\mu}^k) + \alpha_k\nabla J(\bar{\boldsymbol{\mu}}^k))\|^2$
 Step 6: compute $\boldsymbol{\mu}^{k+1} = \boldsymbol{\mu}^k - \gamma M^{-1}(\boldsymbol{\mu}^k - \bar{\boldsymbol{\mu}}^k - \alpha_k\nabla J(\boldsymbol{\mu}^k) + \alpha_k\nabla J(\bar{\boldsymbol{\mu}}^k))$
 Step 7: $rx = \boldsymbol{\mu}^{k+1} - A^k,\ k = k+1$ go to Step 3
End

The Solodov-Tseng method suggests a more general form for the advanced extragradient methods:

$$\begin{aligned}\bar{\boldsymbol{\mu}}^k &= P_{\boldsymbol{\mu}}(\boldsymbol{\mu}^k - \alpha_k \nabla J(\boldsymbol{\mu}^k))\\ \boldsymbol{\mu}^{k+1} &= P_{\boldsymbol{\mu}}(\boldsymbol{\mu}^k - \eta_k \nabla J(\bar{\boldsymbol{\mu}}^k)),\end{aligned}$$

where α_k and η_k are chosen using different rules.

3.7. Goldstein-Type Methods. The classical Goldstein projection method presented in [**29**] is of the form:

$$\boldsymbol{\mu}^{k+1} = P_K(\boldsymbol{\mu}^k - \beta_k \nabla J(\boldsymbol{\mu}^k))$$

The He-Goldstein method, an extragradient method that requires Lipschitz continuity and strong monotonicity of ∇J is of the form:

$$\begin{aligned}\bar{\boldsymbol{\mu}}^k &= P_K(\nabla J(\boldsymbol{\mu}^k) - \beta_k \boldsymbol{\mu}^k)\\ \boldsymbol{\mu}^{k+1} &= \boldsymbol{\mu}^k - \frac{1}{\beta_k}\{\nabla J(\boldsymbol{\mu}^k) - \bar{\boldsymbol{\mu}}^k\}.\end{aligned}$$

It can also be expressed:

$$\begin{aligned}r(\boldsymbol{\mu}^k, \beta_k) &= \frac{1}{\beta_k}\{\nabla J(\boldsymbol{\mu}^k) - P_K[\nabla J(\boldsymbol{\mu}^k) - \beta_k \boldsymbol{\mu}^k]\}\\ \boldsymbol{\mu}^{k+1} &= \boldsymbol{\mu}^k - r(\boldsymbol{\mu}^k, \beta_k).\end{aligned}$$

We implement the more general version above, presented in [**29**], as it allows us to control the second projection (i.e. choosing η_k).

Algorithm: Improved He-Goldstein

Initialize: choose $\beta_U > \beta_L > \frac{1}{(4\tau)}, \gamma \in (0,2), \epsilon > 0, \boldsymbol{\mu}^0, \beta_0 \in [\beta_L, \beta_U], k = 0$

Step 1: Compute:
  $r(\boldsymbol{\mu}^k, \beta_k) = \frac{1}{\beta_k}\{\nabla J(\boldsymbol{\mu}^k) - P_K[\nabla J(\boldsymbol{\mu}^k) - \beta_k \boldsymbol{\mu}^k]\}$
  If $\|r(\boldsymbol{\mu}^k, \beta_k)\| \leq \epsilon$ then Stop

Step 2: $\boldsymbol{\mu}^{k+1} = \boldsymbol{\mu}^k - \gamma \alpha_k r(\boldsymbol{\mu}^k, \beta_k)$ where $\alpha_k := 1 - \frac{1}{4\beta_k \tau}$

Step 3: Update β_k
  $\omega_k = \frac{\|\nabla J(\boldsymbol{\mu}^{k+1}) - \nabla J(\boldsymbol{\mu}^k)\|}{\beta_k \|\boldsymbol{\mu}^{k+1} - \boldsymbol{\mu}^k\|}$
  If $\omega_k < \frac{1}{2}$ Then $\beta_{k+1} = \max\{\beta_L, \frac{1}{2}\beta_k\}$
  Else if $\omega_k > \frac{3}{2}$ Then $\beta_{k+1} = \min\{\beta_U, \frac{6}{5}\beta_k\}$

Step 4: $k = k + 1$, go to Step 1

3.8. Two-step Extragradient Method. Zykina and Melenchuk in [**46**] consider a three step projection method which they called a two-step extragradient method. Numerical experiments with saddle point problem for bilinear function given in [**46**] shows that the convergent of this method is faster compared to the standard extragradient method. The adaptive algorithm is of the form

$$\begin{aligned}\bar{\boldsymbol{\mu}}^k &= P_K(\boldsymbol{\mu}^k - \alpha_k \nabla J(\boldsymbol{\mu}^k))\\ \widetilde{\boldsymbol{\mu}}^k &= P_K(\bar{\boldsymbol{\mu}}^k - \eta_k \nabla J(\bar{\boldsymbol{\mu}}^k))\\ \boldsymbol{\mu}^{k+1} &= P_K(\boldsymbol{\mu}^k - \xi_k \nabla J(\widetilde{\boldsymbol{\mu}}^k))\end{aligned}$$

where α_k, η_k, and ξ_k are all chosen using different rules.

3.9. Hyperplane Extragradient Method. We choose η_k using the following rule from [**40**]:

$$\eta_k = \frac{\langle \nabla J(\bar{\boldsymbol{\mu}}^k), \boldsymbol{\mu}^k - \bar{\boldsymbol{\mu}}^k \rangle}{\|\nabla J(\bar{\boldsymbol{\mu}}^k)\|^2}$$

The idea here is that the hyperplane of all solutions $\boldsymbol{\mu}$ such that

$$\langle \nabla J(\bar{\boldsymbol{\mu}}^k), \bar{\boldsymbol{\mu}}^k - \boldsymbol{\mu} \rangle = 0,$$

separates all the solutions onto one side of the hyperplane. Looking at the variational inequality, we know which side the solutions fall onto:

$$\langle \nabla J(\boldsymbol{\mu}), \bar{\boldsymbol{\mu}}^k - \boldsymbol{\mu} \rangle \geq 0.$$

Consequently, if ∇J is monotone, then we also have

$$\langle \nabla J(\bar{\boldsymbol{\mu}}^k), \bar{\boldsymbol{\mu}}^k - \boldsymbol{\mu} \rangle \geq 0.$$

Thus if

$$\langle \nabla J(\bar{\boldsymbol{\mu}}^k), \bar{\boldsymbol{\mu}}^k - \boldsymbol{\mu}^k \rangle < 0,$$

then we know that we have to look for the solution on the other side of the hyperplane.

The method presented here, first given by Konnov (see [**13**] and [**27**]), follows the development of the method by Iusem and Svaiter [**22**]. It requires three constants, $\epsilon \in (0,1)$ and $\tilde{\alpha} \geq \hat{\alpha} > 0$ such that the sequence α_k is computed with

$$\langle \nabla J(\bar{\boldsymbol{\mu}}^k), \bar{\boldsymbol{\mu}}^k - \boldsymbol{\mu}^k \rangle \leq 0$$

when $\alpha_k \in [\hat{\alpha}, \tilde{\alpha}]$.

Algorithm: Hyperplane (Iusem-Svaiter)

Choose: $\boldsymbol{\mu}^0, \epsilon, \hat{\alpha}, \tilde{\alpha}$
Initialize: $k = 0, rx = \text{ones}(m, 1)$
While $\|rx\| > \text{TOL}$
 Step 1: Choose $\tilde{\alpha}_k$ using a finite bracketing procedure
 Step 2: Compute $K^k = P_K(\boldsymbol{\mu}^k - \tilde{\alpha}_k \nabla J(\boldsymbol{\mu}^k))$ and $\nabla J(K^k)$
 Step 3: If $\nabla J(K^k) = 0$ then Stop
 Step 4: If $\|\nabla J(\tilde{\boldsymbol{\mu}}^k) - \nabla J(\boldsymbol{\mu}^k)\| \leq \frac{\|K^k - \boldsymbol{\mu}^k\|^2}{2\tilde{\alpha}_k^2 \|\nabla J(\boldsymbol{\mu}^k)\|}$
  Then $\bar{\boldsymbol{\mu}}^k = K^k$
  Else find $\alpha_k \in (0, \tilde{\alpha}_k)$ such that
  $\epsilon \frac{\|K^k - \boldsymbol{\mu}^k\|^2}{2\tilde{\alpha}_k^2 \|\nabla J(\boldsymbol{\mu}^k)\|} \leq \|\nabla J(P_K(\boldsymbol{\mu}^k - \alpha_k \nabla J(\boldsymbol{\mu}^k))) - \nabla J(\boldsymbol{\mu}^k)\| \leq \frac{\|K^k - \boldsymbol{\mu}^k\|^2}{2\tilde{\alpha}_k^2 \|\nabla J(\boldsymbol{\mu}^k)\|}$
 Step 5: Compute $\bar{\boldsymbol{\mu}}^k = P_K(\boldsymbol{\mu}^k - \alpha_k \nabla J(\boldsymbol{\mu} k^k))$
 Step 6: If $\nabla J(\bar{\boldsymbol{\mu}}^k) = 0$ then Stop
 Step 7: Compute η_k
 Step 8: Compute $\boldsymbol{\mu}^{k+1} = P_K(\boldsymbol{\mu}^k - \eta_k \nabla J(\bar{\boldsymbol{\mu}}^k))$
 Step 9: $rx = \boldsymbol{\mu}^{k+1} - \boldsymbol{\mu}^k, k = k + 1$; go to Step 3;
End

4. Performance Analysis

In this section, we present a comparison for the following methods:

(1) Gradient Projection Using Armijo Line Search
(2) Fast Gradient Projection Using Armijo Line Search
(3) Scaled Gradient Projection Using Barzilai-Borwein Rules
(4) Khobotov Extragradient Method Using Marcotte Rules (3 versions)
(5) Solodov-Tseng (Projection-Contraction) Method (2 versions)
(6) Improved He-Goldstein Type Extragradient Method
(7) Hyperplane Extragradient Method
(8) Two-Step Method

For means of comparison, we considered two representative examples of elastography inverse problems for the recovery of a variable μ on a two dimensional isotropic domain $\Omega = (0,1) \times (0,1)$ with boundary $\partial\Omega = \Gamma_1 \cup \Gamma_2$. Γ_1, where the Dirichlet boundary conditions hold, is taken as the top boundary of the domain while Γ_2, where the Neumann conditions hold, is taken as the union of the remaining boundary points.

The problem was discretized on a 50×50 quadrangular mesh with 2,601 quadrangles and 8,009 total degrees of freedom.

Maintaining the near incompressibility of the elastography inverse problem, λ was taken as a large constant in both examples: $\lambda = 10^6$.

The functions defining the coefficient, load, and boundary conditions for each example are as follows.

Example 1:

$$\mu(x,y) = 1 + \frac{1}{2}\operatorname{sinc}\left(2\pi\sqrt{\left(x-\frac{1}{2}\right)^2 + \left(y-\frac{1}{2}\right)^2}\right), \qquad f(x,y) = \begin{bmatrix} x \\ \cos(\pi x y) \end{bmatrix},$$

$$g(x,y) = \begin{bmatrix} 0 \\ 0 \end{bmatrix} \text{ on } \Gamma_1, \qquad h(x,y) = \frac{1}{2}\begin{bmatrix} 1+2x^2 \\ 1+2y^2 \end{bmatrix} \text{ on } \Gamma_2.$$

Example 2:

$$\mu(x,y) = 2 + \frac{1}{4}\sin(\pi x)\cos(3\pi x y), \qquad f(x,y) = \begin{bmatrix} 2.3 + \frac{1}{10}x \\ 2.3 + \frac{1}{10}y \end{bmatrix},$$

$$g(x,y) = \begin{bmatrix} 0 \\ 0 \end{bmatrix} \text{ on } \Gamma_1, \qquad h(x,y) = \frac{1}{2}\begin{bmatrix} 1+2x^2 \\ 1+2y^2 \end{bmatrix} \text{ on } \Gamma_2.$$

All of the optimization methods laid out in the previous section were implemented and applied to these two example parameter identification inverse problems. The performance results for all methods are presented in Table 1 and for comparison, are represented by the total number of objective function, J, evaluations, total CPU time, and total number of algorithm iterations. For a selection of the methods (FISTA, Scaled Projected Gradient, Marcotte 2, Two-step, Solodov-Tseng (H)), the convergence history of both J and $\|\nabla J\|$ over each algorithm iteration are provided in Figures 1 & 2 and Figures 3 & 4, respectively. For these same methods, the computed μ for several algorithm steps as well as a comparison of the exact discretized μ with the final computed μ are presented in Figures 5—13. All methods that converged achieved a similar level of accuracy (L^2 error: 3.12E-04±1.25E-04) with similar stopping criteria ($\|\nabla J\|$ <1E-10).

4.1. Discussion. In this subsection, we compare the results for the presented method and discuss their benefits and potential problems as well as the issues related to algorithmic design and parameter selection. We emphasize that our remarks are based on our experiments in connection with the parameter identification problem.

We begin our discussion with the classical extragradient method and its direct extensions: Korpelevich, Khobotov, and the two related Marcotte variants. As the data in Table 1 shows, the Khobotov and Marcotte variants both improve over the performance of the Korpelevich method across the board. Comparing the Marcotte variants against the Khobotov algorithm shows more mixed results. The second Marcotte variant converges in fewer iterations than both the first Marcotte variant and the plain Khobotov, but due to the additional computational steps in the reduction of α_k, this actually results in increased overall CPU time over Khobotov. For Example 2, we also note that the first Marcotte variant performs substantially worse than even the Khobotov method. Overall, these results suggests that, provided a large enough α_0 is selected, the more complicated methods for reducing α_k have a marginal impact at best on performance in the context of the parameter identification problem.

Continuing with the comparison of the extragradient methods, we see a similar pattern with the Two-step and Hyperplane methods. Although these methods both provide convergence in fewer iterations, the additional computational complexity of the algorithms leads to increased CPU time overall and thus have little to offer in terms of "real-world" performance benefits over simpler methods. In the case of the Hyperplane method, this increased complexity results in a more than twofold increase in the runtime over the Khobotov method. This performance discrepancy can partially be explained by the fact that there are instances when the algorithm seemingly gets "stuck" near the hyperplane. In particular, this occurs when $\bar{\boldsymbol{\mu}}^k$ is near the hyperplane, making the difference between $\boldsymbol{\mu}^k$ and $\boldsymbol{\mu}^{k+1}$ small. Sometimes, the sequence of $\alpha's$ or $\eta's$ effectively stops being adaptive and the solution begins to converge slowly. Other extragradient methods such as the He-Goldstein and the unscaled Solodov-Tseng (I) methods perform roundly worse than the Khobotov method and are more in line with the performance of the Korpelevich method.

For the other non-extragradient methods under consideration, the novel FISTA algorithm outperforms the simplistic Projected Gradient method (as do all the other methods) but performs only in the middle of the pack when compared with the other extragradient methods. Again, although improving in total iterations against a method like Khobotov, FISTA's implementation results in more computationally expensive evaluations of the objective function and reduced overall performance.

We also note that, rather than implementing the algorithms to stop only at a fixed iteration, they were also implemented to stop when the norm of gradient of the objective function reached a certain tolerance (generally 10^{-10}). This provided another way to assess the costs and benefits of each method and to fully understand their convergence properties. Figures 3 &4 show the behavior of $\|\nabla J\|$ over the course of several of the algorithm's iterations and both SPG and FISTA display some instability in this regard. Although both still have relatively fast convergence, this would likely also contribute to slower convergence when the gradient of J is used as the stopping criteria.

The most significant performance result, highlighted in Table 1, are those methods that incorporate a scaling factor, and particularly where the scaling information is taken from the Hessian of J. These methods, the scaled Solodov-Tseng (H) and Scaled Projected Gradient, easily outperform all the other methods by nearly tenfold margins and this is perhaps shown most starkly in the contrast between the poorly-performing unscaled Solodov-Tseng (I) method and its Hessian-scaled counterpart. Although this result is not particularly surprising given the accelerated convergence provided by the inclusion of second-order information, it does indicate that proper scaling is perhaps a dominant feature of the solution of the parameter identification problem using optimization methods such as these. That is, where the added computational overhead of the Marcotte variants provides little actual performance benefit, the added computational overhead of computing the Hessian for use in scaling provides a remarkable enhancement.

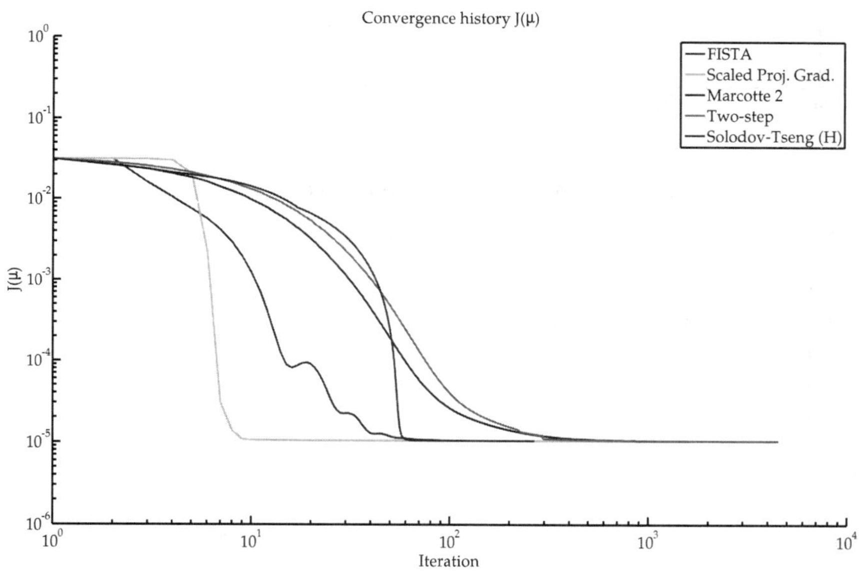

FIGURE 1. Example 1 — Convergence history of $J(\mu)$

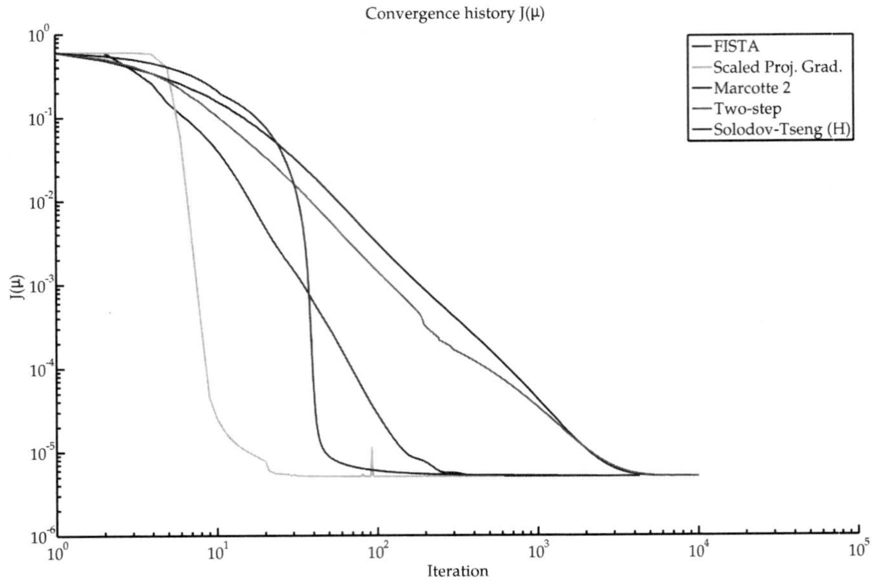

FIGURE 2. Example 2 — Convergence history of $J(\mu)$

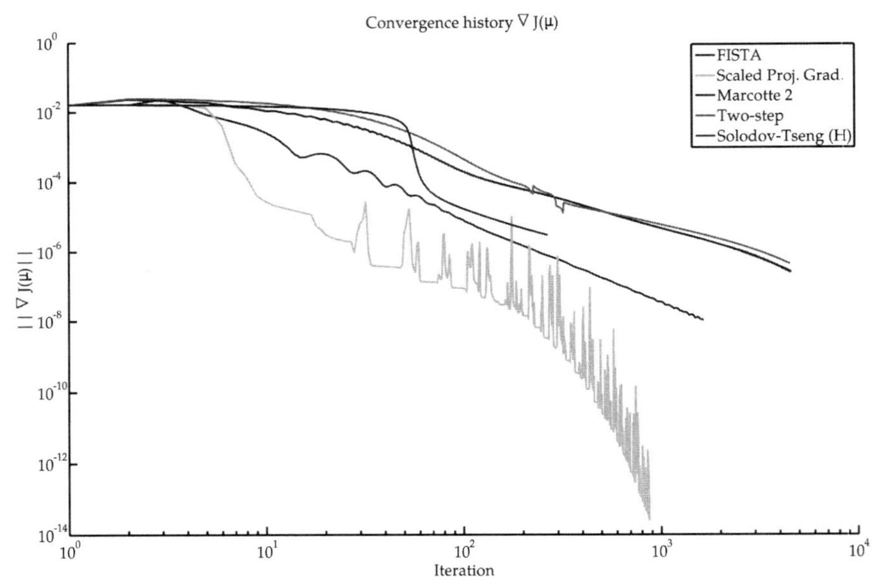

FIGURE 3. Example 1 — Convergence history of $\nabla J(\mu)$

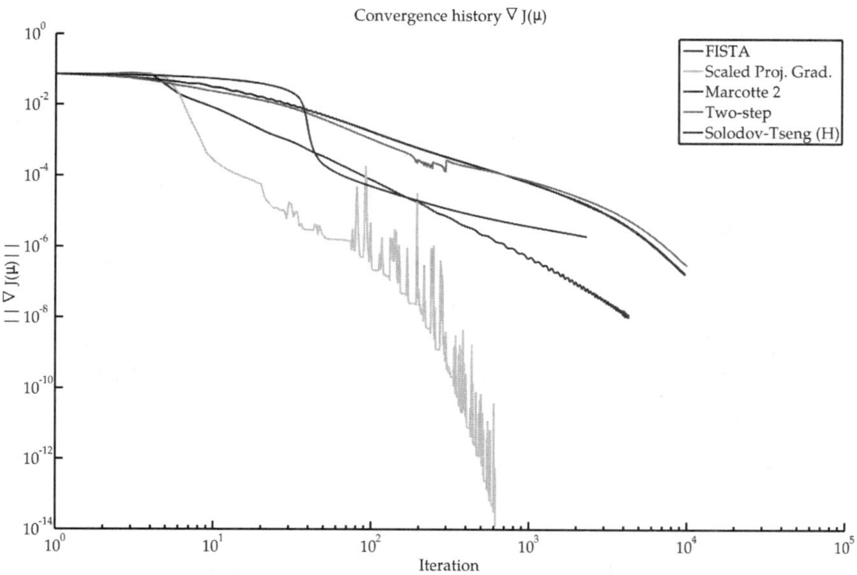

FIGURE 4. Example 2 — Convergence history of $\nabla J(\mu)$

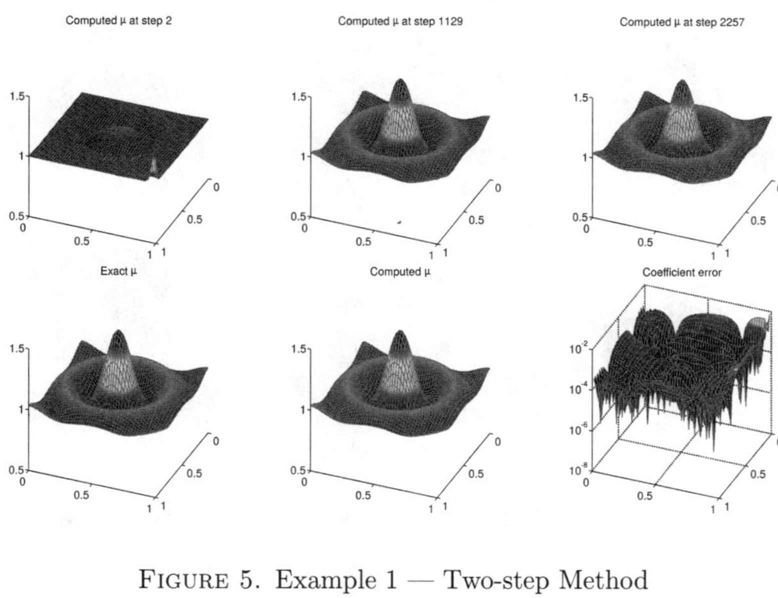

FIGURE 5. Example 1 — Two-step Method

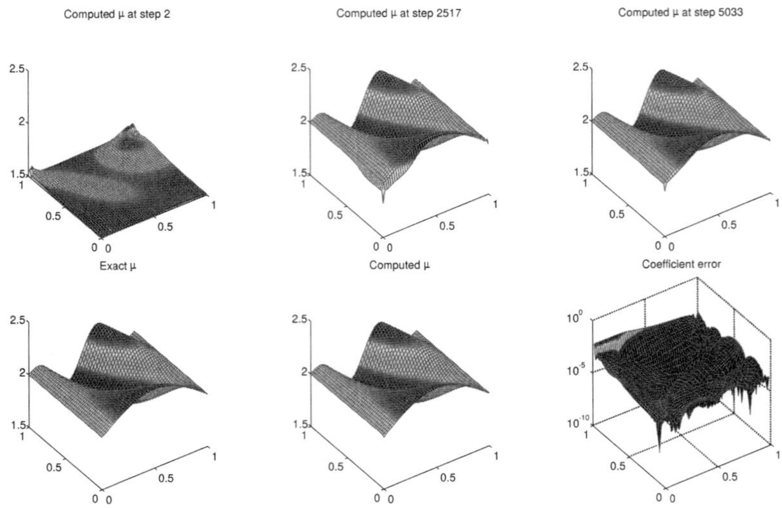

FIGURE 6. Example 2 — Two-step Method

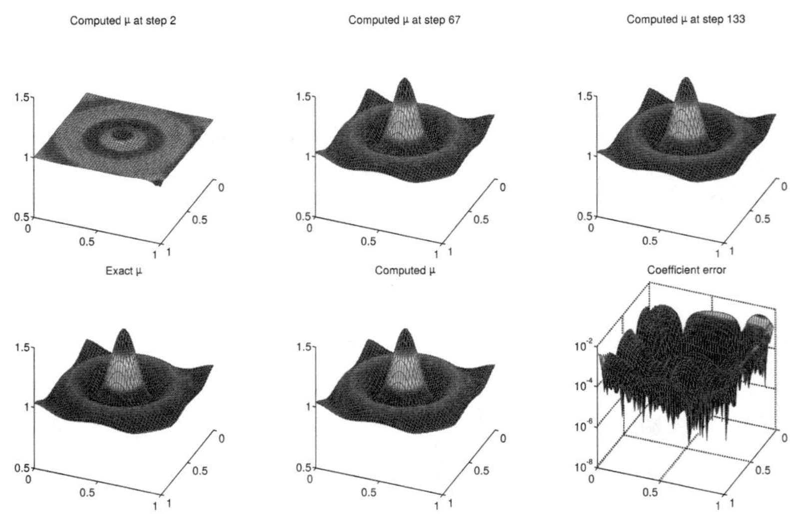

FIGURE 7. Example 1 — Solodov-Tseng Method using Hessian

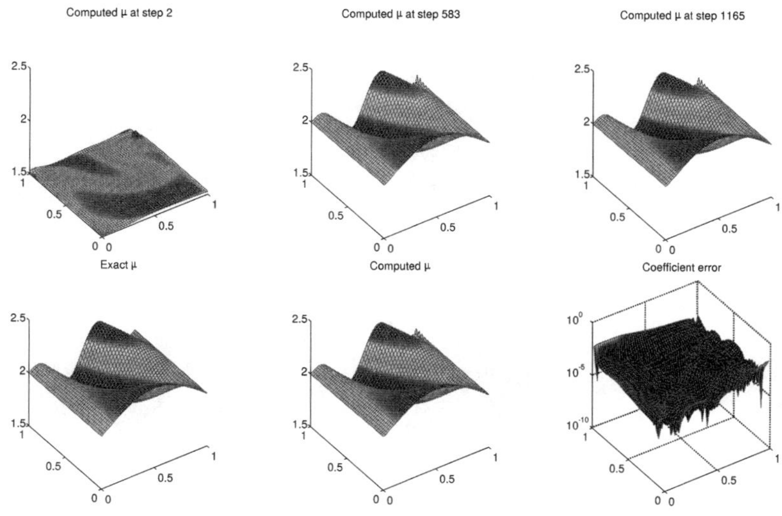

FIGURE 8. Example 2 — Solodov-Tseng Method using Hessian

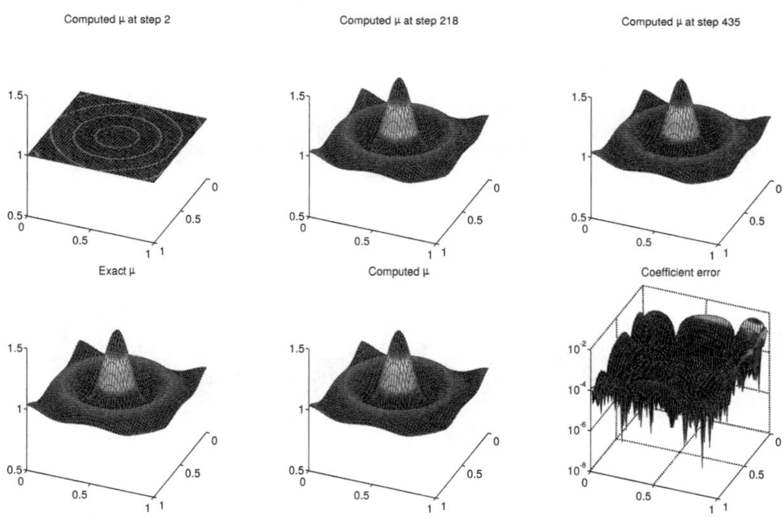

FIGURE 9. Example 1 — Projected Gradient with Hessian

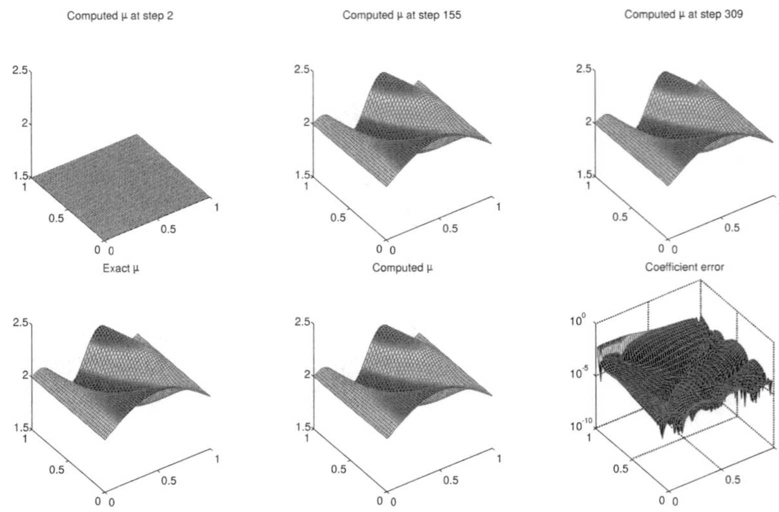

FIGURE 10. Example 2 — Projected Gradient with Hessian

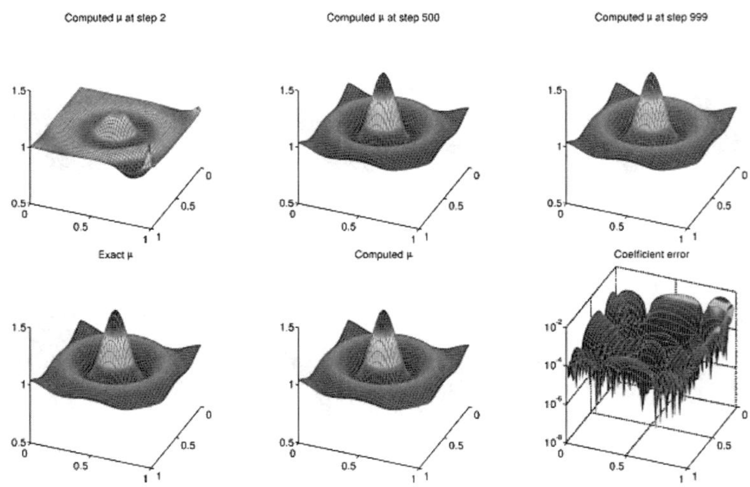

FIGURE 11. Example 1 — Fast Iterative Shrinkage-Thresholding (FISTA)

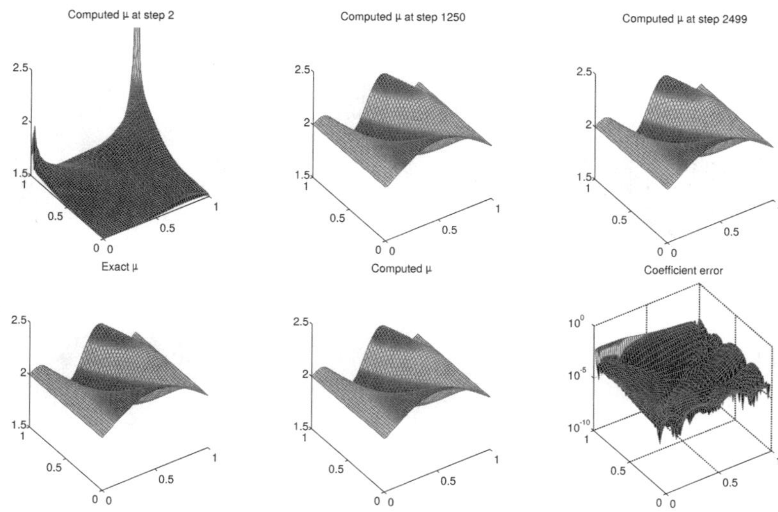

FIGURE 12. Example 2 — Fast Iterative Shrinkage-Thresholding (FISTA)

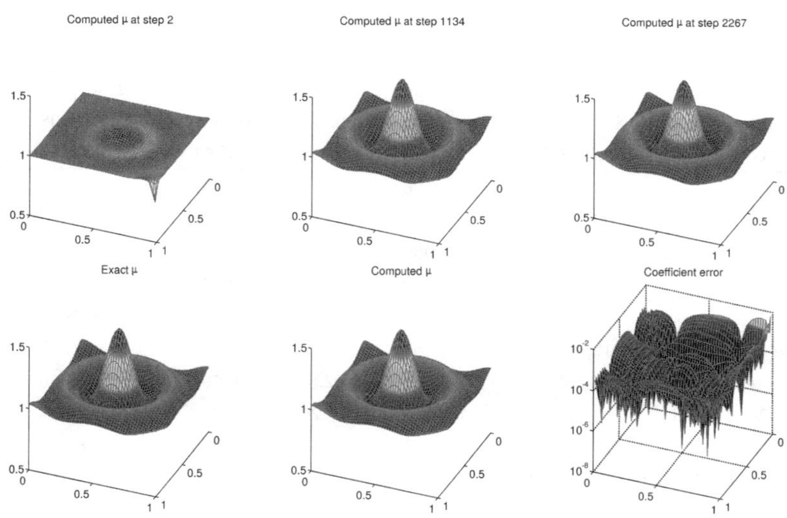

FIGURE 13. Example 1 — Marcotte (Second Variant)

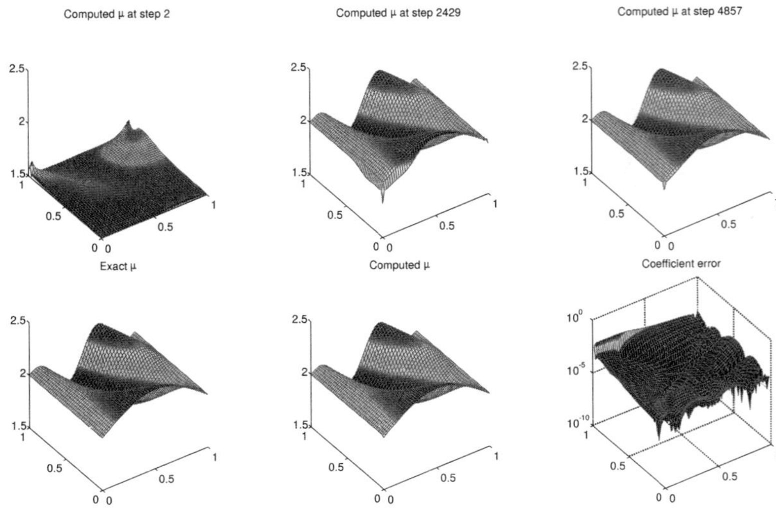

FIGURE 14. Example 2 — Marcotte (Second Variant)

Method	Example 1 J evals	Example 1 CPU time (s)	Example 1 Iterations	Example 2 J evals	Example 2 CPU time (s)	Example 2 Iterations
FISTA	45,704	569.296	1,640	123,568	1,662.358	4,291
Proj. Grad.	891,997	11,846.639	100,000[1]	807,300	13,780.608	100,000[1]
Scaled PG[2]	**2,075**	**29.765**	**868**	**1,561**	**22.342**	**616**
He-Goldstein	49,837	693.426	49,837	86,037	1,200.740	86,037
Hyperplane	30,495	414.684	4,621	59,036	810.453	10,415
Khobotov	10,924	150.366	5,459	22,778	312.540	11,385
Korpelevich	51,696	691.320	25,845	200,000	2,705.447	100,000[1]
Marcotte 1	10,352	143.768	5,171	74,822	1,028.67	37,406
Marcotte 2	12,096	168.053	4,533	25,912	355.013	9,714
Solodov-Tseng (I)	97,785	71,889.027	32,595	210,791	154,232.466	70,263
Solodov-Tseng (H)[2]	**814**	**17.390**	**268**	**6,999**	**105.616**	**2,332**
Two-step	13,764	187.995	4,513	30,405	418.769	10,064

TABLE 1. Performance results for the gradient and extragradient methods

[1]This value indicates the failure of the method to converge.

[2]This method also required one evaluation of the Hessian of J.

Acknowledgments: The work of A. A. K. is supported by a grant from the Simons Foundation (#210443 to Akhtar Khan).

References

[1] A. S. Antipin, L. A. Artem′eva, and F. P. Vasil′ev, *A regularized extragradient method for solving a parametric multicriteria equilibrium programming problem* (Russian, with Russian summary), Zh. Vychisl. Mat. Mat. Fiz. **50** (2010), no. 12, 2083–2098, DOI 10.1134/S096554251012002X; English transl., Comput. Math. Math. Phys. **50** (2010), no. 12, 1975–1989. MR2760071 (2012f:90151)

[2] A. Auslender and M. Teboulle, *Projected subgradient methods with non-Euclidean distances for non-differentiable convex minimization and variational inequalities*, Math. Program. **120** (2009), no. 1, Ser. B, 27–48, DOI 10.1007/s10107-007-0147-z. MR2496425 (2010b:90085)

[3] A. Beck and M. Teboulle, *A fast iterative shrinkage-thresholding algorithm for linear inverse problems*, SIAM J. Imaging Sci. **2** (2009), no. 1, 183–202, DOI 10.1137/080716542. MR2486527 (2010d:35390)

[4] J. Y. Bello Cruz and A. N. Iusem, *Convergence of direct methods for paramonotone variational inequalities*, Comput. Optim. Appl. **46** (2010), no. 2, 247–263, DOI 10.1007/s10589-009-9246-5. MR2644205 (2011c:49009)

[5] A. Bnouhachem, M.-A. Noor, E. Al-Said, M. Khalfaoui, and S. Zhaohan, *Extragradient method for variational inequalities*, Hacet. J. Math. Stat. **40** (2011), no. 6, 839–854. MR2952185

[6] S. Bonettini and V. Ruggiero, *An alternating extragradient method for total variation-based image restoration from Poisson data*, Inverse Problems **27** (2011), no. 9, 095001, 26, DOI 10.1088/0266-5611/27/9/095001. MR2824760 (2012i:94031)

[7] R. S. Burachik, S. Scheimberg, and B. F. Svaiter, *Robustness of the hybrid extragradient proximal-point algorithm*, J. Optim. Theory Appl. **111** (2001), no. 1, 117–136, DOI 10.1023/A:1017523331361. MR1850681 (2002f:90082)

[8] Y. Censor, A. Gibali, and S. Reich, *Strong convergence of subgradient extragradient methods for the variational inequality problem in Hilbert space*, Optim. Methods Softw. **26** (2011), no. 4-5, 827–845, DOI 10.1080/10556788.2010.551536. MR2837800 (2012g:49012)

[9] Y. Censor, A. Gibali, and S. Reich, *The subgradient extragradient method for solving variational inequalities in Hilbert space*, J. Optim. Theory Appl. **148** (2011), no. 2, 318–335, DOI 10.1007/s10957-010-9757-3. MR2780566 (2011k:49014)

[10] Y. Censor, A. Gibali, and S. Reich, *Extensions of Korpelevich's extragradient method for the variational inequality problem in Euclidean space*, Optimization **61** (2012), no. 9, 1119–1132, DOI 10.1080/02331934.2010.539689. MR2966121

[11] E. Crossen, M. S. Gockenbach, B. Jadamba, A. A. Khan, and B. Winkler, *An equation error approach for the elasticity imaging inverse problem for predicting tumor location*, Comput. Math. Appl. **67** (2014), no. 1, 122–135, DOI 10.1016/j.camwa.2013.10.006. MR3141709

[12] M. M. Doyley, B. Jadamba, A. A. Khan, M. Sama, and B. Winkler, *A new energy inversion for parameter identification in saddle point problems with an application to the elasticity imaging inverse problem of predicting tumor location*, Numer. Funct. Anal. Optim. **35** (2014), no. 7-9, 984–1017, DOI 10.1080/01630563.2014.935859. MR3230088

[13] F. Facchinei and J.-S. Pang, *Finite-dimensional variational inequalities and complementarity problems*, vol. 2, Springer, New York, 2003 (English).

[14] M. S. Gockenbach, B. Jadamba, and A. A. Khan, *Equation error approach for elliptic inverse problems with an application to the identification of Lamé parameters*, Inverse Probl. Sci. Eng. **16** (2008), no. 3, 349–367, DOI 10.1080/17415970701602580. MR2427499 (2009e:74048)

[15] M. S. Gockenbach, B. Jadamba, A. A. Khan, C. Tammer, and B. Winkler, *Proximal method for the elastography inverse problem of tumor identification using an equation error approach*, Advances in Variational and Hemivariational Inequalities with Applications, Springer, 2014, pp. 1–25.

[16] M. S. Gockenbach and A. A. Khan, *Identification of Lamé parameters in linear elasticity: a fixed point approach*, J. Ind. Manag. Optim. **1** (2005), no. 4, 487–497, DOI 10.3934/jimo.2005.1.487. MR2186621 (2006f:35290)

[17] M. S. Gockenbach and A. A. Khan, *An abstract framework for elliptic inverse problems. I. An output least-squares approach*, Math. Mech. Solids **12** (2007), no. 3, 259–276, DOI 10.1177/1081286505055758. MR2325602 (2008b:35302)

[18] M. S. Gockenbach and A. A. Khan, *An abstract framework for elliptic inverse problems. II. An augmented Lagrangian approach*, Math. Mech. Solids **14** (2009), no. 6, 517–539, DOI 10.1177/1081286507087150. MR2548203 (2010i:35435)

[19] D. Han and H. K. Lo, *Two new self-adaptive projection methods for variational inequality problems*, Comput. Math. Appl. **43** (2002), no. 12, 1529–1537, DOI 10.1016/S0898-1221(02)00116-5. MR1907686 (2003c:49010)

[20] Y. He, *A new double projection algorithm for variational inequalities*, J. Comput. Appl. Math. **185** (2006), no. 1, 166–173, DOI 10.1016/j.cam.2005.01.031. MR2165359 (2006c:90064)

[21] A. N. Iusem and L. R. Lucambio Pérez, *An extragradient-type algorithm for non-smooth variational inequalities*, Optimization **48** (2000), no. 3, 309–332, DOI 10.1080/02331930008844508. MR1811750 (2001k:49024)

[22] A. N. Iusem and B. F. Svaiter, *A variant of Korpelevich's method for variational inequalities with a new search strategy*, Optimization **42** (1997), no. 4, 309–321, DOI 10.1080/02331939708844365. MR1609571 (98m:49020)

[23] P. D. Khanh and P. T. Vuong, *Modified projection method for strongly pseudomonotone variational inequalities*, J. Global Optim. **58** (2014), no. 2, 341–350, DOI 10.1007/s10898-013-0042-5. MR3158057

[24] E. N. Khobotov, *A modification of the extragradient method for solving variational inequalities and some optimization problems* (Russian), Zh. Vychisl. Mat. i Mat. Fiz. **27** (1987), no. 10, 1462–1473, 1597. MR918546 (88m:49005)

[25] E. V. Khoroshilova, *Extragradient-type method for optimal control problem with linear constraints and convex objective function*, Optim. Lett. **7** (2013), no. 6, 1193–1214, DOI 10.1007/s11590-012-0496-2. MR3081686

[26] I. V. Konnov, *A class of combined relaxation methods for decomposable variational inequalities*, Optimization **51** (2002), no. 1, 109–125, DOI 10.1080/02331930211985. MR1926303 (2003g:49008)

[27] I. V. Konnov, *Combined relaxation methods for the search for equilibrium points and solutions of related problems* (Russian), Izv. Vyssh. Uchebn. Zaved. Mat. **2** (1993), 46–53; English transl., Russian Math. (Iz. VUZ) **37** (1993), no. 2, 44–51. MR1258456 (94k:90106)

[28] G. M. Korpelevič, *An extragradient method for finding saddle points and for other problems* (Russian), Èkonom. i Mat. Metody **12** (1976), no. 4, 747–756. MR0451121 (56 #9408)

[29] M. Li, L.-Z. Liao, and X. M. Yuan, *Some Goldstein's type methods for co-coercive variant variational inequalities*, Appl. Numer. Math. **61** (2011), no. 2, 216–228, DOI 10.1016/j.apnum.2010.10.001. MR2739938 (2011m:65137)

[30] P.-E. Maingé and A. Moudafi, *Coupling viscosity methods with the extragradient algorithm for solving equilibrium problems*, J. Nonlinear Convex Anal. **9** (2008), no. 2, 283–294. MR2444460 (2010f:49018)

[31] P. Marcotte, *Application of khobotov's algorithm to variational inequalities and network equilibrium problems*, INFORM **29** (1991), no. 4, 258–270.

[32] R. D. C. Monteiro and B. F. Svaiter, *Iteration-complexity of a Newton proximal extragradient method for monotone variational inequalities and inclusion problems*, SIAM J. Optim. **22** (2012), no. 3, 914–935, DOI 10.1137/11083085X. MR3023757

[33] R. D. C. Monteiro and B. F. Svaiter, *An accelerated hybrid proximal extragradient method for convex optimization and its implications to second-order methods*, SIAM J. Optim. **23** (2013), no. 2, 1092–1125, DOI 10.1137/110833786. MR3063151

[34] N. Nadezhkina and W. Takahashi, *Modified extragradient method for solving variational inequalities in real Hilbert spaces*, Nonlinear analysis and convex analysis, Yokohama Publ., Yokohama, 2004, pp. 359–366. MR2144057

[35] A. Nemirovski, *Prox-method with rate of convergence $O(1/t)$ for variational inequalities with Lipschitz continuous monotone operators and smooth convex-concave saddle point problems*, SIAM J. Optim. **15** (2004), no. 1, 229–251 (electronic), DOI 10.1137/S1052623403425629. MR2112984 (2006f:90046)

[36] Yu. E. Nesterov, *A method for solving the convex programming problem with convergence rate $O(1/k^2)$* (Russian), Dokl. Akad. Nauk SSSR **269** (1983), no. 3, 543–547. MR701288 (84i:90119)

[37] M. V. Solodov, *Convergence rate analysis of iteractive algorithms for solving variational inquality problems*, Math. Program. **96** (2003), no. 3, Ser. A, 513–528, DOI 10.1007/s10107-002-0369-z. MR1993461 (2004d:49026)

[38] M. V. Solodov and B. F. Svaiter, *A new projection method for variational inequality problems*, SIAM J. Control Optim. **37** (1999), no. 3, 765–776, DOI 10.1137/S0363012997317475. MR1675086 (2000b:49023)

[39] M. V. Solodov and P. Tseng, *Modified projection-type methods for monotone variational inequalities*, SIAM J. Control Optim. **34** (1996), no. 5, 1814–1830, DOI 10.1137/S0363012994268655. MR1404858 (97m:49008)

[40] F. Tinti, *Numerical solution for pseudomonotone variational inequality problems by extragradient methods*, Variational analysis and applications, Nonconvex Optim. Appl., vol. 79, Springer, New York, 2005, pp. 1101–1128, DOI 10.1007/0-387-24276-7_63. MR2160662 (2006d:49021)

[41] P. Tseng, *Alternating projection-proximal methods for convex programming and variational inequalities*, SIAM J. Optim. **7** (1997), no. 4, 951–965, DOI 10.1137/S1052623495279797. MR1479608 (99a:90156)

[42] X. Wang, B. He, and L.-Z. Liao, *Steplengths in the extragradient type methods*, J. Comput. Appl. Math. **233** (2010), no. 11, 2925–2939, DOI 10.1016/j.cam.2009.11.037. MR2592266 (2011a:65165)

[43] Y. J. Wang, N. H. Xiu, and C. Y. Wang, *Unified framework of extragradient-type methods for pseudomonotone variational inequalities*, J. Optim. Theory Appl. **111** (2001), no. 3, 641–656, DOI 10.1023/A:1012606212823. MR1873421 (2002i:90090)

[44] Y. Yao, M. A. Noor, and Y.-C. Liou, *Strong convergence of a modified extragradient method to the minimum-norm solution of variational inequalities*, Abstr. Appl. Anal., posted on 2012, Art. ID 817436, 9, DOI 10.1155/2012/817436. MR2872298

[45] A. J. Zaslavski, *The extragradient method for convex optimization in the presence of computational errors*, Numer. Funct. Anal. Optim. **33** (2012), no. 12, 1399–1412, DOI 10.1080/01630563.2012.706769. MR2995496

[46] A. V. Zykina and N. V. Melen′chuk, *A two-step extragradient method for variational inequalities* (Russian, with English and Russian summaries), Izv. Vyssh. Uchebn. Zaved. Mat. **9** (2010), 82–85, DOI 10.3103/S1066369X10090082; English transl., Russian Math. (Iz. VUZ) **54** (2010), no. 9, 71–73. MR2789310 (2012d:49017)

Mathematics Department, ORT Braude College of Engineering, Karmiel 21982, Israel

E-mail address: avivg@braude.ac.il

Center for Applied and Computational Mathematics, School of Mathematical Sciences, Rochester Institute of Technology, 85 Lomb Memorial Drive, Rochester, New York, 14623

E-mail address: aaksma@rit.edu

Center for Applied and Computational Mathematics, School of Mathematical Sciences, Rochester Institute of Technology, 85 Lomb Memorial Drive, Rochester, New York, 14623

Facoltà di Ingegneria dell'Università di Catania, Dipartimento di Matematica e Informatica dell'Università di Catania, 95125-Catania, Italy

E-mail address: fraciti@dmi.unict.it

Center for Applied and Computational Mathematics, School of Mathematical Sciences, Rochester Institute of Technology, 85 Lomb Memorial Drive, Rochester, New York, 14623

Contemporary Mathematics
Volume **659**, 2016
http://dx.doi.org/10.1090/conm/659/13156

Mappings with upper integral bounds for p-moduli

Anatoly Golberg and Ruslan Salimov

ABSTRACT. We consider a class of homeomorphisms of domains in $\mathbb{R}^n$ with integrally bounded from above p-moduli of the families of curves, which essentially extends the well-known classes of mappings such as quasiconformal, quaiisometric, Lipschitzian, etc. The aim of the paper is to survey the known results in this field, but mainly to establish new geometric and differential features of such mappings. In particular, the estimates for Jacobian and for various dilatation coefficients are obtained.

1. Introduction

The theory of quasiconformal mappings and its wide generalization is an important part of non-linear analysis and has found fundamental applications to real and complex analysis, geometry, the Riemann surface theory, elliptic differential equations, dynamical systems, topology, combinatorial group theory, continuous mechanics and physics. Though this theory is about 80 years old, it still remains a very intensively developing field, with strong intriguing problems. From topological point of view, quasiconformal mappings interpolate between diffeomorphisms and general homeomorphisms. Omitting injectivity leads to quasiregular mappings which provide an essential generalization of multivalent holomorphic functions. Though the theory has been widely expanded to higher dimensions, this case (in view of deep connections with other fields and many non-solved inner problems) still attracts many mathematicians. However, in the case of mappings in $\mathbb{R}^n$, $n \geq 3$, there are no analogues of the fundamental Riemann mapping theorem, and the most of methods applied to the mappings on the plane do not work. Fortunately, there is a universal method of extremal lengths (conformal moduli), which will be used here. The origins of this method go back to classical works of Grötzsch and Ahlfors-Beurling. More precisely, we deal with p-module of k-dimensional surfaces in $\mathbb{R}^n$, $p \geq k$.

Let $\mathcal{S}_k$ be a family of k-dimensional surfaces $\mathcal{S}$ in $\mathbb{R}^n$, $1 \leq k \leq n-1$ (curves for $k = 1$). The *conformal module* of $\mathcal{S}_k$ is defined as

$$\mathcal{M}(\mathcal{S}_k) = \inf \int_{\mathbb{R}^n} \rho^n \, dm(x), \tag{1.1}$$

2010 *Mathematics Subject Classification.* Primary 30C65, 31B15; Secondary 26B30.

Key words and phrases. Weighted p-module, Q-homeomorphisms, Lipschitz mappings, quasiconformal mappings, quasiconformality in the mean, homeomorphisms of finite distortion.

where the infimum is taken over all admissible functions for $\mathcal{S}_k$, i.e. Borel measurable functions $\rho \geq 0$ such that $\int\limits_{\mathcal{S}} \rho^k \, d\sigma_k \geq 1$ for every $\mathcal{S} \in \mathcal{S}_k$.

The quasiinvariance of the confornal module, i.e.

$$K^{\frac{k-n}{n-1}} \mathcal{M}(\mathcal{S}_k) \leq \mathcal{M}(f(\mathcal{S}_k)) \leq K^{\frac{n-k}{n-1}} \mathcal{M}(\mathcal{S}_k),$$

completely characterizes quasiconformality ([**37**]), while the quasiinvariance of p-module ($p \neq n$), where the exponent n in (1.1) is replaced by p, describes quasiisometric (bilipschtitz) mappings.

Let $f : G \to G^*$, $G, G^* \subset \mathbb{R}^n$, be a homeomorphism such that f and f^{-1} are differentiable almost everywhere (a.e.) with nonzero Jacobians in G and G^*, respectively. It was shown in [**10**], that under more restrictive conditions on f the following bounds for the p-module of k-dimensional surface families

$$\begin{aligned} \inf_{\varrho \in \mathrm{ext}_p \mathrm{adm}\, \mathcal{S}_k} \int\limits_G \frac{\varrho^p(x)}{K_{O,p}(x,f)} \, dm(x) &\leq \mathcal{M}_p(f(\mathcal{S}_k)) \\ &\leq \inf_{\rho \in \mathrm{adm}\, \mathcal{S}_k} \int\limits_G \rho^p(x) K_{I,p}(x,f) \, dm(x) \end{aligned} \tag{1.2}$$

are fulfilled. Here $K_{I,p}(x,f)$ and $K_{O,p}(x,f)$ stand for the p-inner and p-outer dilatations of f at $x \in G$ (see, e.g., [**7**]).

As mentioned above, when these dilatations are uniformly bounded by some absolute constant K in G, one obtains the well-known classes of quasiconformal mappings ($p = n$) and bilipschitz mappings ($p \neq n$) in G (cf. [**6, 7, 15**]).

A natural question is to establish various properties of mappings which satisfy either

$$\mathcal{M}_p(f(\mathcal{S}_k)) \leq \inf_{\rho \in \mathrm{adm}\, \mathcal{S}_k} \int\limits_G \rho^p(x) Q(x) \, dm(x), \tag{1.3}$$

or

$$\mathcal{M}_p(f(\mathcal{S}_k)) \geq \inf_{\varrho \in \mathrm{ext}_p \mathrm{adm}\, \mathcal{S}_k} \int\limits_G \frac{\varrho^p(x)}{Q(x)} \, dm(x), \tag{1.4}$$

with a given measurable function $Q : G \to [0, \infty]$. The main underlying idea of our approach consists in establishing the sharp upper and lower bounds for conformal module and p-module of the images of initial surfaces. These bounds provide the keys to an essential development of distortion theory for quasiconformal and quasiregular mappings in higher dimensions and strictly improve many known quantitative and qualitative estimates. A calculation of the inner and outer dilatations in several examples relates to the optimization problems and illustrates the sharpness of the bounds.

This paper can be regarded as a survey of the mostly general Q-homeomorphisms with respect to p-module (topological mappings satisfying the inequality (1.3) for $k = 1$), introduced in [**8**], and on the other hand it contains new results of this theory. Note also that such representation of moduli in (1.3)–(1.4) can be treated as the quasiinvariance of weighted moduli (cf. [**1, 2, 4, 5, 39**]).

2. Dilatations in $\mathbb{R}^n$ and their calculation

The dilatations play a crucial role in studying various classes of mappings. One of classical results in $\mathbb{R}^n$ says that the mappings with the inner or outer dilatations close to one, must be close to the Möbius transformations. Observe that this is not true for p-dilatations which will be shown below. Let us also mention that the calculation of the dilatations usually is not trivial and relates to optimization problems.

Let $A : \mathbb{R}^n \to \mathbb{R}^n$ be a linear bijection. The numbers

$$K_{I,p}(A) = \frac{|\det A|}{l^p(A)}, \quad K_{O,p}(A) = \frac{||A||^p}{|\det A|}, \quad p \geq 1,$$

are called the *p-inner* and *p-outer* dilatations of A, respectively. Here

$$l(A) = \min_{|h|=1} |Ah|, \quad ||A|| = \max_{|h|=1} |Ah|,$$

denote the minimal and maximal stretching of A and $\det A$ is the determinant of A.

Let G and G^* be two bounded domains in $\mathbb{R}^n$, $n \geq 2$, and let a mapping $f : G \to G^*$ be differentiable at a point $x \in G$. This means there exists a linear mapping $f'(x) : \mathbb{R}^n \to \mathbb{R}^n$, called the (strong) derivative of the mapping f at x, such that

$$f(x+h) = f(x) + f'(x)h + \omega(x,h)|h|,$$

where $\omega(x,h) \to 0$ as $h \to 0$.

We denote

$$K_{I,p}(x,f) = K_{I,p}(f'(x)), \quad K_{O,p}(x,f) = K_{O,p}(f'(x)).$$

These quantities naturally extend the classical quasiconformal dilatations (inner and outer)

$$K_I(x,f) = K_{I,n}(x,f), \quad K_O(x,f) = K_{O,n}(x,f).$$

The third dilatation of quasiconformality called *linear*

$$K_l(x,f) = \frac{||f'(x)||}{l(f'(x))}$$

is a direct analog of the classical planar Lavrentiev dilatation.

The relations between the classical quasiconformal dilatations

$$\begin{aligned} K_l(x,f) &\leq \min(K_I(x,f), K_O(x,f)) \leq K_l^{n/2}(x,f) \\ &\leq \max(K_I(x,f), K_O(x,f)) \leq K_l^{n-1}(x,f). \end{aligned} \tag{2.1}$$

show that they are finite or infinity simultaneously; see [**40**]. Moreover, a closeness of one of the dilatations to 1 necessarily implies that all dilatations also are close to 1, and therefore the corresponding mapping is close to a Möbius transformation (see, e.g., [**31**]).

The p-inner and p-outer dilatations provide the study of a class of mappings whose basic properties are close to quasiconformal homeomorphisms. On the other hand, there are some essential differences caused by the fact that the dilatations $K_I(x,f)$ and $K_O(x,f)$ are always greater than or equal to 1, while the p-inner and outer dilatations range between 0 and ∞.

The following class of mappings provides a natural extension of quasiconformal mappings and contains many well-known classes of mappings. Define for the fixed real numbers $\alpha, \beta, \gamma, \delta$ such that $1 \le \alpha < \beta < \infty$, $1 \le \gamma < \delta < \infty$, the *class of mappings with finite mean dilatations* which consists of homeomorphisms $f : G \to G^*$ satisfying:
(i) f and f^{-1} are in $W^{1,1}_{\mathrm{loc}}$ (the well-known Sobolev class);
(ii) f and f^{-1} are differentiable, with Jacobians $J(x,f) \neq 0$ and $J(y, f^{-1}) \neq 0$ a.e. in G and G^*; respectively,
(iii) $K_{I,\alpha}(x,f)$ and $K_{O,\delta}(x,f)$ are integrable in G with exponents $\beta/(\beta-\alpha)$ and $\gamma/(\delta-\gamma)$, respectively.

The mappings with finite mean dilatations were investigated in [**7**].

In contrast with the classical dilation coefficients their integral counterparts defined in (iii) need not satisfy any relation of kind (2.1). The following simple example shows that the unboundedness of one from these dilatations does not depend on the value of another mean dilatation.

As the first example, consider the unit cube

$$G = \{x = (x_1, \dots, x_n) : 0 < x_k < 1, k = 1, \dots, n\}$$

and let

$$f_1(x) = \left(x_1, \dots, x_{n-1}, \frac{x_n^{1-c}}{1-c}\right), \quad 0 < c < 1.$$

An easy computation shows that f belongs to the class of mappings with finite mean dilatations if and only if

$$0 < c < 1 - \alpha/\beta \quad \text{and} \quad 0 < c < 1 - (\gamma - 1)\delta/(\delta - 1)\gamma.$$

If

$$1 - \alpha/\beta \le c < 1 \quad \text{and} \quad 1 - (\gamma - 1)\delta/(\delta - 1)\gamma \le c < 1,$$

then we have

$$\int\limits_G K_{I,\alpha}^{\frac{\beta}{\beta-\alpha}}(x, f_1)\, dm(x) = \infty \quad \text{and} \quad \int\limits_G K_{O,\delta}^{\frac{\gamma}{\delta-\gamma}}(x, f_1)\, dm(x) = \infty.$$

Thus, by a suitable choice of parameters $c, \alpha, \beta, \gamma, \delta$, one obtains any desired relations between the integrability conditions in (iii).

Now let us calculate the dilatations for the homeomorphism

$$f_2(x) = \left(\frac{x_1}{x_n}, \dots, \frac{x_{n-1}}{x_n}, \frac{x_n^{n-p}}{n-p}\right), \qquad 1 \le p < n-1, \tag{2.2}$$

defined in the unbounded cylinder $D = \{x \in \mathbb{R}^n : x_1^2 + \ldots + x_{n-1}^2 < 1,\ x_n > 1\}$ ([**6**]). Its Jacobian matrix has the form

$$f_2'(x) = \begin{pmatrix} \frac{1}{x_n} & 0 & \cdots & 0 & -\frac{x_1}{x_n^2} \\ 0 & \frac{1}{x_n} & \cdots & 0 & -\frac{x_2}{x_n^2} \\ \vdots & \vdots & \ddots & \vdots & \vdots \\ 0 & 0 & \cdots & \frac{1}{x_n} & -\frac{x_{n-1}}{x_n^2} \\ 0 & 0 & \cdots & 0 & x_n^{n-p-1} \end{pmatrix}$$

and the determinant $J(x, f_2) = x_n^{-p}$.

We use the method of Lagrange multipliers to find

$$l(f_2'(x)) = \min_{|h|=1} |f_2'(x)h| \quad \text{and} \quad ||f_2'(x)|| = \max_{|h|=1} |f_2'(x)h|.$$

Letting by $F(x,h) = |f_2'(x)h|^2$, we solve the optimization problem

$$F(x,h) \to \text{extr}$$

subject to the equality $h_1^2 + h_2^2 + \ldots + h_n^2 = 1$. A straightforward calculation yields that one of critical points satisfies

$$h_n = 0, \quad h_1^2 + h_2^2 + \ldots + h_{n-1}^2 = 1, \quad x_1 h_1 + x_2 h_2 + \ldots + x_{n-1} h_{n-1} = 0.$$

It is not hard to verify that for such points $F(x,h)$ achieves the minimum, which is equal to $1/x_n^2$. Thus $l(f_2'(x)) = 1/x_n$. Since f_2 is symmetric with respect to $n-1$ coordinates $x_1, x_2, \ldots, x_{n-1}$, one concludes that $n-1$ stretchings $\lambda_2, \lambda_3, \ldots, \lambda_n$ have the same value. The last one $\lambda_1 = ||f_2'(x)|| = x_n^{n-p-1}$, because $J(x, f_2) = \lambda_1 \lambda_2 \cdot \ldots \cdot \lambda_n$. Finally, the p-inner and p-outer dilatations are equal

$$K_{I,p}(x, f_2) = \frac{J(x, f_2)}{l^p(f_2'(x))} = 1, \qquad K_{O,p}(x, f_2) = \frac{||f_2'(x)||^p}{J(x, f_2)} = x_n^{(n-p)p}.$$

A point is that under the certain regularity conditions for homeomorphisms $f : G \to G^*$ (namely preserving (N) and (N^{-1})-properties for k-dimensional surface families), the function $Q(x)$ in (1.3) and (1.4) can be replaced by $K_{I,p}(x, f)$ and $K_{O,p}(x, f)$, respectively; see [**10**]. In the general case, these bounds for $\mathcal{M}_p(f(\mathcal{S}_k))$ given by (1.3) and (1.4) cannot be improved. In the case of these Q, the radial stretchings $f_3(x) = x|x|^{K-1}$ and $f_4(x) = x|x|^{1/K-1}$ with $K > 1$ (quasiconformal automorphisms of the unit ball $\mathbb{B}^n(0,1)$) yield that the above inequalities are sharp.

However, if we fix a point $x_0 \in G$ and restrict ourselves by considering only the curve families that join boundary components of a ring $A \subset G$ centered at x_0, then the inequality (1.3) possesses a strengthened form (see [**13**]). In this case, the majorant Q can be regarded as the angular dilatation introduced for quasiconformal mappings in $\mathbb{R}^n$ in [**16**].

Recall that a homeomorphic mapping $f : G \to \mathbb{R}^n$, $n \geq 2$, is called K-*quasiconformal*, $K \geq 1$, if $f \in W^{1,n}_{\text{loc}}(G)$ and if $K_I(x, f) \leq K$ and $K_O(x, f) \leq K$ almost everywhere (a.e.) in G; see [**40**], [**22**].

Let $f : G \to \mathbb{R}^n$ be a quasiconformal mapping. The angular dilatation coefficient $D(x, x_0, f)$ of the mapping $f : G \to \mathbb{R}^n$ at the point $x \in G$ with respect to $x_0 \in \mathbb{R}^n$, $x \neq x_0$, is defined by

$$D(x, x_0, f) = \frac{J(x, f)}{\ell^n(x, x_0, f)},$$

where

$$\ell(x, x_0, f) = \min_{|h|=1} \frac{|\partial_h f(x)|}{|\langle h, \frac{x - x_0}{|x - x_0|}\rangle|}. \tag{2.3}$$

Here $\partial_h f(x)$ denotes the derivative of f at x in the direction h and the minimum is taken over all unit vectors $h \in \mathbb{R}^n$. The dilatation $D(x, x_0, f)$ is a measurable function in G and satisfies the inequalities

$$\frac{1}{K_O(x, f)} \leq D(x, x_0, f) \leq K_I(x, f).$$

The following example of the volume preserving quasiconformal mapping

$$f_5(x) = \left(ze^{2i\log|z|}, x_3, \ldots, x_n\right), \quad f_5(0) = 0, \quad |x| < 1,$$

of the unit ball $\mathbb{B}^n(0,1)$ with a "quick rotation" in the plane $x_3 = \ldots = x_n = 0$ is given in [**16**]. Here $x = (x_1, \ldots, x_n)$ and $z = x_1 + ix_2$. For this mapping the angular dilatation $D(x, 0, f_5) \equiv 1$ at all points of the unit ball $\mathbb{B}^n(0,1)$ except for the origin. Note that the mapping is far to be conformal, because $K_I(x, f_5) = K_O(x, f_5) = (1+\sqrt{2})^n$.

For a sense-preserving homeomorphism $f : G \to \mathbb{R}^n$ which is differentiable with non-degenerate Jacobian at a point x, we define p-angular dilatation $D_p(x, x_0, f)$ with respect to a point $x_0 \in \mathbb{R}^n$ by

$$D_p(x, x_0, f) = \frac{J(x,f)}{\ell^p(x, x_0, f)}, \qquad p \geq 1,$$

where $\ell(x, x_0, f)$ is determined by (2.3).

The calculation of the angular dilatations is much more complicated than of the classical dilatation coefficients of quasiconformality. We illustrated it in [**13**] for a "slow rotation" of $\mathbb{B}^n(0,1)$ in the plane $x_3 = \ldots = x_n = 0$. Consider the mapping

$$f_6(x) = \left(ze^{i\sqrt{-\log|z|}}, x_3, \ldots, x_n\right), \quad f_6(0) = 0, \quad |x| < 1.$$

For this mapping,

$$||f_6'(x)|| = \frac{\sqrt{16\log\frac{1}{|z|}+1}+1}{4\sqrt{\log\frac{1}{|z|}}}, \qquad l(f_6'(x)) = \frac{\sqrt{16\log\frac{1}{|z|}+1}-1}{4\sqrt{\log\frac{1}{|z|}}},$$

(all other stretchings are equal to one) and, hence, the Jacobian

$$J(x, f_6) = ||f_6'(x)|| l(x, f_6) \equiv 1$$

at all points x of the unit ball $\mathbb{B}^n$ except for the origin. Thus,

$$K_O(x, f_6) = K_I(x, f_6) = \left(\frac{\sqrt{16\log\frac{1}{|z|}+1}+1}{4\sqrt{\log\frac{1}{|z|}}}\right)^n > 1,$$

however $D(x, 0, f_6) = D_p(x, 0, f_6) = 1$ at all $x \in \mathbb{B}^n(0,1)\backslash\{0\}$ and any $p \geq 1$.

The *quasiregular mappings* are defined by the same way as quasiconformal mappings assuming the continuity of mappings instead of homeomorphy. Such mappings have been introduced by Reshetnyak ([**30**]) (and named *mappings with bounded distortion*) in the middle of 1960s and a few years later intensively studied by Marto, Rickman, Väisälä (see, e.g. [**23**]).

One of the modern extensions of quasiregular mappings is the class of mappings of finite distortion. This class has deep applications in various fields of mathematics, for instance in the elasticity theory. For the features of spatial mappings of finite distortion we refer to [**18**] and [**17**]. Recall that a mapping $f : G \to \mathbb{R}^n$ is of *finite (outer) distortion* if $f \in W^{1,1}_{\text{loc}}(G)$, $J \in L^1_{\text{loc}}(G)$ and there is a function $K : G \to [1, \infty]$ with $K(x) < \infty$ a.e. such that $\|f'(x)\|^n \leq K(x)J(x,f)$ for almost all $x \in G$.

This definition goes back to the analytic definition of quasiregular (quasiconformal) mappings. Next we consider another approach which relates to Geometric Function Theory.

3. Q-homeomorphisms

Now we give a geometric (modular) description of quasiconformality in $\mathbb{R}^n$ starting with the definition of k-dimensional Hausdorff measure H^k, $k = 1, \ldots, n-1$ in $\mathbb{R}^n$. For a given $E \subset \mathbb{R}^n$, put

$$H^k(E) = \sup_{r>0} H^k_r(E),$$

where

$$H^k_r(E) = \Omega_k \inf \sum_i (\delta_i/2)^k.$$

Here the infimum is taken over all countable coverings $\{E_i, i = 1, 2, \ldots\}$ of E with diameters δ_i, and Ω_k is the volume of the unit ball in $\mathbb{R}^k$.

Let $\mathcal{S}$ be a k-dimensional surface, which means that $\mathcal{S} : D_s \to \mathbb{R}^n$ is a continuous image of the closed domain $D_s \subset \mathbb{R}^k$. We denote by

$$N(\mathcal{S}, y) = \operatorname{card} \mathcal{S}^{-1}(y) = \operatorname{card}\{x \in D_s : \mathcal{S}(x) = y\}$$

the multiplicity function of the surface $\mathcal{S}$ at the point $y \in \mathbb{R}^n$. For a given Borel set $B \subseteq \mathbb{R}^n$, the k-dimensional Hausdorff area of B in $\mathbb{R}^n$ associated with the surface $\mathcal{S}$ is determined by

$$\mathcal{H}_{\mathcal{S}}(B) = \mathcal{H}^k_{\mathcal{S}}(B) = \int_B N(\mathcal{S}, y)\, dH^k y.$$

If $\rho : \mathbb{R}^n \to [0, \infty]$ is a Borel function, the integral of ρ over $\mathcal{S}$ is defined by

$$\int_{\mathcal{S}} \rho\, d\sigma_k = \int_{\mathbb{R}^n} \rho(y) N(\mathcal{S}, y)\, dH^k y.$$

Let $\mathcal{S}_k$ be a family of k-dimensional surfaces $\mathcal{S}$ in $\mathbb{R}^n$, $1 \le k \le n-1$ (curves for $k = 1$). The *p-module* of $\mathcal{S}_k$ is defined as

$$\mathcal{M}_p(\mathcal{S}_k) = \inf \int_{\mathbb{R}^n} \rho^p\, dm(x), \quad p \ge k,$$

where the infimum is taken over all Borel measurable functions $\rho \ge 0$ such that

$$\int_{\mathcal{S}} \rho^k\, d\sigma_k \ge 1$$

for every $\mathcal{S} \in \mathcal{S}_k$. We call each such ρ an *admissible function* for $\mathcal{S}_k$ ($\rho \in \operatorname{adm} \mathcal{S}_k$). The n-module $\mathcal{M}_n(\mathcal{S}_k)$ will be denoted by $\mathcal{M}(\mathcal{S}_k)$.

Following [**19**], a metric ρ is said to be *extensively admissible* for $\mathcal{S}_k$ with respect to p-module ($\rho \in \operatorname{ext}_p\operatorname{adm} \mathcal{S}_k$) if $\rho \in \operatorname{adm}(\mathcal{S}_k \backslash \widetilde{\mathcal{S}}_k)$ such that $\mathcal{M}_p(\widetilde{\mathcal{S}}_k) = 0$.

Accordingly, we say that a property P holds for p-almost every (p-a.e.) k-dimensional surface, if P holds for all surfaces except a family of zero p-module.

We also remind that a continuous mapping f satisfies (N)-property with respect to k-dimensional Hausdorff area if $\mathcal{H}^k_{\mathcal{S}}(f(B)) = 0$ whenever $\mathcal{H}^k_{\mathcal{S}}(B) = 0$. Similarly, f has (N^{-1})-property if $\mathcal{H}^k_{\mathcal{S}}(B) = 0$ whenever $\mathcal{H}^k_{\mathcal{S}}(f(B)) = 0$.

Now we provide the bounds for the p-module of k-dimensional surfaces (see [**10**]).

THEOREM 3.1. *Let $f : G \to \mathbb{R}^n$ be a homeomorphism satisfying (i)-(ii) with $K_{I,p}, K_{O,p}^{-1} \in L^1_{\rm loc}(G)$. Suppose that for some k, $1 \le k \le n-1$ $(k \le p)$, and for p-a.e. k-dimensional surface $\mathcal{S}$ the restriction $f|_{\mathcal{S}}$ has the (N) and (N^{-1})-properties with respect to k-dimensional Hausdorff area in G and $G^* = f(G)$, respectively. Then the double inequality (1.2) holds for any family $\mathcal{S}_k$ of k-dimensional surfaces in G, and for each $\rho \in \operatorname{adm} \mathcal{S}_k$ and $\varrho \in \operatorname{ext}_p\operatorname{adm} \mathcal{S}_k$.*

Inequalities of type (1.3) naturally lead to important extensions of quasiconformality and quasiisometry called *(p, Q)-homeomorphisms* or *Q-homeomorphisms with respect to p-module.* Mappings that are subject to (1.4) are called *lower Q-homeomorphisms with respect to p-module.* An essential difference between (1.3) and (1.4) is based on the following property: the infimum in (1.3) can be omitted assuming that ρ is a arbitrary admissible for the appropriate family. However, the dropping infimum is not possible in the inequality (1.4).

As was mentioned above, we shall restrict ourselves by considering only inequalities of kind (1.3) related to the case $k = 1$. More precisely, given a measurable function $Q : G \to [0, \infty]$, we call a homeomorphism $f : G \to \mathbb{R}^n$ the *Q-homeomorphism with respect to p-module* if

$$\mathcal{M}_p(f(\Gamma)) \le \int_G Q(x)\rho^p(x)\, dm(x) \tag{3.1}$$

for every family of curves Γ in G and for every admissible function ρ for Γ.

For $p = n$, the class of Q-homeomorphisms was introduced in [**24**]. It admits various rich properties; see, e.g. [**25**] and [**36**] (the latter for open and discrete mappings). We are focused here to the case $p \ne n$. The ranges $1 \le p < n$ and $p > n$ for p are needed to apply, similar to the classical case $p = n$, the "continuum" analysis, since for $p < n$ and $p > n$ there appear some essential differences from the case $p = n$. In Sections 3–5 we survey some known results starting with differentiability of such mappings.

THEOREM 3.2. *Let $f : G \to G^*$ be a Q-homeomorphism with respect to p-module with $Q \in L^1_{\rm loc}(G)$ and $p > n-1$. Then*
(1) *f is ACL-homeomorphism which is differentiable a.e. in G;*
(2) *$f \in W^{1,1}_{\rm loc}(G)$;*
(3) *$K_{I,p} \in L^1_{\rm loc}(G)$ assuming in addition $J(x, f) \ne 0$ a.e.*

The proof of Theorem 3.2 is given in [**8**] (cf. [**32**] for $p = n$). Note that for $1 \le p \le n-1$ the question on differentiability of Q-mappings with respect to p-module remains still open.

Upper bounds for the norm of Jacobi matrix of f and its determinant have been established in [**11**]. Observe that the restrictions on the parameter p are essential in the proof.

THEOREM 3.3. *Let G and G^* be two bounded domains in $\mathbb{R}^n$, $n \ge 2$, and $f : G \to G^*$ be a sense preserving Q-homeomorphism with respect to p-module, $n-1 < p < n$, such that that $Q \in L^1_{\rm loc}$. Then*

$$\|f'(x)\| \le \lambda_{n,p}\, Q^{\frac{1}{n-p}}(x) \ \text{ a.e.}$$

and

$$J(x,f) \leq \gamma_{n,\alpha}\, Q^{\frac{n}{n-p}}(x) \quad \text{a.e.},$$

where $\lambda_{n,p}$ and $\gamma_{n,p}$ are a positive constants depending only on n and p. If in addition $Q(x) \in L_{\text{loc}}^{\frac{s}{n-p}}$ with $s > n-p$, then $f \in W_{\text{loc}}^{1,s}$.

The following result is a consequence of the previous one.

THEOREM 3.4. Let $f : G \to G^*$ be a Q-homeomorphism with respect to p-module such that $Q(x) \in L_{\text{loc}}^{\frac{n}{n-p}}$, $n-1 < p < n$. Then f satisfies (N)-property and, moreover,

$$mf(E) \leq \gamma_{n,p} \int\limits_E Q^{\frac{n}{n-p}}(x)\, dm(x)$$

for any Lebesgue measurable set E. Here $\gamma_{n,p}$ is a constant depending only on n and p.

One of natural questions concerns the connection between the dilatation coefficients and the majorant Q. The boundedness of p-inner dilatation by Q from above provides one of crucial tools in our approach.

THEOREM 3.5. Let $f : G \to G^*$ be a Q-homeomorphism with respect to p-module, $p > n-1$, with $Q \in L_{\text{loc}}^1(G)$ such that $J(x,f) \neq 0$ a.e. in G. Then

$$K_{I,p}(x,f) \leq Q(x) \quad \text{a.e. in} \quad G.$$

REMARK 3.6. If one omits the restriction $p > n-1$ (i.e. for $1 \leq p \leq n-1$), Theorem 3.5 can be proved assuming additionally that the Q-homeomorphisms f with respect to p-module are differentiable a.e.

Relaxing $Q \in L^\infty$, we can formulate the following statement given in [**6**] in the terms of Q-homeomorphisms. We say that $f : G \to G^*$ belongs to $\text{Lip}(K)$ if there is K, $0 < K < \infty$, satisfying

$$\limsup_{y \to x} \frac{|f(y) - f(x)|}{|y - x|} \leq K$$

for each $x \in G$.

THEOREM 3.7. If $f : G \to G^*$ is a Q-homeomorphism with respect to p-module such that $Q(x) \leq K$ a.e. in G, then
(1) $f \in \text{Lip}(K_0)$ for $n-1 < p < n$;
(2) $f^{-1} \in \text{Lip}(K_0)$ for $p > n$.
In both cases K_0 depends only on p, n, and K.

4. Homeomorphisms Lipschitzian in the mean

Various classes of the spatial mappings, which are quasiconformal in the mean with integrally bounded dilatations have been intensively studied already about 50 years (see, e.g., [**21**], [**20**], [**7**], [**9**]).

As was mentioned above, the conditions of type (3.1) provide an integral form of restriction on p-module, $n-1 < p < n$, and for $Q(x) \leq K$ at all $x \in G$ such mappings are Lipschitzian. It allows to define a class of sense-preserving homeomorphisms $f : G \to G^*$ which depend on given real numbers α and p, $\alpha, p \geq 1$, and satisfying:

1) f and f^{-1} belong to the Sobolev class $W^{1,p}_{\text{loc}}$ of G and G^*, respectively;

2) f and f^{-1} are differentiable with the Jacobians $J(x, f) \neq 0$ and $J(y, f^{-1}) \neq 0$ a.e. in G and G^*, respectively;

3) p-angular dilatation $D_p(x, x_0, f) \in L^\alpha_{\text{loc}}(G)$.

We call such homeomorphisms f *mappings Lipschitzian in (α, p)-mean with respect to the point x_0*; see [**13**].

One of natural questions here is how the differential features of such mappings are far from or close to the corresponding properties of Lipschizian mappings. In this section, we present some distortion results for the mappings which are Lipschitzian in the mean with respect to euclidian distances and their logarithms.

THEOREM 4.1. *Let G and G^* be domains in $\mathbb{R}^n$, $n \geq 2$, and let $f : G \to G^*$ be a mapping Lipschitzian in $(\frac{n}{n-p}, p)$-mean with respect to a point $x_0 \in G$ with $n-1 < p < n$. Then the following estimate*

$$|f(x) - f(x_0)| \leq C_{n,p} \, \|D_p\|^{\frac{1}{n-p}}_{\frac{n}{n-p}} \left(\log \frac{1}{|x - x_0|} \right)^{-\frac{p(n-1)}{n(n-p)}} \tag{4.1}$$

holds for any $x \in G$ satisfying $|x - x_0| = r < \delta = \min\{1, \text{dist}^4(x_0, \partial G)\}$. Here $C_{n,p}$ is a positive constant depending only on n and p, and

$$||D_p(x_0, r)||_{\frac{n}{n-p}} = \left(\int\limits_A D_p^{\frac{n}{n-p}}(x, x_0, f) \, dm(x) \right)^{\frac{n-p}{n}} \tag{4.2}$$

is the $n/(n-p)$-norm over the ring $A = A(x_0, r, \sqrt[4]{r})$.

Theorem 4.1 was proved in [**12**] under much stronger conditions: the weight function Q is itself integrable over the balls centered at x_0. In [**13**], we require the integrability of p-angular dilatations over some rings. It was also shown that the integrability degree $n/(n-p)$ cannot be decreased, and it is impossible to replace the logarithmic Hölder continuity by the usual Hölder continuity; in addition, the exponent $p(n-1)/n(n-p)$ in (4.1) is sharp.

Fix p, $p \in (n-1, n)$, and consider the mapping $f_7 : \mathbb{B}^n \to \mathbb{B}^n$ given by formula

$$y = f_7(x) = \frac{x}{|x|} \left(1 + \frac{n-p}{p-1} \log \frac{1}{|x|} \right)^{-\frac{p-1}{n-p}}, \quad x \neq 0, \quad \text{and} \quad f_7(0) = 0. \tag{4.3}$$

Take the spherical coordinates (ρ, ψ_i) in the image-space y and (r, φ_i) in x, $i = 1, \ldots, n-1$, one can rewrite (4.3) for $x \neq 0$ as

$$f_7(x) = \left\{ \rho = \left(1 + \frac{n-p}{p-1} \log \frac{1}{r} \right)^{-\frac{p-1}{n-p}}, \psi_i = \varphi_i, 0 < r < 1, 0 \leq \varphi_i < 2\pi \right\}.$$

The p-inner dilatation of this mapping can be calculated similarly to [**10**], and one obtains that this dilatation coefficient is equal to r^{p-n}. Obviously,

$$\|D_p(x, 0, f)\|_{\frac{n}{n-p}} \leq \|K_{I,p}(x, f)\|_{\frac{n}{n-p}}.$$

Applying Theorem 4.1 from [10], we obtain from (4.1) the estimate

$$|f_7(x)| \leq C_{n,p} \left(\log\frac{1}{|x|}\right)^{\frac{1}{n}} \left(\log\frac{1}{|x|}\right)^{-\frac{p(n-1)}{n(n-p)}} = C_{n,p} \left(\log\frac{1}{|x|}\right)^{\frac{1-p}{n-p}}.$$

On the other hand, the mapping f_7 is not Hölder continuous at the origin with any exponent $0 < \alpha < 1$, and a direct calculation yields

$$\lim_{x\to 0} |f(x)| \left(\log\frac{1}{|x|}\right)^{\frac{p-1}{n-p}} = \left(\frac{p-1}{n-p}\right)^{\frac{p-1}{n-p}}.$$

This shows that the estimate given by Theorem 4.1 is sharp with respect to the order.

The next statement [13] provides the Hölder continuity condition for mappings Lipschitzian in (α, p)-mean if $\alpha > n/(n-p)$ and p ranges between $n-1$ and n.

THEOREM 4.2. Let G and G^* be domains in $\mathbb{R}^n$, $n \geq 2$, and let $f : G \to G^*$ be a mapping Lipschtzian in (α, p)-mean with respect to a point $x_0 \in G$ with $n - 1 < p < n$ and $\alpha > \frac{n}{n-p}$. Then

$$|f(x) - f(x_0)| \leq \lambda_{n,p} \|D_p\|_{\alpha}^{\frac{1}{n-p}} |x - x_0|^{1-\frac{n}{\alpha(n-p)}},$$

for any x satisfying $|x - x_0| = r < \frac{1}{4}\mathrm{dist}(x_0, \partial G)$. Here $\|D_p\|_\alpha$ is α-norm defined by (4.2) over the ring $A = A(x_0, r, 4r)$ and $\lambda_{n,p}$ is a positive constant depending only on n and p.

The sufficient condition for mappings Lipschitzian in $(1,p)$-mean to be Lipschitzian in a prescribed point x_0 is given also in [13]. This condition involves an integral mean of the form

$$d_{x_0} = \limsup_{r\to 0} \frac{1}{mA(x_0, r, 4r)} \int\limits_{A(x_0,r,4r)} D_p(x, x_0, f)\, dm(x).$$

THEOREM 4.3. Let G and G^* be domains in $\mathbb{R}^n$, $n \geq 2$, and let $f : G \to G^*$ be a mapping Lipschizian in $(1,p)$-mean with respect to a point $x_0 \in G$ for $n-1 < p < n$. If $d_{x_0} < \infty$, then

$$L(x_0, f) = \limsup_{x\to x_0} \frac{|f(x) - f(x_0)|}{|x - x_0|} \leq C_{n,p}\, d_{x_0}^{\frac{1}{n-p}},$$

where $C_{n,p}$ is a positive constant that depends only on n and p.

Close classes of finite Lipschitzian homeomorphisms were studied in [33] and [34].

5. Extensions of the Schwarz Lemma to Q-homeomorphisms

Let G be a domain in $\mathbb{R}^n$ $(n \geq 2)$, and $Q : G \to [0, \infty]$ be a Lebesgue measurable function. Then for almost all $r > 0$ one can define the spherical mean

$$q(r) = \frac{1}{\omega_{n-1} r^{n-1}} \int\limits_{S_r} Q(x)\, d\mathcal{A}, \tag{5.1}$$

where $d\mathcal{A}$ denotes the element of surface area on $S_r = \{x \in \mathbb{R}^n : |x| = r\}$.

The following theorem given in [**14**] provides an upper bound for Q-homeomorphisms with respect to p-module $(1 < p < n)$, which can be regarded as a generalization of Schwarz's lemma. Denote

$$\|Q\|_\alpha = \left(\int\limits_{\mathbb{B}^n} Q^\alpha(x)\, dm(x) \right)^{\frac{1}{\alpha}} .$$

THEOREM 5.1. For any Q-homeomorphism $f : \mathbb{B}^n \to \mathbb{B}^n$ with respect to p-module, $1 < p < n$, with $Q \in L^\alpha_{\mathrm{loc}}(\mathbb{B}^n)$, $\alpha > \frac{n}{n-p}$, and normalized by $f(0) = 0$, we have that

$$\liminf_{x \to 0} \frac{|f(x)|}{|x|^\gamma} \leq c_0\, ||Q_\alpha||^{\frac{1}{n-p}}, \tag{5.2}$$

where $\gamma = 1 - \frac{n}{\alpha(n-p)}$ and c_0 is a constant depending only on α, p and n.

The following theorem in [**14**] estimates the growth of Q-homeomorphisms of the unit ball $\mathbb{B}^n$, in the case when $p = n$, again in the spirit of Schwarz's lemma.

THEOREM 5.2. Let $f : \mathbb{B}^n \to \mathbb{B}^n$ be a Q-homeomorphism preserving the origin. If for some real β,

$$\mathcal{I} = \int\limits_0^1 \left(\beta - \frac{1}{q^{\frac{1}{n-1}}(t)} \right) \frac{dt}{t} < \infty,$$

then

$$\liminf_{x \to 0} \frac{|f(x)|}{|x|^\beta} \leq e^{\mathcal{I}} .$$

It seems likely that the assumption $Q \in L^\alpha$ with $\alpha > \frac{n}{n-p}$ in Theorem 5.1 can not be improved. In fact, if $\alpha \leq n/(n-p)$, the estimate (5.2) becomes trivial because in this case the assumptions of this theorem yields $\gamma \leq 0$, and, obviously,

$$\liminf_{x \to 0} \frac{|f(x)|}{|x|^\gamma} = 0 .$$

The same situation occurs when $||Q||_\alpha = \infty$. This shows that the assumptions on growth of Q are sharp in some sense.

To illustrate the bound (5.2), consider the function $f_4(x) = x|x|^{\frac{1}{K}-1}$, $K > 1$, $f_4(0) = 0$, mapping quasiconformally the ball $\mathbb{B}^n$ into itself. A direct calculation gives that the Jacobian of f_4 is equal to $|x|^{n(\frac{1}{K}-1)}/K$, and $l(f_4'(x)) = |x|^{\frac{1}{K}-1}/K$. Thus, p-inner dilatation $K_{I,p}(x, f_4) = K^{p-1}|x|^{(n-p)(\frac{1}{K}-1)}$. The majorant $Q(x)$ is integrable over the ball $\mathbb{B}^n$ with degree α provided that $\alpha < \frac{K}{K-1}\frac{n}{n-p}$. Restricting α by

$$\frac{n}{n-p} < \alpha < \frac{K}{K-1}\frac{n}{n-p},$$

one obtains that in this case $0 < \gamma < \frac{1}{K}$.

There should be in Theorem 5.2 $\mathcal{I} < +\infty$. In the case $\mathcal{I} = -\infty$, we have

$$\liminf_{x \to 0} \frac{|f(x)|}{|x|^\beta} = 0 .$$

The example given in Section 2 relates to Theorem 5.2. We now rewrite f_5 in the form

$$f_5(x) = (x_1 \cos\theta - x_2 \sin\theta, x_2 \cos\theta + x_1 \sin\theta, x_3, \ldots, x_n), \quad |x| < 1, \quad f_5(0) = 0,$$

where $x = (x_1, \ldots, x_n)$ and $\theta = \log(x_1^2 + x_2^2)$. This is a rotation with the speed depending on the coordinates x_1 and x_2. Due to [**3**] and [**16**], $Q(x) = K_I(x, f_5) \equiv (1+\sqrt{2})^n$. By (5.1), the spherical mean $q(r)$ is also equal to $(1+\sqrt{2})^n$, hence the integral $\mathcal{I}$ is finite if $\beta = (\sqrt{2}-1)^{n/(n-1)}$, which in turn yields

$$\liminf_{x\to 0} \frac{|g(x)|}{|x|^{(\sqrt{2}-1)^{n/(n-1)}}} \leq 1\,.$$

The behavior of Q-homeomorphisms with weighted $|f(x)|$ located in the closed unit ball has been established in [**35**].

6. Capacity of condenser and related estimates

We also use especially another tool which is important in Potential Theory and Mathematical Analysis.

Following in general [**23**], a pair $E = (A, C)$, where $A \subset \mathbb{R}^n$ is an open set and $C \subset A$ is a nonempty compact, is called the *condenser*. We say that the condenser E is the *ring condenser*, if $\mathcal{R} = A \backslash C$ is a ring domain. The condenser E is bounded, if A is bounded. We also say that a condenser $E = (A, C)$ lies in a domain G when $A \subset G$. Obviously, for an open and continuous mapping $f : G \to \mathbb{R}^n$ and for any condenser $E = (A, C) \subset G$, the pair $(f(A), f(C))$ is a condenser in $f(G)$. In this case we shall use the notation $f(E) = (f(A), f(C))$.

Let $E = (A, C)$ be a condenser. Denote by $\mathcal{C}_0(A)$ the set of all continuous functions $u : A \to \mathbb{R}^1$ with compact support in A. Consider the set $\mathcal{W}_0(E) = \mathcal{W}_0(A, C)$ of all nonnegative functions $u : A \to \mathbb{R}^1$ such that
1) $u \in \mathcal{C}_0(A)$, 2) $u(x) \geq 1$ for $x \in C$ and 3) u belongs ACL. Put

$$\operatorname{cap}_p E = \operatorname{cap}_p(A, C) = \inf_{u \in \mathcal{W}_0(E)} \int\limits_A |\nabla u|^p \, dm(x), \quad p \geq 1,$$

where, as usual

$$|\nabla u| = \left(\sum_{i=1}^n (\partial_i u)^2 \right)^{1/2}.$$

This quantity is called *p-capacity of condenser E*.

It was proven in [**38**] that for $p > 1$

$$\operatorname{cap}_p E = \mathcal{M}_p(\Delta(\partial A, \partial C; A \setminus C)), \tag{6.1}$$

where $\Delta(\partial A, \partial C; A \setminus C))$ denotes the set of all continuous curves which join the boundaries ∂A and ∂C in $A \setminus C$. For general properties of p-capacities and their relation to the mapping theory, we refer for instance to [**15**] and [**27**]. In particular, for $p > 1$,

$$\begin{aligned} &\operatorname{cap}_p(A, C) \geq n\Omega_n^{\frac{p}{n}} \left| \frac{p-n}{p-1} \right|^{p-1} \left| (mC)^{\frac{p-n}{n(p-1)}} - (mA)^{\frac{p-n}{n(p-1)}} \right|^{1-p}, \quad p \neq n, \\ &\operatorname{cap}_n(A, C) \geq n^n \Omega_n \left(\log \frac{mA}{mC} \right)^{1-n}, \end{aligned} \tag{6.2}$$

where Ω_n denotes the volume of the unit ball in $\mathbb{R}^n$, and mE is the n-dimensional Lebesgue measure of E.

For $n-1 < p \le n$, there is the following lower estimate

$$(6.3) \qquad \left(\operatorname{cap}_p E\right)^{n-1} \ge \gamma \frac{d(C)^p}{(mA)^{1-n+p}},$$

where $d(C)$ denotes the diameter of the connected compact set C, and γ is a positive constant depending only on n and p (see [**20**]). By applying the Hölder inequality, the estimate (6.3) is extended to any $p > n-1$.

7. Estimates for outer dilatations and volume derivative

The estimates for the p-inner dilatation were given in Section 3. Now we discuss the boundedness of another p-dilatation, namely the outer p-dilatation, and of Jacobian. To this end, we first recall the notion of the volume derivative.

Let G and G^* be domains in $\mathbb{R}^n$ and $f : G \to G^*$ be a homeomorphism. Consider the set function $\Phi(\mathcal{B}) = mf(\mathcal{B})$ defined over the algebra of all Borel sets $\mathcal{B}$ in the domain G. By Lebesgue's theorem on the differentiability of nonnegative subadditive locally finite set functions, there exists a finite limit

$$\varphi(x) = \lim_{\varepsilon\to 0} \frac{\Phi(B(x,\varepsilon))}{\Omega_n \varepsilon^n} \quad \text{for a.e.} \quad x \in G,$$

where $B(x,\varepsilon)$ denotes the ball in $\mathbb{R}^n$ centered at $x \in G$ with radius $\varepsilon > 0$ (see, e.g., [**29**, III.2.4] or [**40**, 23.5]). The quantity $\varphi(x)$ is called the *volume derivative* of f at the point x.

THEOREM 7.1. *Let $f : G \to G^*$ be a Q-homeomorphism with respect to p-module with $Q \in L^1_{\text{loc}}(G)$ and $x \in G$. If $1 < p < n$, then*

$$(7.1) \qquad \varphi(x) \le C_1\, Q^{\frac{n}{n-p}}(x) \quad \text{a.e.},$$

If $n < p < \infty$ and $Q(x) > 0$ a.e., then

$$(7.2) \qquad \varphi(x) \ge C_2\, Q^{\frac{n}{n-p}}(x) \quad \text{a.e.},$$

where the constants C_1 and C_2 depend only on n and p.

PROOF. Given $x \in G$, consider the spherical ring $A_\varepsilon = \{y : \varepsilon < |x-y| < 2\varepsilon\}$, choosing $\varepsilon > 0$ such that $A_\varepsilon \subset G$. Since $\left(f(B(x,2\varepsilon)), \overline{f(B(x,\varepsilon))}\right)$ is a ring-like condenser in G^*, it satisfies by (6.1) the relation

$$\operatorname{cap}_p \left(f(B(x,2\varepsilon)), \overline{f(B(x,\varepsilon))}\right) = \mathcal{M}_p\left(\triangle(\partial f(B(x,2\varepsilon)), \partial f(B(x,\varepsilon)); f(A_\varepsilon))\right).$$

The homeomorphy of f yields

$$\triangle\left(\partial f(B(x,2\varepsilon)), \partial f(B(x,\varepsilon)); f(A_\varepsilon)\right) = f\left(\triangle\left(\partial B(x,2\varepsilon), \partial B(x,\varepsilon); A_\varepsilon\right)\right).$$

Thus, since f is a Q-homeomorphism with respect to p-module, we have

$$\operatorname{cap}_p \left(f(B(x,2\varepsilon)), \overline{f(B(x,\varepsilon))}\right) \le \int_G Q(x)\, \rho^p(x)\, dm(x)$$

for every admissible function ρ for the curve family $\triangle(\partial B(x,2\varepsilon), \partial B(x,\varepsilon); A_\varepsilon)$. Obviously, the function

$$\rho(x) = \begin{cases} \frac{1}{\varepsilon}, & \text{if } x \in A_\varepsilon, \\ 0, & \text{if } x \in G \setminus A_\varepsilon, \end{cases}$$

is admissible for the family and, hence,

$$\operatorname{cap}_p\left(f(B(x,2\varepsilon)),\overline{f(B(x,\varepsilon))}\right) \leq \frac{1}{\varepsilon^p}\int\limits_{B(x,2\varepsilon)} Q(y)\,dm(y). \tag{7.3}$$

Suppose that $p > n$. Then inequality (6.2) can be rewritten as

$$\operatorname{cap}_p(D,C) \geq n\Omega_n^{\frac{p}{n}}\left(\frac{p-n}{p-1}\right)^{p-1}(mD)^{\frac{n-p}{n}}. \tag{7.4}$$

Let E_1 be a set of all points $x \in G$ at which f has the total differential and

$$0 < \lim_{\varepsilon\to 0}\frac{\Phi\left(B(x,\varepsilon)\right)}{mB(x,\varepsilon)} < \infty\,,$$

where Φ is defined in this section. Denote by E_2 a set of points $x \in G$ such that

$$0 < Q(x) = \lim_{\varepsilon\to 0}\frac{1}{mB(x,\varepsilon)}\int\limits_{B(x,\varepsilon)} Q(y)\,dm(y) < \infty.$$

Note that by construction the set $G' = G\setminus(E_1\cup E_2)$ has zero Lebesgue measure. Pick any $x_0 \in G\setminus G'$.

Replacing ε by 2ε and combining the lower bound (7.4) for the condenser $\left(f(B(x_0,4\varepsilon)), f(\overline{B(x_0,2\varepsilon)})\right)$ with the upper estimate (7.3), one gets

$$n\Omega_n^{\frac{p}{n}}\left(\frac{p-n}{p-1}\right)^{p-1}(mf(B(x_0,4\varepsilon)))^{\frac{n-p}{n}} \leq \frac{1}{(2\varepsilon)^p}\int\limits_{B(x_0,4\varepsilon)} Q(x)\,dm(x).$$

This inequality is equivalent to

$$1 \leq C_{n,p}\left(\frac{mf(B(x_0,4\varepsilon))}{mB(x_0,4\varepsilon)}\right)^{\frac{p-n}{n}}\frac{\int\limits_{B(x_0,4\varepsilon)} Q(x)\,dm(x)}{mB(x_0,4\varepsilon)}.$$

Hence,

$$\varphi(x_0) = \lim_{\varepsilon\to 0}\frac{mf(B(x_0,4\varepsilon))}{mB(x_0,4\varepsilon)} \geq C_2 Q^{\frac{n}{n-p}}(x)$$

a.e. in G; here C_2 is a constant depending on n and p.

Let now $1 < p < n$. We rewrite the inequality (6.2) for the condenser (D,C) in the form

$$\operatorname{cap}_p(D,C) \geq n\Omega_n^{\frac{p}{n}}\left(\frac{n-p}{p-1}\right)^{p-1}(mC)^{\frac{n-p}{n}},$$

and, arguing similar to above, we get

$$n\Omega_n^{\frac{p}{n}}\left(\frac{n-p}{p-1}\right)^{p-1}(mf(B(x_0,2\varepsilon)))^{\frac{n-p}{n}} \leq \frac{1}{(2\varepsilon)^p}\int\limits_{B(x_0,4\varepsilon)} Q(x)\,dm(x).$$

Letting again $\limsup$ as $\varepsilon\to 0$ in the inequality

$$\left(\frac{mf(B(x_0,2\varepsilon))}{mB(x_0,2\varepsilon)}\right)^{\frac{n-p}{n}} \leq C_1\frac{\int\limits_{B(x_0,4\varepsilon)} Q(x)\,dm(x)}{mB(x_0,4\varepsilon)},$$

we obtain the upper bound (7.1) for the volume derivative stated in the assertion of the theorem. Here C_1 also is a constant depending only on n and p. □

Since all Q-homeomorphisms with respect to p-module are differentiable a.e. when $p > n-1$ (see Theorem 3.2), the volume derivative $\varphi(x)$ coincides with Jacobian a.e. This implies

COROLLARY 7.1. *Let $f : G \to G^*$ be a Q-homeomorphism with respect to p-module, $p > n-1$, and $Q \in L^1_{\mathrm{loc}}(G)$. Then*

$$J(x,f) \leq C_1\, Q^{\frac{n}{n-p}}(x) \quad \text{a.e. if} \quad n-1 < p < n,$$

and if $Q(x) > 0$ a.e.

$$J(x,f) \geq C_2\, Q^{\frac{n}{n-p}}(x) \quad \text{a.e. if} \quad p > n.$$

Here C_1 and C_2 depend only on n and p.

For $1 < p \leq n-1$ the estimate of Jacobian requires certain differential properties of mappings.

COROLLARY 7.2. *Let $f : G \to G^*$ be a Q-homeomorphism with respect to p-module, $1 < p \leq n-1$, and $Q \in L^1_{\mathrm{loc}}(G)$. If f is ACL and differentiable a.e. in G, then*

$$J(x,f) \leq C_1\, Q^{\frac{n}{n-p}}(x) \quad \text{a.e.},$$

where C_1 depends only on n and p.

THEOREM 7.2. Let $f : G \to G^*$ be a Q-homeomorphism with respect to p-module, $p > n-1$, and $Q \in L^1_{\mathrm{loc}}(G)$. Then

$$K_{O,q}(x,f) \leq C_{n,p}\ [Q(x)]^{q-1} \quad \text{a.e.}$$

where $q = p/(p-n+1)$ and the constant $C_{n,p}$ depends only on n and p.

PROOF. Arguing as in the beginning of the proof of Theorem 7.1, we arrive at the inequality (7.3). Applying the lower bound (6.3) to this p-capacity, one gets

$$\text{(7.5)} \qquad \mathrm{cap}_p\left(f(B(x,2\varepsilon)), \overline{f(B(x,\varepsilon))}\right) \geq \left(\gamma_{n,p}\frac{d^p(f(B(x,\varepsilon)))}{m^{1-n+p}f(B(x,2\varepsilon))}\right)^{\frac{1}{n-1}},$$

where $C_{n,p}$ is a constant depending only on n and p, $d(A)$ and mA denote the diameter and the Lebesgue measure of a set A in $\mathbb{R}^n$, respectively.

Combining (7.3) and (7.5) yields

$$\left(\frac{d(f(B(x,\varepsilon)))}{\varepsilon}\right)^q \frac{mB(x,2\varepsilon)}{mf(B(x,2\varepsilon))} \leq C_{n,p}\left(\frac{1}{mB(x,2\varepsilon)}\int\limits_{B(x,2\varepsilon)} Q(y)\,dm(y)\right)^{q-1},$$

where $q = p/(p-n+1)$ and $C_{n,p}$ is a constant.

Now letting $\varepsilon \to 0$, and using the differentiability of f a.e., which follows from Theorem 3.2, we derive the desired estimate for the outer dilatation $K_{O,q}(x,f)$. □

The following estimate for the outer dilatation K_O is a consequence of Theorem 7.2.

COROLLARY 7.3. *Let G and G^* be domains in $\mathbb{R}^n$, $n \geq 2$, and $f : G \to G^*$ be a Q-homeomorphism with respect to p-module with $Q \in L^1_{\mathrm{loc}}$, $p \geq n$. Then*

$$\text{(7.6)} \qquad K_O(x,f) \leq C_{n,p}\left(Q^n(x)\, J^{p-n}(x,f)\right)^{\frac{n-1}{p}} \quad \text{a.e.},$$

where the constant $C_{n,p}$ depends only on n and p.

Indeed, for all $q \geq 1$,

$$K_O(x,f) = K_{O,q}^{\frac{n}{q}}(x,f)\, J^{\frac{n-q}{q}}(x,f).$$

Replacing q by $p/(p-n+1)$ and applying the assertion of Theorem 7.2, one arrives at the desired estimate (7.6).

The homeomorphism f_2 given by (2.2) shows that for $1 < p \leq n-1$ the p-outer dilatation cannot be bounded from above. Indeed, following Theorem 4.1 from [**10**], one can replace the majorant $Q(x)$ by the p-inner dilatation $K_{I,p}(x, f_2) \equiv 1$ in D. Then, assuming the boundedness of $K_{O,p}(x, f_2)$ from above by some absolute constant $\gamma_{n,p}$ depending only on n and p, we can choose a bounded cylinder $D_{1,d} \subset D$ (taking $1 < x_n < d$ such that $d^{(n-p)p} > \gamma_{n,p}$). This contradicts to the boundedness of $K_{O,p}$ from above.

THEOREM 7.3. Let G and G^* be domains in $\mathbb{R}^n$, $n \geq 2$, and $f : G \to G^*$ be a Q-homeomorphism with respect to p-module and $Q \in L^1_{\mathrm{loc}}$, $p \geq n$. Then f has Lusin's (N^{-1})-property and $J(x,f) \neq 0$ a.e.

PROOF. By Corollary 7.3, we have

$$\begin{aligned}\int\limits_C K_O^{\frac{1}{n-1}}(x,f)\,dm(x) &\leq \gamma_{n,p} \int\limits_C Q^{\frac{n}{p}}(x) J(x,f)^{\frac{p-n}{p}}\,dm(x)\\ &\leq \gamma_{n,p}\left(\int\limits_C Q(x)\,dm(x)\right)^{\frac{p}{n}}\left(\int\limits_C J(x,f)\,dm(x)\right)^{\frac{p-n}{n}} < \infty\end{aligned}$$

for each compact set C in G. Since $f \in W^{1,1}_{\mathrm{loc}}(G)$ and f is a homeomorphism of finite distortion satisfying $K_O \in L^{\frac{1}{n-1}}_{\mathrm{loc}}(G)$, one can concludes by [**17**, Theorem 5.14] that f has Lusin's (N^{-1})-property, i.e. $mE = 0$ whenever $mf(E) = 0$. Due to [**28**], it is equivalent to the condition $J(x,f) \neq 0$ a.e. □

COROLLARY 7.4. *Let $f : G \to G^*$ be a Q-homeomorphism with respect to p-module with $Q(x) > 0$ a.e. and $Q \in L^{\frac{n}{n-p}}_{\mathrm{loc}}$, $p > n$. Then f satisfies (N^{-1})-property, and moreover, for any measurable set $E \subset G$*

$$mf(E) \geq \gamma_{n,p} \int\limits_E Q^{\frac{n}{n-p}}(x)\,dm(x),$$

where $\gamma_{n,p}$ depends only on n and p.

PROOF. Let E be an arbitrary measurable set in $\mathbb{R}^n$, $n \geq 2$. Since $Q \in L^1_{\mathrm{loc}}(G)$ and $p > n$, by Theorem 3.2, $f \in W^{1,1}_{\mathrm{loc}}$. Thus, applying [**17**, Corollary A36] and (7.2), we have

$$mf(E) \geq \int\limits_E J(x,f)\,dm(x) \geq \gamma_{n,p} \int\limits_E Q^{\frac{n}{n-p}}(x)\,dm(x),$$

where $\gamma_{n,p}$ depends only on n and p. □

8. Strengthened differential properties for Q-homeomorphisms

It is well-known that for every quasiconformal mapping there is $p > n$ such that $\|f'(x)\| \in L^p_{\mathrm{loc}}$, i.e. $f \in W^{1,p}_{\mathrm{loc}}$.

We now establish that Q-homeomorphisms with respect to p-module, $1 < p \leq n-1$, belong to the Sobolev class $W^{1,\alpha}_{\mathrm{loc}}$ if the majorant Q and p-outer dilatation are locally integrable with the appropriate degrees.

THEOREM 8.1. Let $f : G \to G^*$ be an ACL Q-homeomorphism with respect to p-module with $Q \in L^1_{\mathrm{loc}}(G)$, $1 < p \leq n-1$, $n \geq 3$, and differentiable with nonvanishing Jacobian a.e. in G. Suppose that α is such that $1 < \alpha < p(n-p)/n$ and $K_{O,p} \in L^{\beta}_{\mathrm{loc}}(G)$ with $\beta = \alpha(n-p)/(p(n-p)-\alpha n)$. Then $f \in W^{1,\alpha}_{\mathrm{loc}}(G)$.

PROOF. Let $C \subset G$ be an arbitrary compact set. By the definition of the p-outer dilatation

$$\int\limits_C \|f'(x)\|^{\alpha}\, dm(x) = \int\limits_C (\|f'(x)\|^p)^{\frac{\alpha}{p}}\, dm(x) = \int\limits_C (K_{O,p}(x,f)J(x,f))^{\frac{\alpha}{p}}\, dm(x).$$

Applying the upper bound (7.1) and Hölder's inequality, one obtains

$$\int\limits_C \|f'(x)\|^{\alpha}\, dm(x) \leq \left(\int\limits_C K^{\beta}_{O,p}(x,f)\, dm(x)\right)^{\frac{p(n-p)-\alpha n}{p(n-p)}} \left(\int\limits_C Q(x)\, dm(x)\right)^{\frac{\alpha n}{p(n-p)}},$$

which provides that $f \in W^{1,\alpha}_{\mathrm{loc}}(G)$, whenever Q and $K_{O,p}$ are locally integrable with the degrees 1 and $\beta = \alpha(n-p)/(p(n-p)-\alpha n)$, respectively. □

Observe that under restriction $1 < p \leq n-1$, the upper bound $p(n-p)/n$ for α ranges between $1/n$ and $n-2+1/n$ and cannot arrive at $n-1$ (note that any homeomorphisms $f \in W^{1,n-1}_{\mathrm{loc}}$ are differentiable a.e.; see e.g. [**17**]).

Assuming the local integrability of the majorant Q and p-inner dilatation (with the certain degrees), we show that the inverse mapping of the Sobolev class $W^{1,1}$ admits better differential properties (belongs to an appropriate Sobolev class $W^{1,\alpha}_{\mathrm{loc}}$ with $\alpha > n$).

THEOREM 8.2. Let $f : G \to G^*$ be a Q-homeomorphism with respect to p-module with $Q \in L^1_{\mathrm{loc}}(G)$, $n < p < \infty$, and $f^{-1} \in W^{1,1}_{\mathrm{loc}}(G^*)$. Suppose that $p < \alpha < p + (p-n)/n$ is such that $K_{I,p} \in L^{\gamma}_{\mathrm{loc}}(G)$ with $\gamma = \alpha(n-p)/(p^2 - \alpha n)$. Then $f^{-1} \in W^{1,\alpha}_{\mathrm{loc}}(G^*)$, and therefore f^{-1} is differentiable a.e. in G^*.

PROOF. Let $C \subset G^*$ be an arbitrary compact set. By the definition of p-inner dilatation, Lusin's (N^{-1})-property and the relationship between the minimal and maximal stretchings and Jacobians in the image and inverse image, we have

$$\int\limits_{C} \|f^{-1'}(y)\|^{\alpha}\, dm(y) = \int\limits_{C} \frac{\|f^{-1'}(y)\|^{\alpha}}{J(y,f^{-1})}\, J(y,f^{-1})\, dm(y)$$

$$= \int\limits_{f^{-1}(C)} \frac{\|f^{-1'}(f(x))\|^{\alpha}}{J(f(x),f^{-1})}\, dm(x) = \int\limits_{f^{-1}(C)} \frac{J(x,f)}{l^{\alpha}(f'(x))}\, dm(x)$$

$$= \int\limits_{f^{-1}(C)} K_{I,p}^{\frac{\alpha}{p}}(x,f)\, J^{\frac{p-\alpha}{p}}(x,f)\, dm(x),$$

see [**17**, Lemma A.33]. Taking into account the range of the parameter p, one can apply the lower bound (7.2). Then by Hölder's inequality,

$$\int\limits_{C} \|f^{-1'}(y)\|^{\alpha}\, dm(y) \leq \int\limits_{f^{-1}(C)} K_{I,p}^{\frac{\alpha}{p}}(x,f)\, Q^{\frac{n(\alpha-p)}{p(p-n)}}(x)\, dm(x)$$

$$\leq \left(\int\limits_{f^{-1}(C)} K_{I,p}^{\gamma}(x,f)\, dm(x) \right)^{\frac{\alpha}{p\gamma}} \left(\int\limits_{f^{-1}(C)} Q(x)\, dm(x) \right)^{\frac{n(\alpha-p)}{p(p-n)}},$$

which provides that $f^{-1} \in W^{1,\alpha}_{\rm loc}(G^*)$, because Q and $K_{I,p}$ are locally integrable with the degrees 1 and $\gamma = \alpha(n-p)/(p^2 - \alpha n)$, respectively. Since $\alpha > p > n$, this implies the differentiability of f^{-1} a.e. in G^*; see, e.g. [**17**, Corollary 2.25]. $\square$

9. (Q_1, Q_2)-homeomorphisms

In this section, we merge Q-homeomorphisms with respect p-moduli in the image and preimage and describe the features for such mappings.

Let $(Q_1(x), Q_2(y))$ be a pair of measurable functions $Q_1 : G \to [0,\infty]$ and $Q_2 : G^* \to [0,\infty]$. We say that $f : G \to \mathbb{R}^n$, $n \geq 2$, $f(G) = G^*$, is (Q_1,Q_2)-*homeomorphism with respect* (p,q)-*moduli* if

$$\mathcal{M}_p(f(\Gamma)) \leq \int\limits_{G} Q_1(x)\rho^p(x) dm(x)$$

and

$$\mathcal{M}_q(\Gamma) \leq \int\limits_{G^*} Q_2(y)\rho_*^q(y) dm(y)$$

for every family Γ of paths in G and all $\rho \in \operatorname{adm}\Gamma$ and $\rho_* \in \operatorname{adm}\ f(\Gamma)$.

All assertions of the following theorem follow from Theorems 3.2 and 7.3 and from the above definition.

THEOREM 9.1. Let $f : G \to G^*$ be an (Q_1, Q_2)-homeomorphism with respect (p,q)-moduli ($p \geq n$, $q \geq n$) and $Q_1 \in L^1_{\rm loc}(G)$, $Q_2 \in L^1_{\rm loc}(G^*)$. Then

1) f and f^{-1} are ACL-homeomorphisms;
2) $f \in W^{1,1}_{\rm loc}(G)$ and $f^{-1} \in W^{1,1}_{\rm loc}(G^*)$;
3) f and f^{-1} are differentiable with the Jacobians $J(x,f) \neq 0$ and $J(y,f^{-1}) \neq 0$ a.e. in G and G^*, respectively;
4) f and f^{-1} has the Lusin (N)-property;

5) $K_{I,p}(x,f) \le Q_1(x)$ and $K_{I,q}(y,f) \le Q_2(y)$ a.e. in G and G^*, respectively;
6) the equality

$$\int\limits_E g(f(x))|J(x,f)|\,dm(x) = \int\limits_{\mathbb{R}^n} g(y)\,dm(y)$$

holds for every measurable function $g : \mathbb{R}^n \to [0,\infty)$ and every measurable set $E \subset G$.

THEOREM 9.2. Let G and G^* be domains in $\mathbb{R}^n$, $n \ge 2$, and $f : G \to G^*$ be a (Q_1, Q_2)-homeomorphism with respect (p,q)-moduli $(p \ge n,\ q \ge n)$. Suppose that Q_1 and Q_2 are locally integrable in G and G^*, respectively. Then $f \in W^{1,q}_{\rm loc}(G)$ and $g = f^{-1} \in W^{1,p}_{\rm loc}(f(G))$. Moreover, for any compact set $C \subset G$,

$$\int\limits_C \|f'(x)\|^q\,dm(x) \le \int\limits_{f(C)} Q_2(y)\,dm(y)\,,$$

and

$$\int\limits_{f(C)} \|g'(y)\|^p\,dm(y) \le \int\limits_C Q_1(x)\,dm(x)\,. \tag{9.1}$$

PROOF. Let C be a compact set in G. Since $Q_1(x) \in L^1_{\rm loc}(G)$ and $Q_2(y) \in L^1_{\rm loc}(G^*)$, one concludes from the previous theorem that f is differentiable a.e. with $J(x,f) \ne 0$ and the mapping f^{-1} is also differentiable a.e. with $J\left(x,f^{-1}\right) \ne 0$. Then

$$f'(x) = \left({f^{-1}}'(f(x))\right)^{-1}$$

in G, and in addition the change variable formula is true. Thus

$$\begin{aligned}\int\limits_C \|f'(x)\|^q\,dm(x) &= \int\limits_C \|\left({f^{-1}}'(f(x))\right)^{-1}\|^q\,dm(x)\\ &= \int\limits_C \frac{1}{l^q\left({f^{-1}}'(f(x))\right)}\,dm(x) = \int\limits_{f(C)} \frac{|J\left(y,f^{-1}\right)|}{l^q\left({f^{-1}}'(y)\right)}\,dm(y)\\ &= \int\limits_{f(C)} K_{I,q}\left(y,f^{-1}\right)\,dm(y) \le \int\limits_{f(C)} Q_2(y)\,dm(y) < \infty,\end{aligned}$$

and, hence, $f \in W^{1,q}_{\rm loc}$, because $f \in ACL$, see e.g. [**26**, Theorem 2, 1.1.3]

Assuming $Q_1(x) \in L^1_{\rm loc}(G)$, we similarly obtain that $f^{-1} \in W^{1,p}_{\rm loc}$. The validity of (9.1) is proven in a similar way. □

References

[1] Cabiria Andreian Cazacu, *Some formulae on the extremal length in n-dimensional case*, Proceedings of the Romanian-Finnish Seminar on Teichmüller Spaces and Quasiconformal Mappings (Braşov, 1969), Publ. House of the Acad. of the Socialist Republic of Romania, Bucharest, 1971, pp. 87–102. (errata insert). MR0301194 (46 #352)

[2] Cabiria Andreian Cazacu, *Module inequalities for quasiregular mappings*, Ann. Acad. Sci. Fenn. Ser. A I Math. **2** (1976), 17–28. MR0470213 (57 #9971)

[3] Christopher J. Bishop, Vladimir Ya. Gutlyanskiĭ, Olli Martio, and Matti Vuorinen, *On conformal dilatation in space*, Int. J. Math. Math. Sci. **22** (2003), 1397–1420, DOI 10.1155/S0161171203110034. MR1980177 (2004c:30038)

[4] Mihai Cristea, *On generalized quasiconformal mappings*, Complex Var. Elliptic Equ. **59** (2014), no. 2, 232–246, DOI 10.1080/17476933.2012.731398. MR3170756

[5] Mihai Cristea, *Local homeomorphisms satisfying generalized modular inequalities*, Complex Var. Elliptic Equ. **59** (2014), no. 10, 1363–1387, DOI 10.1080/17476933.2013.845176. MR3219683

[6] F. W. Gehring, *Lipschitz mappings and the p-capacity of rings in n-space*, Advances in the theory of Riemann surfaces (Proc. Conf., Stony Brook, N.Y., 1969), Princeton Univ. Press, Princeton, N.J., 1971, pp. 175–193. Ann. of Math. Studies, No. 66. MR0310238 (46 #9340)

[7] Anatoly Golberg, *Homeomorphisms with finite mean dilatations*, Complex analysis and dynamical systems II, Contemp. Math., vol. 382, Amer. Math. Soc., Providence, RI, 2005, pp. 177–186, DOI 10.1090/conm/382/07057. MR2175886 (2007b:30020)

[8] Anatoly Golberg, *Differential properties of (α, Q)-homeomorphisms*, Further progress in analysis, World Sci. Publ., Hackensack, NJ, 2009, pp. 218–228, DOI 10.1142/9789812837332_0015. MR2581624 (2011b:30055)

[9] A. Golberg, *Spatial quasiconformal mappings and directional dilatations*, Progress in analysis and its applications, World Sci. Publ., Hackensack, NJ, 2010, pp. 45–52, DOI 10.1142/9789814313179_0006. MR2757987 (2012g:30067)

[10] Anatoly Golberg, *Homeomorphisms with integrally restricted moduli*, Complex analysis and dynamical systems IV. Part 1, Contemp. Math., vol. 553, Amer. Math. Soc., Providence, RI, 2011, pp. 83–98, DOI 10.1090/conm/553/10934. MR2868590

[11] Anatoly Golberg and Ruslan Salimov, *Topological mappings of integrally bounded p-moduli*, Ann. Univ. Buchar. Math. Ser. **3(LXI)** (2012), no. 1, 49–66. MR3034962

[12] Anatoly Golberg and Ruslan Salimov, *Logarithmic Hölder continuity of ring homeomorphisms with controlled p-module*, Complex Var. Elliptic Equ. **59** (2014), no. 1, 91–98, DOI 10.1080/17476933.2013.823164. MR3170745

[13] A. Golberg and R. Salimov, *Homeomorphisms Lipschizian in the mean.* Complex Analysis and Potentioal Theory with Applications, Camb. Sci. Publ., Cambridge, 2014, 95–111.

[14] Anatoly Golberg and Ruslan Salimov, *Extension of the Schwarz lemma to homeomorphisms with controlled p-module*, Georgian Math. J. **21** (2014), no. 3, 273–279. MR3259071

[15] V. M. Gol′dshteĭn and Yu. G. Reshetnyak, *Quasiconformal mappings and Sobolev spaces*, Mathematics and its Applications (Soviet Series), vol. 54, Kluwer Academic Publishers Group, Dordrecht, 1990. Translated and revised from the 1983 Russian original; Translated by O. Korneeva. MR1136035 (92h:46040)

[16] Vladimir Ya. Gutlyanskiĭ and Anatoly Golberg, *On Lipschitz continuity of quasiconformal mappings in space*, J. Anal. Math. **109** (2009), 233–251, DOI 10.1007/s11854-009-0032-1. MR2585395 (2011f:30045)

[17] Stanislav Hencl and Pekka Koskela, *Lectures on mappings of finite distortion*, Lecture Notes in Mathematics, vol. 2096, Springer, Cham, 2014. MR3184742

[18] Tadeusz Iwaniec and Gaven Martin, *Geometric function theory and non-linear analysis*, Oxford Mathematical Monographs, The Clarendon Press, Oxford University Press, New York, 2001. MR1859913 (2003c:30001)

[19] Denis Kovtonyk and Vladimir Ryazanov, *On the theory of mappings with finite area distortion*, J. Anal. Math. **104** (2008), 291–306, DOI 10.1007/s11854-008-0025-5. MR2403438 (2009e:30042)

[20] V. I. Kruglikov, *Capacities of condensors and quasiconformal in the mean mappings in space* (Russian), Mat. Sb. (N.S.) **130(172)** (1986), no. 2, 185–206, 284. MR854971 (88d:30028)

[21] S. L. Kruškal′, *On mean quasiconformal mappings* (Russian), Dokl. Akad. Nauk SSSR **157** (1964), 517–519. MR0176067 (31 #342)

[22] Samuil L. Krushkal′, *Quasiconformal mappings and Riemann surfaces*, V. H. Winston & Sons, Washington, D.C.; John Wiley & Sons, New York-Toronto, Ont.-London, 1979. Edited by Irvin Kra [Irwin Kra]; Translated from the Russian; A Halsted Press Book; Scripta Series in Mathematics; With a foreword by Lipman Bers. MR536488 (80j:30025)

[23] O. Martio, S. Rickman, and J. Väisälä, *Definitions for quasiregular mappings*, Ann. Acad. Sci. Fenn. Ser. A I No. **448** (1969), 40. MR0259114 (41 #3756)

[24] O. Martio, V. Ryazanov, U. Srebro, and E. Yakubov, *Q-homeomorphisms*, Complex analysis and dynamical systems, Contemp. Math., vol. 364, Amer. Math. Soc., Providence, RI, 2004, pp. 193–203, DOI 10.1090/conm/364/06685. MR2099027 (2005h:30046)

[25] Olli Martio, Vladimir Ryazanov, Uri Srebro, and Eduard Yakubov, *Moduli in modern mapping theory*, Springer Monographs in Mathematics, Springer, New York, 2009. MR2466579 (2012g:30004)

[26] Vladimir G. Maz'ja, *Sobolev spaces*, Springer Series in Soviet Mathematics, Springer-Verlag, Berlin, 1985. Translated from the Russian by T. O. Shaposhnikova. MR817985 (87g:46056)

[27] Vladimir Maz′ya, *Lectures on isoperimetric and isocapacitary inequalities in the theory of Sobolev spaces*, Heat kernels and analysis on manifolds, graphs, and metric spaces (Paris, 2002), Contemp. Math., vol. 338, Amer. Math. Soc., Providence, RI, 2003, pp. 307–340, DOI 10.1090/conm/338/06078. MR2039959 (2005e:26031)

[28] S. P. Ponomarev, *The N^{-1}-property of mappings, and Luzin's (N) condition* (Russian, with Russian summary), Mat. Zametki **58** (1995), no. 3, 411–418, 480, DOI 10.1007/BF02304773; English transl., Math. Notes **58** (1995), no. 3-4, 960–965 (1996). MR1368549 (97c:26017)

[29] T. Rado and P. V. Reichelderfer, *Continuous transformations in analysis. With an introduction to algebraic topology*, Die Grundlehren der mathematischen Wissenschaften in Einzeldarstellungen mit besonderer Berücksichtigung der Anwendungsgebiete, Bd. LXXV, Springer-Verlag, Berlin-Göttingen-Heidelberg, 1955. MR0079620 (18,115c)

[30] Yu. G. Reshetnyak, *Space mappings with bounded distortion*, Translations of Mathematical Monographs, vol. 73, American Mathematical Society, Providence, RI, 1989. Translated from the Russian by H. H. McFaden. MR994644 (90d:30067)

[31] Yu. G. Reshetnyak, *Stability theorems in geometry and analysis*, Mathematics and its Applications, vol. 304, Kluwer Academic Publishers Group, Dordrecht, 1994. Translated from the 1982 Russian original by N. S. Dairbekov and V. N. Dyatlov, and revised by the author; Translation edited and with a foreword by S. S. Kutateladze. MR1326375 (96i:30016)

[32] Ruslan Salimov, ACL *and differentiability of Q-homeomorphisms*, Ann. Acad. Sci. Fenn. Math. **33** (2008), no. 1, 295–301. MR2386853 (2009b:30040)

[33] R. R. Salimov, *On finitely Lipschitz space mappings*, Sib. Èlektron. Mat. Izv. **8** (2011), 284–295. MR2876547 (2012m:30036)

[34] Ruslan Salimov, *On Q-homeomorphisms with respect to p-modulus*, Ann. Univ. Buchar. Math. Ser. **2(LX)** (2011), no. 2, 207–213. MR3034384

[35] R. R. Salimov, *Estimation of the measure of the image of the ball* (Russian, with Russian summary), Sibirsk. Mat. Zh. **53** (2012), no. 4, 920–930, DOI 10.1134/S0037446612040155; English transl., Sib. Math. J. **53** (2012), no. 4, 739–747. MR3013536

[36] R. R. Salimov and E. A. Sevost′yanov, *Theory of ring Q-mappings and geometric function theory* (Russian, with Russian summary), Mat. Sb. **201** (2010), no. 6, 131–158, DOI 10.1070/SM2010v201n06ABEH004096; English transl., Sb. Math. **201** (2010), no. 5-6, 909–934. MR2682368 (2011i:30026)

[37] B. V. Šabat, *The modulus method in space*, Soviet Math. Dokl. **1** (1960), 165–168. MR0120346 (22 #11101)

[38] V. A. Shlyk, *On the equality between p-capacity and p-modulus* (Russian, with English and Russian summaries), Sibirsk. Mat. Zh. **34** (1993), no. 6, 216–221, v, x, DOI 10.1007/BF00973485; English transl., Siberian Math. J. **34** (1993), no. 6, 1196–1200. MR1268174 (95d:31003)

[39] P. M. Tamrazov, *Moduli and extremal metrics in nonorientable and twisted Riemannian manifolds* (Russian, with English and Ukrainian summaries), Ukraïn. Mat. Zh. **50** (1998), no. 10, 1388–1398, DOI 10.1007/BF02513490; English transl., Ukrainian Math. J. **50** (1998), no. 10, 1586–1598 (1999). MR1711225 (2000e:53039)

[40] Jussi Väisälä, *Lectures on n-dimensional quasiconformal mappings*, Lecture Notes in Mathematics, Vol. 229, Springer-Verlag, Berlin-New York, 1971. MR0454009 (56 #12260)

Department of Applied Mathematics, Holon Institute of Technology, 52 Golomb St., P.O.B. 305, 5810201 Holon, Israel
E-mail address: golberga@hit.ac.il

Institute of Mathematics, National Academy of Sciences of Ukraine, 3 Tereschenkivska St., Kiev-4 01601, Ukraine
E-mail address: ruslan623@yandex.ru

Contemporary Mathematics
Volume **659**, 2016
http://dx.doi.org/10.1090/conm/659/13193

Finite-parameters feedback control for stabilizing damped nonlinear wave equations

Varga K. Kalantarov and Edriss S. Titi

ABSTRACT. In this paper we introduce a finite-parameters feedback control algorithm for stabilizing solutions of various classes of damped nonlinear wave equations. Specifically, stabilization the zero steady state solution of initial boundary value problems for nonlinear weakly and strongly damped wave equations, nonlinear wave equation with nonlinear damping term and some related nonlinear wave equations, introducing a feedback control terms that employ parameters, such as, finitely many Fourier modes, finitely many volume elements and finitely many nodal observables and controllers. In addition, we also establish the stabilization of the zero steady state solution to initial boundary value problem for the damped nonlinear wave equation with a controller acting in a proper subdomain. Notably, the feedback controllers proposed here can be equally applied for stabilizing other solutions of the underlying equations.

1. Introduction

This paper is devoted to the study of finite-parameters feedback control for stabilizing solutions of initial boundary value problems for nonlinear damped wave equations. Feedback control, stabilization and control of wave equations is a well established area of control theory. Many interesting results were obtained, in the last dacades, on stabilization of linear and nonlinear wave equations (see, e.g., [**2**],[**4**], [**17**], [**20**], [**23**], [**25**], [**27**], [**28**] , [**29**] and references therein). Most of the problems considered are problems of stabilization by interior and boundary controllers involving linear or nonlinear damping terms. However, only few results are known on feedback stabilization of linear hyperbolic equation by employing finite-dimensional controllers (see, e.g., [**3**] and references therein).
We study the problem of feedback stabilization of initial boundary value problem for damped nonlinear wave equation

$$\partial_t^2 u - \Delta u + b\partial_t u - au + f(u) = -\mu w, \ x \in \Omega, t > 0, \tag{1.1}$$

2010 *Mathematics Subject Classification.* Primary 35L71; Secondary 35B40, 35L15.

The work of V. K. Kalantarov was supported in part by The Scientific and Research Council of Turkey, grant no. 112T934.

The work of E.S. Titi was supported in part by the NSF grants DMS–1109640 and DMS–1109645.

nonlinear wave equation with nonlinear damping term

$$\partial_t^2 u - \Delta u + bg(\partial_t u) - au + f(u) = -\mu w, \ x \in \Omega, t > 0, \tag{1.2}$$

and the strongly damped wave equation

$$\partial_t^2 u - \Delta u - b\Delta\partial_t u - \lambda u + f(u) = -\mu w, \ x \in \Omega, t > 0. \tag{1.3}$$

Here and in what follows μ, a, b, ν are given positive parameters, w is a feedback control input (different for different problems), $f(\cdot) : \mathbb{R} \to \mathbb{R}$ is a given continuously differentiable function such that $f(0) = 0$ and

$$f(s)s - \mathcal{F}(s) \geq 0, f'(s) \geq 0, \ \forall s \in \mathbb{R}, \ \ \mathcal{F}(s) := \int_0^s f(\tau)d\tau, \ \ \forall s \in \mathbb{R}. \tag{1.4}$$

We show that the feedback controller proposed in [**1**] for nonlinear parabolic equations can be used for stabilization of the zero steady state solution of the above mentioned wide class of nonlinear dissipative wave equations. It is worth mentioning that the same type of feedback control can be used for stabilizing any other given exact solution of the underlying equation.
We also show stabilization of the initial boundary value problem for equation (1.1), with the control input acting on some proper subdomain of Ω. It is worth mentioning that the same type of feedback control can be used for stabilizing any other given exact solution of the underlying equation.
Our study, as well as the results obtained in [**1**], are inspired by the fact that dissipative dynamical systems generated by initial boundary value problems, such as the 2D Navier - Stokes equations, nonlinear reaction-diffusion equation, Cahn-Hilliard equation, damped nonlinear Schrödinger equation, damped nonlinear Klein - Gordon equation, nonlinear strongly damped wave equation and related equations and systems have a finite-dimensional asymptotic (in time) behavior (see, e.g., [**2**], [**6**]-[**10**], [**12**]-[**14**], [**21**], [**22**], [**29**] and references therein). Specifically, the long-term behavior for such systems is determined by finitely many parameters, which are typically represented in terms of projections onto the large/coarse spatial length scales. As a result of this observation, we show that it is sufficient to stabilize/nudge these large spatial length scales in order to stabilize the full solution. This property has also been implicitly used in the numerical feedback control result [**5**] in which the authors overcome the spillover phenomenon in the Navier-Stokes equations. Motivated by this observation, we specifically show

(1) stabilization of $1D$ damped wave equation (1.1), under Neuman boundary conditions, when the feedback control input employs observables based on measurement of *finite volume elements*,

(2) stabilization of equation (1.1), under homogeneous Dirichlet boundary condition, with one feedback controller supported on some proper subdomain of $\Omega \subset \mathbb{R}^n$,

(3) stabilization of equation (1.1) and equation (1.3), under the Dirichlet boundary condition, when the feedback control involves *finitely many Fourier modes of the solution*, based on eigenfunctions of the Laplacian subject to homogeneous Dirichlet boundary condition,

(4) stabilization of the $1D$ equation (1.3), when the feedback control incorporates observables at *finitely many nodal points.*

In the sequel we will use the notations:

- $(\cdot,\cdot)$ and $\|\cdot\|$ denote the inner product and the norm of $L^2(\Omega)$, and the following inequalities:

- Young's inequality

$$ab \leq \varepsilon a^2 + \frac{\varepsilon}{4} b^2, \tag{1.5}$$

that is valid for all positive numbers a, b and ε
- the Poincaré inequality

$$\|u\|^2 \leq \lambda_1^{-1} \|\nabla u\|^2, \tag{1.6}$$

which holds for each $u \in H_0^1(\Omega)$.

2. Feedback control of damped nonlinear wave equations

In this section, we show that the initial boundary value problem for nonlinear damped wave equation can be stabilized by employing finite volume elements feedback controller, feedback controllers acting in a subdomain of Ω, or feedback controllers involving finitely many Fourier modes.

2.1. Stabilization employing finite volume elements feedback control.

We consider the following feedback control problem

$$\begin{cases} \partial_t^2 u - \nu \partial_x^2 u + b\partial_t u - \lambda u + f(u) = -\mu \sum\limits_{k=1}^{N} \overline{u}_k \chi_{J_k}(x), & x \in (0,L),\ t > 0, \\ \partial_x u(0,t) = \partial_x u(L,t) = 0,\ t > 0, \\ u(x, 0 = u_0(x), \quad \partial_t u(x,0) = u_1(x), \quad x \in (0,L). \end{cases} \tag{2.1}$$

Here $J_k := \left[(k-1)\frac{L}{N}, k\frac{L}{N}\right)$, for $k = 1, 2, \cdots N-1$ and $J_N = [\frac{N-1}{N}L, L]$, $\overline{\phi}_k := \frac{1}{|J_k|} \int\limits_{J_k} \phi(x)dx$, and $\chi_{J_k}(x)$ is the characteristic function of the interval J_k. In what follows we will need the following lemma

LEMMA 2.1. *(see [**1**]) Let $\phi \in H^1(0,L)$. Then*

$$\|\phi - \sum_{k=1}^{N} \overline{\phi}_k \chi_{J_k}(\cdot)\| \leq h \|\phi_x\|, \tag{2.2}$$

and

$$\|\phi\|^2 \leq h \sum_{k=1}^{N} \overline{\phi}_k^2 + \left(\frac{h}{2\pi}\right)^2 \|\phi_x\|^2, \tag{2.3}$$

where $h := \frac{L}{N}$.

By employing this Lemma, we proved the following theorem:

THEOREM 2.2. *Suppose that the nonlinear term $f(\cdot)$ satisfies condition* (1.4) *and that μ and N are large enough satisfying*

$$\mu \geq 2\left(\lambda + \frac{\delta_0 b}{2}\right) \text{ and } N^2 > \frac{L^2}{2\nu\pi^2}\left(\lambda + \frac{\delta_0 b}{2}\right), \tag{2.4}$$

where

$$\delta_0 = \frac{b}{2}\min\{1,\nu\}. \tag{2.5}$$

Then each solution of the problem (2.1) *satisfies the following decay estimate:*

$$\|\partial_t u(t)\|^2 + \|\partial_x u(t)\|^2 \leq K(\|u_1\|^2 + \|\partial_x u_0\|^2)e^{-\delta_0 t}, \tag{2.6}$$

where K is some positive constant depending on b, λ and L.

PROOF. First observe that one can use standard tools of the theory of nonlinear wave equations to show global existence and uniqueness of solution to problem (2.1) (see, e.g., [**24**]).
Taking the $L^2(0,L)$ inner product of (2.1) with $\partial_t u+\varepsilon u$, where $\varepsilon > 0$ is a parameter, to be determined later, gives us the following relation:

(2.7)
$$\frac{d}{dt}\left[\frac{1}{2}\|\partial_t u\|^2 + \frac{\nu}{2}\|\partial_x u\|^2 - \frac{1}{2}(\varepsilon b - \lambda)\|u\|^2 + \int_0^L \mathcal{F}(u)dx + \frac{1}{2}h\mu\sum_{k=1}^N \overline{u}_k^2 + \varepsilon(u,\partial_t u)\right]$$
$$+ (b-\varepsilon)\|\partial_t u\|^2 + \varepsilon\nu\|\partial_x u\|^2 - \varepsilon\lambda\|u\|^2 + \varepsilon(f(u),u) + \varepsilon\mu h\sum_{k=1}^N \overline{u}_k^2 = 0.$$

It follows from (2.7) that

$$\begin{aligned}\frac{d}{dt}\Phi_\varepsilon(t) + \delta\Phi_\varepsilon(t) + (b-\varepsilon)\|\partial_t u\|^2 + \varepsilon\nu\|\partial_x u\|^2 - \varepsilon\lambda\|u\|^2 + \varepsilon(f(u),u)\\ + \varepsilon\mu h\sum_{k=1}^N \overline{u}_k^2 - \frac{\delta}{2}\|\partial_t u\|^2 - \frac{\delta\nu}{2}\|\partial_x u\|^2 - \frac{\delta}{2}(\varepsilon b - \lambda)\|u\|^2 - \delta(\mathcal{F}(u),1)\\ - \frac{\delta}{2}h\mu\sum_{k=1}^N \overline{u}_k^2 - \delta\varepsilon(u,\partial_t u) = 0.\end{aligned} \tag{2.8}$$

Here

$$\Phi_\varepsilon(t) := \frac{1}{2}\|\partial_t u(t)\|^2 + \frac{\nu}{2}\|\partial_x u(t)\|^2 + \frac{1}{2}(\varepsilon b - \lambda)\|u(t)\|^2 + \int_0^L \mathcal{F}(u(x,t))dx$$
$$+ \frac{1}{2}h\mu\sum_{k=1}^N \overline{u}_k^2(t) + \varepsilon(u(t),\partial_t u(t)).$$

Due to condition (1.4) and the Cauchy-Schwarz inequality we have the following lower estimate for $\Phi_\varepsilon(t)$

$$\Phi_\varepsilon(t) \geq \frac{1}{4}\|\partial_t u\|^2 + \frac{\nu}{2}\|\partial_x u\|^2 + (\frac{b\varepsilon}{2} - \frac{\lambda}{2} - \varepsilon^2)\|u\|^2 + \frac{1}{2}h\mu\sum_{k=1}^N \overline{u}_k^2.$$

By choosing $\varepsilon \in (0, \frac{b}{2}]$ and by employing inequality (2.3), we get from the above inequality:

$$\Phi_\varepsilon(t) \geq \frac{1}{4}\|\partial_t u\|^2 + \frac{\nu}{2}\|\partial_x u\|^2 - \lambda\left[h\sum_{k=1}^{N}\overline{u}_k^2 + \left(\frac{h}{2\pi}\right)^2\|\partial_x u\|^2\right] \tag{2.9}$$

$$+\frac{1}{2}\mu h\sum_{k=1}^{N}\overline{u}_k^2 = \frac{1}{4}\|\partial_t u\|^2 + \left(\frac{\nu}{2} - \lambda\frac{L^2}{4\pi^2N^2}\right)\|\partial_x u\|^2 + h(\frac{\mu}{2} - \lambda)\sum_{k=1}^{N}\overline{u}_k^2,$$

Thanks to the condition (2.4) we also have

$$\Phi_\varepsilon(t) \geq \frac{1}{4}\|\partial_t u(t)\|^2 + d_0\|\partial_x u(t)\|^2, \tag{2.10}$$

where $d_0 = \frac{\delta_0 bL^2}{4N^2\nu\pi^2}$, and δ_0 is defined in (4.21).
Let $\delta \in (0, \varepsilon)$, to be chosen below. According to condition (1.4) $f(u)u \geq \mathcal{F}(u), \forall u \in \mathbb{R}$. Therefore (2.8) implies

$$\frac{d}{dt}\Phi_\varepsilon(t) + \delta\Phi_\varepsilon(t) + \frac{b}{2}\|\partial_t u\|^2 + (\frac{b\nu}{2} - \frac{\delta}{2})\|\partial_x u\|^2 \tag{2.11}$$

$$+\left(\frac{\delta\lambda}{2} - \frac{\delta b^2}{4} - \frac{b\lambda}{2}\right)\|u\|^2 + \mu h\left(\frac{b}{2} - \frac{\delta}{2}\right)\sum_{k=1}^{N}\overline{u}_k^2 \leq 0.$$

We choose here $\delta = \delta_0 := \frac{b}{2}\min\{\nu, 1\}$ and employ inequality (2.3) to obtain:

$$\frac{d}{dt}\Phi_\varepsilon(t) + \delta_0\Phi_\varepsilon(t) + \frac{b}{2}\|\partial_t u\|^2 + \left[\frac{b\nu}{4} - \frac{b}{2}\left(\lambda + \frac{\delta_0 b}{2}\right)\frac{L^2}{4\pi^2N^2}\right]\|\partial_x u\|^2$$

$$+\frac{bh}{2}\left[\frac{\mu}{2} - \left(\lambda + \frac{\delta_0 b}{2}\right)\right]\sum_{k=1}^{N}\overline{u}_k^2 \leq 0.$$

Finally by using condition (2.4) we get:

$$\frac{d}{dt}\Phi_\varepsilon(t) + \delta_0\Phi_\varepsilon(t) \leq 0.$$

Thus by Gronwalls inequality and thanks (2.10) we have

$$\|\partial_t u(t)\|^2 + \|\partial_x u(t)\|^2 \leq K(\|u_1\|^2 + \|\partial_x u_0\|^2)e^{-\delta_0 t}, \tag{2.12}$$

where δ_0 is defined in (4.21), and K is a positive constant, depending on b, λ and L. □

2.2. Stabilization with feedback control on a subdomain. In this section we study the problem of internal stabilization of initial boundary value problem for nonlinear damped wave equation on a bounded domain. We show that the problem can be exponentially stabilized by a feedback controller acting on a strict subdomain. So, we consider the following feedback control problem:

$$\partial_t^2 u - \Delta u + b\partial_t u - au + |u|^{p-2}u = -\mu\chi_\omega(x)u, \ x \in \Omega, t > 0, \tag{2.13}$$

$$u(x,t) = 0, \ x \in \partial\Omega, \ t > 0, \tag{2.14}$$

$$u(x,0) = u_0(x), \ \partial_t u(x,0) = u_1(x), \ x \in \Omega. \tag{2.15}$$

Here $a > 0, p \geq 2$ are given numbers, and $\mu > 0$ is a parameter to be determined, $\Omega \subset \mathbb{R}^n$ is a bounded domain with smooth boundary $\partial\Omega$, $\chi_\omega(x)$ is the characteristic function of the subdomain $\omega \subset \Omega$ with smooth boundary and $\overline{\omega} \subset \Omega$.
Let us denote by $\lambda_1(\Omega_\omega)$ the first eigenvalue of the problem

$$-\Delta v = \lambda v, \ x \in \Omega_\omega; \ v = 0, \ x \in \partial\Omega_\omega,$$

where $\Omega_\omega := \Omega \setminus \overline{\omega}$. We will need the following Lemma in the proof of the main result of this section:

LEMMA 2.3. *(see, e.g.,*[**30**]*) For each $d > 0$ there exists a number $\mu_0(d) > 0$ such that the following inequality holds true*

$$\int_\Omega \left(|\nabla v(x)|^2 + \mu\chi_\omega(x)v^2(x)\right)dx \geq (\lambda_1(\Omega_\omega) - d)\int_\Omega v^2(x)dx, \ \forall v \in H_0^1(\Omega), \tag{2.16}$$

whenever $\mu > \mu_0$.

THEOREM 2.4. *Suppose that*

$$\lambda_1(\Omega_\omega) \geq 4a + \frac{3b^2}{2}, \quad \text{and} \quad \mu > \mu_0, \tag{2.17}$$

where μ_0 is the parameter stated in Lemma 2.3 , corresponding to $d = \frac{1}{2}\lambda_1(\Omega_\omega)$. Then the energy norm of each weak solution of the problem (2.13)-(2.15) tends to zero with an exponential rate. More precisely, the following estimate holds true:

$$\|\partial_t u(t)\|^2 + \|\nabla u(t)\|^2 + \|u(t)\|^p_{L^p(\Omega)} \leq C_0 e^{-\frac{b}{2}t}, \tag{2.18}$$

where C_0 is a positive constant depending on initial data.

PROOF. Taking the $L^2(\Omega)$ inner product of (2.13) with $\partial_t u + \frac{b}{2}u$ we get

$$\frac{d}{dt}E_b(t) + \frac{b}{2}\left[\|\partial_t u\|^2 + \|\nabla u\|^2 - a\|u\|^2 + \|u\|^p_{L^p(\Omega)} + \mu\int_\Omega \chi_\omega(x)u^2(x)dx\right] = 0, \tag{2.19}$$

where

$$E_b(t) := \frac{1}{2}\|\partial_t u\|^2 + \frac{1}{2}\|\nabla u\|^2 - \frac{a}{2}\|u\|^2 + \frac{1}{p}\|u\|^p_{L^p(\Omega)} + \frac{\mu}{2}\int_\Omega \chi_\omega(x)u^2(x)dx + \frac{b}{2}(u, \partial_t u) + \frac{b^2}{4}\|u\|^2.$$

By Cauchy-Schwarz and Young inequalities we have

$$\frac{b}{2}|(u, \partial_t u)| \leq \frac{1}{4}\|\partial_t u\|^2 + \frac{b^2}{4}\|u\|^2. \tag{2.20}$$

By employing (2.20), we obtain the lower estimate for $E_b(t)$:

$$E_b(t) \geq \frac{1}{4}\|\partial_t u\|^2 + \frac{1}{4}\|\nabla u\|^2 + \frac{1}{p}\|u\|^p_{L^p(\Omega)} + \frac{1}{4}\left[\|\nabla u\|^2 + 2\mu\|u\|^2_{L^2(\omega)} - 2a\|u\|^2\right]. \tag{2.21}$$

According to Lemma 2.3

$$\|\nabla u\|^2 + \mu\|u\|^2_{L^2(\omega)} \geq \frac{1}{2}\lambda_1(\Omega_\omega)\|u\|^2, \quad \text{for every} \ \ \mu > \mu_0. \tag{2.22}$$

Thus, thanks to condition (2.17), we have

$$E_b(t) \geq \frac{1}{4}\|\partial_t u\|^2 + \frac{1}{4}\|\nabla u\|^2 + \frac{1}{p}\|u\|^p_{L^p(\Omega)}. \tag{2.23}$$

Adding to the left-hand side of (2.19) the expression $E_b(t) - \delta E_b(t)$ with some $\delta > 0$ (to be chosen below), we get

$$\frac{d}{dt}E_b(t) + \delta E_b(t) + \frac{1}{2}(b-\delta)\|\partial_t u\|^2 + \frac{1}{2}(b-\delta)\|\nabla u\|^2$$
$$+\frac{1}{2}(a\delta - ab - \frac{\delta b^2}{2})\|u\|^2 + \frac{\mu}{2}(b-\delta)\|u\|^2_{L^2(\omega)} - \frac{1}{2}\delta b(u,\partial_t u) = 0$$

We use here the inequality

$$\frac{1}{2}\delta b|(u,\partial_t u)| \le \frac{b}{4}\|\partial_t u\|^2 + \frac{b\delta^2}{4}\|u\|^2,$$

then in the resulting inequality we choose $\delta = \frac{b}{2}$, and get:

$$\frac{d}{dt}E_b(t) + \frac{b}{2}E_b(t) + \frac{b}{4}\left[\|\nabla u\|^2 + \mu\|u\|^2_{L^2(\omega)} - (2a + \frac{3}{4}b^2)\|u\|^2\right].$$

Finally, by using the condition (2.17), thanks to Lemma 2.3 we obtain

$$\frac{d}{dt}E_b(t) + \frac{b}{2}E_b(t) \le 0.$$

Integrating the last inequality and taking into account (2.23), we arrive at the desired estimate (2.18). □

2.3. Weakly damped nonlinear wave equation: Stabilization with finitely many Fourier modes. In this section we consider the feedback control problem for damped nonlinear wave equation based on finitely many Fourier modes, i.e. we consider the feedback system of the following form:

$$\partial_t^2 u - \nu\Delta u + b\partial_t u - au + |u|^{p-2}u = -\mu\sum_{k=1}^{N}(u,w_k)w_k,\ x\in\Omega, t>0, \tag{2.24}$$

$$u = 0,\ x\in\partial\Omega,\ t>0, \tag{2.25}$$

$$u(x,0 = u_0(x),\ \partial_t u(x,0) = u_1(x),\ \ x\in\Omega,\ t>0. \tag{2.26}$$

Here $\nu > 0, a > 0, b > 0, \mu > 0, p \ge 2$ are given numbers; $w_1, w_2, \dots, w_n, \dots$ is the set of orthonormal (in $L^2(\Omega)$) eigenfunctions of the Laplace operator $-\Delta$ under the homogeneous Dirichlet boundary condition, corresponding to eigenvalues $0 < \lambda_1 \le \lambda_2 \dots \le \lambda_n \le, \dots$.

THEOREM 2.5. *Suppose that μ and N are large enough such that*

$$\nu \ge (2a + 3b^2/4)\lambda_{N+1}^{-1}, \quad \textit{and} \quad \mu \ge a + 3b^2/4. \tag{2.27}$$

Then the following decay estimate holds true

$$\|\partial_t u(t)\|^2 + \|\nabla u(t)\|^2 + \int_\Omega |u(x,t)|^p dx \le E_0 e^{-\frac{b}{2}t}, \tag{2.28}$$

where

$$E_0 := \frac{1}{2}\|u_1\|^2 + \frac{\nu}{2}\|\nabla u_0\|^2 + (\frac{b^2}{4} - \frac{a}{2})\|u_0\|^2$$
$$+\frac{1}{p}\int_\Omega |u_0(x)|^p dx + \frac{\mu}{2}\sum_{k=1}^{N}(u_0,w_k)^2 + \frac{b}{2}(u_0,u_1).$$

PROOF. Multiplication of (2.24) by $\partial_t u + \frac{b}{2}u$ and integration over Ω gives

$$\frac{d}{dt}E_b(t) + \frac{b}{2}\left[\|\partial_t u\|^2 + \nu\|\nabla u\|^2 - a\|u(t)\|^2 + \int_\Omega |u|^p dx + \mu\sum_{k=1}^N (u, w_k)^2\right] = 0, \tag{2.29}$$

where

$$E_b(t) := \frac{1}{2}\|\partial_t u(t)\|^2 + \frac{\nu}{2}\|\nabla u(t)\|^2$$
$$+ (\frac{b^2}{4} - \frac{a}{2})\|u(t)\|^2 + \frac{1}{p}\int_\Omega |u(x,t)|^p dx + \frac{\mu}{2}\sum_{k=1}^N (u(t), w_k)^2 + \frac{b}{2}(u(t), \partial_t u(t)).$$

Thanks to the Cauchy-Schwarz and Young inequalities we have

$$\frac{b}{2}|(u(t), \partial_t u(t))| \leq \frac{1}{4}\|\partial_t u\|^2 + \frac{b^2}{4}\|u\|^2.$$

Consequently,

$$E_b(t) \geq \frac{1}{4}\|\partial_t u\|^2 + \frac{\nu}{2}\|\nabla u\|^2 - \frac{a}{2}\|u\|^2 + \frac{1}{p}\int_\Omega |u|^p dx + \frac{\mu}{2}\sum_{k=1}^N (u, w_k)^2.$$

Since

$$-\frac{a}{2}\|u\|^2 + \frac{\mu}{2}\sum_{k=1}^N (u, w_k)^2 = \frac{1}{2}(\mu - a)\sum_{k=1}^N (u, w_k)^2 - \frac{a}{2}\sum_{k=N+1}^\infty (u, w_k)^2,$$

by using the Poincaré -like inequality

$$\|\phi - \sum_{k=1}^N (\phi, w_k)w_k\|^2 \leq \lambda_{N+1}^{-1}\|\nabla\phi\|^2, \tag{2.30}$$

which is valid for each $\phi \in H_0^1(\Omega)$, we get

$$E_b(t) \geq \frac{1}{4}\|\partial_t u\|^2 + \frac{1}{2}\left(\nu - a\lambda_{N+1}^{-1}\right)\|\nabla u\|^2 + \frac{1}{2}(\mu - a)\sum_{k=1}^N (u, w_k)^2 + \frac{1}{p}\int_\Omega |u|^p dx. \tag{2.31}$$

Thus,

$$E_b(t) \geq \frac{1}{4}\|\partial_t u\|^2 + \frac{\nu}{4}\|\nabla u\|^2 + \frac{1}{p}\int_\Omega |u|^p dx. \tag{2.32}$$

Adding to the left-hand side of (2.29) the expression $\delta E_b(t) - \delta E_b(t)$ (with δ to be chosen later), we obtain

$$\frac{d}{dt}E_b(t) + \delta E_b(t) + \frac{1}{2}(b - \delta)\|\partial_t u\|^2 + \frac{\nu}{2}(b - \delta)\|\nabla u\|^2$$
$$+ \left(-\frac{ba}{2} + \frac{\delta a}{2} - \frac{\delta b^2}{4}\right)\|u\|^2 + \left(\frac{b}{2} - \frac{\delta}{2}\right)\|u\|_{L^p(\Omega)}^p$$
$$+ \frac{\mu}{2}(b - \delta)\sum_{k=1}^N (u, w_k)^2 - \frac{1}{2}b\delta(u, \partial_t u) = 0.$$

We choose here $\delta = \frac{b}{2}$, and obtain

$$\frac{d}{dt}E_b(t) + \delta E_b(t) + \frac{b}{4}\|\partial_t u\|^2 + \frac{\nu b}{4}\|\nabla u\|^2 - \left(\frac{ba}{4} + \frac{b^3}{8}\right)\|u\|^2 + \frac{\mu b}{4}\sum_{k=1}^{N}(u, w_k)^2 - \frac{b^2}{4}(u, \partial_t u) = 0. \tag{2.33}$$

Employing the inequality

$$\frac{b^2}{4}|(u, \partial_t u)| \leq \frac{b}{4}\|\partial_t u\|^2 + \frac{b^3}{16}\|u\|^2,$$

and inequality (2.30), we obtain from (2.34):

$$\frac{d}{dt}E_b(t) + \delta E_b(t) + \frac{b}{4}\left[\nu - (a + 3b^2/4)\lambda_{N+1}^{-1}\right]\|\nabla u\|^2 + \frac{b}{4}\left[\mu - (a + 3b^2/4)\right]\sum_{k=1}^{N}(u, w_k)^2. \tag{2.34}$$

$$\frac{d}{dt}E_b(t) + \delta E_b(t) + \left(\frac{b}{2} - \frac{\delta}{2}\right)\|\partial_t u\|^2 + \left(\frac{b}{2}(\nu - a\lambda_{N+1}^{-1}) - \frac{\delta\nu}{2}\right)\|\nabla u\|^2 + \left(\frac{b}{2}(\mu - a) - \frac{\delta\mu}{2}\right)\sum_{k=1}^{N}(u, w_k)^2 + \left(\frac{b}{2} - \frac{\delta}{p}\right)\int_\Omega |u|^p dx - \delta\left(\frac{b^2}{4} - \frac{a}{2}\right)\|u\|^2 + \frac{\delta b}{2}(u, \partial_t u) \leq 0. \tag{2.35}$$

By using the inequalities

$$\frac{\delta b}{2}|(u, \partial_t u)| \leq \frac{b}{4}\|\partial_t u\|^2 + \frac{\delta^2 b}{4}\|u\|^2,$$

and (2.30) in (2.35) we get

$$\frac{d}{dt}E_b(t) + \delta E_b(t) + \left(\frac{b}{4} - \frac{\delta}{2}\right)\|\partial_t u\|^2 + \left(\frac{b}{2}(\nu - a\lambda_{N+1}^{-1}) - \frac{\delta^2 b}{4}\lambda_{N+1}^{-1} - \frac{\delta\nu}{2}\right)\|\nabla u\|^2 + \left(\frac{b}{2}(\mu - a) - \frac{\delta\mu}{2}\right)\sum_{k=1}^{N}(u, w_k)^2 + \left(\frac{b}{2} - \frac{\delta}{p}\right)\int_\Omega |u|^p dx - \delta\left(\frac{b^2}{4} - \frac{a}{2} - \frac{\delta^2 b}{4}\right)\|u\|^2 \leq 0.$$

By choosing $\delta = \frac{b}{2}$ we obtain :

$$\frac{d}{dt}E_b(t) + \frac{b}{2}E_b(t) + \frac{b}{2}\left[\nu - (\frac{b^2}{8} + \frac{3}{2}a)\lambda_{N+1}^{-1}\right]\|\nabla u\|^2 + \frac{b}{2}\left[\frac{\mu}{2} - \frac{3}{2}a - \frac{b^2}{8}\right]\sum_{k=1}^{N}(u, w_k)^2 \leq 0.$$

Taking into account conditions (2.27) we deduce from the last inequality the inequality

$$\frac{d}{dt}E_b(t) + \frac{b}{2}E_b(t) \leq 0.$$

Integrating the last inequality we get the desired estimate (2.28) thanks to (2.32). □

REMARK 2.6. We would like to note that estimate (2.28) allows us to prove existence of a weak solution to the problem (2.24), (2.25) such that (see [**24**])

$$u \in L^\infty\left(\mathbb{R}^+; H_0^1(\Omega) \cap L^p(\Omega)\right), \ u \in L^\infty\left(\mathbb{R}^+; L^2(\Omega)\right)$$

Note that there are no restrictions on the spatial dimension of the domain Ω or the growth of nonlinearity.

3. Nonlinear wave equation with nonlinear damping term: stabilization with finitely many Fourier modes

In this section we consider the initial boundary value problem for a nonlinear wave equation with nonlinear damping term with a feedback controller involving finitely many Fourier modes:

$$\partial_t^2 u - \nu\Delta u + b|\partial_t u|^{m-2}\partial_t u - au + |u|^{p-2}u = -\mu\sum_{k=1}^N (u, w_k)w_k, \ x \in \Omega, t > 0, \tag{3.1}$$

$$u = 0, \ x \in \partial\Omega, \ t > 0, \tag{3.2}$$

$$u(x,0) = u_0(x), \quad \partial_t u(x,0) = u_1(x), \quad x \in \Omega, \tag{3.3}$$

where $\nu > 0, b > 0, a > 0, p \geq m > 2$ are given parameters.
We show stabilization of solutions of this problem to the zero with a polynomial rate. Our main result is the following theorem:

THEOREM 3.1. *Suppose that μ ad N are large enough such that*

$$\nu > 2a\lambda_{N+1}^{-1}, \quad \text{and} \quad \mu > a. \tag{3.4}$$

Then for each solution of the problem (3.1)-(3.3) *following estimate holds true*

$$\|\partial_t u(t)\|^2 + \|\nabla u(t)\|^2 + \int_\Omega |u(x,t)|^p dx \leq Ct^{-\frac{m-1}{m}}, \tag{3.5}$$

where C is a positive constant depending on initial data.

PROOF. Taking the inner product of equation (3.1) in $L^2(\Omega)$ with $\partial_t u$ we get

$$\frac{d}{dt}\mathcal{E}(t) + b\|\partial_t u(t)\|^m = 0, \tag{3.6}$$

where

$$\mathcal{E}(t) := \frac{1}{2}\|\partial_t u(t)\|^2 + \frac{1}{2}\|\nabla u(t)\|^2 - \frac{a}{2}\|u(t)\|^2 + \frac{1}{p}\|u(t)\|^p_{L^p(\Omega)} + \frac{\mu}{2}\sum_{k=1}^N (u(t), w_k)^2. \tag{3.7}$$

Similar to (2.31) we have

(3.8)
$$\begin{aligned}\mathcal{E}(t) &\geq \frac{1}{2}\|\partial_t u\|^2 + \frac{1}{2}\left(\nu - a\lambda_{N+1}^{-1}\right)\|\nabla u\|^2 + \frac{1}{2}(\mu - a)\sum_{k=1}^N (u, w_k)^2 + \frac{1}{p}\int_\Omega |u|^p dx \\ &\geq \frac{1}{2}\|\partial_t u\|^2 + \frac{\nu}{2}\|\nabla u\|^2 + \frac{1}{p}\|u(t)\|^p_{L^p(\Omega)}.\end{aligned}$$

This estimate implies that the function $\mathcal{E}(t)$ is non-negative, for $t \geq 0$. Let us integrate (3.6) over the interval $(0,t)$:

$$\mathcal{E}(0) - \mathcal{E}(t) = \int_0^t \|\partial_t u(\tau)\|_m^m d\tau. \tag{3.9}$$

Taking the inner product of (3.1) in $L^2(\Omega)$ with u gives

$$\frac{d}{dt}(u, \partial_t u) =$$

$$\|\partial_t u\|^2 - \nu\|\nabla u\|^2 - b\int_\Omega u|\partial_t u|^{m-2}\partial_t u dx + a\|u\|^2 - \int_\Omega |u|^p dx - \mu\sum_{k=1}^N (u, w_k)^2.$$

By using notation (3.7) we can rewrite the last relation in the following form:

$$\begin{aligned}\frac{d}{dt}(u, \partial_t u) &= \frac{3}{2}\|\partial_t u\|^2 - \mathcal{E}(t) - \frac{\nu}{2}\|\nabla u\|^2 \\ &+ \frac{a}{2}\|u\|^2 - \frac{\mu}{2}\sum_{k=1}^N (u, w_k)^2 - \frac{p-1}{p}\int_\Omega |u|^p dx - b\int_\Omega u|\partial_t u|^{m-2}\partial_t u dx.\end{aligned} \tag{3.10}$$

By using inequality (2.30) we obtain from (3.10)

$$\begin{aligned}\frac{d}{dt}(u, \partial_t u) &\leq -\mathcal{E}(t) + \frac{3}{2}\|\partial_t u\|^2 - \frac{1}{2}(\mu - a)\sum_{k=1}^N (u, w_k)^2 \\ &+ \frac{a}{2}\sum_{k=N+1}^\infty (u, w_k)^2 - \frac{\nu}{2}\|\nabla u\|^2 + b\int_\Omega |u||\partial_t u|^{m-1}dx \leq -\mathcal{E}(t) + \frac{3}{2}\|\partial_t u\|^2 \\ &- \frac{1}{2}(\mu - a)\sum_{k=1}^N (u, w_k)^2 - \frac{1}{2}(\nu - a\lambda_{N+1}^{-1})\|\nabla u\|^2 + b\int_\Omega |u||\partial_t u|^{m-1}dx.\end{aligned}$$

Taking into account condition (3.4) we deduce the inequality

$$\mathcal{E}(t) \leq -\frac{d}{dt}(u(t), \partial_t u(t)) + \frac{3}{2}\|\partial_t u(t)\|^2 + b\int_\Omega |u(x,t)|\partial_t u(x,t)|^{m-1}dx.$$

After integration over the interval $(0,t)$, we obtain

$$\begin{aligned}\int_0^t \mathcal{E}(\tau)d\tau &\leq (u(0), \partial_t u(0)) - (u(t), \partial_t u(t)) \\ &+ \frac{3}{2}\int_0^t \|\partial_t u(\tau)\|^2 d\tau + b\int_0^t\int_\Omega |u(x,\tau)|\partial_t u(x,\tau)|^{m-1}dxd\tau.\end{aligned} \tag{3.11}$$

Since $\mathcal{E}(t) \leq \mathcal{E}(0)$ and $p \geq 2$, then due to (3.8), we have

$$|(u(0), \partial_t u(0)) - (u(t), \partial_t u(t))| \leq C, \tag{3.12}$$

where C depends on the initial data.

By using the Hölder inequality and estimate (3.9), we estimate the second term on the right-hand side of (3.11):

$$\int_0^t \|\partial_t u(\tau)\|^2 d\tau = \int_0^t \int_\Omega |\partial_t u(x,\tau)|^2 \cdot 1 dx d\tau \le \left(\int_0^t \|\partial_t u(\tau)\|^m_{L^m(\Omega)} d\tau\right)^{\frac{2}{m}} (|\Omega| t)^{\frac{m-2}{m}} \le C t^{\frac{m-2}{m}}. \tag{3.13}$$

The third term on the right-hand side of (3.11) we estimate again by using the estimates (3.9) similarly (recalling that $m \le p$):

$$\int_0^t \int_\Omega |\partial_t u(x,\tau)|^{m-1} |u(x,\tau)| dx d\tau \le \left(\int_0^t \int_\Omega |\partial_t u(x,\tau)|^m dx d\tau\right)^{\frac{m-1}{m}} \left(\int_0^t \int_\Omega |u(x,\tau)|^m dx d\tau\right)^{\frac{1}{m}} \le C t^{\frac{1}{m}}. \tag{3.14}$$

Since $\mathcal{E}(t)$ is non-increasing, positive function we have:

$$t\mathcal{E}(t) \le \int_0^t \mathcal{E}(\tau) d\tau. \tag{3.15}$$

Thus employing (3.12)-(3.15) we obtain from (3.11)

$$\mathcal{E}(t) \le C t^{-\frac{m-1}{m}}.$$

Hence

$$\|\partial_t u(t)\|^2 + \|\nabla u(t)\|^2 + \|u(t)\|^p_{L^p(\Omega)} \le C t^{-\frac{m-1}{m}}. \tag{3.16}$$

□

REMARK 3.2. We would like to note that unlike the result on finite set of functionals determining long time behavior of solutions to equation

$$\partial_t^2 u - \nu \Delta u + b|\partial_t u|^{m-2} \partial_t u - au + |u|^{p-2} u = 0, \ x \in \Omega, t > 0,$$

under the homogeneous Dirichlet's boundary condition, established in [**7**], (as in the case of the equation (2.24)), we do not require restrictions neither on the dimension of the domain Ω nor on the parameters $m > 0, p > 0$. It suffices to know that problem (2.24)-(2.26) has a global solution such that (see, e.g., [**24**]):

$$u \in L^\infty\left(\mathbb{R}^+; H_0^1(\Omega) \cap L^p(\Omega)\right) \cap L^{m+2}(\mathbb{R}^+; L^{m+2}(\Omega)), \ u \in L^\infty\left(\mathbb{R}^+; L^2(\Omega)\right).$$

4. Nonlinear strongly damped wave equation

In this section, we study the problem of feedback control of initial boundary value problem for nonlinear strongly damped equation with controllers involving finitely many Fourier modes and by nodal observables.

4.1. Feedback control employing finitely many nodal valued observables.

First we consider first the following problem

$$\partial_t^2 u - \partial_x^2 u - b\partial_x^2 \partial_t u - au + f(u) = -\mu \sum_{k=1}^N h u(\bar{x}_k) \delta(x - x_k), \ x \in (0,L), t > 0, \tag{4.1}$$

$$u(0,t) = u(L,t) = 0, \ x \in (0,L), t > 0, \tag{4.2}$$

where $x_k, \bar{x}_k \in J_k = [(k-1)\frac{L}{N}, k\frac{L}{N}], k = 1, \dots, N, h = \frac{L}{N}$, $f(\cdot)$ is continuously differentiable function that satisfies the conditions (1.4), $\delta(x - x_k)$ is the Dirac delta function, a, b and μ are given positive parameters.

Our estimates will be based on the following lemma:

LEMMA 4.1. *(see,e.g., [**1**]) Let $x_k, \overline{x}_k \in J_k = [(k-1)\, h, k\, h], k = 1, .., N$, where $h = \frac{L}{N}$, $N \in \mathbb{Z}^+$. Then for every $\varphi \in H^1(0, L)$ the following inequalities hold true*

$$\sum_{k=1}^{N} |\varphi(x_k) - \varphi(\overline{x}_k)|^2 \le h \, \|\varphi_x\|_{L^2}^2, \tag{4.3}$$

and

$$\|\varphi\|^2 \le 2 \left[h \sum_{k=1}^{N} |\varphi(x_k)|^2 + h^2 \, \|\varphi_x\|^2 \right]. \tag{4.4}$$

Taking the H^{-1} action of (4.1) on $\partial_t u + \varepsilon u \in H^1$, where $\varepsilon > 0$, to be determined, we get

$$\begin{aligned} \frac{d}{dt} \left[\frac{1}{2} \|\partial_t u\|^2 + \frac{1}{2}(1 + \varepsilon b) \|\partial_x u\|^2 - a \|u\|^2 + (F(u), 1) + \varepsilon(u, \partial_t u) \right] + \\ b \|\partial_t \partial_x u\|^2 + \varepsilon \|\partial_x u\| + \varepsilon(f(u), u) - a\varepsilon \|u\|^2 - \varepsilon \|\partial_t u\|^2 = \\ - \mu h \sum_{k=1}^{N} u(\bar{x}_k) \partial_t u(x_k) - \varepsilon \mu h \sum_{k=1}^{N} u(\bar{x}_k) u(x_k). \end{aligned} \tag{4.5}$$

By using the equalities

$$\sum_{k=1}^{N} u(\bar{x}_k) \partial_t u(x_k) = \frac{1}{2} \frac{d}{dt} \sum_{k=1}^{N} u^2(\bar{x}_k) + \sum_{k=1}^{N} u(\bar{x}_k) \left(\partial_t u(x_k) - \partial_t u(\bar{x}_k) \right)$$

and

$$\sum_{k=1}^{N} u(\bar{x}_k) u(x_k) = \sum_{k=1}^{N} u^2(\bar{x}_k) + \sum_{k=1}^{N} u(\bar{x}_k) \left(\partial_t u(x_k) - u(\bar{x}_k) \right),$$

we can rewrite (4.5) in the following form

(4.6)

$$\begin{aligned} \frac{d}{dt} E_\varepsilon(t) + b \|u_{xt}\|^2 + \varepsilon \|\partial_x u\| + \varepsilon(f(u), u) - a\varepsilon \|u\|^2 - \varepsilon \|\partial_t u\|^2 + \varepsilon \mu h \sum_{k=1}^{N} u^2(\bar{x}_k) = \\ - \mu h \sum_{k=1}^{N} u(\bar{x}_k) \left(\partial_t u(x_k) - \partial_t u(\bar{x}_k) \right) - \varepsilon \mu h \sum_{k=1}^{N} u(\bar{x}_k) \left(u(x_k) - u(\bar{x}_k) \right), \end{aligned}$$

where

$$E_\varepsilon(t) := \frac{1}{2} \|\partial_t u\|^2 + \frac{1}{2}(1 + \varepsilon b) \|\partial_x u\|^2 - a \|u\|^2 + (F(u), 1) + \varepsilon(u, \partial_t u) + \frac{\mu h}{2} \sum_{k=1}^{N} u^2(\bar{x}_k).$$

By using inequality (4.4) we obtain

$$E_\varepsilon(t) \geq \frac{1}{4}\|\partial_t u\|^2 + \left[\frac{1}{2}(1+\varepsilon b) - 2(a+\varepsilon^2)h^2\right]\|\partial_x u\|^2 + \varepsilon(F(u),1) + \left[\frac{\mu h}{2} - 2(a+\varepsilon^2)h\right]\sum_{k=1}^N u^2(\bar{x}_k). \tag{4.7}$$

Employing (4.3) we get

$$\mu h \sum_{k=1}^N u(\bar{x}_k)\left(\partial_t u(x_k) - \partial_t u(\bar{x}_k)\right) \leq \frac{\varepsilon\mu h}{4}\sum_{k=1}^N u^2(\bar{x}_k) + \frac{\mu h^2}{\varepsilon}\|\partial_x\partial_t u\|^2, \tag{4.8}$$

$$\mu h \sum_{k=1}^N u(\bar{x}_k)\left(u(x_k) - u(\bar{x}_k)\right) \leq \frac{\varepsilon\mu h}{4}\sum_{k=1}^N u^2(\bar{x}_k) + \frac{\mu h^2}{\varepsilon}\|\partial_x u\|^2. \tag{4.9}$$

Now we choose $\varepsilon = \frac{b\lambda_1}{2}$, use the Poincaré inequality, (1.6), and inequalities (4.8) and (4.9) in (4.6) and obtain:

$$\frac{d}{dt}E_\varepsilon(t) + \frac{b}{2}\|\partial_x\partial_t u\|^2 + \left(\varepsilon - \frac{4}{\varepsilon}\mu h^2\right)\|\partial_x u\|^2 + \varepsilon(F(u),1) - a\varepsilon\|u\|^2 + \frac{1}{2}\varepsilon\mu h\sum_{k=1}^N u^2(\bar{x}_k) \leq 0.$$

We employ here inequality (4.4) to obtain

$$\frac{d}{dt}E_\varepsilon(t) + \frac{b}{2}\|u_{xt}\|^2 + \left(\varepsilon - \frac{4}{\varepsilon}\mu h^2 - 4a\varepsilon h^2\right)\|\partial_x u\|^2 + \varepsilon(F(u),1) + \left(\frac{1}{2}\varepsilon\mu h - 2a\varepsilon h\right)\sum_{k=1}^N u^2(\bar{x}_k) \leq 0. \tag{4.10}$$

It is not difficult to see that if μ is large enough such that

$$\mu > 4\left(a + \frac{\lambda_1^2 b^2}{4}\right), \tag{4.11}$$

and $N = \frac{L}{h}$ is large enough such that

$$\frac{\lambda_1 b}{2} - 2h^2\left(\frac{\mu}{\lambda_1 b} - a\lambda_1 b\right) > 0, \tag{4.12}$$

$$\frac{b^2\lambda_1^2}{4} - a^2\lambda_1^2 b^2 h^2 - \mu h^2 > 0, \tag{4.13}$$

then there exists $d_1 > 0$ such that

$$E_\varepsilon(t) \geq d_1\left(\|\partial_t u\|^2 + \|\partial_x u\|^2\right), \tag{4.14}$$

and that there exists a positive number δ such that

$$\frac{b}{2}\|u_{xt}\|^2 + \left(\varepsilon - \frac{4}{\varepsilon}\mu h^2\right)\|\partial_x u\|^2 + \varepsilon(F(u),1) - a\varepsilon\|u\|^2 + \frac{1}{2}\varepsilon\mu h\sum_{k=1}^N u^2(\bar{x}_k) \geq \delta E_\varepsilon(t). \tag{4.15}$$

By virtue of (4.15) we deduce from (4.10) the inequality

$$\frac{d}{dt}E_\varepsilon(t) + \delta E_\varepsilon(t) \le 0.$$

This inequality and inequality (4.14) imply the exponential stabilization estimate

$$\|\partial_t u(t)\|^2 + \|\partial_x u(t)\|^2 \le D_0 e^{-\delta t}. \tag{4.16}$$

Consequently we have proved the following:

THEOREM 4.2. *Suppose that conditions (4.11)-(4.13) are satisfied. Then all solutions of the problem (4.1)-(4.2) tend to zero with an exponential rate, as $t \to \infty$.*

REMARK 4.3. By using similar arguments we can prove an analog to Theorem 4.2 for solutions of the semilinear pseudo-hyperbolic equation

(4.17)

$$\partial_t^2 u - u_{xxtt} - \nu\partial_x^2 u - bu_{xxt} - au + f(u) = -\mu\sum_{k=1}^{N} hu(\bar{x}_k)\delta(x - x_k), \ x \in (0, L), t > 0,$$

under the boundary conditions (4.2) or under the periodic boundary conditions. Here a, b, ν are positive parameters, and f satisfies conditions (1.4), μ and N should be chosen appropriately large enough.

4.2. Feedback control employing finitely many Fourier modes. The second problem we are going to study in this section is the following feedback problem

$$\begin{cases} \partial_t^2 u - \nu\Delta u - b\Delta\partial_t u - au + |u|^{p-2}u = -\mu\sum_{k=1}^{N}(u, w_k)w_k, \ x \in \Omega, t > 0, \\ u(x, 0) = u_0(x), \ \partial_t u(x, 0) = u_1(x), \ x \in \Omega, \\ u = 0, \ x \in \partial\Omega, \ t > 0, \end{cases} \tag{4.18}$$

where $\nu > 0, b > 0, a > 0, p > 2$ are given parameters.
The following theorem guarantees the exponential feedback stabilization of solutions of problem (4.18)

THEOREM 4.4. *Suppose that μ is large enough such that*

$$\mu > 2a + \frac{1}{4}\delta_0\lambda_1 b, \tag{4.19}$$

and N is large enough such that

$$\nu \ge \left(2a + \frac{1}{4}b\lambda_1\delta_0\right)\lambda_{N+1}^{-1}, \tag{4.20}$$

where

$$\delta_0 := \frac{b\lambda_1\nu}{2\nu + b^2\lambda_1}. \tag{4.21}$$

Then the solution of (4.18) satisfies the following exponential decay estimate:

$$\|\partial_t u(t)\|^2 + \|\nabla u(t)\|^2 + \|u(t)\|^p_{L^p(\Omega)} \le E_0 e^{-\delta_0 t}, \ \textit{for all} t > 0. \tag{4.22}$$

with a constant E_0 depending on initial data.

PROOF. The proof of this theorem is similar to the proof of Theorem 2.5. The energy equality in this case has the form

$$\frac{d}{dt}E_\varepsilon(t) + b\|\nabla\partial_t u(t)\|^2 - \varepsilon\|\partial_t u(t)\|^2 + \varepsilon\nu\|\nabla u(t)\|^2 + \varepsilon\int_\Omega |u(x,t)|^p dx - \varepsilon a\|u\|^2 + \varepsilon\mu\sum_{k=1}^N (u(t), w_k)^2 = 0, \tag{4.23}$$

where $\varepsilon = \frac{b\lambda_1}{2}$, and

$$E_\varepsilon(t) := \frac{1}{2}\|\partial_t u\|^2 + \frac{1}{2}(\nu+\varepsilon b)\|\nabla u\|^2 - \frac{a}{2}\|u\|^2 + \frac{1}{p}\|u\|^p_{L^p(\Omega)} + \frac{\mu}{2}\sum_{k=1}^n (u, w_k)^2 + \varepsilon(u, \partial_t u).$$

Employing the Young inequality (1.5), the Poincaré inequality (1.6), the Poincaré-like inequality (1.6), and the conditions (4.19) and (4.20) we get:

$$\begin{aligned} E_\varepsilon(t) &\geq \frac{1}{4}\|\partial_t u\|^2 + \frac{1}{2}\left(\nu + \frac{b^2\lambda_1}{2}\right)\|\nabla u\|^2 - \left(\frac{a}{2} + \frac{b^2\lambda_1^2}{4}\right)\|u\|^2 + \frac{1}{p}\|u\|^p_{L^p(\Omega)} \\ &+ \frac{\mu}{2}\sum_{k=1}^n (u, w_k)^2 \geq \frac{1}{4}\|\partial_t u\|^2 + \frac{\nu}{2}\|\nabla u\|^2 - \frac{a}{2}\|u\|^2 + \frac{\mu}{2}\sum_{k=1}^n (u, w_k)^2 + \frac{1}{p}\|u\|^p_{L^p(\Omega)} \\ &\geq \frac{1}{4}\|\partial_t u\|^2 + \frac{\nu}{4}\|\nabla u\|^2 + \left(\frac{\nu}{4} - \frac{a}{2}\lambda_{N+1}^{-1}\right)\|\nabla u\|^2 + \left(\frac{\mu}{2} - \frac{a}{2}\right)\sum_{k=1}^n (u, w_k)^2 + \frac{1}{p}\|u\|^p_{L^p(\Omega)} \\ &\geq \frac{1}{4}\|\partial_t u\|^2 + \frac{\nu}{4}\|\nabla u\|^2 + \frac{1}{p}\|u\|^p_{L^p(\Omega)}. \end{aligned} \tag{4.24}$$

Let $\delta \in (0, \varepsilon)$, be another parameter to be chosen below. Then employing the Poincaré inequality, and the fact that $\delta < \varepsilon = \lambda_1 b/2$, and $p > 2$, we obtain from (4.23)

$$\begin{aligned} &\frac{d}{dt}E_\varepsilon(t) + \delta E_\varepsilon(t) + \left(\frac{b}{2} - \frac{\delta}{2\lambda_1}\right)\|\nabla\partial_t u\|^2 \\ &+ \left(\frac{b\lambda_1\nu}{2} - \frac{\delta}{2}(\nu + \frac{b^2\lambda_1}{2})\right)\|\nabla u\|^2 - \frac{ab\lambda_1}{4}\|u\|^2 + \left(\frac{b\lambda_1}{2} - \frac{\delta}{p}\right)\|u\|^p_{L^p(\Omega)} \\ &+ \frac{\mu}{2}(b\lambda_1 - \delta)\sum_{k=1}^n (u, w_k)^2 - \frac{\delta\lambda_1 b}{2}(u, \partial_t u) \leq 0. \end{aligned} \tag{4.25}$$

By using the inequality

$$\frac{1}{2}\delta\lambda_1 b|(u, \partial_t u)| \leq \frac{1}{2}\delta\lambda_1^{\frac{1}{2}} b\|\nabla\partial_t u\|\|u\| \leq \frac{\delta}{2\lambda_1}\|\nabla\partial_t u\|^2 + \frac{\delta}{8}\lambda_1^2 b^2\|u\|^2,$$

and assuming here that $\delta \in (0, \varepsilon)$, we get

$$\begin{aligned} &\frac{d}{dt}E_\varepsilon(t) + \delta E_\varepsilon(t) + \left(\frac{b}{2} - \frac{\delta}{\lambda_1}\right)\|\nabla\partial_t u\|^2 + \frac{1}{2}\left[\nu b\lambda_1 - \delta(\nu + \frac{b^2\lambda_1}{2})\right]\|\nabla u\|^2 \\ &- \frac{1}{2}\left(ab\lambda_1 + \frac{\delta}{8}\lambda_1^2 b^2\right)\|u\|^2 + \frac{\mu}{2}(b\lambda_1 - \delta)\sum_{k=1}^n (u, w_k)^2. \end{aligned} \tag{4.26}$$

We choose in the last inequality $\delta = \delta_0$, defined in (4.21), and obtain the inequality
(4.27)

$$\frac{d}{dt}E_\varepsilon(t) + \delta_0 E_\varepsilon(t) + \frac{1}{4}\nu b\lambda_1\|\nabla u\|^2 - \frac{b\lambda_1}{2}\left(a + \frac{\delta}{8}\lambda_1 b\right)\|u\|^2 + \frac{\mu}{4}b\lambda_1\sum_{k=1}^{n}(u, w_k)^2.$$

Finally, employing in (4.27) inequality (2.30), and the conditions (4.19) and (4.20) we obtain the desired inequality

$$\frac{d}{dt}E_\varepsilon(t) + \delta_0 E_\varepsilon(t) \leq 0.$$

Thanks to (4.24), this inequality implies the desired decay estimate (4.22). □

REMARK 4.5. In a similar way we can prove exponential stabilization of solutions to strongly damped Boussinesq equation, with homogeneous Dirichlet boundary conditions

$$\begin{cases} \partial_t^2 u - \nu\partial_x^4 u - b\partial_x^2\partial_t u + \partial_x^2\left(au - |u|^{p-2}u\right) = -\mu w, \ x \in (0, L), t > 0, \\ u(0,t) = u(L,t) = \partial_x^2 u(0,t) = \partial_x^2 u(L,t) = 0, \quad t > 0, \end{cases} \tag{4.28}$$

where a, ν, b are given positive parameters, and w is a controller of the

$$w = \sum_{k=1}^{N}\lambda_k(u, w_k)w_k, \quad \lambda_k = \frac{k^2\pi^2}{L^2}, \quad w_k(x) = \sin\frac{k\pi}{L}.$$

Here also we can find N and μ large enough such that

$$\|\partial_t u(t)\|^2_{H^{-1}(0,L)} + \|\partial_x^2 u(t)\|^2 \leq E_0 e^{-\delta t}$$

with some $\delta > 0$, depending on parameters of the problem, i.e. ν, b, p and L.

REMARK 4.6. It is worth noting that the estimates we obtained in Theorem 2.5 and Theorem 4.4 are valid for weak solutions of the corresponding problems from the class of functions such that

$$\partial_t u \in L^\infty(\mathbb{R}^+; L^2(\Omega)), \ u \in L^\infty(\mathbb{R}^+; H_0^1(\Omega) \cap L^p(\Omega)).$$

Existence of a weak solution for each of these problems, as well as justification of our estimates can be done by using the Galerkin method (see e.g. [**24**]). We would like also to note that the feedback stabilization estimate for problem (2.24)-(2.25) is established, for arbitrary $p > 2$, without any restrictions on the spatial dimension of the domain Ω. As far as we know, even uniqueness of a weak solution in the case $p > 5, \Omega \subset \mathbb{R}^3$, is an open problem.

REMARK 4.7. It is also worth mentioning that the feedback stabilization of nonlinear damped wave equation, nonlinear strongly damped wave equation, nonlinear wave equation with nonlinear damping term with controllers involving finitely many parameters, can be shown by employing the concept of general determining functionals (projections) introduced in [**8**],[**9**], then exploited and developed for the study of dissipative wave equations [**6**].
Recently it was shown in [**15**], [**16**] and [**18**] that the approach of a new feedback controlling of dissipative PDEs using finitely many determining parameters can be used to show that the global (in time) dynamics of the 2D Navier - Stokes equations, and of that of the 1D damped driven nonlinear Schrödinger equation

$$i\partial_t u - \partial_x^2 u + i\gamma u + |u|^2 u = f, \quad b > 0,$$

can be embedded in an infinite-dimensional dynamical system induced by an ordinary differential equation, called *determining form*, governed by a global Lipschitz vector field.
The existence of determining form for the long-time dynamics of nonlinear damped wave equation and nonlinear strongly damped wave equation is a subject of a future research.

References

[1] Abderrahim Azouani and Edriss S. Titi, *Feedback control of nonlinear dissipative systems by finite determining parameters—a reaction-diffusion paradigm*, Evol. Equ. Control Theory **3** (2014), no. 4, 579–594, DOI 10.3934/eect.2014.3.579. MR3274649

[2] A. V. Babin and M. I. Vishik, *Attractors of evolution equations*, Studies in Mathematics and its Applications, vol. 25, North-Holland Publishing Co., Amsterdam, 1992. Translated and revised from the 1989 Russian original by Babin. MR1156492 (93d:58090)

[3] Mark J. Balas, *Feedback control of dissipative hyperbolic distributed parameter systems with finite-dimensional controllers*, J. Math. Anal. Appl. **98** (1984), no. 1, 1–24, DOI 10.1016/0022-247X(84)90275-0. MR728514 (84m:93055)

[4] J. M. Ball and M. Slemrod, *Feedback stabilization of distributed semilinear control systems*, Appl. Math. Optim. **5** (1979), no. 2, 169–179, DOI 10.1007/BF01442552. MR533618 (81a:93070)

[5] Chongsheng Cao, Ioannis G. Kevrekidis, and Edriss S. Titi, *Numerical criterion for the stabilization of steady states of the Navier-Stokes equations*, Indiana Univ. Math. J. **50** (2001), no. Special Issue, 37–96, DOI 10.1512/iumj.2001.50.2154. Dedicated to Professors Ciprian Foias and Roger Temam (Bloomington, IN, 2000). MR1855664 (2002h:35232)

[6] I. D. Chueshov, *The theory of functionals that uniquely determine the asymptotic dynamics of infinite-dimensional dissipative systems* (Russian), Uspekhi Mat. Nauk **53** (1998), no. 4(322), 77–124, DOI 10.1070/rm1998v053n04ABEH000057; English transl., Russian Math. Surveys **53** (1998), no. 4, 731–776. MR1668050 (2000b:37089)

[7] Igor D. Chueshov and Varga K. Kalantarov, *Determining functionals for nonlinear damped wave equations* (English, with English, Russian and Ukrainian summaries), Mat. Fiz. Anal. Geom. **8** (2001), no. 2, 215–227. MR1859881 (2002h:35192)

[8] Bernardo Cockburn, Don A. Jones, and Edriss S. Titi, *Determining degrees of freedom for nonlinear dissipative equations* (English, with English and French summaries), C. R. Acad. Sci. Paris Sér. I Math. **321** (1995), no. 5, 563–568. MR1356554 (96j:34106)

[9] Bernardo Cockburn, Don A. Jones, and Edriss S. Titi, *Estimating the number of asymptotic degrees of freedom for nonlinear dissipative systems*, Math. Comp. **66** (1997), no. 219, 1073–1087, DOI 10.1090/S0025-5718-97-00850-8. MR1415799 (98a:35130)

[10] Peter Constantin, Charles R. Doering, and Edriss S. Titi, *Rigorous estimates of small scales in turbulent flows*, J. Math. Phys. **37** (1996), no. 12, 6152–6156, DOI 10.1063/1.531769. MR1419159 (97j:35121)

[11] J.-M. Coron and E.Trélat, Feedback stabilization along a path of steady-states for 1-D semilinear heat and wave equations, Proceedings of the 44th IEEE Conference on Decision and Control, and the European Control Conference 2005 Seville, Spain, December 12-15, 2005.

[12] C. Foiaş and G. Prodi, *Sur le comportement global des solutions non-stationnaires des équations de Navier-Stokes en dimension* 2 (French), Rend. Sem. Mat. Univ. Padova **39** (1967), 1–34. MR0223716 (36 #6764)

[13] Jack K. Hale, *Asymptotic behavior of dissipative systems*, Mathematical Surveys and Monographs, vol. 25, American Mathematical Society, Providence, RI, 1988. MR941371 (89g:58059)

[14] Ciprian Foias and Edriss S. Titi, *Determining nodes, finite difference schemes and inertial manifolds*, Nonlinearity **4** (1991), no. 1, 135–153. MR1092888 (92a:65241)

[15] Ciprian Foias, Michael S. Jolly, Rostyslav Kravchenko, and Edriss S. Titi, *A determining form for the two-dimensional Navier-Stokes equations: the Fourier modes case*, J. Math. Phys. **53** (2012), no. 11, 115623, 30, DOI 10.1063/1.4766459. MR3026568

[16] K. Foyash, M. S. Dzholli, R. Kravchenko, and È. S. Titi, *A unified approach to the construction of defining forms for a two-dimensional system of Navier-Stokes equations: the case of general interpolating operators* (Russian, with Russian summary), Uspekhi Mat. Nauk **69** (2014), no. 2(416), 177–200; English transl., Russian Math. Surveys **69** (2014), no. 2, 359–381. MR3236940

[17] A. Haraux, Systemés dynamiques dissipatifs et applications, R. M. A. 17, ed. Ph. Ciarlet and J.-L. Lions. Masson, Paris (1990).

[18] Michael S. Jolly, Tural Sadigov, and Edriss S. Titi, *A determining form for the damped driven nonlinear Schrödinger equation—Fourier modes case*, J. Differential Equations **258** (2015), no. 8, 2711–2744, DOI 10.1016/j.jde.2014.12.023. MR3312642

[19] Don A. Jones and Edriss S. Titi, *Upper bounds on the number of determining modes, nodes, and volume elements for the Navier-Stokes equations*, Indiana Univ. Math. J. **42** (1993), no. 3, 875–887, DOI 10.1512/iumj.1993.42.42039. MR1254122 (94k:35249)

[20] V. Komornik, *Exact controllability and stabilization*, RAM: Research in Applied Mathematics, Masson, Paris; John Wiley & Sons, Ltd., Chichester, 1994. The multiplier method. MR1359765 (96m:93003)

[21] O. A. Ladyženskaja, *The dynamical system that is generated by the Navier-Stokes equations* (Russian), Zap. Naučn. Sem. Leningrad. Otdel. Mat. Inst. Steklov. (LOMI) **27** (1972), 91–115. Boundary value problems of mathematical physics and related questions in the theory of functions, 6. MR0328378 (48 #6720)

[22] O. A. Ladyzhenskaya, *Attractors for Semi-groups and Evolution Equations*, Cambridge University Press (1991).

[23] I. Lasiecka and R. Triggiani, *Uniform stabilization of the wave equation with Dirichlet or Neumann feedback control without geometrical conditions*, Appl. Math. Optim. **25** (1992), no. 2, 189–224, DOI 10.1007/BF01182480. MR1142681 (93b:93099)

[24] J.-L. Lions, *Quelques méthodes de résolution des problèmes aux limites non linéaires* (French), Dunod; Gauthier-Villars, Paris, 1969. MR0259693 (41 #4326)

[25] Weijiu Liu, *Elementary feedback stabilization of the linear reaction-convection-diffusion equation and the wave equation*, Mathématiques & Applications (Berlin) [Mathematics & Applications], vol. 66, Springer-Verlag, Berlin, 2010. MR2573579 (2010m:93004)

[26] Andrey Smyshlyaev, Eduardo Cerpa, and Miroslav Krstic, *Boundary stabilization of a 1-D wave equation with in-domain antidamping*, SIAM J. Control Optim. **48** (2010), no. 6, 4014–4031, DOI 10.1137/080742646. MR2645471 (2011d:93142)

[27] Pierangelo Marcati, *Decay and stability for nonlinear hyperbolic equations*, J. Differential Equations **55** (1984), no. 1, 30–58, DOI 10.1016/0022-0396(84)90087-1. MR759826 (86e:35019)

[28] G. Perla Menzala, C. F. Vasconcellos, and E. Zuazua, *Stabilization of the Korteweg-de Vries equation with localized damping*, Quart. Appl. Math. **60** (2002), no. 1, 111–129. MR1878262 (2002j:35273)

[29] Roger Temam, *Infinite-dimensional dynamical systems in mechanics and physics*, 2nd ed., Applied Mathematical Sciences, vol. 68, Springer-Verlag, New York, 1997. MR1441312 (98b:58056)

[30] Yu Xin and Li Yongyong, The Stabilization of FitzHugh-Nagumo Systems with One Feedback Controller, Proceedings of the 27th Chinese Control Conference July 16–18, 2008.

[31] Enrike Zuazua, *Uniform stabilization of the wave equation by nonlinear boundary feedback*, SIAM J. Control Optim. **28** (1990), no. 2, 466–477, DOI 10.1137/0328025. MR1040470 (92c:35021)

Department of mathematics, Koç University, Rumelifeneri Yolu, Sariyer 34450, Sariyer, Istanbul, Turkey
E-mail address: vkalantarov@ku.edu.tr

Department of Mathematics, Texas A&M University, 3368 – TAMU, College Station, Texas 77843-3368. **Also**, Department of Computer Science and Applied Mathematics, Weizmann Institute of Science, Rehovot 76100, Israel
E-mail address: titi@math.tamu.edu
E-mail address: edriss.titi@weizmann.ac.il

Contemporary Mathematics
Volume **659**, 2016
http://dx.doi.org/10.1090/conm/659/13147

Multiple Fourier transforms and trigonometric series in line with Hardy's variation

E. Liflyand

ABSTRACT. For functions of bounded Hardy's variation, integrability results for their Fourier transforms are presented and proved in a somewhat different way than before. In order to obtain integrability theorems for multiple trigonometric series with coefficients also of bounded Hardy's variation, passage relations are elaborated. Unlike those known earlier, they deal only with one variation, the Hardy variation.

1. Introduction

Known results on the integrability of the Fourier transform of a function with bounded variation can naturally be extended to several dimensions in various ways. Here we fulfil this task in the category of functions with bounded Hardy variation. Some results of that kind are more or less known but here we present them in a more convenient and transparent form and cause the proofs to fit this presentation. Similar estimates are known for multiple trigonometric series with coefficients also with bounded variation of Hardy type. It is worth mentioning that there is a variety of the notions of bounded variation in several dimensions but for numerous reasons Hardy's variation is the most natural in the considered problems. One of them is that Fourier integrals and trigonometric series are well defined and converge almost everywhere in this setting (see, e.g., [**44**, Ch.1]). A simple confusion is as follows. In dimension one there is a relation that allows one to derive integrability results for trigonometric series with coefficients of bounded variation from known results for the Fourier transform of a function with bounded variation. In several dimensions, though various extensions of such results to the scope of Hardy's variation are known, no tools has existed in terms of Hardy's variation only. To wit, we are concerned with the following problem. Let

$$\widehat{f}(-x) = \int_{\mathbb{R}^n} f(u)e^{i\langle x,u\rangle}du,$$

2010 *Mathematics Subject Classification.* Primary 42B10; Secondary 42B20, 42B30, 42B35, 42A50, 42A20, 42A38, 26B30.

Key words and phrases. Fourier transform, Hilbert transform, bounded variation, Hardy space, Hardy variation, integrability of trigonometric series.

where $u = (u_1, \ldots, u_n) \in \mathbb{R}^n$, similarly $x \in \mathbb{R}^n$, and $\langle x, u\rangle = x_1u_1 + \cdots + x_nu_n$, be the Fourier integral (Fourier transform) of a suitable function f and let

$$\sum_{k=(k_1,\ldots,k_n)\in\mathbb{Z}^n} f(k)e^{i\langle x,k\rangle},$$

be the trigonometric series generated by f, more precisely, with $f(k)$ taken as the k-th coefficients of this trigonometric series. We wish to compare the two values, or so to say build a "bridge" between them. Because of the periodicity of the series it is meaningful for $0 < |x_j| \le \pi$, $j = 1, \ldots, n$ only. To construct something multi-dimensional, one should have a clear picture of how the one-dimensional prototype looks like. Though one may try to trace certain relations back to 30-s - 40-s of the 20-th century, the first result within the scope of our study is due to Belinsky. In his paper [**3**], the function f is assumed to be of bounded variation (and of compact support which is an unnecessary restriction), while the estimate is provided in the L^p norm with $p < \infty$. Further results are due to Trigub. In dimension one his estimates are of great generality and sharpness, see [**43**] and [**44**, Th.4.1.2] first of all. Let f be of bounded (usual one-dimensional) total variation $BV(f)$ and vanishing at infinity, we have

$$\sup_{|x|\le\pi}\left|\int_{\mathbb{R}} f(t)e^{ixt}dt - \sum_{k\in\mathbb{Z}} f(k)e^{ixk}\right| \lesssim BV(f). \tag{1}$$

Here and in what follows we shall use the notation "$\lesssim$" and "$\gtrsim$" as abbreviations for "$\le C$" and "$\ge C$", with C being an absolute positive constant, maybe different in different occurrences.

In dimension one, (1) allows one to immediately obtain many known results on the integrability of trigonometric series (see, e.g., [**40**]) from estimates for the Fourier transforms, like those in [**27**].

In higher dimension his results were not in such a final form; an accessible references are [**28**, Sect.3], [**32**], [**33**]. The problem is that either two different variations are assumed to be bounded simultaneously or, like in [**32**], a new variation is introduced, which in fact can be treated as a mixture of two variations or a supplement to one variation.

However, in many applications only one, the so-called Hardy variation seems to be natural. We are going to prove a generalization of (1) in which only Hardy's variation is involved. In this paper we present such tools. Of course, and this is known long ago, no way to get this under the only assumption of the boundedness of Hardy's variation, but an additional condition we use does not involve any other variation and is by no means restrictive in the problems considered.

The outline of the paper is as follows. In the next section we briefly introduce the needed facts about Hardy's variation (via Vitali's variation). In Section 3 a variety of spaces involved in the integrability problems for the Fourier transform of a function of bounded variation is considered, in one and several dimensions. In Section 4 we present the results on the integrability of the Fourier transform of a function of bounded Hardy's variation. Being known in fact, (see, e.g., [**27**] or [**34**], [**22**]) they are given in a more convenient form than before. In Section 5 we apply these results to integrability problems for multiple trigonometric series. The passage from the Fourier transform to trigonometric series, an analog of (1), is our main advancement.

2. Hardy variation

One of the simplest and direct generalization of the one-dimensional variation, the Vitali variation, is defined as follows (cf., e.g., [**1**, **10**]). Let f be a complex-valued function and

$$\Delta_u f(x) = \left(\prod_{j=1}^{n} \Delta_{u_j}\right) f(x),$$

with

$$\Delta_{u_j} f(x) = f(x + u_j e_j) - f(x),$$

be a "mixed" difference with respect to the parallelepiped $[x, x+u] = [x_1, x_1 + u_1] \times \cdots \times [x_n, x_n + u_n]$. We will need the following notations. Let $\eta = (\eta_1, \ldots, \eta_n)$ be an n-dimensional vector with the entries either 0 or 1 only. Correspondingly, $|\eta| = \eta_1 + \cdots + \eta_n$. The inequality of vectors is meant coordinate wise. Denote by $\Delta_{u_\eta} f(x)$ the partial difference

$$\Delta_{u_\eta} f(x) = \left(\prod_{j:\eta_j=1} \Delta_{u_j}\right) f(x).$$

Here and in what follows $D^\eta f$ for $\eta = \mathbf{0} = (0, 0, \ldots, 0)$ or $\eta = \mathbf{1} = (1, 1, \ldots, 1)$ mean the function itself and the partial derivative repeatedly in each variable, respectively, where

$$D^\eta f(x) = \left(\prod_{j:\,\eta_j=1} \frac{\partial}{\partial x_j}\right) f(x).$$

By $\mathbf{x}_\eta$ and $d\mathbf{x}_\eta$ we denote the $|\eta|$-tuple consisting only of x_j such that $\eta_j = 1$ and $\prod\limits_{j:\eta_j=1} dx_j$, respectively.

Let us take an arbitrary number of non-overlapping parallelepipeds, and form a mixed difference with respect to each of them. Then the Vitali variation is

$$VV(f) = \sup \sum |\Delta_u f(x)|,$$

where the sum and then the least upper bound are taken over all the sets of such nonoverlapping parallelepipeds. For smooth functions f, the Vitali variation is expressed as the following integral

$$VV(f) = \int\limits_{\mathbb{R}^n} \left| \frac{\partial^n f(x)}{\partial x_1 \ldots \partial x_n} \right| dx. \tag{2}$$

Already this variation has numerous applications; besides those in [**1**], see also [**44**, Sec.3.3.9-3.3.10] and [**42**]. However, in many problems Vitali's variation is helpless, because marginal functions of a smaller number of variables, as bad as one can imagine, may be added to a function of bounded Vitali's variation. The next notion is free of this disadvantage.

A function f is said to be of bounded Hardy variation, written $f \in VH(f)$, if it is of bounded Vitali variation and is of bounded variation with respect to any smaller number of variables (in fact, Vitali with respect to these variables or just BV when with respect to a single variable; see, e.g., [**19**], [**10**]; sometimes this notion is also attributed to Krause, see, e.g., [**21**, p. 345]). The latter will be denoted by $VV_\eta(f) < \infty$, with $\eta \neq \mathbf{1}, \mathbf{0}$. Correspondingly, $VV(f) := VV_{\mathbf{1}}(f)$. In other words, $VH(f) < \infty$ if and only if $VV_\eta(f) < \infty$ for all η, except $\eta = \mathbf{0}$ which is meaningless. However, just for convenience, we can understand $VV_{\mathbf{0}}(f) := f$.

If f is of bounded Vitali variation on $\mathbb{R}^n$ and $\lim\limits_{|x|\to\infty} f(x) = 0$, then functions depending on a smaller number of variables than n are excluded. Such a function is of bounded Hardy variation.

3. Spaces

3.1. One-dimensional preliminaries. The Fourier transform $\widehat{g}$ of a (complex-valued) function g in $L^1(\mathbb{R})$ is defined by

$$\widehat{g}(t) := \int_{\mathbb{R}} g(x)e^{-itx}\,dx, \quad t \in \mathbb{R},$$

while its Hilbert transform $\tilde{g}$ is defined by

$$\begin{aligned}\mathcal{H}g(x) &:= \frac{1}{\pi}\,(\text{P.V.})\int_{\mathbb{R}} g(x-u)\frac{du}{u} = \frac{1}{\pi}\,(\text{P.V.})\int_{\mathbb{R}} \frac{g(u)}{x-u}\,du \\ &= \frac{1}{\pi}\lim_{\delta\downarrow 0}\int_\delta^\infty \{g(x-u) - g(x+u)\}\frac{du}{u}, \quad x \in \mathbb{R}. \end{aligned} \tag{3}$$

As is well known, for $g \in L^1(\mathbb{R})$ this limit exists for almost all x in $\mathbb{R}$. Of course, near infinity this integral is understood in the standard way as an improper integral but we do not rest upon this. The real Hardy space $H^1(\mathbb{R})$ is defined to be

$$H^1(\mathbb{R}) := \{g \in L^1(\mathbb{R}) : \mathcal{H}g \in L^1(\mathbb{R})\},$$

where $L^1(\mathbb{R})$ is the usual space of integrable functions with norm

$$\|g\|_{L^1} := \int_{\mathbb{R}} |g(x)|\,dx.$$

The Hardy space is endowed with the norm

$$\|g\|_{H^1} := \|g\|_{L^1} + \|\mathcal{H}g\|_{L^1}. \tag{4}$$

If $g \in H^1(\mathbb{R})$, then

$$\int_{\mathbb{R}} g(t)\,dt = 0. \tag{5}$$

It was apparently first mentioned in [**26**].

An odd function always satisfies (5). However, not every odd integrable function belongs to $H^1(\mathbb{R})$, for counterexamples, see [**35**] and [**30**]. When in the definition of the Hilbert transform (49) the function g is odd, we will denote this transform by $\mathcal{H}^o$, and it is equal to

$$\mathcal{H}^o g(x) = \frac{2}{\pi}\int_0^\infty \frac{tg(t)}{x^2 - t^2}\,dt. \tag{6}$$

If it is integrable, we will denote the corresponding Hardy space by $H_0^1(\mathbb{R})$. Correspondingly, when in the definition of the Hilbert transform the function g is even, we will denote this transform by $\mathcal{H}^e$, and it is equal to

$$\mathcal{H}^e g(x) = \frac{2}{\pi}\int_0^\infty \frac{xg(t)}{x^2 - t^2}\,dt. \tag{7}$$

The cosine Fourier transform of f is

$$\widehat{f}_c(x) = \int_0^\infty f(t) \cos xt \, dt \tag{8}$$

and the sine Fourier transform is

$$\widehat{f}_s(x) = \int_0^\infty f(t) \sin xt \, dt. \tag{9}$$

We may equivalently study, for $\gamma = 0$ or 1, the Fourier transforms

$$\widehat{f}_\gamma(x) = \int_0^\infty f(t) \cos(xt - \frac{\pi\gamma}{2}) \, dt. \tag{10}$$

It is clear that $\widehat{f}_\gamma$ represents the cosine Fourier transform in the case $\gamma = 0$, while taking $\gamma = 1$ gives the sine Fourier transform.

The following result can be found in [**27**] (cf. [**14**] and [**16**]; see also [**22**, Ch.3]).

THEOREM 3.1. *Let $f : \mathbb{R}_+ \to \mathbb{C}$ be locally absolutely continuous, of bounded variation and $\lim\limits_{t\to\infty} f(t) = 0$. Let also $\mathcal{H}^o f' \in L^1(\mathbb{R}_+)$. Then the cosine Fourier transform of f (8) is Lebesgue integrable on $\mathbb{R}_+$, with*

$$\|\widehat{f}_c\|_{L^1(\mathbb{R}_+)} \lesssim \|f'\|_{L^1(\mathbb{R}_+)} + \|\mathcal{H}^o f'\|_{L^1(\mathbb{R}_+)} = \|f'\|_{H_0^1(\mathbb{R})}, \tag{11}$$

and for the sine Fourier transform, we have, for $x > 0$,

$$\widehat{f}_s(x) = \frac{1}{x} f\left(\frac{\pi}{2x}\right) + F(x), \tag{12}$$

where

$$\|F\|_{L^1(\mathbb{R}_+)} \lesssim \|f'\|_{L^1(\mathbb{R}_+)} + \|\mathcal{H}^o f'\|_{L^1(\mathbb{R}_+)} = \|f'\|_{H_0^1(\mathbb{R})}. \tag{13}$$

There is a scale of handy subspaces of $H_0^1(\mathbb{R}_+)$. For $1 < q < \infty$, set

$$\|g\|_{A_q} = \int_0^\infty \left(\frac{1}{x} \int_{x \le t \le 2x} |g(t)|^q dt\right)^{1/q} dx.$$

These spaces and their sequence analogs first appeared in the paper by D. Borwein [**8**], but became - for sequences - widely known after the paper by G. A. Fomin [**13**]; see also [**16, 17**]. On the other hand, these spaces are a partial case of the so-called Herz spaces (see first of all the initial paper by C. Herz [**20**], and also a relevant paper of Flett [**12**]).

Further, for $q = \infty$ let

$$\|g\|_{A_\infty} = \int_0^\infty \operatorname*{ess\,sup}_{x \le t \le 2x} |g(t)| \, dx.$$

The role of integrable monotone majorant for problems of almost everywhere convergence of singular integrals is known from the work of D.K. Faddeev (see, e.g., [**2**, Ch.IV, §4]; also [**38**, Ch.I]); for spectral synthesis problems it was used by A. Beurling [**6**], for more details see [**4**].

Known are (see, e.g., [**27**]) the following relations.

LEMMA 3.2. *The following embeddings hold:*

$$A_\infty \hookrightarrow A_{p_1} \hookrightarrow A_{p_2} \hookrightarrow H_0^1 \hookrightarrow L^1 \ (p_1 > p_2 > 1). \tag{14}$$

For examples on the difference (proper embedding) between these spaces, see [**29**].

In fact, in the previous works integrability properties of the Fourier transform have been considered for functions of bounded variation with the derivative in one of such spaces.

3.2. Widest integrability spaces. The space $H_0^1(\mathbb{R}_+)$, is one of the widest spaces the belonging of the derivative f' to which ensures the integrability of the cosine Fourier transform of f. However, the possibility of a wider space of such type to exist is of interest and importance.

Let us show that the maximal space of integrability does exist. In fact, it has in essence been introduced (for different purposes) in [**24**] as

$$Q = \{g : g \in L^1(\mathbb{R}), \int_{\mathbb{R}} \frac{|\widehat{g}(x)|}{|x|}\,dx < \infty\}. \tag{15}$$

The following result is a simple effect of the structure of this space (see [**31**]).

THEOREM 3.3. *Let $f : \mathbb{R}_+ \to \mathbb{C}$ be locally absolutely continuous on $(0,\infty)$, of bounded variation and $\lim_{t\to\infty} f(t) = 0$.*

a) The cosine Fourier transform of f given by (8) is Lebesgue integrable on $\mathbb{R}_+$ if and only if $f' \in Q$.

b) The sine Fourier transform of f given by (9) is Lebesgue integrable on $\mathbb{R}_+$ if and only if $f' \in Q$.

To be precise, in a) and b) the conditions are given in terms of different subspaces of Q. We shall use the space Q_o of the odd functions from Q rather than Q:

$$Q_o = \{g : g \in L^1(\mathbb{R}), g(-t) = -g(t), \int_0^\infty \frac{|\widehat{g_s}(x)|}{x}\,dx < \infty\}; \tag{16}$$

such functions naturally satisfy (5). This is exactly the space that a) is concerned with. As for b), an even counterpart of Q_o is used, more precisely,

$$Q_e = \{g : g \in L^1(\mathbb{R}), g(-t) = g(t), \int_0^\infty \frac{|\widehat{g_c}(x)|}{x}\,dx < \infty\}. \tag{17}$$

This makes sense only if

$$\int_0^\infty g(t)\,dt = 0.$$

This is the case for the considered function class.

3.3. Multidimensional spaces. Hardy's variation, at least in our study, is closely related with the product Hardy space $H_m^1 = H^1(\mathbb{R}^{n_1} \times \cdots \times \mathbb{R}^{n_m})$. These spaces are of interest and importance in certain questions of Fourier Analysis (see, e.g., [**9**], [**11**], [**18**]).

We are going to deal with a special case of these spaces when $m = n$ and, correspondingly, $n_1 = \cdots = n_n = 1$. In this case we can use the notation

$$H^1(\mathbb{R} \times \cdots \times \mathbb{R}) := H_n^1(\mathbb{R} \times \cdots \times \mathbb{R})$$

and a modified definition on the basis of the Hilbert transforms applied to each variable as in [**18**]. Such a transform applied to the j-th variable will be defined $\mathcal{H}_j$ and, consequently, $\mathcal{H}_j\mathcal{H}_k \ldots \mathcal{H}_l := \mathcal{H}_{jk\ldots l}$. However, there is a more convenient notation for this. We shall define $H_\eta^1(\mathbb{R} \times \cdots \times \mathbb{R})$ if the Hilbert transform with respect to each of the j-th variables, for which $\eta_j = 1$, is the even Hilbert transform, while for the rest it is odd. We will be especially interested in the case where $\eta = \mathbf{0}$, that is, all the Hilbert transforms are odd. The corresponding notation is similar to that in dimension one: $H^1_{\mathbf{0}}(\mathbb{R} \times \cdots \times \mathbb{R})$.

By this, the norm of a function $g(x)$ in $H^1(\mathbb{R} \times \cdots \times \mathbb{R})$ will be

$$\|g\|_{H^1(\mathbb{R}\times\cdots\times\mathbb{R})} = \sum_{\mathbf{0}\le\eta\le\mathbf{1}} \|\mathcal{H}_\eta g\|_{L^1(\mathbb{R}^n)}. \tag{18}$$

One-dimensional Hardy's inequality [**15**, (7.24)]

$$\int_{\mathbb{R}} \frac{|\widehat{g}(x)|}{|x|}\, dx \lesssim \|g\|_{H^1(\mathbb{R})} \tag{19}$$

is for this space naturally rewritten as

$$\int_{\mathbb{R}^n} \frac{|\widehat{g}(x)|}{|x_1|\ldots|x_n|}\, dx \lesssim \|g\|_{H^1(\mathbb{R}\times\cdots\times\mathbb{R})}. \tag{20}$$

It is given in [**18**] in dimension 2, with the reference to [**23**] for the proof. However, in [**23**] the result is also two-dimensional, which may lead to certain misunderstandings for higher dimensions. A simple inductive argument proves (20) in full generality. Indeed, using (19) as a basis of induction and assuming (20) to hold in dimension $n-1$, let us prove it for dimension n. Denoting, like above, F_η to be the Fourier transform with respect to the variables x_j for which $\eta_j = 1$, we have

$$\begin{aligned}\int_{\mathbb{R}^n} \frac{|\widehat{g}(x)|}{|x_1|\ldots|x_n|}\, dx &= \int_{\mathbb{R}} \frac{1}{|x_1|} \left[\int_{\mathbb{R}^{n-1}} \frac{|\widehat{g}(x)|}{|x_2|\ldots|x_n|}\, dx_2\ldots dx_n\right] dx_1\\ &\lesssim \int_{\mathbb{R}} \frac{1}{|x_1|} \sum_{\substack{\mathbf{0}\le\eta\le\mathbf{1},\\ \eta_1=0}} \|\mathcal{H}_\eta F_{(1,0,\ldots,0)} g(x_1,\cdot,\ldots,\cdot)\|_{L^1(\mathbb{R}^{n-1})}\, dx_1\\ = \sum_{\substack{\mathbf{0}\le\eta\le\mathbf{1},\\ \eta_1=0}} \int_{\mathbb{R}^{n-1}} \int_{\mathbb{R}} &\frac{|F_{(1,0,\ldots,0)}\mathcal{H}_\eta g(x_1,x_2,\ldots,x_n)|}{|x_1|}\, dx_1\, dx_2\ldots dx_n.\end{aligned}$$

Applying now (19) with respect to x_1, we complete the proof.

Note that

$$H^1(\mathbb{R} \times \cdots \times \mathbb{R}) \subset H^1(\mathbb{R}^n), \tag{21}$$

see, e.g., [**18**, Th.1]. In the same paper [**18**], one can find A. Uchiyama's example that this inclusion is proper.

Let us now introduce a version of Q which corresponds to our setting:

$$Q_H = \{g \in L^1(\mathbb{R}^n) : \|g\|_{Q_H} = \|g\|_{L^1(\mathbb{R}^n)} + \int_{\mathbb{R}^n} \frac{|\widehat{g}(x)|}{|x_1\ldots x_n|}\, dx < \infty\}. \tag{22}$$

3.4. Absolute continuity. In order to present multidimensional versions of Theorems 3.1 and 3.3, we should discuss a multidimensional notion of absolute continuity; see, e.g., [**5**] or [**17**]. There are several equivalent definitions. It suffices to define such functions to be those representable as

$$f(x) = \int_{-\infty}^{x_1} \cdots \int_{-\infty}^{x_n} h(u)\, du + \sum_{\eta\ne\mathbf{1}} h_\eta(x_\eta), \tag{23}$$

where marginal functions h_η depending on a smaller number of variables than n, in fact, $|\eta| < n$, since $|\eta| = n$ only if $\eta = \mathbf{1}$, are absolutely continuous on $\mathbb{R}^{|\eta|}$.

This inductive definition is correct since reduces to the usual absolute continuity on $\mathbb{R}$ for marginal functions of one variable. Locally absolute continuity means absolute continuity on every finite rectangle $[a,b] = [a_1,b_1]\times\cdots\times[a_n,b_n]$. In this case, $a_1,\ldots,a_n$, respectively, should replace $-\infty$ in (23). In [**39**] a less restrictive version of absolute continuity is given in a similar way, without assuming the absolute continuity of the marginal functions in (23). However, this is not applicable in our considerations.

The next statement "justifies" the study of the Fourier transform of a function with bounded Hardy's variation.

PROPOSITION 3.4. *Let the integral on the left-hand side of (20) be finite, or, equivalently, let $g\in Q_H$. Then g is the D^1 derivative of a function of bounded Hardy variation f which is absolutely continuous, $\lim\limits_{|x|\to\infty} f(x) = 0$ and its Fourier transform is integrable.*

PROOF. Let

$$f(x) = \int_{-\infty}^{x_1}\cdots\int_{-\infty}^{x_n} g(u)\,du. \tag{24}$$

First of all, this immediately gives absolute continuity. For simplicity, let us continue the proof in dimension two. We have

$$\begin{aligned}\int_{-\infty}^{\infty}\int_{-\infty}^{\infty} g(s,t)e^{-ixs}e^{-iyt}dsdt &= \int_{-\infty}^{\infty}\left[e^{-iyt}\int_{-\infty}^{t} g(s,v)\,dv\right]_{-\infty}^{\infty} e^{-ixs}ds\\ &\quad+iy\int_{-\infty}^{\infty}\left[\int_{-\infty}^{\infty}\int_{-\infty}^{t} g(s,v)\,dve^{-iyt}\,dt\right]e^{-ixs}ds\\ &= iy\int_{-\infty}^{\infty}\left[\int_{-\infty}^{\infty}\int_{-\infty}^{t} g(s,v)\,dve^{-iyt}\,dt\right]e^{-ixs}ds.\end{aligned}$$

Here the integrated term at infinity and those similar in the sequel vanish because of the assumption of the proposition on the finiteness of the integral on the left-hand side of (20). We then repeat the same procedure with respect to the other variable. Note that we are always able to change the order of operations because of the integrability of g.

$$\begin{aligned}&iy\int_{-\infty}^{\infty}\left[\int_{-\infty}^{\infty}\int_{-\infty}^{t} g(s,v)\,dve^{-iyt}\,dt\right]e^{-ixs}ds\\ &= iy\int_{-\infty}^{\infty}\left[\int_{-\infty}^{\infty}\int_{-\infty}^{t} g(s,v)\,dve^{-ixs}\,ds\right]e^{-iyt}dt\\ &= iy\int_{-\infty}^{\infty}\left[e^{-ixs}\int_{-\infty}^{s}\int_{-\infty}^{t} g(u,v)\,dv\,du\right]_{-\infty}^{\infty}e^{-iyt}dt\\ &+ixiy\int_{-\infty}^{\infty}\left[\int_{-\infty}^{\infty}\int_{-\infty}^{s}\int_{-\infty}^{t} g(u,v)\,du\,dve^{-ixs}\,ds\right]e^{-iyt}dt\\ &= ixiy\int_{-\infty}^{\infty}\int_{-\infty}^{\infty}\left[\int_{-\infty}^{s}\int_{-\infty}^{t} g(u,v)\,du\,dv\right]e^{-ixs}e^{-iyt}ds\,dt\\ &= ixiy\int_{-\infty}^{\infty}\int_{-\infty}^{\infty} f(s,t)e^{-ixs}e^{-iyt}ds\,dt.\end{aligned}$$

It follows now from $g \in Q_H$ that the L^1 norm of $\widehat{f}$ coincides modulo with the Q_H norm of the derivative of f. □

For the difference between the space of functions with bounded variation (the derivative of which is integrable) and the (Sobolev) space of functions with integrable derivative, see, e.g., [**41**].

4. Integrability of the Fourier transforms

In this section we obtain multidimensional results similar to, say, Theorem 3.1 and those related for the integrability of the Fourier transform

$$\widehat{f}(x) = \int_{\mathbb{R}^n} f(u)e^{-i\langle x,u\rangle}\,du. \tag{25}$$

Let us mention that various results of that kind can be found in the survey paper [**34**]. For example, in Section 7 of that paper many results are extensions of Zygmund's test for the absolute convergence of Fourier series of a function of bounded variation.

We begin with a simple extension of Theorem 3.3.

THEOREM 4.1. *Let f be of bounded Hardy's variation and f vanishes at infinity along with all $D^\eta f$ except that for $\eta = \mathbf{1}$. Let also f and the same $D^\eta f$ be locally absolutely continuous in the above sense. Then $\widehat{f} \in L^1(\mathbb{R}^n)$ if and only if $D^{\mathbf{1}} f \in Q_H$.*

PROOF. In (25), we simply integrate by parts in each of the variables. Observe that we do need the absolute continuity with respect to a part of variables here. The integrated terms vanish each time due to the assumptions of the theorem. After n steps we integrate modulo

$$\frac{\widehat{D^{\mathbf{1}} f}(x)}{|x_1 \dots x_n|}$$

in place of $\widehat{f}(x)$ which completes the proof. □

COROLLARY 4.2. *Let f be of bounded Hardy's variation and f vanishes at infinity along with all $D^\eta f$ except that for $\eta = \mathbf{1}$. Let also f and the same $D^\eta f$ be locally absolutely continuous in the above sense. If $D^{\mathbf{1}} f \in H^1(\mathbb{R} \times \dots \times \mathbb{R})$, then $\widehat{f} \in L^1(\mathbb{R}^n)$.*

PROOF. We just apply Hardy's inequality (20) to the integral that assigns, by virtue ofthe above theorem, the belonging of $D^{\mathbf{1}} f$ to Q_H. □

Such an extension, a direct consequence of Hardy's inequality is not that developed as Theorem 3.1 for the sine case. We now give an analog of Theorem 3.1. It is not every bit new, one can find similar versions in [**27**] (the same one is given in [**22**, Ch.3]) and in [**36**]. We hope that our present attempt is more transparent.

We consider (cf. (10))

$$\widehat{f}_\eta(x) = \int_{\mathbb{R}^n_+} f(u) \prod_{i:\eta_i=1} \cos x_i u_i \prod_{i:\eta_i=0} \sin x_i u_i\,du. \tag{26}$$

If $\eta = \mathbf{1}$ we have the purely cosine transform, while if $\eta = \mathbf{0}$ we have the purely sine transform, otherwise we have a mixed transform with both cosines and sines.

THEOREM 4.3. *Let $f : \mathbb{R}^n_+ \to \mathbb{C}$ be of bounded Hardy's variation on $\mathbb{R}^n_+$ and f vanishes at infinity along with all $D^\eta f$ except $\eta = \mathbf{1}$. Let also f and the same $D^\eta f$ be locally absolutely continuous in the above sense. If $D^{\mathbf{1}} f \in H^1_{\mathbf{0}}(\mathbb{R} \times \cdots \times \mathbb{R})$, then for $\eta = \mathbf{1}$ we have*

$$\int_{\mathbb{R}^n_+} |\widehat{f_{\mathbf{1}}}(x)|\, dx \lesssim \|D^{\mathbf{1}} f\|_{H^1_{\mathbf{0}}(\mathbb{R}\times\cdots\times\mathbb{R})}; \tag{27}$$

for $\eta \ne \mathbf{1}, \mathbf{0}$,

$$\int_{\mathbb{R}^n_+} |\widehat{f_\eta}(x)|\, dx \lesssim \sum_{\substack{\eta\ne\mathbf{1},\mathbf{0}\\ \mathbf{0}\le\kappa\le\eta}} \sum_{\substack{\mathbf{0}\le\chi\le\mathbf{1}-\eta\\ \mathbf{0}\le\zeta\le\chi}} \int_{\mathbb{R}^n_+} \frac{|\mathcal{H}^o_{\kappa+\zeta} D^{\eta+\chi} f(x)|}{\prod\limits_{j:\chi_j=\eta_j=0} x_j}\, dx; \tag{28}$$

and for $\eta = \mathbf{0}$,

$$\widehat{f_{\mathbf{0}}}(x) = f(\frac{\pi}{2x_1}, \dots, \frac{\pi}{2x_n}) \prod_{j=1}^n \frac{1}{x_j} + F(x), \tag{29}$$

with

$$\int_{\mathbb{R}^n_+} |F(x)|\, dx \lesssim \sum_{\substack{\mathbf{0}\le\chi\le\mathbf{1},\chi\ne\mathbf{0}\\ \mathbf{0}\le\zeta\le\chi}} \int_{\mathbb{R}^n_+} \frac{|\mathcal{H}^o_\zeta D^\chi f(x)|}{\prod\limits_{j:\chi_j=0} x_j}\, dx. \tag{30}$$

PROOF. We start with the proof in dimension two. The only cosine case is simple for any dimension: we integrate by parts $|\eta|$ times, in each of the variables and then apply (20).

The following two cases in dimension two illustrate along which lines the proof runs in other situations. In the mixed case, integrating by parts, we obtain

$$\begin{aligned}\widehat{f}_{(1,0)}(x,y) &= \int_{\mathbb{R}^2_+} f(u,v) \cos xu \sin yv\, du\, dv \\ &= -\frac{1}{x} \int_{\mathbb{R}^2_+} D^{(1,0)} f(u,v) \sin xu \sin yv\, du\, dv.\end{aligned}$$

Applying now (19) in x and then (11), we have

$$\begin{aligned}\int_0^\infty |\widehat{f}_{(1,0)}(x,y)|\, dx &\lesssim \int_0^\infty \left| \int_0^\infty D^{(1,0)} f(x,v) \sin yv\, dv \right| dx \\ &+ \int_0^\infty \left| \int_0^\infty \mathcal{H}^o_{(1,0)} D^{(1,0)} f(x,v) \sin yv\, dv \right| dx.\end{aligned}$$

Applying (12) and (13) to the inner integrals on the right-hand side, we finally get

$$\begin{aligned}\int_0^\infty\int_0^\infty |\widehat{f}_{(1,0)}(x,y)|\,dx\,dy \lesssim \int_{\mathbb{R}^2_+} &\frac{|D^{(1,0)}f(x,y)|}{y}\,dx\,dy\\ &+\int_{\mathbb{R}^2_+}|D^{(1,1)}f(x,y)|\,dx\,dy\\ &+\int_{\mathbb{R}^2_+}\frac{|\mathcal{H}^o_{(1,0)}D^{(1,0)}f(x,y)|}{y}\,dx\,dy\\ &+\int_{\mathbb{R}^2_+}|\mathcal{H}^o_{(0,1)}D^{(1,1)}f(x,y)|\,dx\,dy\\ &+\int_{\mathbb{R}^2_+}|\mathcal{H}^o_{(1,0)}D^{(1,1)}f(x,y)|\,dx\,dy\\ &+\int_{\mathbb{R}^2_+}|\mathcal{H}^o_{(1,1)}D^{(1,1)}f(x,y)|\,dx\,dy.\end{aligned}$$

Let us proceed to the only sine case. By (12) and (13) we get

$$\begin{aligned}\widehat{f}_{(0,0)}(x) &= \int_{\mathbb{R}^2_+} f(u,v)\sin xu\sin yv\,du\,dv\\ &= \frac{1}{y}\int_0^\infty f(u,\frac{\pi}{2y})\sin xu\,du + F_2(x,y),\end{aligned}\tag{31}$$

with

$$\begin{aligned}\int_0^\infty |F_2(x,y)|\,dy \lesssim \int_0^\infty &\left|\int_0^\infty D^{(0,1)}f(u,y)\sin xu\,du\right|dy\\ &+\int_0^\infty\left|\mathcal{H}^o_{(0,1)}\int_0^\infty D^{(0,1)}f(u,y)\sin xu\,du\right|dy.\end{aligned}$$

Applying again (12) and (13) to the leading term on the right-hand side of (31) and to the inner integrals that give estimates for F_2, we finally obtain

$$\begin{aligned}\widehat{f}_{(0,0)}(x,y) &= \int_{\mathbb{R}^2_+} f(u,v)\sin xu\sin yv\,du\,dv\\ &= \frac{1}{x}\frac{1}{y}f(\frac{\pi}{2x},\frac{\pi}{2y}) + F(x,y),\end{aligned}\tag{32}$$

with

$$\int_0^\infty\int_0^\infty |F(x,y)|\,dx\,dy \lesssim \int_{\mathbb{R}^2_+}\frac{|D^{(1,0)}f(x,y)|}{y}\,dx\,dy + \int_{\mathbb{R}^2_+}|D^{(1,1)}f(x,y)|\,dx\,dy$$
$$+\int_{\mathbb{R}^2_+}\frac{|D^{(0,1)}f(x,y)|}{x}\,dx\,dy + \int_{\mathbb{R}^2_+}\frac{|\mathcal{H}^o_{(1,0)}D^{(1,0)}f(x,y)|}{y}\,dx\,dy + \int_{\mathbb{R}^2_+}\frac{|\mathcal{H}^o_{(0,1)}D^{(0,1)}f(x,y)|}{y}\,dx\,dy$$
$$+\int_{\mathbb{R}^2_+}|\mathcal{H}^o_{(0,1)}D^{(1,1)}f(x,y)|\,dx\,dy + \int_{\mathbb{R}^2_+}|\mathcal{H}^o_{(1,0)}D^{(1,1)}f(x,y)|\,dx\,dy + \int_{\mathbb{R}^2_+}|\mathcal{H}^o_{(1,1)}D^{(1,1)}f(x,y)|\,dx\,dy.$$

Now, only some additional explanation about the general case are in order. If $\eta\ne\mathbf{1},\mathbf{0}$, then we integrate by parts $|\eta|$ times, in each of the variables u_j where $\eta_j=1$ and apply (20). For the rest of the variables, we apply, step by step, (12) and (13), as above in the two-dimensional purely sine case. The latter works in the purely sine case, that is, if $\eta=\mathbf{0}$, as well, but in addition to the estimates we obtain the leading term in (29). In the estimates (30) as well as in the "sine" part of (27) each of the variables is subject to one of the following three operations: either division by the argument with no differentiation, or differentiation, or the (odd) Hilbert transform of the derivative. The only difference from the given above two-dimensional case with details is that two variables is not enough for all three operations together, they appear at most in pairs, while for the larger dimension all the combinations may appear simultaneously. However, the proof itself goes along the same lines. □

Applying Lemma 3.2 in each of the needed variables, we can establish the following consequences of the above theorems. Written in a somewhat different form (less convenient in our opinion) they can be found in [**27**], [**34**, Th.12.10] and in [**22**, Ch.3]; for those in full detail in dimension two, see [**16**], [**17**], [**18**].

Corollary 4.4. *Let $f:\mathbb{R}^n\to\mathbb{C}$ be of bounded Hardy's variation on $\mathbb{R}^n$ and f vanishes at infinity along with all $D^\eta f$ except $\eta=\mathbf{1}$. Let also f and the same $D^\eta f$ be locally absolutely continuous in the above sense. Then for $\eta\ne\mathbf{0}$, we have*

$$\int_{\mathbb{R}^n_+}|\widehat{f}_\eta(x)|\,dx \lesssim \sum_{\mathbf{0}\le\kappa\le\mathbf{1}-\eta}\int_{\mathbb{R}^n_+}\prod_{i:\kappa_i=1}\frac{1}{x_i}$$
$$\left[\left(\prod_{j:\kappa_j=0}\frac{1}{x_j}\int_{x_j\le u_j\le 2x_j}du_j\right)|D^{\mathbf{1}-\kappa}f(u_{\mathbf{1}-\kappa},x_\kappa)|^q\right]^{\frac1q}dx, \tag{33}$$

and for $\eta=\mathbf{0}$, we have (29), with

$$\int_{\mathbb{R}^n_+} |F(x)|\,dx \lesssim \int_{\mathbb{R}^n_+} \left[\left(\prod_{j=1}^{n} \frac{1}{x_j} \int_{x_j \le u_j \le 2x_j} \right) |D^{\mathbf{1}} f(u)|^q \,du \right]^{\frac{1}{q}} dx, \tag{34}$$

for some $1 < q < \infty$.

COROLLARY 4.5. *Let* $f : \mathbb{R}^n \to \mathbb{C}$ *be of bounded Hardy's variation on* $\mathbb{R}^n$ *and* f *vanishes at infinity along with all* $D^\eta f$ *except* $\eta = \mathbf{1}$. *Let also* f *and the same* $D^\eta f$ *be locally absolutely continuous in the above sense. Then for* $\eta \neq \mathbf{0}$, *we have*

$$\int_{\mathbb{R}^n_+} |\widehat{f}_\eta(x)|\,dx$$

$$\lesssim \sum_{\mathbf{0} \le \kappa \le \mathbf{1}-\eta} \int_{\mathbb{R}^n_+} \prod_{i:\kappa_i=1} \frac{1}{x_i} \operatorname*{ess\,sup}_{\substack{x_j \le u_j \le 2x_j \\ j:\kappa_j=0}} |D^{\mathbf{1}-\kappa} f(u_{\mathbf{1}-\kappa}, x_\kappa)|\,dx, \tag{35}$$

and for $\eta = \mathbf{0}$, *we have* (29), *with*

$$\int_{\mathbb{R}^n_+} |F(x)|\,dx \lesssim \int_{\mathbb{R}^n_+} \operatorname*{ess\,sup}_{\substack{x_j \le u_j \le 2x_j \\ j=1,2,\dots,n}} |D^{\mathbf{1}} f(u)|\,dx. \tag{36}$$

5. Integrability of trigonometric series

The main problem in the integrability of trigonometric series reads as follows. We begin with the one-dimensional setting. Given a trigonometric series

$$a_0/2 + \sum_{k=1}^{\infty} (a_k \cos kx + b_k \sin kx),$$

find assumptions on the sequences of coefficients $\{a_k\}$, $\{b_k\}$ under which the series is the Fourier series of an integrable function. Frequently, the series

$$a_0/2 + \sum_{k=1}^{\infty} a_k \cos kx \tag{37}$$

and

$$\sum_{k=1}^{\infty} b_k \sin kx \tag{38}$$

are investigated separately, since there is a difference in their behavior. Usually, integrability of (38) requires additional assumptions. However, one of the basic assumptions for both (37) and (38) is that the sequence $\{a_k\}$ or $\{b_k\}$ is of bounded variation, written $\{a_k\} \in bv$ or $\{b_k\} \in bv$, that is, satisfies the condition

$$\sum_{k=1}^{\infty} |\Delta a_k| < \infty,$$

where $\Delta a_k = a_k - a_{k+1}$, and similarly for Δb_k.

As an application of integrability results for the Fourier transform, we can obtain integrability results for (37) and (38) given above, for (38) even in a stronger, asymptotic form, see [**27**].

We can relate this problem to a similar one for Fourier transforms as follows. First, given series (37) or (38) with the null sequence of coefficients being in an appropriate sequence space, set for $x \in [k, k+1)$

$$\begin{aligned} A(x) &= a_k + (k-x)\Delta a_k, \qquad a_0 = 0, \\ B(x) &= b_k + (k-x)\Delta b_k. \end{aligned}$$

So, we construct a corresponding function by means of linear interpolation of the sequence of coefficients. Of course, one may interpolate not only linearly, but there are no problems where this might be of importance so far.

Further, for functions of bounded variation φ, to pass from series to integrals and vice versa, we will make use of (1).

We are going to provide the same in several dimensions, that is, to generalize (1), dealing with functions and sequences with bounded Hardy variation only.

5.1. Temporary bridge. We are now in a position to formulate and prove our main results.

THEOREM 5.1. *Let $f(x)$ be a function of bounded Hardy variation on $\mathbb{R}^n$. Let, in addition, VV_η vanish at infinity for all η except $\eta = \mathbf{1}$. Then*

$$\sup_{|x_j|\le\pi, j=1,2,\dots,n} \Bigg| \sum_{k\in\mathbb{Z}^n} f(k)e^{i\langle x,k\rangle} \tag{39}$$

$$- \sum_{\eta\ne\mathbf{0},\mathbf{1}} (-1)^{|\eta|} \sum_{k_\eta} \int_{\mathbb{R}^{n-|\eta|}} f(k_\eta, u_{\mathbf{1}-\eta}) e^{i\langle x_{\mathbf{1}-\eta}, u_{\mathbf{1}-\eta}\rangle} du_{\mathbf{1}-\eta}\, e^{i\langle x_\eta, k_\eta\rangle} \tag{40}$$

$$+ (-1)^d \int_{\mathbb{R}^n} f(u) e^{i\langle x,u\rangle} du \Bigg| \lesssim \|f\|_{VH(f)}. \tag{41}$$

For the sake of simplicity, we formulate and prove the two-dimensional version. The general case is treated in the completely same way.

THEOREM 5.2. *Let $f(x,y)$ be a function of bounded Hardy variation on $\mathbb{R}^2$. Let, in addition, $\lim\limits_{|x|+|y|\to\infty} f(x,y), VV_{(1,0)}(y), VV_{(0,1)}(x) = 0$. Then*

$$\begin{aligned} \sup_{|x|,|y|\le\pi} \Bigg| &\sum_{k,l} f(k,l)e^{i(kx+ly)} - \sum_k \int_{\mathbb{R}} f(k,v)e^{iyv} dv\, e^{ikx} \\ &- \sum_l \int_{\mathbb{R}} f(u,l) e^{ixu} du\, e^{ily} + \int_{\mathbb{R}^2} f(u,v) e^{i(xu+yv)} du\, dv \Bigg| \lesssim \|f\|_{VH(f)}. \end{aligned} \tag{42}$$

PROOF. Denoting $s(t) = \frac{t}{2\sin\frac{t}{2}}$ and using a standard trick (see, e.g., [**45**, Ch.5, Th.2.29])

$$e^{ikx} = s(x) \int_{k-\frac12}^{k+\frac12} e^{ixt} dt, \tag{43}$$

we have

$$\sum_{k,l} f(k,l)e^{i(kx+ly)} - \sum_k \int_{\mathbb{R}} f(k,v)e^{iyv}dv\, e^{ikx}$$

$$= \sum_k s(y) \sum_l \int_{l-\frac{1}{2}}^{l+\frac{1}{2}} [f(k,l) - f(k,v)]e^{iyv}dv\, e^{ikx}$$

$$+ \sum_k (s(y)-1) \int_{\mathbb{R}} f(k,v)e^{iyv}dv\, e^{ikx}. \tag{44}$$

We now calculate

$$\sum_{k,l} f(k,l)e^{i(kx+ly)} - \sum_k \int_{\mathbb{R}} f(k,v)e^{iyv}dv\, e^{ikx}$$

$$-\sum_l \int_{\mathbb{R}} f(u,l)e^{ixu}du\, e^{ily} + \int_{\mathbb{R}^2} f(u,v)e^{i(xu+yv)}du\,dv. \tag{45}$$

We can represent two last summands in (45) as

$$- s(y) \sum_l \int_{l-\frac{1}{2}}^{l+\frac{1}{2}} [f(u,l) - f(u,v)]e^{i(xu+yv)}du\,dv$$

$$- (s(y)-1) \int_{\mathbb{R}^2} f(u,v)e^{i(xu+yv)}du\,dv.$$

This means that (45) is just equivalent to the application of the procedure used in (44) in l to the left-hand side of (44) but in k. In other words, we can deal with

$$\sum_{k,l} f(k,l)e^{i(kx+ly)} - \sum_k \int_{\mathbb{R}} f(k,v)e^{iyv}dv\, e^{ikx}$$

$$- s(y) \sum_l \int_{l-\frac{1}{2}}^{l+\frac{1}{2}} [f(u,l) - f(u,v)]e^{i(xu+yv)}du\,dv$$

$$- (s(y)-1) \int_{\mathbb{R}^2} f(u,v)e^{i(xu+yv)}du\,dv$$

rather than with (45). This value is equal to

$$s(x)s(y) \sum_{k,l} \int_{k-\frac{1}{2}}^{k+\frac{1}{2}} \int_{l-\frac{1}{2}}^{l+\frac{1}{2}} [f(k,l) - f(k,v) - f(u,l) + f(u,v)]e^{i(xu+yv)}du\,dv$$

$$+s(y)(s(x)-1) \sum_l \int_{l-\frac{1}{2}}^{l+\frac{1}{2}} \int_{\mathbb{R}} [f(u,l) - f(u,v)]e^{i(xu+yv)}du\,dv$$

$$+s(x)(s(y)-1) \sum_k \int_{k-\frac{1}{2}}^{k+\frac{1}{2}} \int_{\mathbb{R}} [f(k,v) - f(u,v)]e^{i(xu+yv)}du\,dv$$

$$+(s(x)-1)(s(y)-1) \int_{\mathbb{R}^2} f(u,v)e^{i(xu+yv)}du\,dv. \tag{46}$$

Observe that $1 \le |s(t)| \le \pi/2$ for $0 < |t| < \pi$. This immediately implies that the first summand in (46) is dominated by $\|f\|_{VH(f)}$. Taking into account that $\frac{s(t)-1}{t}$ is also bounded, we get the same estimate for the last summand in (46) by integrating by parts repeatedly in each variable as a Stieltjes integral. The intermediate two summands are treated in the same way, by integrating by parts in x where we have the factor $s(x)-1$ and in y where we have the factor $s(y)-1$. The proof is complete. □

5.2. Stable bridge. To be able to apply the results for the Fourier transform to trigonometric series, it is desirable to make use of Theorem 5.1. In its assertion (39) and (41) fit this claim. However, (40) is a mixture of sums and integrals, and it is desirable to provide further calculations in order to get rid of the sums in (40). The price for it is twofold. First, we let the function be from a specific function space in addition to the boundedness of Hardy's variation. Such a space is natural in the problems of integrability of the Fourier transform of a function of bounded Hardy's variation. Secondly, we estimate the deviation between the series and the integral in the L^1 norm over $\mathbb{T}^n$, with $\mathbb{T} = [-\pi, \pi)$, rather than in the sup-norm. This is, in a sense, resetting, since in the first result of that sort by Belinsky L^p norms were estimated.

THEOREM 5.3. *Let f be of bounded Hardy's variation and f vanishes at infinity along with all $D^\eta f$ except that for $\eta = \mathbf{1}$. Let also f and the same $D^\eta f$ be locally absolutely continuous in the above sense. Then*

$$\sum_{k\in\mathbb{Z}^n} f(k)e^{i\langle x,k\rangle} = \int_{\mathbb{R}^n} f(u)e^{i\langle x,u\rangle}du + \Phi(x), \tag{47}$$

where

$$\int_{\mathbb{T}^n} |\Phi(x)|\,dx \lesssim \sum_{\eta\neq\mathbf{0}} \int_{\mathbb{R}^{|\eta|}} \Big(\prod_{j:\eta_j=0} \int_{\mathbb{T}} \frac{1}{|x_j|}\Big) |F_{\mathbf{1}-\eta}(D^{\mathbf{1}}f)(x_{\mathbf{1}-\eta}, u_\eta)|\,dx_{\mathbf{1}-\eta}\,du_\eta. \tag{48}$$

PROOF. Thus, we deal with every of the summands, corresponding to $\eta \neq \mathbf{0}, \mathbf{1}$, in (40):

$$I_\eta = \sum_{k_\eta} \int_{\mathbb{R}^{n-|\eta|}} f(k_\eta, u_{\mathbf{1}-\eta})e^{i\langle x_{\mathbf{1}-\eta}, u_{\mathbf{1}-\eta}\rangle}du_{\mathbf{1}-\eta}\, e^{i\langle x_\eta, k_\eta\rangle}.$$

We then apply to the $|\eta|$-dimensional sum in I_η the corresponding assertion (39)-(41) in Lemma 5.1. It will be controlled by Hardy's variation of $F_{\mathbf{1}-\eta}f$ on $\mathbb{R}^{|\eta|}$.

In the integral, integrate by parts $n - |\eta|$ times, repeatedly in each of the variables u_j for which $\eta_j = 0$. Since integrated terms vanish, we obtain

$$I_\eta = \sum_{k_\eta} \prod_{j:\eta=0} \frac{1}{ix_j} F_{\mathbf{1}-\eta}f(k_\eta, -x_{\mathbf{1}-\eta})\, e^{i\langle x_\eta, k_\eta\rangle}.$$

The L^1 norm of this value gives a part of (48). It remains to observe that due to the signs $\sum\limits_{\eta\neq\mathbf{0}} (-1)^{|\eta|} = -1$.

Mixed relations of type I_η but of lower dimension are treated in the same manner, which completes the proof. □

5.3. Multidimensional sequence spaces. For the sequence $a = \{a_k\} \in \ell^1$, the discrete Hilbert transform is defined for $m \in \mathbb{Z}$ as (see, e.g., [**25**, (13.127)], [**7**])

$$\hbar a(m) = \sum_{\substack{k=-\infty \\ k \neq m}}^{\infty} \frac{a_k}{m-k}. \tag{49}$$

If the sequence a is either even or odd, the corresponding Hilbert transforms $\hbar_e$ and $\hbar_o$ may be expressed in a special form (see, e.g., [**25**, (13.130) and (13.131)]). More precisely, if a is even, with $a_0 = 0$, we have $\hbar_e(0) = 0$ and for $m = 1, 2, \ldots$

$$\hbar_e a(m) = \sum_{\substack{k=1 \\ k \neq m}}^{\infty} \frac{2ma_k}{m^2 - k^2} + \frac{a_m}{2m}. \tag{50}$$

If a is odd, with $a_0 = 0$, we have for $m = 0, 1, 2, \ldots$

$$\hbar_o a(m) = \sum_{\substack{k=1 \\ k \neq m}}^{\infty} \frac{2ka_k}{m^2 - k^2} - \frac{a_m}{2m}. \tag{51}$$

Of course, $\frac{a_0}{0}$ is considered to be zero.

In this case, similarly to that for functions, we can use the notation

$$h^1(\mathbb{Z} \times \cdots \times \mathbb{Z}) := h^1_n(\mathbb{Z} \times \cdots \times \mathbb{Z})$$

and a modified definition on the basis of the discrete Hilbert transforms applied to each variable. Such a transform applied to the j-th variable will be defined $\hbar_j$ and, consequently, $\hbar_j \hbar_k \ldots \hbar_l := \hbar_{jk\ldots l}$. However, there is a more convenient notation for this. We shall define $h^1_\eta(\mathbb{Z} \times \cdots \times \mathbb{Z})$ if the discrete Hilbert transform with respect to each of the j-th variables, for which $\eta_j = 1$, is the even discrete Hilbert transform, while for the rest it is odd. We will be especially interested in the case where $\eta = \mathbf{0}$, that is, all the discrete Hilbert transforms are odd. The corresponding notation is similar to that in dimension one: $h^1_{\mathbf{0}}(\mathbb{Z} \times \cdots \times \mathbb{Z})$.

By this, the norm of a sequence a in $h^1(\mathbb{Z} \times \cdots \times \mathbb{Z})$ will be

$$\|a\|_{h^1(\mathbb{Z} \times \cdots \times \mathbb{Z})} = \sum_{\mathbf{0} \leq \eta \leq \mathbf{1}} \|\hbar_\eta a\|_{\ell^1(\mathbb{Z}^n)}. \tag{52}$$

5.4. Integrability of multiple trigonometric series. We study the trigonometric series

$$\sum_{k \in \mathbb{Z}^n_+} a_k \prod_{i:\eta_i = 1} \cos x_i k_i \prod_{i:\eta_i = 0} \sin x_i k_i, \tag{53}$$

with the null sequence of coefficients a_k being in an appropriate sequence space, set for $k \in [k_1, k_1 + 1) \times \cdots \times [k_n, k_n + 1)$

$$A(x) = \sum_{\mathbf{0} \leq \eta \leq \mathbf{1}} \prod_{j:\eta_j = 1} (k_j - x_j) \, \Delta_\eta a_k, \tag{54}$$

where

$$\Delta_\eta a_k = \prod_{j:\eta_j = 1} (a_{k_1,\ldots,k_{j-1},k_j,k_{j+1},\ldots,k_n} - a_{k_1,\ldots,k_{j-1},k_j+1,k_{j+1},\ldots,k_n}).$$

In the multivariate case, it is natural to consider trigonometric series with the sequence of coefficients of bounded Hardy variation, that is, such that

$$\sum_{k\in\mathbb{Z}_+^n} |\Delta_\eta a_k| < \infty$$

for each $\eta \neq \mathbf{0}$. Indeed, such a series converges almost everywhere in the Pringsheim sense (see, e.g., [**37**]).

THEOREM 5.4. *Let a be a sequence with bounded Hardy's variation on $\mathbb{Z}_+^n$ and a vanishes at infinity along with all $\Delta_\eta a$ except $\eta = \mathbf{1}$. If $\Delta_{\mathbf{1}} a \in h_{\mathbf{0}}^1(\mathbb{Z} \times \cdots \times \mathbb{Z})$, then for $\eta = \mathbf{1}$ we have*

$$\int_{\mathbb{T}_+^n} |\sum_{\mathbb{Z}_+^n} a_k \prod_{i=1}^n \cos x_i k_i|\, dx \lesssim \|\Delta_{\mathbf{1}} a\|_{h_{\mathbf{0}}^1(\mathbb{Z}\times\cdots\times\mathbb{Z})}; \tag{55}$$

for $\eta \neq \mathbf{1}, \mathbf{0}$,

$$\begin{aligned} &\int_{\mathbb{T}_+^n} |\sum_{k\in\mathbb{Z}_+^n} a_k \prod_{i:\eta_i=1} \cos x_i k_i \prod_{i:\eta_i=0} \sin x_i k_i|\, dx \\ &\lesssim \sum_{\substack{\eta\neq\mathbf{1},\mathbf{0}\\ \mathbf{0}\le\kappa\le\eta}} \sum_{\substack{\mathbf{0}\le\chi\le\mathbf{1}-\eta\\ \mathbf{0}\le\zeta\le\chi}} \sum_{k\in\mathbb{Z}_+^n} \frac{|\hbar_{\kappa+\zeta}^o \Delta_{\eta+\chi} a_k|}{\prod\limits_{j:\chi_j=\eta_j=0} (k_j+1)}; \end{aligned} \tag{56}$$

and for $\eta = \mathbf{0}$,

$$\sum_{k\in\mathbb{Z}_+^n} a_k \prod_{i=1}^n \sin x_i k_i = A(\frac{\pi}{2x_1}, \dots, \frac{\pi}{2x_n}) \prod_{j=1}^n \frac{1}{x_j} + F(x), \tag{57}$$

with

$$\int_{\mathbb{T}_+^n} |F(x)|\, dx \lesssim \sum_{\substack{\mathbf{0}\le\chi\le\mathbf{1},\chi\neq\mathbf{0}\\ \mathbf{0}\le\zeta\le\chi}} \sum_{k\in\mathbb{Z}_+^n} \frac{|\hbar_\zeta^o \Delta_\chi a_k|}{\prod\limits_{j:\chi_j=0} (k_j+1)}. \tag{58}$$

PROOF. Given the sequence from the theorem, we define the function $A(x)$ by (54). This function naturally satisfies the assumptions of Theorem 4.3. To see that (48) is satisfied, we apply to the integrals in it Hardy's inequality (20) but for lower dimension (or (19) in dimension one). We then are within the scope of the integrability of the Fourier transforms of Theorem 4.3. The bridge (47) allows us to conclude that (55)-(58) take place, in each of the appropriate cases, just following from (27)-(30). □

As above for functions, using similar embeddings for sequences, we can establish the following consequences of Theorem 5.4.

COROLLARY 5.5. *Let a be a sequence with bounded Hardy's variation on $\mathbb{Z}_+^n$ and a vanishes at infinity along with all $\Delta^\eta a$ except $\eta = \mathbf{1}$. Then for $\eta \neq \mathbf{0}$, we have*

$$\begin{aligned} &\int_{\mathbb{T}_+^n} |\sum_{k\in\mathbb{Z}_+^n} a_k \prod_{i:\eta_i=1} \cos x_i k_i \prod_{i:\eta_i=0} \sin x_i k_i|\, dx \lesssim \sum_{\mathbf{0}\le\kappa\le\mathbf{1}-\eta} \sum_{k\in\mathbb{Z}_+^n} \prod_{i:\kappa_i=1} \frac{1}{k_i} \\ &\left[\Big(\prod_{j:\kappa_j=0} \frac{1}{k_j} \sum_{k_j\le m_j<2k_j}\Big) |\Delta_{\mathbf{1}-\kappa} a_{m_{\mathbf{1}-\kappa},k_\kappa}|^q\right]^{\frac{1}{q}}, \end{aligned} \tag{59}$$

and for $\eta = \mathbf{0}$, *we have* (57), *with*

$$\int_{\mathbb{T}_+^n} |F(x)|\,dx \lesssim \sum_{k\in\mathbb{Z}_+^n} \left[\left(\prod_{j=1}^{n} \frac{1}{k_j} \sum_{k_j\le m_j<2k_j} \right) |\Delta_{\mathbf{1}} a_m|^q \right]^{\frac{1}{q}}, \tag{60}$$

for some $1 < q < \infty$.

COROLLARY 5.6. *Let* a *be a sequence with bounded Hardy's variation on* $\mathbb{Z}_+^n$ *and* a *vanishes at infinity along with all* $\Delta^\eta a$ *except* $\eta = \mathbf{1}$. *Then for* $\eta \ne \mathbf{0}$, *we have*

$$\int_{\mathbb{R}_+^n} |\sum_{k\in\mathbb{Z}_+^n} a_k \prod_{i:\eta_i=1} \cos x_i k_i \prod_{i:\eta_i=0} \sin x_i k_i|\,dx$$

$$\lesssim \sum_{\mathbf{0}\le\kappa\le\mathbf{1}-\eta} \sum_{k\in\mathbb{Z}_+^n} \prod_{i:\kappa_i=1} \frac{1}{k_i} \max_{\substack{k_j\le m_j<2k_j\\ j:\kappa_j=0}} |\Delta_{\mathbf{1}-\kappa} a_{m_{\mathbf{1}-\kappa},k_\kappa}|, \tag{61}$$

and for $\eta = \mathbf{0}$, *we have* (57), *with*

$$\int_{\mathbb{T}_+^n} |F(x)|\,dx \lesssim \sum_{k\in\mathbb{Z}_+^n} \max_{\substack{k_j\le m_j<2k_j\\ j=1,2,\ldots,n}} |\Delta_{\mathbf{1}} a_m|. \tag{62}$$

Acknowledgements

The matter presented became a self-contained work as a preparation to the talk given by the author at a workshop "Complex and Harmonic Analysis" held in the Holon Inst. Tech. in June, 2014, which, in turn, was a part (satellite conference) of the joint AMS-IMU meeting. The author is indebted to the organizers of that workshop, first of all to A. Golberg.

The author thanks the referee of the paper for valuable remarks and suggestions.

References

[1] C. Raymond Adams and James A. Clarkson, *Properties of functions $f(x,y)$ of bounded variation*, Trans. Amer. Math. Soc. **36** (1934), no. 4, 711–730, DOI 10.2307/1989819. MR1501762

[2] G. Alexits, Convergence problems of orthogonal series, Akadémiai Kiadó, Budapest.

[3] E. S. Belinsky, *On asymptotic behavior of integral norms of trigonometric polynomials*, Metric Questions of the Theory of Functions and Mappings, Nauk. Dumka, Kiev **6**(1975), 15–24 (Russian).

[4] E. S. Belinskii, E. R. Liflyand, and R. M. Trigub, *The Banach algebra A^* and its properties*, J. Fourier Anal. Appl. **3** (1997), no. 2, 103–129, DOI 10.1007/s00041-001-4052-1. MR1438893 (98a:42003)

[5] E. Berkson and T. A. Gillespie, *Absolutely continuous functions of two variables and well-bounded operators*, J. London Math. Soc. (2) **30** (1984), no. 2, 305–321, DOI 10.1112/jlms/s2-30.2.305. MR771426 (86c:47044)

[6] Arne Beurling, *On the spectral synthesis of bounded functions*, Acta Math. **81** (1948), 14. MR0027891 (10,371c)

[7] Santiago Boza and María J. Carro, *Discrete Hardy spaces*, Studia Math. **129** (1998), no. 1, 31–50. MR1611926 (99c:42025)

[8] D. Borwein, *Linear functionals connected with strong Cesàro summability*, J. London Math. Soc. **40** (1965), 628–634. MR0185426 (32 #2893)

[9] Sun-Yung A. Chang and Robert Fefferman, *Some recent developments in Fourier analysis and H^p-theory on product domains*, Bull. Amer. Math. Soc. (N.S.) **12** (1985), no. 1, 1–43, DOI 10.1090/S0273-0979-1985-15291-7. MR766959 (86g:42038)

[10] James A. Clarkson and C. Raymond Adams, *On definitions of bounded variation for functions of two variables*, Trans. Amer. Math. Soc. **35** (1933), no. 4, 824–854, DOI 10.2307/1989593. MR1501718

[11] Robert Fefferman, *Some recent developments in Fourier analysis and H^p theory on product domains. II*, Function spaces and applications (Lund, 1986), Lecture Notes in Math., vol. 1302, Springer, Berlin, 1988, pp. 44–51, DOI 10.1007/BFb0078862. MR942256 (89g:42036)

[12] T. M. Flett, *Some elementary inequalities for integrals with applications to Fourier transforms*, Proc. London Math. Soc. (3) **29** (1974), 538–556. MR0360959 (50 #13406)

[13] G. A. Fomin, *A class of trigonometric series* (Russian), Mat. Zametki **23** (1978), no. 2, 213–222. MR0487218 (58 #6878)

[14] S. Fridli, *Hardy spaces generated by an integrability condition*, J. Approx. Theory **113** (2001), no. 1, 91–109, DOI 10.1006/jath.2001.3614. MR1866249 (2003d:42041)

[15] José García-Cuerva and José L. Rubio de Francia, *Weighted norm inequalities and related topics*, North-Holland Mathematics Studies, vol. 116, North-Holland Publishing Co., Amsterdam, 1985. Notas de Matemática [Mathematical Notes], 104. MR807149 (87d:42023)

[16] Dang Vu Giang and Ferenc Móricz, *On the L^1 theory of Fourier transforms and multipliers*, Acta Sci. Math. (Szeged) **61** (1995), no. 1-4, 293–304. MR1377366 (97i:42006)

[17] Dăng Vũ Giang and Ferenc Móricz, *Lebesgue integrability of double Fourier transforms*, Acta Sci. Math. (Szeged) **58** (1993), no. 1-4, 299–328. MR1264239 (95c:42014)

[18] Dang Vu Giang and Ferenc Móricz, *Hardy spaces on the plane and double Fourier transforms*, J. Fourier Anal. Appl. **2** (1996), no. 5, 487–505, DOI 10.1007/s00041-001-4040-5. MR1412065 (98b:42037)

[19] G. H. Hardy, *On double Fourier series*, Quart. J. Math. **37** (1906), 53–79.

[20] C. S. Herz, *Lipschitz spaces and Bernstein's theorem on absolutely convergent Fourier transforms*, J. Math. Mech. **18** (1968/69), 283–323. MR0438109 (55 #11028)

[21] E. W. Hobson, *The theory of functions of a real variable and the theory of Fourier's series. Vol. 1*, third edition (University Press, Cambridge, 1927; Dover, New York, 1957).

[22] Alex Iosevich and Elijah Liflyand, *Decay of the Fourier transform*, Birkhäuser/Springer, Basel, 2014. Analytic and geometric aspects. MR3308120

[23] Björn Jawerth and Alberto Torchinsky, *A note on real interpolation of Hardy spaces in the polydisk*, Proc. Amer. Math. Soc. **96** (1986), no. 2, 227–232, DOI 10.2307/2046158. MR818449 (87h:42033)

[24] R. L. Johnson and C. R. Warner, *The convolution algebra $H^1(R)$*, J. Funct. Spaces Appl. **8** (2010), no. 2, 167–179, DOI 10.1155/2010/524036. MR2667874 (2011i:46053)

[25] Frederick W. King, *Hilbert transforms. Vol. 1*, Encyclopedia of Mathematics and its Applications, vol. 124, Cambridge University Press, Cambridge, 2009. MR2542214 (2011c:42003)

[26] H. Kober, *A note on Hilbert's operator*, Bull. Amer. Math. Soc. **48** (1942), 421–427. MR0006775 (4,40d)

[27] E. R. Liflyand, *On asymptotics of Fourier transform for functions of certain classes* (English, with Russian summary), Anal. Math. **19** (1993), no. 2, 151–168, DOI 10.1007/BF02905077. MR1239677 (94j:42014)

[28] E. R. Liflyand, *Lebesgue constants of multiple Fourier series*, Online J. Anal. Comb. **1** (2006), Art. 5, 112 pp. (electronic). MR2249993 (2007g:42017)

[29] Elijah Liflyand, *A counterexample of embedding of spaces*, Georgian Math. J. **16** (2009), no. 4, 757–760. MR2640803 (2011b:44002)

[30] Elijah Liflyand, *Fourier transform versus Hilbert transform*, Ukr. Mat. Visn. **9** (2012), no. 2, 209–218, DOI 10.1007/s10958-012-1048-0; English transl., J. Math. Sci. (N. Y.) **187** (2012), no. 1, 49–56. MR2986607

[31] E. Liflyand, *Interaction between the Fourier transform and the Hilbert transform*, Acta Comm. Uni. Tartu. Math. **18** (2014), 19–32.

[32] E. Liflyand, U. Stadtmüller, and R. Trigub, *An interplay of multidimensional variations in Fourier analysis*, J. Fourier Anal. Appl. **17** (2011), no. 2, 226–239, DOI 10.1007/s00041-010-9136-3. MR2776774 (2012a:42015)

[33] E. Liflyand and U. Stadtmüller, *A multidimensional Euler-Maclaurin formula and an application*, Complex analysis and dynamical systems V, Contemp. Math., vol. 591, Amer. Math. Soc., Providence, RI, 2013, pp. 183–194, DOI 10.1090/conm/591/11836. MR3155687

[34] E. Liflyand, S. Samko, and R. Trigub, *The Wiener algebra of absolutely convergent Fourier integrals: an overview*, Anal. Math. Phys. **2** (2012), no. 1, 1–68, DOI 10.1007/s13324-012-0025-6. MR2891740 (2012m:42009)

[35] Elijah Liflyand and Sergey Tikhonov, *Weighted Paley-Wiener theorem on the Hilbert transform* (English, with English and French summaries), C. R. Math. Acad. Sci. Paris **348** (2010), no. 23-24, 1253–1258, DOI 10.1016/j.crma.2010.10.028. MR2745334 (2011j:42017)

[36] E. Liflyand and R. Trigub, *Conditions for the absolute convergence of Fourier integrals*, J. Approx. Theory **163** (2011), no. 4, 438–459, DOI 10.1016/j.jat.2010.11.001. MR2775139 (2012c:42020)

[37] Ferenc Móricz, *On the integrability and L^1-convergence of double trigonometric series*, Studia Math. **98** (1991), no. 3, 203–225. MR1115190 (92h:42022)

[38] Elias M. Stein and Guido Weiss, *Introduction to Fourier analysis on Euclidean spaces*, Princeton University Press, Princeton, N.J., 1971. Princeton Mathematical Series, No. 32. MR0304972 (46 #4102)

[39] A. A. Talalyan and G. G. Gevorkyan, *Representation of absolutely continuous functions of several variables* (Russian), Acta Sci. Math. (Szeged) **54** (1990), no. 3-4, 277–283. MR1096808 (92b:26025)

[40] S. A. Teljakovskiĭ, *Integrability conditions for trigonometrical series and their application to the study of linear summation methods of Fourier series* (Russian), Izv. Akad. Nauk SSSR Ser. Mat. **28** (1964), 1209–1236. MR0172067 (30 #2293)

[41] S. A. Telyakovskiĭ and V. N. Temlyakov, *On the convergence of Fourier series of functions of several variables of bounded variation* (Russian, with Russian summary), Mat. Zametki **61** (1997), no. 4, 583–595, DOI 10.1007/BF02354993; English transl., Math. Notes **61** (1997), no. 3-4, 484–494. MR1620007 (99e:42020)

[42] R. M. Trigub, *Absolute convergence of Fourier integrals, summability of Fourier series and approximation by polynomials of functions on a torus* (Russian), Izv. Akad. Nauk SSSR Ser. Mat. **44** (1980), no. 6, 1378–1409, 1439. MR603581 (82d:42011)

[43] R. M. Trigub, *A generalization of the Euler-Maclaurin formula* (Russian), Mat. Zametki **61** (1997), no. 2, 312–316, DOI 10.1007/BF02355737; English transl., Math. Notes **61** (1997), no. 1-2, 253–257. MR1619963 (99d:42006)

[44] R. M. Trigub, E. S. Belinsky, *Fourier Analysis and Appoximation of Functions*, Kluwer, 2004.

[45] A. Zygmund, *Trigonometric series, Vol. I, II*, Cambridge Univ. Press, Cambridge, U.K., 1966.

Department of Mathematics, Bar-Ilan University, 52900 Ramat-Gan, Israel
E-mail address: liflyand@math.biu.ac.il

Contemporary Mathematics
Volume **659**, 2016
http://dx.doi.org/10.1090/conm/659/13148

Optimal irrigation scheduling for wheat production in the Canadian prairies: A modelling study

Raphael Linker and Ilya Ioslovich

ABSTRACT. Food security and reducing malnutrition of the growing world population is a permanent issue in agricultural research. Wheat (Triticum aestivum L.) is an important part of the world food market. The Canadian Prairies comprise the provinces of Alberta, Saskatchewan, and Manitoba which produce very large quantities of wheat, mostly for export. While in many other countries, and in particular in the Middle East and North Africa (MENA) region, wheat is grown with supplemental irrigation, the common practice in Canadian Prairies is to grow wheat as a rain-fed crop. Taking into account the growing pressure on fresh water resources demand and factors of the profitability the possibly optimal use of the irrigation water should be determined according to the its deficit in the region. Thus the set of the optimization problems must be solved: find maximal wheat yield with respect to the limited irrigation water quota and given weather and hydrological data. Systematically solving this problem for different values of the water quota (W) allows to create an irrigation water use efficiency function $IWUE = Y(W)$ which presents the yield as a function of the total irrigation water applied optimally during the season. We demonstrate this approach using the FAO model AquaCrop in conjunction with the TOMLAB optimization library. The results of the model-based optimization show that for this specific case study (Carman,Manitoba 2006) wheat yield could be roughly doubled with limited amount of irrigation water. The average increase of the yield in the range of $0 - 100$ mm of irrigation water was more than 20 kg/ha per mm of water.

1. INTRODUCTION

The growing demand for fresh water for direct human consumption is generating increasing pressure on industry, and especially agriculture, to become more efficient in terms of water use, especially in regions plagued with water scarcity and water-related conflicts.

The Middle East and North Africa Region (MENA) is one of the driest regions in the world and the lack of rainfall severely limits food production in many of MENA countries. According to UN population projections, this region has one of the highest population growth rates with a base population of 517 million people in 2010, which is expected to reach 818 million people by 2050. This represents a 38% increase in 40 years, [**1**]. Somewhat surprisingly, wheat is a traditional source

2010 *Mathematics Subject Classification.* Primary 54C40, 14E20; Secondary 46E25, 20C20.
The second author was supported in part by the EU project FIGARO.

of food in many MENA countries. Although in some cases wheat is imported (e.g. in Egypt), in the majority of MENA countries it is grown locally using irrigation despite the chronic lack of water in these countries.

In 2010 the MENA region imported 41 million tons of wheat which, assuming a water requirement of 1300 m^3/ton per year, translates into a saving of 53 billion m^3 of so-called 'virtual water'. The countries that save the largest amount of water in this way are Algeria, Iraq, Iran, Morocco and Yemen, where the water requirement for local production is more than 2000 m^3/ton, [**1**]. At the same time [**2**] note that "serious cultural problem dominates in the minds and actions of farmers' culture of abundance of water. They deal with water as if it was inexhaustible resource, are wasteful in irrigation and non-compliant with regulations for voracious water crops". These authors also point out illogical practices such as the export of rice by water-scarce Egypt to countries which do not face water scarcity, which means that Egypt is actually exporting its much-needed water to these countries.

One of the largest wheat exporters is Canada, where wheat is grown as a rain-fed crop. In 2010, the area planted with wheat (Triticum aestivum L) in western Canada (Canadian Prairies) was 8.1 million ha, with a total production of 21 million metric tons, [**3**]. About 75% - 80% of the wheat grown in the Prairie Provinces is exported. However, wheat yields in western Canada fluctuate from year to year due primarily to variable weather conditions and especially precipitation, see [**4–6**]. In this region wheat is not considered as a good candidate for irrigation because other crops, such as e.g. potato, demand less water and are more profitable. The present paper shows that adopting a more global framework for planning food production could lead to revising such conceptions.

The expansion of international trade and the process of globalization enable the development of sustainable policies that take into account the acute local loads on water resources. Virtual water trade is the water embodied in the internationally exchanged products. In particular we may consider strategies that increase wheat production in water-rich countries like Canada by implementing irrigation technologies, and increase wheat export (probably at subsidized prices) to water-scarce regions like MENA. The net result of such a policy would be that more fresh water is available for direct human consumption in the water-scarce countries. Devising such a policy requires adequate model-based evaluations and numerical analysis. The most important factors are the relationships between crop yield and water use, taking into account the timing of the water application. The present paper focuses on the derivation of such relationships using optimal control framework. The paper is organized as follows. In section 2 we outline the model for optimal crop production and allocation. In section 3 we describe the AquaCrop model and its use with TOMLAB optimization solver. In section 4 the results of a simulation study conducted for irrigated wheat production in Carman (Manitoba) in 2006 are presented. Section 5 shows the conclusions.

2. Model of the optimal crop production and allocation

There are many models that describe the relationship between water and crop production in the literature, e.g. [**7–10**] and many others. Here we describe a framework for investigating the concept of global management of crop production and water allocation. The aim of this framework is to enable a quantitative evaluation of the effects of international trade of virtual water, taking into account local constraints and quotas for agricultural production.

2.1. Objective and constraints. The objective function is the total net income of a number of different crops and their international allocation (taking into account the costs of water and transportation). The constraints of the model are as follows.

$$\begin{gathered} Z = \sum_{ikj}[y_{ik}^j c_{ik}^j - p_k x_{ik} w_{ik} - t_{ik}^j y_{ik}^j] \to \max \\ \sum_i x_{ik} \le x_k^u,\ x_{ik}^l \le x_{ik} \le x_{ik}^u, \\ 0 \le w_{ik} \le w_{ik}^u, \sum_i w_{ik} \le w_k^u, \\ \sum_j y_{ik}^{lj} \le \sum_j y_{ik}^j \le \sum_j y_{ik}^{uj},\ \sum_j y_{ik}^j = x_{ik} f_{ik}(w_{ik}), \\ \sum_k y_{ik}^j \le \sum_k y_{ik}^j \le \sum_j y_{ik}^{uj}. \end{gathered} \tag{2.1}$$

Here the following notations are used:
Indices and indicators:
i - index of the crop;
k - index of the country (production and import);
j - index of the country (export to);
l - indicator of the lower bound;
u - indicator of the upper bound.

Variables, functions and coefficients:
x_{ik} - land area [ha] in the country k used for crop i;
w_{ik} - irrigation water [mm] used in the country k for crop i;
$f_{ik}(w_{ik})$ - irrigation water use efficiency (water production function) [t/ha] of crop i in the country k;
y_{ik}^j - amount of crop i [t] imported into country j from the country k. y_{ij}^j refers to local consumption of the crop i;
c_{ik}^j- price of the crop i produced in the country k and exported to the country j. This price should include special export subsidies to stimulate the export of virtual water;
p_k - cost of the irrigated water [\$/m^3] in the country k;
t_{ik}^j - cost of transportation of the crop i from country k to country j.

The constraints show that all demands in different crops should be optimally met by local production and import while satisfying the constraints related to available agricultural land and irrigation water. The balance between production, local consumption and export must be maintained for each crop and each country. The total production of crop i in the country k is limited by lower and upper constraints. The last two constraints indicate limits for overall demand (local production and import) for crop i in the country j.

This optimization problem can be solved taking into account that all the functions are known to be concave, see [**9**]. The main problem is to obtain the necessary information, mainly to determine the functions $f_{ik}(w_{ik})$ which we shall call irrigation water use efficiency $IWUE$ and which describe the dependence of the yield of crop i in a given country k on the seasonal amount of irrigation water (w_{ik}).

3. AquaCrop model and its use with TOMLAB optimization library

3.1. AquaCrop model. Depending on the purpose and objectives of the crop model, one can distinguish two main modeling approaches: *scientific* and *engineering*. The first aims mainly at improving the understanding of crop behavior, its physiology, and its responses to environmental changes. The second attempts to provide reasonable management advice to farmers or predictions to policy-makers. Such models are aimed mainly at practitioner-type end-users such as those working for extension services, consulting engineers, governmental agencies, nongovernmental organizations, or farmers associations. They are also designed to fit the need of economists and policy specialists who use relatively simple models for planning and scenario analysis. The AquaCrop model used in this study is a canopy-level engineering type of model that simulates the development of crop biomass and harvestable yield in response to water availability, see [**11, 12**]. It was designed with support of the FAO and simulates yields and other parameters of major field crops as a function of water availability under rain-fed, supplemental, deficit, and full irrigation. An executable version of this model can be downloaded freely from the FAO website. The crop development in AquaCrop is *water-driven*, so that transpiration is calculated first and translated into biomass using a conservative, crop-specific parameter: the biomass water productivity, normalized for atmospheric evaporative demand and air CO_2 concentration. The normalization is used to make AquaCrop applicable to diverse locations and seasons. Simulations are performed on thermal time (degree-days) or calendar time in daily time-steps. The model uses canopy ground cover instead of leaf area index (LAI) as the basis to calculate transpiration and to separate soil evaporation from transpiration. Crop yield is calculated as the product of biomass and a harvest index (HI), which can be affected negatively or positively by water stress depending on its level, timing, and duration. At the beginning of the yield formation period, HI increases linearly with time, until near physiological maturity. Other than yield, there is no biomass partitioning into the various organs. Crop response to water deficit is simulated via four modifiers which are functions of the available soil water modulated by evaporative demand, based on the differential sensitivity to water stress of four key plant processes: canopy expansion, stomatal control of transpiration, canopy senescence, and conversion of biomass into harvestable product.

The AquaCrop model includes (i) the soil, with its water balance; (ii) the plant, with its growth, development, and yield processes; and (iii) the atmosphere, with its thermal regime, rainfall, evaporative demand, and carbon dioxide concentration. Additionally, some management aspects are considered, with emphasis on irrigation and soil fertility as they affect crop development, water productivity, and therefore the final yield.

The majority of studies which tested and validated Aquacrop for different crops and locations came to conclusion that AquaCrop was able to simulate water content in the root zone, crop biomass and grain yield quite accurately, e.g. [**13–16**]. With respect to the type of work presented in the present paper a major advantage of Aquacrop is the large number of crops and locations for which its has been calibrated so that the procedure illustrated below with wheat could be readily applied to other crops.

Aquacrop can be used to simulate both rain-fed and irrigated crops. When irrigation is used, the user defines the irrigation schedule by specifying the time and amount of each application, or let the model generate automatically the schedule on the basis of fixed time interval, fixed amount per application, or fixed percentage of allowable water depletion. The user-defined time/amount option, together with the option to run the simulation manually day by day and applying irrigation at will in chosen amounts while seeing immediately the effect on crop canopy and transpiration, are particularly suited for analyzing and developing irrigation schedules and analyzing yield response.

3.2. TOMLAB optimization library. TOMLAB /OQNLP is a multistart heuristic algorithm designed to find global optima of smooth constrained nonlinear programs (NLPs) and mixed-integer nonlinear programs (MINLPs), [**17**]. TOMLAB /MSNLP is also a multistart heuristic algorithm designed to find global optima of smooth constrained nonlinear programs (NLPs). Up to 10,000 variables and constraints are supported. The multistart feature calls an NLP solver with a different set of initial values and return the feasible solutions as well as the optimal point. The starting points are calculated from scatter search algorithm. The user may also choose to use uniformly distributed initial values. Neither of the two options guarantee that a global optimum is obtained, however the likelihood is high. The main advantage with OQNLP/MSNLP for smooth problems is that good local solutions are easily obtained, and that integer variables are handled. OQNLP solves problems of the form

$$\begin{aligned} &f(x) \to \min \\ &x_L \leq x \leq x_U \\ &b_L \leq Ax \leq b_U \\ &c_L \leq c(x) \leq c_U \end{aligned} \tag{3.1}$$

Some of variables x may be restricted to be integer. The OQNLP solver was chosen from many other solvers implemented in TOMLAB because it has demonstrated the best ability to find good solutions.

3.3. The problem. The considered formulation of the problem of optimal deficit irrigation of wheat is as follows:

$$\begin{aligned} &J = IWE(w_1, w_2, \ldots, w_i, \ldots, w_n)\ [t/ha] \to \max \\ &w_1 + w_2 + \ldots + w_i + \ldots + w_n \leq w\ [mm],\ 0 \leq w_i \leq w_i^U. \end{aligned} \tag{3.2}$$

Here IWE is the yield, w_i is the daily irrigation for day i, and w is a seasonal irrigation water quota. The variables w_i are non-negative and have an upper bound. The function $IWUE(w)$ is obtained by performing the optimization a number of times with various values of the water quota w.

Actually this is a discrete optimal control problem where IWE is result of the multi-step process with an isoperimetric constraint on w_i. However this process is not described explicitly and is considered to be a black box. The function IWE has to be determined by running a simulation model which predicts the crop development in response to the environmental variables. We have used here the AquaCrop model described in the previous section, together with appropriate plug-ins to write irrigation files and read the output files of AquaCrop. The approach to get a good starting point for optimization has been discussed in [**19**]. Having these optimal solutions for a set of different values of w (seasonal water quota) we can construct and plot $IWUE = Y(w)$. This graph can be used for planning purposes both by farmers and by water authorities. For each point of the plot of $IWUE$ we have the vector of the values of w_i that details the irrigation schedule corresponding to this particular solution.

The function $IWUE = Y(w)$ is obtained in the following way. A convex hull of all stored suboptimal points $Y(w)$ is found and a set of its vertices identified. Then a third order polynomial is used to fit these vertices and the resulting third order curve in the interval from minimal w_L to maximal w_U is used as $IWUE(w)$.

4. Results for Canadian wheat in Carman (Manitoba) for year 2006

Weather and soil data from Carman (Manitoba, Canada), 2006, have been used for construction of the corresponding $IWUE$ function. The data were taken from [**18**] of Mkhabela and Bullock, and these authors have kindly provided us with the files for the site-calibrated AquaCrop model. The season was 97 days long starting on May 12, 2006. The total precipitation was 140 mm. The yield simulated without irrigation was 2.770 t/ha. The optimization procedure was performed several times assuming seasonal water quotas ranging from 0 to 250 mm. The $IWUE$ function obtained is shown in Figure 1. In this plot the points of vertices of the convex hull are marked as '$*$', their fitted values on the IWUE curve are marked as 'o' and the suboptimal points are marked as '+'. The average increase of the yield in the range of $0 - 100$ mm irrigation water was more than 20 kg/ha per mm of water. The maximum yield of 5.018 t/ha (almost twice the yield of the rain-fed crop) was achieved with 154 mm of irrigation.

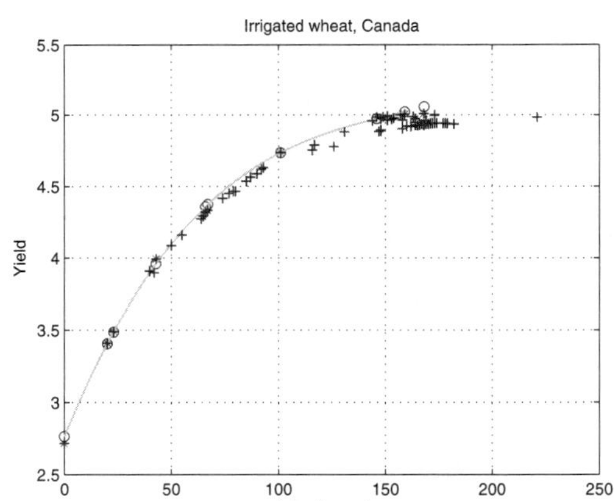

FIGURE 1. IWUE(w) function for wheat in Carman (Manitoba, Canada), 2006. Yield [t/ha] vs irrigation water [mm].

5. Conclusions

In this paper we have outlined a framework for optimal planning of global agricultural production using the concept of virtual water. This framework requires irrigation water use efficiency ($IWUE$) functions which describe the relationship between irrigation and the yield of a specific crop at a specific location. We used the example of wheat production in Carman (Manitoba, Canada) in 2006 to show how the $IWUE$ function can be obtained using the FAO model AquaCrop in conjunction with TOMLAB optimization library. Our results show that considering crop production as a global, rather than local, problem could promote significant changes compared to current common practices. According to the specific example presented in this paper, implementing optimal irrigation could almost double wheat yield compared to the common practice of rain-fed production. The small quantity of irrigation water required (150 mm for 2 t/ha increase, compared to requirement of 2000 m^3/ha for local production in countries such as Algeria or Marocco, [**1**], indicates that significant water savings could be achieved by adopting a global approach for planning food production.

Acknowledgements

The research leading to these results has received funding from the European Community's Seventh Framework Programme (FP7/2007-2013) under grant agreement n° 311903–FIGARO ("Flexible and Precise Irrigation Platform to Improve Farm-Scale Water Productivity") (http://www.figaro-irrigation.net/). The contents of this document are the sole responsibility of the FIGARO Consortium and can under no circumstances be regarded as reflecting the position of the European Union. This Research was supported by Technion General Research Fund. The authors are grateful to P. Bullock and M. Mkhabela from the University of Manitoba for providing files with the calibrated AquaCrop model and weather data.

References

[1] Schrier, H., Garwood, P., China and Middle East. Virtual water, 2012, Univ. of British Columbia, http://wmc.landfood.ubc.ca/webapp/VWM/course/china-and-middle-east/middle-east-north-africa/

[2] Mostafa, H., Fujimoto, N. Water saving scenarios for effective irrigation management in Egyptian rice cultivation, *Ecol. Engin.*, **70** (2014), 11–15.

[3] Statistics Canada, Production of principal field crops, (2010), http://www.statcan.gc.ca/daily-quotidien/110426/t110426a1-eng.htm (accessed June 2012)

[4] Williams, G., D., V. Weather and Prairie wheat production. *Can. J. Agric. Econ.* **17**, (1969), 99–109.

[5] Walker, G., K. Model for operational forecasting of Western Canada wheat yield. *Agric. Forest Meteorol.* **44**, (1989), 339–351.

[6] DePauw, R., M., Malhi, S., S., Bullock, P., R., Gan, Y., T., McKenzie, R., H., Larney, F., Janzen, H., H., Cutforth, H., W., Wang, H. Wheat production in Northern high latitudes - Canadian example. In: Bonjean, A., Angus, W., van Ginkel, M. (Eds.), World Wheat Book 2. Lavoisier Publishing, France, (2011), 607–651.

[7] Amir, I., and Fisher, F. Response of near optimal agricultural production to water policies. *Agricultural Systems*, **64** (2000), 115–130.

[8] Uri Shani, Yacov Tsur, and Amos Zemel, *Optimal dynamic irrigation schemes*, Optimal Control Appl. Methods **25** (2004), no. 2, 91–106, DOI 10.1002/oca.740. MR2063460

[9] Shani, U., Tsur, Y., Zemel, A., Zilberman, D. Irrigation production functions with water-capital substitution. *Agric. Economics*, **40** (2009), 55–66.

[10] Ilya Ioslovich, Mikhail Borshchevsky, and Per-Olof Gutman, *On optimal irrigation scheduling*, Dyn. Contin. Discrete Impuls. Syst. Ser. B Appl. Algorithms **19** (2012), no. 1-2, 303–310. MR2918258

[11] FAO, AquaCrop: The FAO Crop-Model to Simulate Yield Response to Water, http://www.fao.org/nr/water/aquacrop.html, 2009.

[12] Steduto, P., Hsiao, T., C., Raes, D., Ferereset, E., AquaCrop - the FAO crop model to simulate yield response to water: I. Concepts and underlying principles. *Agron. J.*, **101** (2009), 426–437.

[13] Geerts, S., Raes, D., Garcia, M., Miranda, R., Cusicanqui, J.A., Taboada, C., Mendoza, J., Huanca, R., Mamani, A., Condori, O., Mamani, J., Morales, B., Osco, V., Steduto, P., Simulating yield response of Quinoa to water availability with AquaCrop. *Agron. J.*, **101** (2009), 499–508.

[14] Geerts, S., Raes, D., Garcia, M., Using AquaCrop to derive deficit irrigation schedules. *Agric. Water Manage.*, **98** (2010), 213–216.

[15] Heng, L., K., Hsiao, T., Evett, S., Howell, T., Steduto, P., Validating the FAO AquaCrop model for irrigated and water deficient field maize. *Agron. J.*, **101** (2009), 488–498.

[16] Andarzian, B., Bannayan, M., Steduto, P., Mazraeh, H., Barati, M.E., Barati, M.A., Rahnama, A. Validation and testing of the AquaCrop model under full and deficit irrigated wheat production in Iran. *Agric. Water Manage.* **100** (2011), 1–8.

[17] Holmstrom, K., Goran, A., O., Edvall, M., M. User's guide for TOMLAB /OQNLP, http://tomopt.com/tomlab/products/oqnlp/, 2007.

[18] Mkhabela, M., S., Bullock P., R. Performance of the FAO AquaCrop model for wheat grain yield and soil moisture simulation in Western Canada, *Agric. Water Manage.* **110** (2012), 16–24.

[19] Linker, R., Sylaios, G., and Ioslovich, I. Optimization of irrigation scheduling using genetic algorithms and AquaCrop: a case sudy for cotton in Northern Greece. *International Conference on Agriculture Science and Environment Engineering (ICASEE 2013), Beijing, China, December 19-20, 2013* , Conference proceedings (DVD), (2013), paper ICASEE 132117.

The Faculty of Civil and Environmental Engineering, Technion–Israel Institute of Technology, 32000 Haifa, Israel

E-mail address: linkerr@tx.technion.ac.il

The Faculty of Civil and Environmental Engineering, Technion–Israel Institute of Technology, 32000 Haifa, Israel

E-mail address: agrilya@tx.technion.ac.il

Contemporary Mathematics
Volume **659**, 2016
http://dx.doi.org/10.1090/conm/659/13152

Implicit Euler approximation and optimization of one-sided Lipschitzian differential inclusions

Boris S. Mordukhovich and Yuan Tian

ABSTRACT. This paper concerns the study of the generalized Bolza problem governed by differential inclusions satisfying the so-called "relaxed one-sided Lipschitzian" (ROSL) condition with respect to the state variables subject to various types of nonsmooth endpoint constraints. We construct discrete approximations of differential inclusions with ROSL right-hand sides by using the implicit Euler scheme for approximating time derivatives, and then we justify an appropriate well-posedness of such approximations. Our principal result establishes the uniform approximation of strong local minimizers for the continuous-time Bolza problem by optimal solutions to the implicitly discretized finite-difference systems in the general ROSL setting and even by the strengthen $W^{1,2}$-norm approximation of this type in the case "intermediate" (between strong and weak minimizers) local minimizers under additional assumptions. Finally, we derive necessary optimality conditions for the discretized Bolza problems via suitable generalized differential constructions of variational analysis. The obtained results on the well-posedness of discrete approximations and necessary optimality conditions allow us to justify a numerical approach to solve the generalized Bolza problem for one-sided Lipschitzian differential inclusions by using discrete approximations constructed via the implicit Euler scheme.

1. Introduction and Problem Formulation

This paper addresses the following optimization problem (P) of the *generalized Bolza type* for dynamic systems governed by constrained differential inclusions:

$$\text{minimize } J[x] := \varphi_0\big(x(T)\big) + \int_0^T f\big(x(t), \dot{x}(t), t\big)dt \tag{1.1}$$

over absolutely continuous trajectories $x : [0,T] \to \mathbb{R}^n$ satisfying the differential inclusion

$$\dot{x}(t) \in F\big(x(t), t\big) \text{ a.e. } t \in [0,T] \text{ with } x(0) = x_0 \in \mathbb{R}^n \tag{1.2}$$

subject to the geometric and functional endpoint constraints given by, respectively,

$$x(T) \in \Omega \subset \mathbb{R}^n, \tag{1.3}$$

2010 *Mathematics Subject Classification.* Primary 49J53, 49J24, 49M25; Secondary 49J52, 90C30.

This research was partly supported by the National Science Foundation under grant DMS-1007132.

$$\varphi_i(x(T)) \le 0 \ \text{ for } \ i = 1, \ldots, m, \tag{1.4}$$

$$\varphi_i(x(T)) = 0 \ \text{ for } \ i = m+1, \ldots, m+r. \tag{1.5}$$

Here x_0 is a fixed n-vector, $F\colon \mathbb{R}^n \times [0,T] \rightrightarrows \mathbb{R}^n$ is a set-valued mapping/multifunction, Ω is an nonempty set, f and φ_i for $i = 0, \ldots, m+r$ are real-valued functions. Differential inclusion problems of type (P) have been well recognized in dynamic optimization and control theory as a convenient framework to cover the vast majority of conventional and nonconventional models arising in optimization and control of dynamical systems described via time derivatives. We refer the reader to the books [**Q**, **T**] and the bibliographies therein for more discussions, historical overviews as well as applied models governed by differential inclusions. In particular, the differential inclusion model in (1.2) encompasses ODE control systems represented in the parameterized control form

$$\dot{x}(t) = g(x(t), u(t), t) \ \text{ a.e. on } \ [0,T] \ \text{ with } \ u(t) \in U(x(t), t), \tag{1.6}$$

where $g\colon \mathbb{R}^n \times U \times [0,T] \to \mathbb{R}^n$ and $U(x,t) \subset U$ is a variable control set belonging to the given control space U. Indeed, we can take

$$F(x,t) := \{v \in \mathbb{R}^n |\ v = g(x,u,t) \ \text{ for some } \ u \in U(x,t)\} \tag{1.7}$$

in (1.2) to describe (1.6), omitting here details regarding measurable selections. Observe that the differential inclusion framework (1.2) with the velocity mapping F in (1.7) covers not only standard control systems with constant control sets but also significantly more challenging problems with *feedback* reflected by the dependence of the control sets in (1.6) on state variables. On the other hand, the optimization problem (P) is *intrinsically nonsmooth*, (even when $\Omega = \mathbb{R}^n$ and all the functions in (1.1), (1.4), (1.5), and (1.6) are smooth) due to set-valued dynamic constraints in (1.2). Thus the usage and development of appropriate tools of variational analysis and generalized differentiation are required for the study and applications of (P) and related problems governed by differential inclusions.

The *method of discrete approximations* has been well recognized as an efficient approach to investigate differential inclusions and optimization problems for them from both qualitative and quantitative/numerical viewpoints; see, e.g., the surveys and books [**K**, **L**, **Q**, **S**] and the bibliographies therein. A principal question arising in all the aspects and modifications of this method (even without applications to optimization) is about the possibility to approximate, in a suitable sense, feasible trajectories of the given differential inclusion by those for finite-difference inclusions that appear by using one or another scheme to replace time derivatives. The majority of the results in this direction concern explicit Euler schemes under the Lipschitz continuity of velocity mappings with respect to state variables; see [**A**, **K**, **L**, **N**, **Q**, **S**] for more details and references.

The other lines of research on discrete approximations of differential inclusions via the explicit Euler schemes invoke the replacement of the Lipschitz continuity of velocity mappings by various *one-sided Lipschitzian* conditions; see, e.g., [**F**, **G**, **H**, **I**, **L**]. Conditions of this type essentially weaken, from one side, the classical Lipschitz continuity, while from the other side they encompass *dissipativity* properties widely used in nonlinear analysis and the theory of monotone operators. Note to this end the so-called *modified one-sided Lipschitzian* (MOSL) condition introduced and applied in [**I**] to justify a certain strong approximation of solution sets for differential inclusions by finite-difference ones obtained via the explicit Euler scheme and to derive in this way a Bogolyubov-type density theorem for the Bolza problem (P) and the corresponding convergence of discrete optimal solutions.

In this paper we exploit a weaker property than MOSL known as the *relaxed one-sided Lipschitz* (ROSL) condition; see below. The ROSL property of set-valued mappings was introduced by Tzanko Donchev in [**F**] under a different name and has already been employed in the studies of various aspects of analysis of set-valued mappings, differential inclusions, and their discrete approximations; see, e.g., [**C**, **G**, **H**, **J**]. In particular, the paper [**H**] contains an extension to the ROSL case of the fundamental Filippov theorem about relationships between trajectories and "quasitrajectories" of Lipschitzian differential inclusions and provides applications of this result to the stability analysis of the explicit Euler scheme. In [**C**], similar and related solvability and stability results were developed for the parameterized *implicit Euler scheme*

$$\Phi_h(x) := \{y \in \mathbb{R}^n |\, y \in x + hF(y)\}, \quad h > 0, \tag{1.8}$$

generated by ROSL mappings F with compact and convex values. Note that the implicit framework of (1.8) is essentially more involved in comparison with the explicit one

$$\Phi_h(x) := \{y \in \mathbb{R}^n |\, y \in x + hF(x)\}, \quad h > 0, \tag{1.9}$$

studied and applied in [**G**, **H**, **J**] and other publications.

The main goal of this paper is to use the implicit Euler scheme (1.8) to construct and investigate the *discrete approximations*

$$x_{j+1}^k \in x_j^k + h_k F(x_{j+1}^k, t_{j+1}), \quad k \in \mathbb{N} := \{1, 2, \ldots\} \text{ with } h_k \downarrow 0 \text{ as } k \to \infty, \tag{1.10}$$

of the ROSL differential inclusion (1.2) and the generalized Bolza problem (P) for it with establishing the strong convergence of discrete approximations (in the sense specified below) and deriving necessary optimality conditions for their optimal solutions. To the best of our knowledge, the results obtained in what follows are new for discrete approximations constructed via the implicit Euler scheme even for the case of unconstrained differential inclusions satisfying the classical Lipschitz condition with respect to state variables.

After recalling the basic definitions and some background material in Section 2, we develop new results in the aforementioned directions outlined in what follows.

Section 3 presents a constructive procedure allowing us to *strongly approximate* in the norm topology of $C[0,T]$ a given feasible trajectory $\bar{x}(\cdot)$ of the differential inclusion (1.2) by feasible solutions to the *implicit Euler* finite-difference inclusions (1.10) piecewise linearly extended to $[0,T]$. Furthermore, we justify here even *stronger* $W^{1,2}[0,T]$-norm approximation of $\bar{x}(\cdot)$ by feasible extended discrete trajectories in the following two major cases: either F is *ROSL and locally graph-convex*, or F is *locally Lipschitzian.* Some counterparts of this result involving the *explicit* Euler scheme (1.9) can be found (with different proofs) in [**N**, **Q**] for Lipschitzian differential inclusions and in [**I**] for those satisfying the MOSL condition. We are not familiar with any results of this type (involving either the $C[0,T]$ or $W^{1,2}[0,T]$ convergence) for discrete approximations of differential inclusions based on the implicit Euler scheme.

In Section 4 we construct a sequence of finite-difference Bolza type problems (P_k) as $k \in \mathbb{N}$ with the dynamic constraints given by the implicit Euler scheme

(1.10) under appropriate approximations of the cost functional (1.1) and the endpoint constraints in (1.3)–(1.5). Then we show that optimal solutions to (P_k) and their slight modifications exist for all large $k \in \mathbb{N}$ and *norm converge* in the $C[0,T]$ topology for the case of *strong local minimizers* of (P) in the general ROSL setting and the $W^{1,2}[0,T]$ topology in the case of *intermediate local minimizers* of (P) under the additional assumptions on these minimizers imposed in Section 3. The obtained results seem to be the first achievements in this direction for the *implicit* Euler scheme (1.10). It is worth mentioning however that our approach to the strong approximation and convergence results obtained in Sections 3 and 4 require, along with the ROSL condition on the differential inclusion, the *unform boundedness* of the velocity sets. This does not allow us to cover the corresponding developments presented of [**D**, **E**] for discrete approximations of control problems governed by Moreau's sweeping process, which is described by a dissipative while intrinsically unbounded differential inclusion studied in [**D**] via the explicit Euler scheme by exploiting certain specific features of the sweeping process generated by controlled moving sets.

In addition to the well-posedness results for discrete approximations of (P) via the implicit Euler scheme obtained in Sections 3 and 4, we derive in Section 5 under fairly mild assumptions necessary conditions for optimal solutions to the nonsmooth discrete approximations problems (P_k) associated with the implicit discrete inclusions (1.10) that are different from necessary optimality conditions for the corresponding problems associated with explicit Euler counterparts. These conditions are expressed in terms of the advanced tools of generalized differentiation in variational analysis discussed in Section 2. Due to the established convergence of discrete optimal solutions, the necessary optimality conditions for problems (P_k) obtained in this way can be treated as *suboptimality* (almost optimality) conditions for the original Bolza problem (P) and can be also viewed as a certain justification of *numerical algorithms* based on discrete approximations. The final Section 6 presents concluding remarks on some topics of further research including deriving necessary optimality conditions for the one-sided Lipschitzian generalized Bolza problem (P) by using the method of discrete approximations.

2. Basic Definitions and Preliminaries

Throughout the paper we use standard notation and terminology of variational analysis and generalized differentiation; see, e.g., [**P**, **R**]. Recall that $\mathbb{R}^n$ denotes the n-dimensional space with the Euclidean norm $|\cdot|$ and the closed unit ball $I\!B$ and that $\mathcal{CC}(\mathbb{R}^n)$ signifies the space of convex and compact subsets of $\mathbb{R}^n$ endowed with the Pompieu-Hausdorff metric. The distance function associated with an nonempty closed set $\Omega \subset \mathbb{R}^n$ is denoted by

$$\operatorname{dist}(x,\Omega) := \min_{y\in\Omega} |x-y|, \quad x \in \mathbb{R}^n,$$

and the distance between two closed sets $\Omega_1, \Omega_2 \subset \mathbb{R}^n$ is given by

$$\operatorname{dist}(\Omega_1,\Omega_2) := \max\Big\{\max_{x\in\Omega_1}\operatorname{dist}(x,\Omega_2),\ \max_{y\in\Omega_2}\operatorname{dist}(y,\Omega_1)\Big\}. \tag{2.1}$$

Finally, for an arbitrary a set-valued mapping $F\colon \mathbb{R}^n \rightrightarrows \mathbb{R}^m$, the Painlevé-Kuratowski outer limit of F as $x \to \bar{x}$ is defined by

$$\operatorname*{Lim\,sup}_{x\to\bar{x}} F(x) := \Big\{ y \in \mathbb{R}^m \Big|\ \exists\, x_k \to \bar{x},\ y_k \to y \ \text{ with } \ y_k \in F(x_k),\ k \in \mathbb{N}\Big\}. \tag{2.2}$$

The following property introduced in [**F**] is our standing assumption on the right-hand side $F(\cdot,t)$ of the differential inclusion in (1.2) playing a crucial role in this paper.

DEFINITION 2.1 (relaxed one-sided Lipschitzian condition). A set-valued mapping $F:\mathbb{R}^n\to\mathcal{CC}(\mathbb{R}^n)$ is called to be RELAXED ONE-SIDED LIPSCHITZIAN (ROSL) with constant $l\in\mathbb{R}$ if for any given $x_1,x_2\in\mathbb{R}^n$ and $y_1\in F(x_1)$ there exits $y_2\in F(x_2)$ such that

$$\langle y_1-y_2,x_1-x_2\rangle\le l|x_1-x_2|^2. \tag{2.3}$$

Note that the number/modulus l in (2.3) is not required to be positive as in the classical Lipschitz continuity. The ROSL condition is dramatically weaker the standard Lipschitz continuity and essentially relaxes dissipativity and other one-sided Lipschitzian properties; see more discussions and examples in [**F**, **G**, **H**, **L**].

The next result on the solvability of the implicit Euler scheme (1.8) under the ROSL condition is taken from [**C**, Theorem 4] (the proof of which is based on the Kakutani fixed-point theorem) and is useful in what follows. Recall that a set-valued mapping F is upper semicontinuous (usc) on $\mathbb{R}^n$ if for any $\bar{x}\in\mathbb{R}^n$ and any $\varepsilon>0$ there exists $\gamma>0$ such that $F(x)\subset F(\bar{x})+\varepsilon I\!B$ whenever $|x-\bar{x}|\le\gamma$.

LEMMA 2.2 (solvability of the implicit Euler scheme). *Let $F\colon\mathbb{R}^n\to\mathcal{CC}(\mathbb{R}^n)$ be usc and ROSL on $\mathbb{R}^n$ with constraint $l\in\mathbb{R}$ such that $lh<1$. Then for any $x,y\in\mathbb{R}^n$, there exists a solution $\bar{y}\in\Phi_h(x)$ of the implicit Euler scheme* (1.8) *satisfying the estimate*

$$|\bar{y}-y|\le\frac{1}{1-lh}\operatorname{dist}(y,x+hF(y)).$$

Following [**N**], we say that a feasible solution $\bar{x}(\cdot)$ to (P) is an *intermediate local minimizer* (i.l.m.) of rank $p\in[1,\infty)$ for this problem if there are positive numbers ε,α such that $J[\bar{x}]\le J[x]$ for any other feasible solutions $x(\cdot)$ to (P) satisfying the conditions

$$|x(t)-\bar{x}(t)|<\varepsilon\ \text{ as }\ t\in[0,T]\quad\text{and}\quad\alpha\int_0^T|\dot{x}(t)-\dot{\bar{x}}(t)|^pdt<\varepsilon. \tag{2.4}$$

The case of $\alpha=0$ in (2.4) corresponds to the classical notion of *strong* local minimum and surely includes global solutions to (P). The notion of *weak* local minimum corresponds to (2.4) with $\alpha\ne0$ and $p=\infty$; see [**N**, **Q**] for detailed discussions and examples.

In what follows we need a certain modification of the i.l.m. notion formulated above, which related to some local relaxation stability of the initial problem (P). Along with (P), consider its extended/relaxed version constructed in the line well understood in the calculus of variations and optimal control. Let

$$f_F(x,v,t):=f(x,v,t)+\delta(v,F(x,t)), \tag{2.5}$$

where $\delta(\cdot,\Lambda)$ is the indicator function of the set Λ equal to 0 on Λ and to ∞ otherwise. Denote by $\widehat{f}_F(x,v,t)$ the convexification for f_F in the v variable, i.e., the largest convex function majorized by $f_F(x,\cdot,t)$ for each x and t. The *relaxed generalized Bolza problem* (R) consists of minimizing the functional

$$\widehat{J}[x]:=\varphi(x(T))+\int_0^T\widehat{f}_F(x(t),\dot{x}(t),t)dt \tag{2.6}$$

over absolutely continuous trajectories $x\colon [0,T] \to \mathbb{R}^n$ of the differential inclusion (1.2) with the endpoint constraints (1.3)–(1.5).

Note that, due to our standing assumption on the convexity and compactness of the values $F(x,t)$ in the differential inclusion (1.2), the validity of the dynamic constraint (1.2) is automatic for any absolutely continuous function $x(\cdot)$ with $\widehat{J}[x] < \infty$ in (2.6). Thus the relaxed problem (R) reduces to the original one (P) if the integrand f is convex with respect to the velocity variable v; in particular, when f does not depend on v. Furthermore, a remarkable fact for the continuous-time problems under consideration consists of the equality between the infimum values of the coast functionals in (P) and (R), without taking endpoint constraints into account, even when f is not convex in v. This fact is known as "hidden convexity" of continuous-valued variational and control problems and relates to the fundamental results of Bogolyubov's and Lyapunov's types; see, e.g., the books [**B**, **Q**, **T**] for exact formulations and more discussions. The most recent extended version of the Bogolyubov theorem for differential inclusion problems of type (P) was obtained in [**I**] under the MOSL condition on $F(\cdot,t)$ mentioned in Section 1. This discussion makes more natural the following notion taken from [**N**].

DEFINITION 2.3 (relaxed intermediate local minimizers). A feasible solution $\bar{x}(\cdot)$ to the original problem (P) is called a RELAXED INTERMEDIATE LOCAL MINIMUM (*r.i.l.m.*) of rank $p \in [1,\infty)$ for (P) if it provides an intermediate local minimum of rank p to the relaxed problem (R) and satisfies the condition $J[\bar{x}] = \widehat{J}[\bar{x}]$.

Finally in this section, we recall and briefly discuss the generalized differential constructions of variational analysis introduced by the first author [**M**] and employed in Section 5 for deriving necessary optimality conditions; see the books [**P**, **R**] for more details and references on these and related constructions. Given a set $\Omega \subset \mathbb{R}^n$ locally closed around $\bar{x} \in \Omega$, the *normal cone* to Ω at $\bar{x}$ is defined via the outer limit (2.2) by

$$N(\bar{x};\Omega) = \operatorname*{Lim\,sup}_{x \to \bar{x}} \Big[\operatorname{cone}(x - \Pi(x,\Omega))\Big], \tag{2.7}$$

where $\Pi(x,\Omega) = \big\{w \in \Omega \text{ s.t. } |x-w| = \operatorname{dist}(x,\Omega)\big\}$ is the Euclidean projector of x on Ω, and where the symbol "cone" stands for the conic hull of the set in question. This normal cone reduces to the classical normal cone of convex analysis when Ω is convex, which it may take nonconvex values in rather simple situations as, e.g., for the graph of the function $|x|$ and the epigraph of the function $-|x|$ on $\mathbb{R}$. Nevertheless the normal cone (2.7) and the related generalized differential constructions for functions and mappings enjoy comprehensive calculus rules based on variational/extremal principles of variational analysis; see [**P**, **R**] and the references therein.

Given now an extended-real-valued and lower semicontinuous function $\varphi\colon \mathbb{R}^n \to \overline{\mathbb{R}} := (-\infty,\infty]$ finite at $\bar{x}$, we define its *subdifferential* at $\bar{x}$ geometrically

$$\partial\varphi(\bar{x}) := \Big\{v \in \mathbb{R}^n \Big|\ (v,-1) \in N((\bar{x},\varphi(\bar{x}));\operatorname{epi}\varphi)\Big\} \tag{2.8}$$

via the normal cone (2.7) to the epigraphical set

$$\operatorname{epi}\varphi := \{(x,\alpha) \in \mathbb{R}^n \times \mathbb{R}|\ \alpha \ge \varphi(x)\}$$

of φ. The reader can find in [**P**, **R**] various analytical representations and properties of the subgradient mapping $\partial\varphi\colon \mathbb{R}^n \rightrightarrows \mathbb{R}^n$ used in what follows.

We recall also the *symmetric subdifferential* construction for a continuous function $\varphi\colon \mathbb{R}^n \to \mathbb{R}$ at $\bar{x}$ defined by

$$\partial^0\varphi(\bar{x}) := \partial\varphi(\bar{x}) \cup (-\partial(-\varphi)(\bar{x})) \tag{2.9}$$

and employed in Section 5 for expressing necessary optimality conditions for equality constraints. Note the symmetry relation

$$\partial^0(-\varphi)(\bar{x}) = -\partial^0\varphi(\bar{x}),$$

which does not hold for the unilateral subdifferential construction (2.8).

3. Strong Approximation via Implicit Euler Scheme

In this section we justify the possibility to *strongly approximate* (in the norm topology of either $C[0,T]$ or $W^{1,2}[0,T]$) feasible trajectories of the ROSL inclusion (1.2) constructed via the implicit Euler scheme. Given an *arbitrary* trajectory $\bar{x}(\cdot)$ of (1.2), we impose the following assumptions of F near $\bar{x}(\cdot)$ standing throughout Sections 3 and 4. For simplicity, suppose that the uniform boundedness and ROSL moduli below are constant on $[0,T]$. They can obviously be replaced by the continuous functions $m_F(t)$ and $l(t)$ on this compact interval while the proofs of the main results presented in Sections 3 and 4 can be modified to more general cases of the Riemann and Lebesgue integrability.

(H1) There exists an open set $U \subset \mathbb{R}^n$ and a number $m_F > 0$ such that $\bar{x}(t) \in U$ for all $t \in [0,T]$ and the multifunction $F\colon U \times [0,T] \to \mathcal{CC}(\mathbb{R}^n)$ from (1.2) satisfies the *uniform boundedness condition*

$$F(x,t) \subset m_F I\!\!B \text{ for all } x \in U, \quad \text{a.e. } t \in [0,T].$$

(H2) Given U from (H1), for all $x_1, x_2 \in U$, a.e. $t \in [0,T]$, and $y_1 \in F(x_1,t)$ there exists $y_2 \in F(x_2,t)$ such that we have the *relaxed one-sided Lipschitzian condition*

$$\langle y_1 - y_2, x_1 - x_2 \rangle \le l|x_1 - x_2|^2.$$

(H3) The multifunction $F(\cdot,t)$ is *continuous* on the neighborhood U from (H1) for a.e. $t \in [0,T]$ while $F(x,\cdot)$ is *a.e. continuous* on $[0,T]$ uniformly in $x \in U$ with respect to the Pompieu-Hausdorff metric.

We now construct a finite-difference approximation of the differential inclusion in (1.2) by using the *implicit Euler method* to replace the time derivative by

$$x(t+h) \in x(t) + hF(x(t+h),t) \text{ as } h \downarrow 0.$$

To formalize this process, for any $k \in \mathbb{N}$ define the discrete grid/mesh on $[0,T]$ by $T_k := \big(t_j \big| \, j = 0,1,\ldots,k\big)$ with $t_0 := 0$, $t_k := T$, and stepsize

$$h_k := T/k = t_{j+1} - t_j.$$

Then the corresponding discrete inclusions associated with (1.2) via the implicit Euler scheme are constructed as follows:

$$x_{j+1}^k \in x_j^k + h_k F(x_{j+1}^k, t_{j+1}) \text{ for } j = 0,\ldots,k-1, \tag{3.1}$$

where the starting vector x_0 in (3.1) is taken from (1.2).

The next theorem justifies the aforementioned strong $W^{1,2}[0,T]$-approximation of feasible solutions to (1.2) by those for the discrete inclusions (3.1).

THEOREM 3.1 (discrete approximation of ROSL differential inclusions). *Let $\bar{x}(\cdot)$ be a feasible trajectory for* (1.2) *such that $\dot{\bar{x}}(t)$ is Riemann integrable on $[0,T]$ and the standing assumptions* (H1)–(H3) *are satisfied. Then the following assertions hold:*

(1) *There is a sequence $\{z_j^k | j = 0,\ldots,k\}$ of feasible solutions to the discrete inclusions* (3.1) *such that their piecewise linearly extensions to $[0,T]$ converge to $\bar{x}(t)$ uniformly on $[0,T]$, i.e., in the norm topology of $C[0,T]$.*

(2) *Assume in addition that either the graph of $F(\cdot,t)$ is locally convex around $(\bar{x}(t),\dot{\bar{x}}(t))$, or $F(\cdot,t)$ is locally Lipschitzian around $\bar{x}(t)$ for a.e. $t \in [0,T]$. Then there is a sequence $\{z_j^k | j = 0,\ldots,k\}$ of feasible solutions to* (3.1) *such that the piecewise constantly extended to $[0,T]$ discrete velocity functions*

$$v^k(t) := \frac{z_{j+1}^k - z_j^k}{h_k}, \quad t \in (t_j, t_{j+1}], \quad j = 0,\ldots,k-1, \tag{3.2}$$

converge to $\dot{\bar{x}}(\cdot)$ as $k \to \infty$ in the norm topology of $L^2[0,T]$, which is equivalent to the $W^{1,2}[0,T]$-norm convergence of the piecewise linear functions $z^k(t)$ represented by

$$z^k(t) = x_0 + \int_0^t v^k(s)\,ds \quad \textit{for all } t \in [0,T], \quad k = 1,2,\ldots. \tag{3.3}$$

PROOF. Fix an arbitrary feasible trajectory $\bar{x}(t)$ for (1.2) from the formulation of the theorem and denote $\bar{x}_j := \bar{x}(t_j)$. Taking into account the density of step functions in $L^1[0,T]$, we can find without loss of generality a sequence of functions $w^k(\cdot)$ on $[0,T]$ such that $w^k(t)$ are constant on $(t_j, t_{j+1}]$ and $w^k(t)$ converge to $\dot{\bar{x}}(t)$ as $k \to \infty$ in the norm topology of $L^1[0,T]$. It follows from (H1) that

$$|w^k(t)| \le m_F + 1 \quad \text{for all } t \in [0,T] \text{ and } k \in \mathbb{N}.$$

Define further the sequences

$$w_j^k := w^k(t_j) \ \text{ for } j = 1,\ldots,k \ \text{ and } \ \xi_k := \int_0^T |\dot{\bar{x}}(t) - w^k(t)|\,dt \to 0 \text{ as } k \to \infty \tag{3.4}$$

and for each $k \in \mathbb{N}$ form recurrently the collection of vectors $\{y_0^k,\ldots,y_k^k\}$ by

$$y_{j+1}^k := y_j^k + h_k w_{j+1}^k \ \text{ for } j = 0,\ldots,k-1 \ \text{ with } y_0^k = x_0. \tag{3.5}$$

Note that the continuous-time vector functions

$$y^k(t) := x_0 + \int_0^t w^k(s)\,ds, \quad 0 \le t \le T,$$

are piecewise linear extensions of the discrete ones (3.5) on $[0,T]$ satisfying the estimate

$$|y^k(t) - \bar{x}(t)| \le \int_0^t |w^k(s) - \dot{\bar{x}}(s)|\,ds \le \xi_k \ \text{ for all } t \in [0,T],\ k \in \mathbb{N}, \tag{3.6}$$

where ξ_k is taken from (3.4). Now we construct a sequence of discrete trajectories for (3.1) by the following *algorithmic procedure*.

To define such trajectories $z^k = (z_0^k,\ldots,z_k^k)$ of (3.1), put $z_0^k := x_0$ and suppose that the vectors z_j^k have been already found. Then for any $k \in \mathbb{N}$ sufficiently large (i.e., when h_k is small) we use the solvability result from Lemma 2.2 valid under assumptions (H2) and (H3) and solve the discrete inclusions (3.1) for z^k_{j+1}. Taking into account the error estimate in Lemma 2.2, the construction of y^k_j in (3.5), and the corresponding properties of the distance (2.1), we deduce that the vector z^k_{j+1} satisfies the discrete inclusion

$$z^k_{j+1} \in z^k_j + h_k F(z^k_{j+1}, t_{j+1}) \tag{3.7}$$

and the following relationships for each $j \in \{1, \ldots, k-1\}$ and all (large) $k \in \mathbb{N}$:

$$\begin{aligned}
|z^k_{j+1} - y^k_{j+1}| &\le \frac{1}{1 - lh_k}\text{dist}\big(y^k_{j+1}, z^k_j + h_k F(y^k_{j+1}, t_{j+1})\big)\\
&\le \frac{1}{1 - lh_k}\text{dist}(y^k_{j+1}, y^k_j + h_k F(y^k_{j+1}, t_{j+1}))\\
&+ \frac{1}{1 - lh_k}\text{dist}\big(y^k_j + h_k F(y^k_{j+1}, t_{j+1}), z^k_j + h_k F(y^k_{j+1}, t_{j+1})\big)\\
&\le \frac{|z^k_j - y^k_j|}{1 - lh_k} + \frac{h_k}{1 - lh_k}\text{dist}\Big(\frac{y^k_{j+1} - y^k_j}{h_k}, F(y^k_{j+1}, t_{j+1})\Big)\\
&= \frac{|z^k_j - y^k_j|}{1 - lh_k} + \frac{h_k}{1 - lh_k}\text{dist}\big(w^k_{j+1}, F(y^k_{j+1}, t_{j+1})\big).
\end{aligned}$$

Thus we arrive at the estimate valid for all $j = 0, \ldots, k-1$ and $k \in \mathbb{N}$:

$$|z^k_{j+1} - y^k_{j+1}| \le \frac{|z^k_j - y^k_j|}{1 - lh_k} + \frac{h_k}{1 - lh_k}\text{dist}\big(w^k_{j+1}, F(y^k_{j+1}, t_{j+1})\big). \tag{3.8}$$

Proceeding further by induction implies that

$$|z^k_{j+1} - y^k_{j+1}| \le h_k \sum_{m=1}^{j+1} \Big(\frac{1}{1 - lh_k}\Big)^{j+2-m} \text{dist}\big(w^k_m, F(y^k_m, t_m)\big),$$

which yields by choosing $k \in \mathbb{N}$ with $lh_k < 1/2$ that

$$\begin{aligned}
|z^k_{j+1} - y^k_{j+1}| &\le h_k \sum_{m=1}^{j+1} (1 + 2lh_k)^{j+2-m} \text{dist}\big(w^k_m, F(y^k_m, t_m)\big)\\
&\le h_k e^{2lT} \sum_{m=1}^{j+1} \text{dist}(w^k_m, F(y^k_m, t_m)).
\end{aligned} \tag{3.9}$$

We recall next the average modulus of continuity of F defined by

$$\tau(F; h) := \sup{}_{x \in U} \int_0^T \sup\Big\{\text{dist}(F(x, t'), F(x, t''))\Big|\, t', t'' \in \Big[t - \frac{h}{2}, t + \frac{h}{2}\Big]\Big\}\, dt$$

and consider the quantities ζ_k with the estimates

$$\begin{aligned}
\zeta_k &:= \sum_{m=1}^{k} h_k \text{dist}\big(w^k_m, F(y^k_m, t_m)\big) = \sum_{m=1}^{k} \int_{t_{m-1}}^{t_m} \text{dist}(w^k_m, F(y^k_m, t_m))\\
&\le \sum_{m=1}^{k} \int_{t_{m-1}}^{t_m} \text{dist}\big(w^k_m, F(y^k_m, t)\big) + \tau(F; h_k), \quad k \in \mathbb{N}.
\end{aligned} \tag{3.10}$$

It is well known (see, e.g., [**Q**, Proposition 6.3]) that the a.e. continuity of $F(x, \cdot)$ on $[0, T]$ uniformly in $x \in U$ assumed in (H3) is equivalent to the convergence $\tau(F; h_k) \to 0$.

Let us show next that $\displaystyle\sum_{m=1}^{k}\int_{t_{m-1}}^{t_m}\text{dist}(w_m^k,F(y_m^k,t))\to 0$ as $k\to\infty$. Taking into account that $\bar{x}(\cdot)$ is a feasible trajectory for the differential inclusion (1.2) and that $w^k(\cdot)\to\dot{\bar{x}}(\cdot)$ strongly in $L^1[0,T]$ and remembering also that for each $k\in\mathbb{N}$ the functions $w^k(t)$ are constant on the intervals $(t_{m-1},t_m]$, $m=1,2,\ldots$, and that $\dot{\bar{x}}(t)$ is Riemann integrable (that is, a.e. continuous) on $[0,T]$, we can find $\widetilde{t}_m\in(t_{m-1},t_m]$ such that

$$\dot{\bar{x}}(\widetilde{t}_m)\in F(\bar{x}(\widetilde{t}_m),\widetilde{t}_m)\ \text{ and }\ \sum_{m=1}^{k}\int_0^T|\dot{\bar{x}}(\widetilde{t}_m)-w^k(t)|\,dt\le 2\xi_k.$$

This readily leads us to the following inequalities:

$$\begin{aligned}
&\sum_{m=1}^{k}\int_{t_{m-1}}^{t_m}\text{dist}(w_m^k,F(y_m^k,t))dt\\
&\le\sum_{m=1}^{k}\int_{t_{m-1}}^{t_m}\Big[\text{dist}(w_m^k,F(\bar{x}_m,t))+\text{dist}(F(\bar{x}_m,t),F(y_m^k,t))\Big]dt\\
&\le\sum_{m=1}^{k}\int_{t_{m-1}}^{t_m}\Big[\text{dist}(w_m^k,F(\bar{x}(\widetilde{t}_m),\widetilde{t}_m))+\text{dist}(F(\bar{x}_m,t),F(\bar{x}(\widetilde{t}_m),\widetilde{t}_m))\\
&\qquad\qquad+\text{dist}(F(\bar{x}_m,t),F(y_m^k,t))\Big]dt\\
&\le\sum_{m=1}^{k}\int_{t_{m-1}}^{t_m}\Big[|w^k(t)-\dot{\bar{x}}(\widetilde{t}_m)|+\text{dist}(F(\bar{x}_m,t),F(\bar{x}(\widetilde{t}_m),\widetilde{t}_m))\\
&\qquad\qquad+\text{dist}(F(\bar{x}_m,t),F(y_m^k,t))\Big]dt\\
&\le 2\xi_k+\sum_{m=1}^{k}\int_{t_{m-1}}^{t_m}\Big[\text{dist}(F(\bar{x}_m,t),F(\bar{x}(\widetilde{t}_m),\widetilde{t}_m))+\text{dist}(F(\bar{x}_m,t),F(y_m^k,t))\Big]dt\\
&\le 2\xi_k+\tau(F;h_k)+\sum_{m=1}^{k}\int_{t_{m-1}}^{t_m}\Big[\text{dist}(F(\bar{x}(t_m),t_m),F(\bar{x}(\widetilde{t}_m),\widetilde{t}_m))\\
&\qquad\qquad+\text{dist}(F(\bar{x}_m,t),F(y_m^k,t))\Big]dt.
\end{aligned}$$

Under the assumption (H3) we have $\displaystyle\sum_{m=1}^{k}\int_{t_{m-1}}^{t_m}\text{dist}(w_m^k,F(y_m^k,t))\to 0$. By employing (3.4) and the definition of ζ_k in (3.10), it gives us the convergence $\zeta_k\to 0$ as $k\to\infty$.

Using this and the last inequality in (3.9) allows us to conclude that

$$(3.11)\quad |z_{j+1}^k-y_{j+1}^k|\le\zeta_k e^{2lT}\ \text{ for all }\ j=0,\ldots,k-1\ \text{ and all }\ k\in\mathbb{N}.$$

Furthermore, we easily get the estimates

$$(3.12)\qquad |z_{j+1}^k-\bar{x}_{j+1}|\le\zeta_k e^{2lT}+|y_{j+1}^k-\bar{x}_{j+1}|\le\zeta_k e^{2lT}+\xi_k=:\eta_k,$$

where $\eta_k\to 0$ due to (3.4) and $\zeta_k\to 0$ as $k\to\infty$.

Considering next the the piecewise linear functions $z^k(\cdot)$ built in (3.3) by using the discrete velocity $v^k(\cdot)$ from (3.2), we get from (3.2) and (3.7) that

$$\dot{z}^k(t)=v_j^k=\frac{z_j^k-z_{j-1}^k}{h_k}\in F\big(z^k(t_j),t_j\big)\ \text{ on }\ (t_{j-1},t_j],\quad j=1,\ldots,k.$$

It follows from the uniform boundedness of F in (H1) that there is a subsequence of $\{\dot{z}^k(\cdot)\}$ (without relabeling) that converges to some function in $L^1[0,T]$, which cannot be anything but $\dot{\bar{x}}(t)$ due to the relationships in (3.12) established above. Thus $\dot{z}^k(\cdot) \to \dot{\bar{x}}(\cdot)$ *weakly* in $L^1[0,T]$ as $k \to \infty$. The latter is equivalent to the *uniform convergence* $z^k(\cdot) \to \bar{x}(t)$ by the Newton-Leibniz formula (3.3), and so we get (i).

Now we justify assertion (ii) proving that in fact $\dot{z}^k(\cdot) \to \dot{\bar{x}}(\cdot)$ as $k \to \infty$ *strongly* in $L^1[0,T]$ provided that either the graph of $F(\cdot,t)$ is convex around $(\bar{x}(t), \dot{\bar{x}}(t))$, or the mapping $x \mapsto F(x,t)$ is locally Lipschitzian around $\bar{x}(t)$ for a.e. $t \in [0,T]$.

First we examine the case when the *graph* of $F(\cdot,t)$ is *locally convex.* The classical Mazur's weak closure theorem tells us that there is a sequence of *convex combinations* of $\dot{z}^k(\cdot)$, which converges to $\dot{\bar{x}}(\cdot)$ in the norm topology of $L^1[0,T]$ and thus contains a subsequence (no relabeling) converging to $\dot{\bar{x}}(t)$ for a.e. $t \in [0,T]$. Taking into account the graph convexity of $F(\cdot,t)$ and the piecewise constant nature of $\dot{z}^k(t)$ and assuming without loss of generality that for each element of the sequence of convex combinations the corresponding partition of the interval is a subpartition of the previous one, we conclude that all the elements of the aforementioned sequence of convex combinations are *feasible* trajectories of the discrete approximation systems for any $k \in \mathbb{N}$. Therefore we get a sequence of feasible solutions to the discrete inclusions (3.1) whose piecewise linear extensions on $[0,T]$ converges to $\dot{\bar{x}}(\cdot)$ strongly in $L^1[0,T]$. Keeping for simplicity the notation $\dot{z}^k(\cdot)$ for the elements of this sequence allows us to write

$$\alpha_k := \int_0^T |\dot{z}^k(t) - \dot{\bar{x}}(t)|\,dt \to 0 \;\text{ as }\; k \to \infty. \tag{3.13}$$

To complete the proof of the theorem in the convex graph case, it remains to verify the the convergence of $\{\dot{z}^k(\cdot)\}$ to $\dot{\bar{x}}(\cdot)$ in the norm topology of $L^2[0,T]$. By the constructions above and assumption (H1), it is implied by the following relationships:

$$\begin{aligned} &\sum_{j=1}^{k} \int_{t_{j-1}}^{t_j} \left| \frac{z^k(t_j) - z^k(t_{j-1})}{h_k} - \dot{\bar{x}}(t) \right|^2 dt \\ &\qquad = \sum_{j=1}^{k} \max\left(|v_j^k| + |\dot{\bar{x}}(t)|\right) \int_{t_{j-1}}^{t_j} |v_j^k - \dot{\bar{x}}(t)|\,dt \\ &\qquad \le 2m_F \sum_{j=1}^{k} \int_{t_{j-1}}^{t_j} |v_j^k - \dot{\bar{x}}(t)|\,dt = 2m_F\alpha_k, \end{aligned} \tag{3.14}$$

where α_k is taken from (3.13). This justifies the $W^{1,2}[0,T]$-norm convergence of the extended discrete trajectories $\{z^k(\cdot)\}$ from (3.3) in the first case under consideration.

Let us finally consider the other case in (ii) when $F(\cdot,t)$ is *locally Lipschitzian* around $\bar{x}(t)$ for a.e. $t\in[0,T]$. Then for all $j\in\{1,\ldots,k-1\}$ we have the estimates

(3.15)
$$\begin{aligned}
|z^k_{j+1}-z^k_j-h_k w_{j+1}| &\le \frac{1}{1-lh_k}\text{dist}(z^k_j+h_k w_{j+1}, z^k_j+h_kF(z^k_j+h_kw_{j+1},t_{j+1}))\\
&\le \frac{1}{1-lh_k}\text{dist}(h_kw_{j+1},h_kF(z^k_j+h_kw_{j+1},t_{j+1}))\\
&\le \frac{h_k}{1-lh_k}\Big[\text{dist}(w_{j+1},F(y^k_j+h_kw_{j+1},t_{j+1}))\\
&+\text{dist}(F(y^k_j+h_kw_{j+1},t_{j+1}),F(z^k_j+h_kw_{j+1},t_{j+1}))\Big]\\
&\le \frac{h_k}{1-lh_k}\Big[l|z^k_j-y^k_j|+\text{dist}(w_{j+1},F(y_{j+1},t_{j+1}))\Big].
\end{aligned}$$

Combining (3.15) with (3.11) gives us

$$\begin{aligned}
\int_0^T|\dot{z}^k(t)-\dot{y}^k(t)|\,dt &= \sum_{j=1}^k h_k|v^k_j-w^k_j|\\
&\le \frac{1}{1-lh_k}\Big[lh_k\sum_{j=1}^k|z^k_{j-1}-y^k_{j-1}|+\sum_{j=1}^k h_k\text{dist}(w^k_j,F(y^k_j,t_j))\Big]\\
&\le 2(lT\zeta_ke^{2lT}+\zeta_k)=2\zeta_k(lTe^{2lT}+1)\to 0\quad\text{as}\quad k\to\infty,
\end{aligned}$$

where $\zeta_k\to 0$ is taken from (3.10). Taking further into account that

$$\int_0^T|\dot{z}^k(t)-\dot{\bar{x}}(t)|\,dt\le\int_0^T|\dot{z}^k(t)-\dot{y}^k(t)|\,dt+\int_0^T|\dot{y}^k(t)-\dot{\bar{x}}(t)|\,dt$$

and using the convergence $\xi_k\to 0$ in (3.4) with $\dot{y}^k(t)=w^k(t)$ tell us that the $L^1[0,1]$-norm convergence of $\{\dot{z}^k(\cdot)\}$ in (3.13) holds in the second case under consideration. Applying now (3.14) justifies (ii) in this case and thus completes proof of the theorem. □

4. Strong Convergence of Discrete Optimal Solutions

In this section we construct a sequence of well-posed discrete approximations of the Bolza problem (P) for ROSL differential inclusions and justify the norm convergence in either $C[0,T]$ or $W^{1,2}[0,T]$ topology of their optimal solutions to either a strong local minimizer or an intermediate relaxed local minimizer $\bar{x}(\cdot)$ of (P), respectively. In addition to our standing assumptions (H1)–(H3) on the right-hand side F in (1.2) and those (if needed) from Theorem 3.1 formulated now around the given local minimizer, the following ones are imposed here on the functions f and φ_0 in the Bolza cost functional (1.1) as well as on the functions φ_i, $i=1,\ldots,m+r$, and the set Ω in the endpoint constraints (1.3)–(1.5).

(H4) The function $f(x,v,\cdot)$ is a.e. continuous on $[0,T]$ and bounded uniformly in $(x,v)\in U\times(m_FB)$. Furthermore, there exists $\nu>0$ such that the function $f(\cdot,\cdot,t)$ is continuous on the set

$$A_\nu(t)=\big\{(x,v)\in U\times(m_F+\nu)I\!B\big|\ v\in F(x,t')\ \text{ for some }\ t'\in(t-\nu,t]\big\}$$

uniformly in t on the interval $[0,T]$.

(H5) The cost function φ_0 is continuous on U, while the constraint functions φ_i are Lipschitz continuous on U for all $i=1,\ldots,m+r$. Furthermore, the endpoint constraint set Ω is locally closed around $\bar{x}(T)$.

Given a *r.i.l.m.* $\bar{x}(\cdot)$ in (P), suppose without loss of generality (due to (H1)) that $\alpha = 1$ and $p = 2$ in (2.4) and Definition 2.3. Denote by $L > 0$ a common Lipschitz constant for the functions φ_i, $i = 1, \dots, m+r$, on U and take the sequence $\{\eta_k\}$ in (3.12) constructed via the approximation of the local optimal solution $\bar{x}(\cdot)$ under consideration. Then we define a sequence of discrete approximation problems (P_k), $k \in \mathbb{N}$, as follows:

$$
\begin{aligned}
\text{minimize } J_k[x^k] : &= \varphi_0(x^k(t_k)) + h_k \sum_{j=1}^{k} f\Big(x^k(t_j), \frac{x^k(t_j) - x^k(t_{j-1})}{h_k}, t_j\Big) \\
&+ \sum_{j=1}^{k} \int_{t_{j-1}}^{t_j} \left| \frac{x^k(t_j) - x^k(t_{j-1})}{h_k} - \dot{\bar{x}}(t) \right|^2 dt
\end{aligned}
\tag{4.1}
$$

over trajectories $x^k = (x_0^k, \dots, x_k^k)$ of the discrete inclusions (3.1) subject to the constraints

$$
|x^k(t_j) - \bar{x}(t_j)|^2 \le \frac{\varepsilon^2}{4} \quad \text{for } j = 1, \dots, k, \tag{4.2}
$$

$$
\sum_{j=1}^{k} \int_{t_{j-1}}^{t_j} \left| \frac{x^k(t_j) - x^k(t_{j-1})}{h_k} - \dot{\bar{x}}(t) \right|^2 dt \le \frac{\varepsilon}{2}, \tag{4.3}
$$

$$
x_k^k \in \Omega_k := \Omega + \eta_k I\!B, \tag{4.4}
$$

$$
\varphi_i(x_k^k) \le L\eta_k \quad \text{for } i = 1, \dots, m, \tag{4.5}
$$

$$
-L\eta_k \le \varphi_i(x_k^k) \le L\eta_k \quad \text{for } i = m+1, \dots, m+r, \tag{4.6}
$$

where $\varepsilon > 0$ is fixed and taken from (2.4) for the given r.i.l.m. $\bar{x}(\cdot)$.

If $\bar{x}(\cdot)$ is a given *strong local minimizer* for (P) with $f = f(x, t)$, we construct a simplified sequence of discrete approximations problems $(\widetilde{P}_k)$ as follows:

$$
\text{minimize } \widetilde{J}_k[x^k] := \varphi_0(x^k(t_k)) + h_k \sum_{j=1}^{k} f(x^k(t_j), t_j) + \sum_{j=1}^{k} |x^k(t_j) - \bar{x}(t_j)|^2 \tag{4.7}
$$

subject to the constraints (4.2)–(4.6) with η_k taken from (3.12).

The next theorem shows that problems (P_k) and $(\widetilde{P}_k)$ admit optimal solutions for all large $k \in \mathbb{N}$ and that extended discrete optimal solutions to these problems converge to $\bar{x}(\cdot)$ in the corresponding norm topology of either $C[0,T]$ or $W^{1,2}[0,T]$ depending on the type of local minima (strong or intermediate) which (P) achieves at $\bar{x}(\cdot)$.

THEOREM 4.1 (strong convergence of discrete optimal solutions). *Let $\bar{x}(\cdot)$ be a Riemann integrable local optimal solution to the original Bolza problem (P) under the validity of assumptions* (H1)-(H5) *around $\bar{x}(\cdot)$. The following assertions hold:*

(1) *If $\bar{x}(\cdot)$ is a strong local minimizer for (P) with $f = f(x,t)$, then each problem $(\widetilde{P}_k)$ admits an optimal solution $\bar{x}^k(\cdot)$ for large $k \in \mathbb{N}$ and the sequence $\{\bar{x}^k(\cdot)\}$ piecewise linearly extended to $[0,T]$ converges to $\bar{x}(\cdot)$ as $k \to \infty$ in the norm topology of $C[0,T]$.*

(2) *If $\bar{x}(\cdot)$ is a r.i.l.m. in (P) and the assumptions of Theorem* 3.1(ii) *are satisfied for $\bar{x}(\cdot)$, then each problem (P_k) admits an optimal solution $\bar{x}^k(\cdot)$ whenever $k \in \mathbb{N}$ is sufficiently large and the sequence $\{\bar{x}^k(\cdot)\}$ piecewise linearly extended to $[0,T]$ converges to $\bar{x}(\cdot)$ as $k \to \infty$ in the norm topology of $W^{1,2}[0,T]$.*

PROOF. We verify the existence of optimal solutions to problems (P_k) and $(\widetilde{P}_k)$ in a parallel way. Observe first that both (P_k) and $(\widetilde{P}_k)$ admit feasible solutions for all $k \in \mathbb{N}$ sufficiently large. Indeed, take for each k the discrete trajectories $z^k := (z_0^k, \ldots, z_k^k)$ constructed in Theorem 3.1(i) to approximate the r.i.l.m. $\bar{x}(\cdot)$ and in Theorem 3.1(ii) to approximate the r.i.l.m. $\bar{x}(\cdot)$. Then both these functions satisfy the discrete inclusion (3.1), and it remains to verify that the corresponding z^k fulfills the constraints in (4.2), (4.4)–(4.6) in the case of $(\widetilde{P}_k)$ and those in (4.2)–(4.6) in the case of (P_k). The validity of (4.2) and (4.4) in both cases follows from (3.12) for large k, while the validity of the additional constraint (4.3) for (P_k) follows from (3.13). The fulfillment of the inequality constraints in (4.5) and (4.6) for z_k^k follows by these arguments from the validity of (1.4) and (1.5) for $\bar{x}(T)$, respectively, and the local Lipschitz continuity of the endpoint functions

$$|\varphi_i(z_k^k) - \varphi_i(\bar{x}(T))| \le L|z_k^k - \bar{x}(T)| \le L\eta_k, \quad i = 1, \ldots, m+r.$$

Thus for each $k \in \mathbb{N}$ (omitting the expression "for all large k" in what follows) the sets of feasible solutions to (P_k) and $(\widetilde{P}_k)$ are nonempty. It is clear from the construction of (P_k) and $(\widetilde{P}_k)$ and the assumptions made that each of these sets is closed and bounded. This ensures the existence of optimal solutions to (P_k) by the classical Weierstrass existence theorem due to the continuity of the functions φ_0 and f in (4.1) and (4.7) .

Next we proceed with the proof of the strong $W^{1,2}[0,T]$-convergence in (ii) for any sequence of the discrete optimal solutions $\{\bar{x}^k(\cdot))\}$ in (P_k) piecewise linearly extended to the continuous-time interval $[0,T]$. To this end let us first show that

$$\liminf_{k\to\infty} J_k[\bar{x}^k] \le J[\bar{x}] \tag{4.8}$$

for the optimal values of the cost functional in (4.1). It follows from the optimality of $\bar{x}^k(\cdot)$ for (P_k) and the feasibility of $z^k(\cdot)$ taken from the proof of (i) for this problem that $J_k[\bar{x}^k] \le J_k[z^k]$ for each k. To get (4.8), it suffices to show therefore that

$$J_k[z^k] \to J[\bar{x}] \ \text{ as } \ k \to \infty \tag{4.9}$$

including the verification of the existence of the limit. We have from (4.1) that

$$\begin{aligned} J_k[z^k] &= \varphi_0(z^k(t_k)) + h_k \sum_{j=1}^{k} f\Big(z^k(t_j), \frac{z^k(t_j) - z^k(t_{j-1})}{h_k}, t_j\Big) \\ &\quad + \sum_{j=1}^{k} \int_{t_{j-1}}^{t_j} \Big|\frac{z^k(t_j) - z^k(t_{j-1})}{h_k} - \dot{\bar{x}}(t)\Big|^2 dt \end{aligned}$$

and deduce from the convergence $z^k(t_k) \to \bar{x}(T)$ and the continuity assumption on φ_0 in (H5) the convergence $\varphi_0(z^k(t_k)) \to \varphi_0(\bar{x}(T))$ as $k \to \infty$ of the terminal cost function in (4.1). Furthermore, it follows from (3.13) that

$$\sum_{j=1}^{k}\int_{t_{j-1}}^{t_j}\left|\frac{z^k(t_j)-z^k(t_{j-1})}{h_k}-\dot{\bar{x}}(t)\right|^2dt\to 0 \text{ as } k\to\infty.$$

To justify (4.9), we only need to check that

$$h_k\sum_{j=1}^{k}f\Big(z^k(t_j),\frac{z^k(t_j)-z^k(t_{j-1})}{h_k},t_j\Big)\to\int_0^T f(\bar{x}(t),\dot{\bar{x}}(t),t)dt \text{ as } k\to\infty.$$

The continuity assumptions on f in (H4) imply without loss of generality that

$$\Big|f(z^k(t_j),\frac{z^k(t_j)-z^k(t_{j-1})}{h_k},t_j)-f(z^k(t_j),\frac{z^k(t_j)-z_k(t_{j-1})}{h_k},t)\Big|\le\frac{\varepsilon}{T}$$

for all $k\in\mathbb{N}$ and a.e. $t\in[0,T]$. Employing now Lebesgue's dominated convergence theorem together with Theorem 3.1(ii) tells us that

$$\begin{aligned}
&h_k\sum_{j=1}^{k}f\Big(z^k(t_j),\frac{z^k(t_j)-z^k(t_{j-1})}{h_k},t_j\Big)=\sum_{j=1}^{k}\int_{t_{j-1}}^{t_j}f(z^k(t_j),v^k(t),t_j)\,dt\\
&\sim\sum_{j=1}^{k}\int_{t_{j-1}}^{t_j}f(z^k(t_j),v^k(t),t)\,dt\sim\sum_{j=1}^{k}\int_{t_{j-1}}^{t_j}f(\bar{x}(t),v^k(t),t)\,dt\\
&=\int_0^T f(\bar{x}(t),v^k(t),t)\,dt\sim\int_0^T f(\bar{x}(t),\dot{\bar{x}}(t),t)\,dt,
\end{aligned}$$

where the sign '$\sim$' is used to indicate the equivalence as $k\to\infty$. Thus we get (4.9).

To proceed further, consider the numerical sequence

$$c_k:=\int_0^T|\dot{\bar{x}}^k(t)-\dot{\bar{x}}(t)|^2dt,\quad k\in\mathbb{N}, \tag{4.10}$$

and verify that $c_k\to 0$ as $k\to\infty$. Since the numerical sequence in (4.10) is obviously bounded, it has limiting points. Denote by $c\ge 0$ any of them and show that $c=0$. Arguing by contradiction, suppose that $c>0$. It follows from the uniform boundedness assumption (H1) and basic functional analysis that the sequence $\{\dot{\bar{x}}^k(\cdot)\}$ contains a subsequence (without relabeling), which converges in the weak topology of $L^2[0,T]$ to some $v(\cdot)\in L^2[0,T]$. Considering the absolutely continuous function

$$\widetilde{x}(t):=x_0+\int_0^t v(s)\,ds,\quad 0\le t\le T,$$

we deduce from the Newton-Leibniz formula that the sequence of the extended discrete trajectories $\bar{x}^k(\cdot)$ converges to $\widetilde{x}(\cdot)$ in the weak topology of $W^{1,2}[0,T]$, for which we have $\dot{\widetilde{x}}(t)=v(t)$ for a.e. $t\in[0,T]$. By invoking Mazur's weak closure theorem, it follows from the convexity of the sets $F(x,t)$ and the continuity of $F(\cdot,t)$ that the limiting function $\widetilde{x}(\cdot)$ satisfies the differential inclusion (1.2). Furthermore, the construction of the discrete approximation problems (P_k) with $\eta_k\to 0$ therein ensures that $\widetilde{x}(\cdot)$ is a feasible trajectory for the original Bolza problem (P), and therefore for the relaxed (R) as well.

Employing again Mazur's weak closure theorem, we find a sequence of convex combination of $\dot{\bar{x}}^k(\cdot)$ converging to $\dot{\widetilde{x}}(\cdot)$ in the norm topology of $L^2[0,T]$ and hence a.e. on $[0,T]$ along some subsequence. Taking into account the construction of $\widehat{f}_F$

as the convexification of f_F in (2.5) with respect to the velocity variable, we arrive at the inequality

$$\int_0^T \widehat{f}_F(\widetilde{x}(t),\dot{\widetilde{x}}(t),t)\,dt \le \liminf_{k\to\infty} h_k\sum_{j=1}^k f\Big(\bar{x}_j^k,\frac{\bar{x}_j^k-\bar{x}_{j-1}^k}{h_k},t_j\Big). \tag{4.11}$$

Define now the integral functional on $L^2[0,T]$ by

$$I[v] := \int_0^T |v(t)-\dot{\bar{x}}|^2 dt \tag{4.12}$$

and show it is convex on this space. Indeed, picking any $v(\cdot),w(\cdot)\in L^2[0,T]$ and $\lambda\in[0,1]$ and using the Cauchy-Schwartz inequality gives us

$$\begin{aligned} I[\lambda v+(1-\lambda)w] &= \int_0^T |\lambda(v(t)-\dot{\bar{x}}(t))+(1-\lambda)(w(t)-\dot{\bar{x}}(t))|^2 dt\\ &\le \int_0^T \Big[\lambda|v(t)-\dot{\bar{x}}(t)|+(1-\lambda)|w(t)-\dot{\bar{x}}(t)|\Big]^2 dt\\ &= \lambda\int_0^T |v(t)-\dot{\bar{x}}(t)|^2dt+(1-\lambda)\int_0^T |w(t)-\dot{\bar{x}}(t)|^2 dt\\ &= \lambda I[v]+(1-\lambda)I[w], \end{aligned}$$

which justifies the convexity and hence the lower semicontinuity of (4.12) in the weak topology of $L^2[0,T]$. It allows us to conclude that

$$\begin{aligned} \int_0^T |\dot{\widetilde{x}}(t)-\dot{\bar{x}}(t)|^2 dt &\le \liminf_{k\to\infty}\int_0^T |\dot{\bar{x}}^k(t)-\dot{\bar{x}}(t)|^2dt\\ &= \liminf_{k\to\infty}\sum_{j=1}^k\int_{t_{j-1}}^{t_j}\Big|\frac{\bar{x}^k(t_j)-\bar{x}^k(t_{j-1})}{h_k}-\dot{\bar{x}}(t)\Big|^2 dt. \end{aligned}$$

Employing this and passing to the limit in the constraints (4.2) and (4.3) for $\bar{x}^k(\cdot)$ yield

$$|\widetilde{x}(t)-\bar{x}(t)|\le\frac{\varepsilon}{2}\ \text{ for }\ t\in[0,T]\ \text{ and }\ \int_0^T|\dot{\widetilde{x}}(t)-\dot{\bar{x}}(t)|^2dt\le\frac{\varepsilon}{2},$$

which verifies that the feasible trajectory $\widetilde{x}(\cdot)$ for (R) belongs to the prescribed $W^{1,2}[0,T]$ neighborhood of the r.i.l.m. $\bar{x}(\cdot)$ from Definition 2.3.

Now we are able to pass to the limit in the cost functional formula (4.1) in (P_k) for $\bar{x}^k(\cdot)$ by using (4.8), (4.11), and the assumption on $c_k\to c>0$ in (4.10). It gives us

$$\widehat{J}[\widetilde{x}]=\varphi(\widetilde{x}(T))+\int_0^T\widehat{f}_F(\widetilde{x}(t),\dot{\widetilde{x}}(t),t)\,dt\le\liminf_{k\to\infty}J_k[\bar{x}^k]+c<J[\bar{x}]=\widehat{J}[\bar{x}],$$

which contradicts the choice of $\bar{x}(\cdot)$ as a r.i.l.m. for the original Bolza problem (P). Thus we have $c_k\to 0$ as $k\to\infty$ showing in this way that $\bar{x}^k(\cdot)\to\bar{x}(\cdot)$ strongly in $W^{1,2}[0,T]$.

To complete the proof of the theorem, it remains to justify the strong $C[0,T]$ convergence in (i) of discrete optimal trajectories for $(\widetilde{P}_k)$ in the case when $\bar{x}(\cdot)$ is a strong local minimizers of the continuous-time Bolza problem (P). Note that due to the convexity of $F(x,t)$ and the independence of the integrand f on the velocity variable, problem (P) agrees with its relaxation (R). Taking into account the form of the cost functional (4.7) and Theorem 3.1(i) on the strong discrete approximation of $\bar{x}(\cdot)$ in $C[0,T]$, we arrive at the claimed convergence result in assertion (i) of this theorem by just simplifying the above proof of assertion (ii) and replacing the cost functional J_k with $\widetilde{J}_k$. $\square$

5. Optimality Conditions for Discrete Approximations

In this section we derive necessary optimality conditions for each problem (P_k), $k \in \mathbb{N}$, in the sequence of discrete approximations formulated in Section 4. In the same way we can proceed with deriving necessary optimality conditions in the simplified problems $(\widetilde{P}_k)$; we do not present them here due to the full similarity and size limitation.

Note that problems of this type intrinsically belong to nonsmooth optimization even when all the functions f and φ_i for $i = 0, \ldots, m + r$ are smooth and $\Omega = \mathbb{R}^n$. The nonsmoothness comes from the dynamic constraints in (3.1) given by the discretization of the differential inclusion (1.2); the number of these constraints is increasing along with decreasing the step of discretization. To derive necessary optimality conditions for problems (P_k), we employ advanced tools of variational analysis and generalized differentiation discussed in Section 2.

Consider first the following problem of mathematical programming (MP) with finitely many functional and geometric constraints. Given $\phi_j \colon \mathbb{R}^d \to \mathbb{R}$ for $j = 0, \ldots, s$, $g_j : \mathbb{R}^d \to \mathbb{R}^n$ for $j = 0, \ldots, p$, and $\Delta_j \subset \mathbb{R}^d$ for $j = 0, \ldots, q$, we define (MP) by

$$\begin{aligned} &\text{minimize } \ \phi_0(z) \ \text{ subject to} \\ &\phi_j(z) \le 0 \ \text{ for } \ j = 0, \ldots, s, \\ &g_j(z) = 0 \ \text{ for } \ j = 0, \ldots, p, \\ &z \in \Delta_j \ \text{ for } \ j = 0, \ldots, q. \end{aligned}$$

The next result gives us necessary optimality conditions for local minimizers of problem (MP) in the setting needed for the subsequent application to deriving optimality conditions in the discrete approximation problems (P_k). We express these conditions via our basic normal cone (2.7) and subdifferential (2.8) constructions from Section 2.

LEMMA 5.1. **(generalized Lagrange multiplier rule for mathematical programs).** *Let $\bar{z}$ be a local optimal solution to problem (MP). Assume that the functions ϕ_j are Lipschitz continuous around $\bar{z}$, the mappings g_j are continuous differentiable around $\bar{z}$, and the sets Δ_j are locally closed around this point. Then there exist nonnegative numbers μ_j for $j = 0, \ldots, s$ as well as vectors $\psi_j \in \mathbb{R}^n$ for $j = 0, \ldots, p$ and $z_j^* \in \mathbb{R}^d$ for $j = 0, \ldots, q$, not equal to zero simultaneously, such that we have the conditions*

$$z_j^* \in N(\bar{z}; \Delta_j), \quad j = 0, \ldots, q,$$

$$\mu_j \phi_j(\bar{z}) = 0, \quad j = 1, \ldots, s,$$

$$-z_0^* - \ldots - z_q^* \in \partial\Big(\sum_{j=0}^{s} \mu_j \phi_j\Big)(\bar{z}) + \sum_{j=0}^{p} (\nabla g_j(\bar{z}))^T \psi_j,$$

where the symbol "A^T" indicates the matrix transposition.

PROOF. This result follows from necessary optimality conditions given [**Q**, Theorem 5.21] for problems with a single geometric constraint and the basic intersection rule for the normal cone (2.7) taken from [**P**, Theorem 3.4]. □

Now we employ Lemma 5.1 and calculus rules for generalized normals and subgradients to derive necessary optimality conditions for the structural dynamic problems of discrete approximation (P_k) in the extended Euler-Lagrange form. Note that for this purpose we need less assumptions that those imposed in (H1)–(H5). Observe also that the form of the Euler-Lagrange inclusion below reflects the essence of the implicit Euler scheme being significantly different from the adjoint system corresponding to the explicit Euler counterpart from [**N**, **Q**]. The solvability of the new implicit adjoint system is ensures by Lemma 5.1 due the given proof of the this theorem.

THEOREM 5.2 (extended Euler-Lagrange conditions for discrete approximations). *Fix any $k \in \mathbb{N}$ and let $\bar{x}^k = (\bar{x}_0^k, \ldots, \bar{x}_k^k)$ with $\bar{x}_0^k = x_0$ in (3.1) be an optimal solution to problem (P_k) constructed in Section 4. Assume that the sets Ω and $\operatorname{gph} F_j$ with $F_j := F(\cdot, t_j)$ are closed and the functions φ_i for $i = 0, \ldots, m+r$ and $f_j := f(\cdot, \cdot, t_j)$ for $j = 0, \ldots, k$ are Lipschitz continuous around the corresponding points.*

Then there exist real numbers λ_i^k for $i = 0, \ldots, m+r$ and a vector $p^k := (p_0^k, \ldots, p_k^k) \in \mathbb{R}^{(k+1)n}$, which are not equal to zero simultaneously and satisfy the following relationships:

- *The sign conditions*

$$\lambda_i^k \geq 0 \ \text{ for } \ i = 0, \ldots, m;$$

- *the complementary slackness conditions*

$$\lambda_i^k [\varphi_i(\bar{x}_k^k) - L\eta_k] = 0 \ \text{ for } \ i = 1, \ldots, m;$$

- *the extended Euler-Lagrange inclusion held for $j = 1, \ldots, k$:*

$$\Big(\frac{p_j^k - p_{j-1}^k}{h_k}, p_{j-1}^k - \frac{\lambda_0^k \theta_j^k}{h_k}\Big) \in \lambda_0^k \partial f_j\Big(\bar{x}_j^k, \frac{\bar{x}_j^k - \bar{x}_{j-1}^k}{h_k}\Big) + N\Big(\Big(\bar{x}_j^k, \frac{\bar{x}_j^k - \bar{x}_{j-1}^k}{h_k}\Big); \operatorname{gph} F_j\Big);$$

- *the transversality inclusion*

$$-p_k^k \in \sum_{i=0}^{m} \lambda_i^k \partial \varphi_i(\bar{x}_k^k) + \sum_{i=m+1}^{m+r} \lambda_i^k \partial^0 \varphi_i(\bar{x}_k^k) + N(\bar{x}_k^k; \Omega_k),$$

where $\partial^0 \varphi_i$ stands for the symmetric subdifferential (2.9) of φ_i, and where

$$\theta_j^k := -2 \int_{t_{j-1}}^{t_j} \Big(\dot{\bar{x}}(t) - \frac{\bar{x}_j^k - \bar{x}_{j-1}^k}{h_k}\Big) dt. \tag{5.1}$$

PROOF. Skipping for notational simplicity the upper index "k" if no confusions arise, consider the new "long" variable

$$z := (x_0, \ldots, x_k, y_1, \ldots, y_k) \in \mathbb{R}^{(2k+1)n} \ \text{ with the fixed initial vector } \ x_0$$

and for each $k \in \mathbb{N}$ reformulate the discrete approximation problem (P_k) as a mathematical program of the above type (MP) with the following data:

$$\min \phi_0(z) := \varphi_0(x_k) + h_k \sum_{j=1}^{k} f(x_j, y_j, t_j) + \sum_{j=1}^{k} \int_{t_{j-1}}^{t_j} |y_j - \dot{\bar{x}}(t)|^2 dt \tag{5.2}$$

subject to the functional and geometric constraints

$$\phi_j(z) := |x_j - \bar{x}(t_j)|^2 - \frac{\varepsilon^2}{4} \le 0 \ \text{ for } \ j = 1, \ldots, k, \tag{5.3}$$

$$\phi_{k+1}(z) := \sum_{j=1}^{k} \int_{t_j-1}^{t_j} |y_j - \dot{\bar{x}}(t)|^2 dt - \frac{\varepsilon}{2} \le 0, \tag{5.4}$$

$$\phi_{k+1+j}(z) = \varphi_j(x_k) - L\eta_k \le 0 \ \text{ for } \ j = 1, \ldots, m + r, \tag{5.5}$$

$$\phi_{k+1+m+r+j}(z) := -\varphi_{m+j}(x_k) - L\eta_k \le 0 \ \text{ for } \ j = 1, \ldots, r, \tag{5.6}$$

$$g_j(z) := x_j - x_{j-1} - h_k y_j = 0 \ \text{ for } \ j = 1, \ldots, k, \ \ g_0(z) = x(0) - x_0 \equiv 0, \tag{5.7}$$

$$z \in \Delta_0 = \{(x_0, \ldots, x_k, y_1, \ldots, y_k) \in \mathbb{R}^{(2k+1)n} |\ x_k \in \Omega\}, \tag{5.8}$$

$$z \in \Delta_j = \{(x_0, \ldots, x_k, y_1, \ldots, y_k) \in \mathbb{R}^{(2k+1)n} |\ y_j \in F_j(x_j)\}, \quad j = 1, \ldots, k. \tag{5.9}$$

Let $\bar{x}^k = (x_0, \bar{x}_1^k, \ldots, \bar{x}_k^k)$ be a given local optimal solution to problem (P_k), and thus the corresponding extended variable

$$\bar{z} := (x_0, \ldots, \bar{x}_k, (\bar{x}_1 - \bar{x}_0)/h_k, \ldots, (\bar{x}_k - \bar{x}_{k-1})/h_k),$$

where the upper index "k" is omitted, gives a local minimum to the mathematical program (MP) with the data defined in (5.2)–(5.9). Applying now to $\bar{z}$ the generalized Lagrange multiplier rule from Lemma 5.1, we find normal collections

$$z_j^* = (x_{0j}^*, \ldots, x_{kj}^*, y_{1j}^*, \ldots, y_{kj}^*) \in N(\bar{z}; \Delta_j) \ \text{ for } \ j = 0, \ldots, k \tag{5.10}$$

and well as nonnegative multipliers $(\mu_0, \ldots, \mu_{k+1+m+2r})$ and vectors $\psi_j \in \mathbb{R}^n$ for $j = 0, \ldots, k$ such that we have the conditions

$$\mu_j \phi_j(\bar{z}) = 0 \ \text{ for } \ j = 1, \ldots, k + 1 + m + 2r, \tag{5.11}$$

$$-z_0^* - \ldots - z_k^* \in \partial\Big(\sum_{j=0}^{k+1+m+2r} \mu_j \phi_j\Big)(\bar{z}) + \sum_{j=0}^{k} (\nabla g_j(\bar{z}))^T \psi_j. \tag{5.12}$$

It follows from (5.10) and the structure of Δ_0 in (5.8) that

$$x_{k0}^* \in N(\bar{x}_k; \Omega_k), \ y_{00}^* = y_{i0}^* = x_{i0}^* = 0 \ \text{ for } i = 1, \ldots, k-1, \ \text{ and } \ x_{00}^* \ \text{ is free;}$$

the latter is due to the fact that x_0 is fixed. Furthermore, inclusion (5.10) for $j = 1, \ldots, k$ gives us by the structure of Δ_j that

$$(x_{jj}^*, y_{jj}^*) \in N\Big(\Big(\bar{x}_j, \frac{\bar{x}_j - \bar{x}_{j-1}}{h_k}\Big); \text{gph}\, F_j\Big) \ \text{ and } \ x_{ij}^* = y_{ij}^* = 0 \text{ if } \ i \ne j, \ j = 1, \ldots, k.$$

Employing the above conditions together with the subdifferential sum rule from [**P**, Theorem 2.33] with taking into the nonnegativity of μ_j, we get from (5.12) that

$$\begin{aligned}
&\partial\Big(\sum_{j=0}^{k+1+m+2r}\mu_j\phi_j\Big)(\bar{z})+\sum_{j=0}^{k}(\nabla g_j(\bar{z}))^T\psi_j\\
&\subset\sum_{j=0}^{k+1+m+2r}\mu_j\partial\phi_j(\bar{z})+\sum_{j=0}^{k}(\nabla g_j(\bar{z}))^T\psi_j\\
&=\mu_0\nabla\Big[\varphi(x_k)+h_k\sum_{j=1}^{k}f(x_j,y_j,t_j)+\sum_{j=1}^{k}\int_{t_j-1}^{t_j}|y_j-\dot{\bar{x}}(t)|^2dt\Big]\\
&\quad+\sum_{j=1}^{k}\mu_j\nabla(|x_j-\bar{x}(t_j)|^2)+\mu_{k+1}\nabla\Big(\sum_{j=1}^{k}\int_{t_j-1}^{t_j}|y_j-\dot{\bar{x}}(t)|^2dt\Big)\\
&\quad+\sum_{j=1}^{m+r}\mu_{k+1+j}\nabla\varphi_i(\bar{x}_k)-\sum_{j=1}^{r}\mu_{k+1+m+r+j}\nabla\varphi_j(\bar{x}_k)\\
&\quad+\sum_{j=1}^{k}\nabla(x_j-x_{j-1}-h_ky_j)^T\psi_j+\nabla(x(0)-x_0)^T\psi_0,
\end{aligned}$$

where the derivatives (gradients, Jacobians) of all the *composite/sum* functions involves with respect of all their variables of are taken at the optimal point $\bar{z}$. It follows from Theorem 3.1 that for $k\in\mathbb{N}$ sufficiently large we have $\phi_j(\bar{z}^k)<0$ for $\bar{z}=\bar{z}^k$ and $j=1,\ldots,k+1$ due to the structures of the functions ϕ in (5.3) and (5.4) and the complementary slackness conditions in (5.11). This implies $\mu_j=0$ for $j=1,\ldots,k+1$. Considering now the Lagrange multipliers

$$\lambda_0^k:=\mu_0\ \text{ and }\ \lambda_i^k:=\mu_{k+1+i}\ \text{ for }\ i=1,\ldots,m$$

and using the expressions for θ_j^k in (5.1), we find from the above subgradients

$$(v_j,w_j)\in\partial f_j(\bar{x}_j,\bar{y}_j),\ j=1,\ldots,k,\quad u_i^k\in\partial\varphi_i(\bar{x}_k),\ i=0,\ldots,m+r,$$

$$\text{and }\ u_i'^k\in\partial(-\varphi_i)(\bar{x}_k),\ i=m+1,\ldots,m+r,$$

for which we have the conditions

$$-x_{jj}^*=\lambda_0^kh_kv_j+\psi_j-\psi_{j+1},\quad j=1,\ldots,k-1,$$

$$-x_{k0}^*-x_{kk}^*=\lambda_0^kh_kv_k+\psi_k+\sum_{i=0}^{m}\lambda_0^ku_i^k+\sum_{i=m+1}^{m+r}\mu_{k+1+i}u_i^k+\sum_{i=m+1}^{m+r}\mu_{k+1+r+i}u_i'^k,$$

$$-y_{jj}^*=\lambda_0^kh_kw_j+\lambda_0^k\theta_j^k-h_k\psi_j,\quad j=1,\ldots,k.$$

Next we introduce for each $k\in\mathbb{N}$ the adjoint discrete trajectories by

$$p_{j-1}^k:=\psi_j^k\ \text{ for }\ j=1,\ldots,k\ \text{ and}$$

$$p_k^k:=-x_{k0}^*-\sum_{i=0}^{m}\lambda_i^ku_i^k-\sum_{i=m+1}^{m+r}\mu_{k+1+i}u_i^k-\sum_{i=m+1}^{m+r}\mu_{k+1+r+i}u_i'^k.$$

Then we get the relationships

$$\frac{p_j^k-p_{j-1}^k}{h_k}=\frac{\psi_{j+1}^k-\psi_j^k}{h_k}=\lambda_0^kv_j+\frac{x_{jj}^*}{h_k},$$

$$p_{j-1}^k - \frac{\lambda_0^k \theta_j^k}{h_k} = \psi_j^k - \frac{\lambda_0^k \theta_j^k}{h_k} = \lambda_0^k w_j + \frac{y_{jj}^*}{h_k},$$

which ensure the validity of the extended Euler-Lagrange inclusion of the theorem for each $j = 1, \ldots, k$. Furthermore, it follows from (5.5), (5.6) and the complementary slackness conditions in (5.11) that for, $j = m+1, \ldots, m+r$, we have

$$\mu_{k+1+j}(\varphi_j(x_k^k) - L\eta_k) = 0 \quad \text{and} \quad \mu_{k+1+r+j}(-\varphi_j(x_k^k) - L\eta_k) = 0,$$

which implies that either $\mu_{k+1+j} = 0$ or $\mu_{k+1+r+j} = 0$ must be equal to zero for all $j = m+1, \ldots, m+r$. Denoting finally

$$\lambda_i^k := \begin{cases} \mu_{k+1+i} & \text{if } \mu_{k+1+r+i} = 0, \\ -\mu_{k+1+r+i} & \text{if } \mu_{k+1+i} = 0 \end{cases} \tag{5.13}$$

for each $i = m+1, \ldots, m+r$, we get

$$\begin{aligned} -p_k^k &= x_{k0}^* + \sum_{i=0}^{m} \lambda_i^k u_i^k + \sum_{i=m+1}^{m+r} \mu_{k+1+i} u_i^k + \sum_{i=m+1}^{m+r} \mu_{k+1+r+i} u_i'^k \\ &\in N(\bar{x}_k^k; \Omega_k) + \sum_{i=0}^{m} \lambda_i^k \partial\varphi_i(\bar{x}_k^k) \\ &\quad + \sum_{i=m+1}^{m+r} \mu_{k+1+i} \partial\varphi_i(\bar{x}_k^k) + \sum_{i=m+1}^{m+r} \mu_{k+1+r+i} \partial(-\varphi_i)(\bar{x}_k^k) \\ &\subset \sum_{i=0}^{m} \lambda_i^k \partial\varphi_i(\bar{x}_k^k) + \sum_{i=m+1}^{m+r} \lambda_i^k \partial^0 \varphi_i(\bar{x}_k^k) + N(\bar{x}_k^k; \Omega_k). \end{aligned}$$

This justifies the transversality inclusion completes the proof of the theorem. □

The last result of this section specifies the nontriviality condition of Theorem 5.2 (meaning that all the dual elements therein, i.e., λ_i^k for $i = 0, \ldots, m+r$ and p_j^k for $j = 0, \ldots, k$, are not equal to zero simultaneously) for the important class of multifunctions $F_j = F(\cdot, t_j)$ in the discrete inclusions (3.1) of the implicit Euler scheme satisfying the so-called Lipschitz-like (known also as Aubin's pseudo-Lipschitz) property around the optimal solution $\bar{x}^k$ for (P_k). Recall that a set-valued mapping $F\colon \mathbb{R}^n \rightrightarrows \mathbb{R}^m$ is *Lipschitz-like* around $(\bar{x}, \bar{y}) \in \operatorname{gph} F$ if there exist neighborhoods U of $\bar{x}$ and V of $\bar{y}$ as well as a constant $\kappa \geq 0$ such that we have the inclusion

$$F(u) \cap V \subset F(x) + \kappa |x - u| \mathbb{B} \quad \text{for all} \quad x, u \in U.$$

A crucial advantage of the nonconvex normal cone (2.7) is the possibility to obtain in its terms a complete characterization of the Lipschitz-like property of arbitrary closed-graph multifunctions. To formulate this result, we recall coderivative notion for set-valued mappings generated by the normal cone (2.7). Given $F\colon \mathbb{R}^n \rightrightarrows \mathbb{R}^m$ and $(\bar{x}, \bar{y}) \in \operatorname{gph} F$, the *coderivative* of F at $(\bar{x}, \bar{y})$ is a set-valued mapping $D^*F(\bar{x}, \bar{y})\colon \mathbb{R}^m \rightrightarrows \mathbb{R}^n$ defined by

$$D^*F(\bar{x}, \bar{y})(v) := \Big\{ u \in \mathbb{R}^n \Big| \ (u, -v) \in N((\bar{x}, \bar{y}); \operatorname{gph} F) \Big\} \quad \text{for all} \quad v \in \mathbb{R}^m. \tag{5.14}$$

When F is single-valued and smooth around $\bar{x}$ (then we drop $\bar{y} = F(\bar{x})$), its coderivative reduces to the adjoint/transpose Jacobian

$$D^*F(\bar{x})(v) = \{\nabla F(\bar{x})^T v\}, \quad v \in \mathbb{R}^m.$$

In the general nonsmooth and/or set-valued case, the coderivative (5.14) is a positive homogeneous multifunction, which enjoys comprehensive calculus rules based on the variational and extremal principle of variational analysis; see [**P**, **R**].

The results we need in what follows in known as the *coderivative/Mordukhovich criterion* (see [**O**, Theorem 5.7] and [**R**, Theorem 9.40] with the references therein): If F is close-graph around $(\bar{x},\bar{y})$, then it is Lipschitz-like around this point if and only if

$$D^*F(\bar{x},\bar{y})(0)=\{0\}. \tag{5.15}$$

Now we are ready to derive the aforementioned consequence of Theorem 5.2.

COROLLARY 5.3 (enhanced nontriviality condition). In addition to the assumptions of Theorem (5.2), suppose that for each $j=1,\ldots,k$, the multifunction F_j is Lipschitz-like around the optimal point $(\bar{x}_j^k,(\bar{x}_j^k-\bar{x}_{j-1}^k)/h_k)$. Then all the necessary optimality conditions of this theorem hold at $\bar{x}^k$ with the enhanced nontriviality

$$\sum_{i=0}^{m+r}|\lambda_i^k|+|p_0^k|=1 \;\text{ for all }\; k\in\mathbb{N}. \tag{5.16}$$

PROOF. If $\lambda_0^k=0$, then it follows from the Euler-Lagrange inclusion of the theorem that

$$\Big(\frac{p_j^k-p_{j-1}^k}{h_k},p_{j-1}^k\Big)\in N\Big(\Big(\bar{x}_j^k,\frac{\bar{x}_j^k-\bar{x}_{j-1}^k}{h_k}\Big);\operatorname{gph}F_j\Big)$$

for all $j=1,\ldots,k$, which tells us by the coderivative definition (5.14) that

$$\frac{p_j^k-p_{j-1}^k}{h_k}\in D^*F_j\Big(\bar{x}_j^k,\frac{\bar{x}_j^k-\bar{x}_{j-1}^k}{h_k}\Big)(-p_{j-1}^k),\quad j=1,\ldots,k.$$

Employing finally the coderivative criterion (5.15) with taking into account the transversality condition of the theorem as well as the normalization of

$$(\lambda_0,\ldots,\lambda_{m+r},p_0^k)$$

without changing other conditions, we arrive at (5.16) and thus completes the proof. □

6. Concluding Remarks

This paper develops a constructive approach to investigate the generalized Bolza problem of optimizing constrained differential inclusions satisfying the relaxed one-sided Lipschitzian condition by using the implicit Euler scheme of discrete approximations. In this way we not only justify the well-posedness of the suggested discrete approximation procedures in the sense of either the uniform or $W^{1,2}$-convergence of discrete optimal solutions to a given local (strong or intermediate) minimizer of the original nonsmooth Bolza problem, but also derive necessary optimality conditions to solve each problem of the implicit Euler discrete approximations. As mentioned in the introductory Section 1, the results obtained are new even in the case of the implicit Euler scheme for unconstrained differential inclusions satisfying the classical Lipschitz condition.

A natural question arises about the possibility to derive *necessary optimality conditions* for the given intermediate or strong local minimizer of the original problem (P) for favorable classes of *ROSL differential inclusions* by passing to the limit from those obtained for (P_k) and $(\widetilde{P}_k)$, respectively, as $k \to \infty$. It can surely be done in the case when, in the setting of Theorem 4.1, the velocity function F is Lipschitzian around the local minimizer under consideration; cf. [**N**, Theorem 6.1] and [**Q**, Theorem 6.22], where the case of the explicit Euler scheme was investigated. Note to this end that the ROSL and Lipschitz-like properties of F used in Corollary 5.3 are generally independent (even for bounded mappings), and they both are implied by the classical Lipschitz condition.

On the other hand, the method of discrete approximations has been successfully employed in [**E**] to derive necessary optimality conditions for the Bolza problem governed by a dissipative (hence ROSL while unbounded and heavily non-Lipschitzian) differential inclusion that arises in optimal control of Moreau's *sweeping process* with mechanical applications. The procedure in [**E**] exploits some specific features of the controlled sweeping process over convex polyhedral sets, and thus a principal issue of the our further research is about the possibility to extend these results to more general ROSL differential inclusions.

References

[A] Zvi Artstein, *First-order approximations for differential inclusions*, Set-Valued Anal. **2** (1994), no. 1-2, 7–17, DOI 10.1007/BF01027089. Set convergence in nonlinear analysis and optimization. MR1285817 (95d:34025)

[B] Hedy Attouch, Giuseppe Buttazzo, and Gérard Michaille, *Variational analysis in Sobolev and BV spaces*, MPS/SIAM Series on Optimization, vol. 6, Society for Industrial and Applied Mathematics (SIAM), Philadelphia, PA; Mathematical Programming Society (MPS), Philadelphia, PA, 2006. Applications to PDEs and optimization. MR2192832 (2006j:49001)

[C] Wolf-Juergen Beyn and Janosch Rieger, *The implicit Euler scheme for one-sided Lipschitz differential inclusions*, Discrete Contin. Dyn. Syst. Ser. B **14** (2010), no. 2, 409–428, DOI 10.3934/dcdsb.2010.14.409. MR2660865 (2012d:34033)

[D] G. Colombo, R. Henrion, N. D. Hoang, and B. S. Mordukhovich, *Discrete approximations of a controlled sweeping process*, Set-Valued Var. Anal. **23** (2015), no. 1, 69–86, DOI 10.1007/s11228-014-0299-y. MR3305685

[E] G. Colombo, R. Henrion, N. D. Hoang, and B. S. Mordukhovich, *Optimal control of the sweeping process*, Dyn. Contin. Discrete Impuls. Syst. Ser. B Appl. Algorithms **19** (2012), no. 1-2, 117–159. MR2918251

[F] T. Donchev, *Functional differential inclusions with monotone right-hand side*, Nonlinear Anal. **16** (1991), 543–552.

[G] Tzanko Donchev, *Properties of one-sided Lipschitz multivalued maps*, Nonlinear Anal. **49** (2002), no. 1, Ser. A: Theory Methods, 13–20, DOI 10.1016/S0362-546X(00)00244-3. MR1887909 (2003a:47107)

[H] Tzanko Donchev and Elza Farkhi, *Stability and Euler approximation of one-sided Lipschitz differential inclusions*, SIAM J. Control Optim. **36** (1998), no. 2, 780–796, DOI 10.1137/S0363012995293694. MR1616554 (99c:34024)

[I] Tzanko Donchev, Elza Farkhi, and Boris S. Mordukhovich, *Discrete approximations, relaxation, and optimization of one-sided Lipschitzian differential inclusions in Hilbert spaces*, J. Differential Equations **243** (2007), no. 2, 301–328, DOI 10.1016/j.jde.2007.05.011. MR2371790 (2008j:49040)

[J] Tzanko Donchev, Elza Farkhi, and Simeon Reich, *Fixed set iterations for relaxed Lipschitz multimaps*, Nonlinear Anal. **53** (2003), no. 7-8, 997–1015, DOI 10.1016/S0362-546X(03)00036-1. MR1978031 (2004h:34023)

[K] Asen Dontchev and Frank Lempio, *Difference methods for differential inclusions: a survey*, SIAM Rev. **34** (1992), no. 2, 263–294, DOI 10.1137/1034050. MR1166177 (93g:34024)

[L] Frank Lempio and Vladimir Veliov, *Discrete approximations of differential inclusions*, Bayreuth. Math. Schr. **54** (1998), 149–232. MR1643941 (99j:49067)

[M] B. Sh. Mordukhovich, *Maximum principle in the problem of time optimal response with nonsmooth constraints* (Russian), Prikl. Mat. Meh. **40** (1976), no. 6, 1014–1023; English transl., J. Appl. Math. Mech. **40** (1976), no. 6, 960–969 (1977). MR0487669 (58 #7284)

[N] Boris S. Mordukhovich, *Discrete approximations and refined Euler-Lagrange conditions for nonconvex differential inclusions*, SIAM J. Control Optim. **33** (1995), no. 3, 882–915, DOI 10.1137/S0363012993245665. MR1327242 (96d:49028)

[O] Boris Mordukhovich, *Complete characterization of openness, metric regularity, and Lipschitzian properties of multifunctions*, Trans. Amer. Math. Soc. **340** (1993), no. 1, 1–35, DOI 10.2307/2154544. MR1156300 (94a:49011)

[P] Boris S. Mordukhovich, *Variational analysis and generalized differentiation. I*, Grundlehren der Mathematischen Wissenschaften [Fundamental Principles of Mathematical Sciences], vol. 330, Springer-Verlag, Berlin, 2006. Basic theory. MR2191744 (2007b:49003a)

[Q] Boris S. Mordukhovich, *Variational analysis and generalized differentiation. II*, Grundlehren der Mathematischen Wissenschaften [Fundamental Principles of Mathematical Sciences], vol. 331, Springer-Verlag, Berlin, 2006. Applications. MR2191745 (2007b:49003b)

[R] R. Tyrrell Rockafellar and Roger J.-B. Wets, *Variational analysis*, Grundlehren der Mathematischen Wissenschaften [Fundamental Principles of Mathematical Sciences], vol. 317, Springer-Verlag, Berlin, 1998. MR1491362 (98m:49001)

[S] Georgi V. Smirnov, *Introduction to the theory of differential inclusions*, Graduate Studies in Mathematics, vol. 41, American Mathematical Society, Providence, RI, 2002. MR1867542 (2002h:34016)

[T] Richard Vinter, *Optimal control*, Systems & Control: Foundations & Applications, Birkhäuser Boston, Inc., Boston, MA, 2000. MR1756410 (2001c:49001)

Department of Mathematics, Wayne State University, Detroit, Michigan 48202
E-mail address: `boris@math.wayne.edu`

Department of Mathematics, Wayne State University, Detroit, Michigan 48202
E-mail address: `yuantian@wayne.edu`

Contemporary Mathematics
Volume **659**, 2016
http://dx.doi.org/10.1090/conm/659/13153

Excessive revenue model of competitive markets

Yurii Nesterov and Vladimir Shikhman

ABSTRACT. We present an excessive revenue model of a competitive market. Its partial equilibrium is characterized as a saddle point in a convex-concave game. Convex variables (for minimization) represent prices of goods, and concave variables (for maximization) stand for production and consumption bundles. The analysis of the market is crucially based on Convex Analysis. Namely, our market model admits a joint convex potential function. In particular, equilibrium prices have a natural explanation: they minimize the total excessive revenue of the market's participants. Due to this characterization, existence and efficiency of partial equilibrium are derived. Additionally, we decentralize prices by introducing the trade design. The total excessive revenue preserves its convexity w.r.t. decentralized prices. This feature allows us to suggest a decentralized price dynamics based on subgradient-type information. We show that unique equilibrium prices can be approached as limiting points of this price dynamics. From the technical point of view, the most unusual feature of our approach is the absence of the budget constraint in its classical form. As a consequence, some consumers may get bankrupt and eventually leave the market. In order to incorporate wealth effects into our model, we make further assumptions on consumers' behavior. Namely, the consumers are assumed to be responsible and long-term behaved.

1. Introduction

In the microeconomic theory, the concept of market equilibrium is may be the most fundamental. It equalizes the supply and demand of goods by appropriate equilibrium prices. Naturally, the question arises on how the prices can be adjusted in order to reach an equilibrium. In particular, there are two related questions to be addressed within a market model:

1) How is it possible for a central authority to find equilibrium prices efficiently?
2) How do market participants actually find equilibrium prices by themselves?

2010 *Mathematics Subject Classification.* Primary 90C25, 91B52; Secondary 90C33, 91B26.

The research of the first author has been supported by a grant "Action de recherche concertè ARC 04/09-315" from the "Direction de la recherche scientifique - Communautè française de Belgique". The scientific responsibility rests with its author.

The research of the second author has been supported by a grant "Chargé de recherches 98444" from the "Fonds de la Recherche Scientifique - FNRS, Belgique". The scientific responsibility rests with its author.

These issues on the *computability* and the *decentralization of prices* motivate our approach to the market modelling. In order to provide computability and decentralization of prices, we present a new partial equilibrium model for a market of goods.

Our market model is crucially based on Convex Analysis. Recall that in optimization theory there are good reasons to believe that the maximal class of tractable models is formed by problems with convex structure (e.g. see discussion in Section 2.1 in [**17**]). Some of the most general problems of this type are the *saddle-point problems*, where we have a set of primal variables (for minimization) and dual variables (for maximization). The potential function of this problem must be convex in primal variables and concave in the dual ones. We use this framework to describe the economic equilibrium with prices as primal variables and production/consumption bundles as the dual ones. Our model admits a joint potential function, which we call the *total excessive revenue* (TER) of the market. This function is simply the sum of excessive revenues of all market participants. Note that it also represents the growth rate of the capital. Total excessive revenue has the prices as primal variables (for minimization), and the production and consumption bundles as dual variables (for maximization). The characterization theorem reads:

Equilibrium prices minimize the total excessive revenue of the market.

Hence, we refer to our market model as to the excessive revenue model.

The characterization of equilibrium prices immediately suggests some tractable price dynamics. In fact, to obtain equilibrium prices we just need to minimize the convex function TER. Its convex subdifferential w.r.t. prices represents the excess-supply operator. This gives rise to study subgradient-type price dynamics by applying the powerful methodology of convex optimization. Additionally, the characterization of equilibrium prices as minimizers of TER reveals the nature of prices. To see this, let us exchange minimization w.r.t. prices and maximization w.r.t. production and consumption bundles in the above saddle-point problem. Then, it turns out that prices can be viewed as Lagrange multipliers for the market clearance condition. Hence, they guarantee that at equilibrium supply meets demand.

We point out that for the solvability of the convex-concave saddle-point problem it is necessary to have *separable* convex constraints for primal and dual variables. This is exactly where we need to make crucial assumptions on the consumers' behavior in our model. Recall that due to the classical demand theory a consumer is maximizing his utility function subject to the budget constraint. The budget constraint states that the value of a feasible consumption bundle at current prices cannot exceed the available wealth. Hence, the budget constraint may destroy any hope for existence of an efficient strategy, which could approach the equilibrium prices and consumption/production bundles. To avoid the untractable budget constraint, we assume the so-called *responsible consumer*. The responsible consumer minimizes his expenditure while guaranteeing the fixed level of desirable utility. Here, the utility level is not adjusted as it is the case in utility maximization. Thus, it does not implicitly depend on the wealth, but corresponds to the consumer's known standards on goods' qualities. In some relevant situations, these standards are indeed explicitly available to consumers. We may just think of an experienced consumer who is not willing to spend more money than necessary to satisfy his standards on daily goods' qualities. See also Section 7 on the algorithmic justification of the expenditure minimization problem with Leontief-type utility function. There, we introduce and explain a natural consumption strategy for buying daily goods. This strategy is based on the best quality/price ratio and updates individual prices for qualities. We show that it successively leads to a consumption bundle which minimizes the expenditure and guarantees the quality standards.

Now, we need to incorporate *wealth effects* into the model of a responsible consumer. Here, another novelty comes into play, namely the assumption on the *long-term behavior* of a consumer. Namely, the consumption bundles and the wealths are considered as *flows*. This means that we focus on the corresponding objects in each standard interval of time (say, one week). During a price adjustment, a responsible consumer compares his available wealth with the minimal expenditure in each interval of time. Depending on this comparison, he can implement his tentative consumption or not. We express these successes and failures of consumption attempts by one and zero, respectively. On average, this individual history provides a consumption *frequency*. Note that the latter fraction measures consumer's average involvement into the market activities. As the price adjustment evolves, the consumption frequencies approximate finally the consumer's *proper participation level*. We assume that the demand consists of the tentative consumption scaled by the proper participation level. Thus, at given prices a consumer faces three alternatives:

(i) profitability with participation level one,
(ii) bankruptcy with participation level zero,
(iii) marginality with participation level between zero and one.

In particular, profitable consumers fully implement their tentative consumption bundles. Bankrupt consumers suffer a sequence of failed consumption attempts, hence, they must leave the market sooner or later. Marginal consumers succeed from time to time in implementing their tentative consumption bundles, but occasionally they fail. On average, marginal consumers happen to implement just a fraction of their potential activity. This fraction is modelled by the participation level between zero and one. We point out that particular participation levels of a marginal consumer depend on the whole history of price adjustment, actions of other market agents, simply the good or bad luck, say to be the first or the last in line, etc. We note that the marginality of some consumers can be viewed as a source of *poverty* inside the market.

In our model, the same argumentation applies to producers whose production bundles and expenses are also considered as *flows*. Producers compare their maximal profit with fixed costs (see Section 2 for details). Analogously, profitable, bankrupt and marginal producers can be distinguished.

We claim that marginal agents play a crucial role in our approach to market modelling. An important consequence of the marginal behavior is a possibility to clear the market. It is due to the fact that scaled tentative production/consumption bundles form feasible supplies/demands of marginal producers/consumers. From the technical point of view, introduction of poverty allows lifting the budget constraint into a joint potential function TER, ensuring the convexity of the whole model. Overall, the introduction of proper participation levels means that producers and consumers actually maximize their excessive revenues.

Now, we turn our attention to the decentralization of prices. Our goal is to explain how market participants can by themselves tackle the minimization of the nonsmooth convex function TER by successively updating prices. The decentralization of prices can be implemented by the introduction of various price designs. In this article, we focus just on the *trade design*:

producers settle and update their individual prices,
and consumers buy at the lowest purchase price.

By plugging this price design into the producers' and consumers' excessive revenues, we obtain a decentralized total excessive revenue (TER). Now, the decentralized function TER depends on producers' individual prices. It is crucial that the introduction of the trade design preserves convexity of TER. Moreover, its convex subdifferential w.r.t. a producer's price becomes his *individual* excess supply. The latter is the difference between producer's individual supply and all consumers' demands from him. Obviously, this information is available to a producer. Hence, he is able to update his individual prices according to some subgradient-type dynamics. We show that the corresponding sequence of lowest purchase prices approximates the equilibrium price.

Let us mention some literature related to our modeling approach. The study of market models, where prices play the role of Lagrange multipliers, has been initiated in [**6**, **20**]. Based on that, convex market models are considered from the computational point of view in [**1**–**3**]. There, the focus is mainly on the polynomial-time algorithms for equilibrium pricing. Recently, convex analysis was used to model financial equilibria, see [**9**]. Decentralization of prices via the trade design is suggested already in [**5**], where a disequilibrium price dynamics is considered. The corresponding disequilibrium Hahn-process, see [**8**], resembles our assumption on the long-term behavior. We also refer to [**14**] where a simultaneous ascending auction is used to construct a decentralized price dynamics. Starting with [**23**], the theory of bargaining has been successively applied for market decentralization (see e.g. [**7**, **15**]). Concerning our modeling of the demand operator, note that there is evidence from behavioral economics that consumer's choices need not be consistent with the maximization of a preference relation (see [**16**] and references therein). The reason for that is usually referred to as consumers' bounded rationality. Classic examples include status-quo biases, attraction, compromise and framing effects, temptation and self-control, consideration sets, and choice overload. In our approach, the demand operator is consistent with the long-term behavior of responsible consumers.

The article is organized as follows. In Section 2 we present the excessive revenue model and discuss its main assumptions. Section 3 is devoted to the analysis of the excessive revenue model, including characterization, existence, and efficiency of market equilibrium. In Section 4 we describe in detail the decentralization of prices. Based on this, we present and prove convergence of decentralized subgradient-type price dynamics in Section 5. In Section 6 we extend our model to accommodate the labor market and the market of resources. Section 7 introduces natural consumption strategies in order to algorithmically tackle the expenditure minimization problem.

Notation. Our notation is quite standard. We denote by $\mathbb{R}^n$ the space of n-dimensional column vectors $x = (x^{(1)}, \dots, x^{(n)})^T$, and by $\mathbb{R}^n_+$ the set of all vectors with nonnegative components. For x and y from $\mathbb{R}^n$, we introduce the standard scalar product and the Hadamard product

$$\langle x, y \rangle = \sum_{i=1}^{n} x^{(i)} y^{(i)}, \quad x \circ y = \left(x^{(i)} y^{(i)} \right)_{i=1}^{n} \in \mathbb{R}^n.$$

We denote by $e_j \in \mathbb{R}^k$ the j-th coordinate vector of the corresponding vector space. The same notation is used for different spaces, which are always determined by the context. Finally, $\mathbb{R}^{k \times m}$ denotes the space of $k \times m$-matrices, and $(a)_+$ denotes the positive part of the real value $a \in \mathbb{R}$: $(a)_+ = \max\{a, 0\}$.

2. Excessive revenue model

We present the *excessive revenue model* of a competitive market with multiple goods. For that, we need to describe the behavior of producers and consumers. Given prices of goods, they maximize their profits and minimize their expenditures by deciding on tentative production and consumption patterns. The latter must be compatible with their needs and technological constraints. Additionally, producers compare their profits with fixed costs, and consumers compare their expenditures with available wealths. This comparison is needed to adjust real production and consumption bundles accordingly. Then, we define the equilibrium production and consumption flows, which clear the market of goods by some equilibrium prices.

2.1. Producers and consumers. Consider a market with K producers, which are able to produce n different goods. Given a vector of prices $p \in \mathbb{R}^n_+$, the k-th producer forms his supply operator $S_k(p)$ of real production bundles $\widetilde{y}_k \in \mathbb{R}^n_+$. For that, the k-th producer maximizes the profit with respect to his variable cost, subsequently he tries to cover his fixed cost. Namely,

- The k-th producer chooses first the *tentative production* bundle $y_k \in \mathbb{R}^n_+$ by solving the *profit maximization* problem:

$$\pi_k(p) \stackrel{\text{def}}{=} \max_{y_k \in \mathcal{Y}_k} \left[\langle p, y_k \rangle - c_k(y_k) \right]. \tag{2.1}$$

  Here, $\mathcal{Y}_k \subset \mathbb{R}^n_+$ is the production set, assumed to be nonempty, compact and convex. The producer's yield is $\langle p, y_k \rangle$. The variable cost of producing y_k is denoted by $c_k(y_k)$. We assume that c_k is a convex function on $\mathbb{R}^n_+$. Clearly, the profit $\pi_k(p)$ is convex in p as the maximum of linear functions. By $\mathcal{Y}^*_k(p)$ we denote the set of optimal solutions of (2.1), i.e. $y_k \in \mathcal{Y}^*_k(p)$. Note that the profit maximization problem (2.1) appears already in Marshallian partial equilibrium analysis (see e.g. [**13**]). Producers usually effectively solve (2.1) within their own planning departments or by getting support from consulting firms. In what follows, we assume that a solution of the profit maximization problem (2.1) can be obtained by a producer.

- Secondly, the k-th producer compares his profit $\pi_k(p)$ with his fixed cost of maintaining the technological set $\mathcal{Y}_k$, denoted by $\kappa_k \equiv \kappa_k(\mathcal{Y}_k) \in \mathbb{R}_+$. The latter can include the interest paid to the bank, different charges for renting the equipment, land use, etc. By this comparison, a participation level $\alpha_k \equiv \alpha_k(p) \in [0, 1]$ of k-th producer is *properly* adjusted:

$$\alpha_k(p) \stackrel{\text{def}}{=} \begin{cases} 1, & \text{if } \pi_k(p) > \kappa_k, \\ 0, & \text{if } \pi_k(p) < \kappa_k. \end{cases} \tag{2.2}$$

- Finally, the *supply* operator $S_k : \mathbb{R}^n_+ \rightrightarrows \mathbb{R}^n_+$ of the k-th producer is given by

$$S_k(p) \stackrel{\text{def}}{=} \{\widetilde{y}_k = \alpha_k y_k \,|\, \alpha_k \equiv \alpha_k(p) \text{ and } y_k \in \mathcal{Y}^*_k(p)\}\,. \tag{2.3}$$

Here, the *real production* bundles are

$$\widetilde{y}_k \stackrel{\text{def}}{=} \alpha_k y_k,$$

where $\alpha_k \equiv \alpha_k(p)$ is a proper participation level of the k-th producer, and $y_k \in \mathcal{Y}^*_k(p)$ is his tentative production.

Let I consumers be active at the market. The i-th consumer has to decide on his real consumption bundle $\widetilde{x}_i \in \mathbb{R}^n_+$. These real consumption bundles form his demand operator $D_i(p)$, given the price $p \in \mathbb{R}^n_+$. The i-th consumer minimizes the expenditure with an aim to guarantee his desirable utility level. Then he tries to cover this expenditure by the available wealth. Namely,

- The i-th producer decides first on the *tentative consumption* bundle $x_i \in \mathbb{R}^n_+$ by minimizing his expenditure:

$$e_i(p) \stackrel{\text{def}}{=} \min_{\substack{x_i \in X_i \\ u_i(x_i) \geq u_i}} \langle p, x_i \rangle = \min_{x_i \in \mathcal{X}_i} \langle p, x_i \rangle\,, \tag{2.4}$$

where the i-th consumption set is

$$\mathcal{X}_i \stackrel{\text{def}}{=} \{x_i \in X_i \,|\, u_i(x_i) \geq u_i\}\,.$$

Here, $X_i \subset \mathbb{R}^n_+$ is assumed to be nonempty, compact and convex. By $u_i : X_i \to \mathbb{R}_+$ we denote the utility function of the i-th consumer, assumed to be concave. The utility level $u_i \in \mathbb{R}_+$ is desirable by i-th consumer. The consumer's expenditure $e_i(p)$ is concave in p as the minimum of linear functions. By $\mathcal{X}^*_i(p)$ we denote the set of optimal solutions of (2.4), i.e. $x_i \in \mathcal{X}^*_i(p)$.

- Secondly, the i-th consumer compares his expenditure $e_i(p)$ with his available wealth $w_i \in \mathbb{R}_+$. The latter can include the budget, salary and rent payments, etc. By this comparison, a participation level $\beta_i \equiv \beta_i(p) \in [0,1]$ of i-th consumer is *properly* adjusted:

$$\beta_i(p) \stackrel{\text{def}}{=} \begin{cases} 1, & \text{if } e_i(p) < w_i, \\ 0, & \text{if } e_i(p) > w_i. \end{cases} \tag{2.5}$$

- Finally, the *demand* operator $D_i : \mathbb{R}^n_+ \rightrightarrows \mathbb{R}^n_+$ of the i-th consumer is given by

$$D_i(p) \stackrel{\text{def}}{=} \{\widetilde{x}_i = \beta_i x_i \,|\, \beta_i \equiv \beta_i(p) \text{ and } x_i \in \mathcal{X}^*_i(p)\}\,. \tag{2.6}$$

Here, the *real consumption* bundle are

$$\widetilde{x}_i \stackrel{\text{def}}{=} \beta_i x_i,$$

where $\beta_i \equiv \beta_i(p)$ is a proper participation level of the i-th consumer, and $x_i \in \mathcal{X}^*_i(p)$ is his tentative consumption.

There are two non-standard ingredients in our model which need to be explained and thoroughly justified. The first concerns the expenditure minimization problem (2.4) with the given level u_i of desirable utility. The second deals with the proper adjustment of participation levels α_k and β_i in (2.2) and (2.5), respectively.

1) Expenditure minimization and responsible consumer

The minimization of expenditure in (2.4) is well-known in economics as a dual problem for utility maximization (e.g. [**13**]):

$$v_i(p, w_i) \stackrel{\text{def}}{=} \max_{\substack{x_i \in X_i \\ \langle p, x_i \rangle \le w_i}} u_i(x). \tag{2.7}$$

Namely, under some technical assumptions if x_i solves (2.7) then also (2.4) with $u_i = v_i(p, w_i)$; if x_i solves (2.4) then also (2.7) with $w_i = \langle p, x_i \rangle$. In our setting the desirable utility level u_i is given, thus, is a primitive of the model. It mainly reflects the consumer's standards on qualities of goods. Hence, it does not implicitly depend on the wealth w_i as in the classical setting. Note that we model wealth effects by subsequent comparison of w_i with expenditure $e_i(p)$ rather than by the usual budget constraint $\langle p, x_i \rangle \le w_i$ as in (2.7) (cf. also the discussion in 2) below).

Now, let us explain why in some interesting situations the desirable utility level u_i is explicitly available to consumers. For many daily goods there are physical standards to be satisfied. They constitute the so-called minimum standards of life. Additionally, some consumers often accept standards imposed by the society, e.g. through advertisement, their friends or family members. E.g., it became evident that some consumers use to shrink their usual consumption motivated by ecological reasons. Also experienced consumers, who go shopping in a supermarket say on a weekly basis, know the standards of their living. Overall, we may think of a *responsible consumer* who does care about his expenditure. Namely, he is not willing to spend more than necessary to satisfy his given standards. Thus, such consumer tries to minimize his expenditure while guaranteeing the standards. In [**21**] the agent who faces the expenditure minimization problem (2.4) is called the dual consumer. We also refer to [**10**, Chapter 10] and [**11**] for more details on the dual theory of consumption. In Section 7 we show how the expenditure minimization problem (2.4) with Leontief-type utility function can be algorithmically tackled by simple and natural consumption strategies.

Further, the compactness assumption on X_i refers to the fact that the consumption is bounded. Naturally, there are physical limits to what people can consume in order to satisfy their basic needs. The unbounded desire for wealth is not an issue here, since the wealth w_i is a primitive in our model (cf. the discussion on this assumption in [**22**]).

2) Adjustment of participation levels and long-term behavior

In our model, the production and consumption bundles, the consumer's wealths, and producers' costs are considered as *constant flows*. This means that we get the same amount of corresponding objects in each standard interval of time (say, one week). Therefore, if the income of a person or a firm during this interval is greater than the expenses, then he/she can ensure a constant *rate of growth* of the own capital. In this *profitable* case, we have: $\alpha_k(p) = \beta_i(p) = 1$, i.e. producers and consumers implement their tentative bundles. If the income is strictly less than expenses, then the producer/consumer must leave the market sooner or later, i.e. $\alpha_k(p) = \beta_i(p) = 0$. This is true both for producers (bankruptcy), and for consumers (emigration from this market). We refer to those agents as being *bankrupt*. If the regular income is equal to the regular expenses, then tentative production and consumption bundles may shrink due to participation levels $\alpha_k(p), \beta_i(p) \in [0, 1]$. In this *marginal* case, participation levels $\alpha_k(p), \beta_i(p)$ may be any numbers between

0 and 1. We give a behavioral explanation on how $\alpha_k(p), \beta_i(p)$ are adjusted. Note that marginal agents reach their break-even point, hence, they make neither a profit nor a loss. For a marginal producer it means that his profit is equals to the fixed costs: $\pi_k(p) = \kappa_k$. Net saving of a marginal consumer is zero, i.e. his wealth is equal to the minimal possible expenditure: $w_i = e_i(p)$. Hence, for $\hat{p} \approx p$ the break-even point will be mainly shifted either to profitability or bankruptcy. Marginal producers face

$$\text{either} \quad \pi_k(\hat{p}) > \kappa_k \quad \text{or} \quad \pi_k(\hat{p}) < \kappa_k,$$

and marginal consumers face

$$\text{either} \quad w_i > e_i(\hat{p}) \quad \text{or} \quad w_i > e_i(\hat{p}).$$

Hence, sometimes marginal producers/consumers can implement their tentative bundles, i.e.

$$\alpha_k(\hat{p}) = 1, \quad \beta_i(\hat{p}) = 1,$$

and sometimes it is not possible, i.e.

$$\alpha_k(\hat{p}) = 0, \quad \beta_i(\hat{p}) = 0.$$

The particular 0-1 values of $\alpha_k(\hat{p})$ and $\beta_i(\hat{p})$ depend on the individual history of successes and failures for producers and consumers, respectively. To be more concrete, let us consider some price adjustment process $\hat{p}(t) \to p$ with discrete time $t \to \infty$. Now, the participation levels $\alpha_k(p), \beta_i(p)$ can be viewed as *frequencies* of agents' successful and failed attempts. Indeed, by averaging and taking the limit we obtain the participation levels:

$$\frac{1}{t}\sum_{s=1}^{t} \alpha_k(\hat{p}(s)) \to \alpha_k(p), \quad \frac{1}{t}\sum_{s=1}^{t} \beta_i(\hat{p}(s)) \to \beta_i(p) \quad \text{for } t \to \infty.$$

This interpretation of participation levels as frequencies is based on the *long-term behavior* of the agents. By the way, the same argumentation applies for profitable and bankrupt agents. In fact, profitable agents (with $\pi_k(p) > \kappa_k$ and $w_i > e_i(p)$) will enjoy a sequence of successful production/consumption attempts. It means that $\alpha_k(\hat{p}(s)) = 1$, $\beta_i(\hat{p}(s)) = 1$ for almost all s. Thus, their averages approximate the participation levels $\alpha_k(p) = 1$ and $\beta_i(p) = 1$. The latter is in accordance with the definition of a profitable agent. Analogously, bankrupt agents (with $\pi_k(p) < \kappa_k$ and $w_i < e_i(p)$) will suffer a sequence of failed production/consumption attempts. Hence, $\alpha_k(\hat{p}(s)) = 0$, $\beta_i(\hat{p}(s)) = 0$ for almost all s. The corresponding averages approximate the participation levels $\alpha_k(p) = 0$ and $\beta_i(p) = 0$. Again, this is in accordance with the definition of a bankrupt agent. In addition, we discuss the case of seemingly neglectable excess of expenses over agents' income, i.e. $\pi_k(p) = \kappa_k + \varepsilon$ and $w_i = e_i(p) + \varepsilon$ with small $\varepsilon > 0$. Due to our assumption on long-term agents' behavior, producers/consumers face the constant *rate of decrease* ε of the own capital. Hence, they shall decide to leave the market, i.e. $\alpha_k(p) = \beta_i(p) = 0$, in order not to suffer small, but lasting constant losses. Otherwise, they would not be able to guarantee their full standards ever.

Let us address the question why marginal agents do not quit the market although they eventually implement only a share of their tentative production or consumption bundles. As a consequence marginal producers do not fully exploit their available capacities and cannot cover the fixed costs. Marginal consumers do not spend all their available wealths and cannot reach the desirable levels of utility.

Nevertheless, these agents stay at the market since they actually *do not know* that they are marginal. During the price adjustment, the only available information is their individual history of successes and failures while attempting to produce and to consume. With above notation, at time t they know a 0-1 sequence of $\alpha_k(\hat{p}(s))$ and $\beta_i(\hat{p}(s))$, $s = 1, \dots, t$. This particular history depends on many factors, as their luck, current prices, particular actions of other agents, etc. From time to time, marginal agents succeed to implement their tentative production/consumption bundles, but occasionally they fail. This unsure market environment causes marginal agents to temporally reduce consumption and to wait for "fair prices". Such a behavior is typical for poor people, and we can treat their fractional participation levels $\alpha_k(p)$ and $\beta_i(p)$ as a measure of *poverty*. A hidden consequence of this marginal behavior is a possibility to clear the market as we shall see in Section 2.2. We conclude that the marginal agents play a crucial role in our approach to market modelling.

2.2. Equilibrium market flows. In accordance to the previous notations, we eventually say that the *real market flow*

$$\widetilde{F} = \left((\widetilde{y}_k = \alpha_k y_k)_{k=1}^K, (\widetilde{x}_i = \beta_i x_i)_{i=1}^I \right)$$

is defined by the triple (p, F, γ). Here, $p \in \mathbb{R}^n_+$ is the vector of prices,

$$F = \left((y_k)_{k=1}^K, (x_i)_{i=1}^I \right) \in \prod_{k=1}^K \mathcal{Y}_k \times \prod_{i=1}^I \mathcal{X}_i$$

is the *tentative market flow*, and

$$\gamma = \left((\alpha_k)_{k=1}^K, (\beta_i)_{i=1}^I \right) \in [0,1]^{K+I}$$

is the *proper system of participation levels* (w.r.t. p and F), i.e.

$$\alpha_k = \begin{cases} 1, & \text{if } \langle p, y_k \rangle - c_k(y_k) > \kappa_k, \\ 0, & \text{if } \langle p, y_k \rangle - c_k(y_k) < \kappa_k \end{cases}, \quad \beta_i = \begin{cases} 1, & \text{if } \langle p, x_i \rangle < w_i, \\ 0, & \text{if } \langle p, x_i \rangle > w_i. \end{cases}$$

Now we define the market equilibrium in the standard way.

DEFINITION 2.1 (Equilibrium market flow). We say that $p^* \in \mathbb{R}^n$ is the *equilibrium price* if there exists a real market flow

$$\widetilde{F}^* = \left((\widetilde{y}_k^* = \alpha_k^* y_k^*)_{k=1}^K, (\widetilde{x}_i^* = \beta_i^* x_i^*)_{i=1}^I \right) \in \prod_{k=1}^K S_k(p^*) \times \prod_{i=1}^I D_i(p^*),$$

satisfying the *market clearing condition*

$$(2.8) \qquad p^* \geq 0, \quad \sum_{k=1}^K \widetilde{y}_k^* - \sum_{i=1}^I \widetilde{x}_i^* \geq 0, \quad \left\langle p^*, \sum_{k=1}^K \widetilde{y}_k^* - \sum_{i=1}^I \widetilde{x}_i^* \right\rangle = 0.$$

In this case, $\widetilde{F}^*$ is called the equilibrium market flow.

The market clearing condition (2.8) states that the real consumption never exceed the real production, and the markets of goods with positive prices ($p^{(j)} > 0$) are perfectly cleared:

$$\sum_{k=1}^K \widetilde{y}_k^{*(j)} = \sum_{i=1}^I \widetilde{x}_i^{*(j)}.$$

3. Analysis of the excessive revenue model

3.1. Characterization. Given a vector of prices $p \in \mathbb{R}^n_+$, producers maximize their profits and consumers minimize their expenditures. Afterwards, both properly adjust their participation levels by comparing the profits with the fixed costs, in case of producers, or by comparing the expenditures with the wealths, in case of consumers. Exactly the same behavior can be obtained by maximizing their excessive revenues.

The *excessive revenue* of the k-th producer is set as

$$ERP_k(p) \stackrel{\text{def}}{=} (\pi_k(p) - \kappa_k)_+ = \max_{y_k \in \mathcal{Y}_k} \left(\langle p, y_k\rangle - c_k(y_k) - \kappa_k\right)_+ . \tag{3.1}$$

Using the substitution $\widetilde{y}_k = \alpha_k y_k$, we obtain

$$ERP_k(p) = (\pi_k(p) - \kappa_k)_+ = \max_{\alpha_k \in [0,1]} \alpha_k \left(\pi_k(p) - \kappa_k\right) =$$

$$\max_{\substack{\alpha_k \in [0,1] \\ y_k \in \mathcal{Y}_k}} \alpha_k \left(\langle p, y_k\rangle - c_k(y_k) - \kappa_k\right) = \max_{\substack{\alpha_k \in [0,1] \\ \widetilde{y}_k \in \alpha_k \mathcal{Y}_k}} \langle p, \widetilde{y}_k\rangle - \alpha_k c_k\left(\widetilde{y}_k/\alpha_k\right) - \alpha_k \kappa_k.$$

Note that the maximization problem

$$ERP_k(p) = \max_{\substack{\alpha_k \in [0,1] \\ \widetilde{y}_k \in \alpha_k \mathcal{Y}_k}} \langle p, \widetilde{y}_k\rangle - \alpha_k c_k\left(\widetilde{y}_k/\alpha_k\right) - \alpha_k \kappa_k$$

is convex, and its set of optimal solutions consists of proper participation levels α_k and real production bundles $\widetilde{y}_k$. Moreover, $ERP_k(p)$ is convex in p as the maximum of linear functions. The convex subdifferential of the excessive revenue ERP_k gives the supply S_k of the k-th producer, i.e.

$$\partial ERP_k(p) = S_k(p). \tag{3.2}$$

The latter follows e.g. from [**24**, Theorem 2.4.18] on the convex subdifferential of a max-type function.

Analogously, we define the *excessive revenue* of the i-th consumer as follows:

$$ERC_i(p) \stackrel{\text{def}}{=} (w_i - e_i(p))_+ = \max_{x_i \in \mathcal{X}_i} \left(w_i - \langle p, x_i\rangle\right)_+ , \tag{3.3}$$

where

$$\mathcal{X}_i \stackrel{\text{def}}{=} \left\{x_i \in X_i \,|\, u_i(x_i) \geq u_i\right\}.$$

Using the substitution $\widetilde{x}_i = \beta_i x_i$, we obtain

$$ERC_i(p) = (w_i - e_i(p))_+ = \max_{\beta_i \in [0,1]} \beta_i \left(w_i - e_i(p)\right) =$$

$$\max_{\substack{\beta_i \in [0,1] \\ x_i \in \mathcal{X}_i}} \beta_i \left(w_i - \langle p, x_i\rangle\right) = \max_{\substack{\beta_i \in [0,1] \\ \widetilde{x}_i \in \beta_i \mathcal{X}_i}} \beta_i w_i - \langle p, \widetilde{x}_i\rangle .$$

Note that $\widetilde{x}_i \in \beta_i \mathcal{X}_i$ means

$$\widetilde{x}_i \in \beta_i X_i \text{ and } \beta_i u_i\left(\widetilde{x}_i/\beta_i\right) \geq \beta_i u_i.$$

In particular, $\beta_i u_i\left(\widetilde{x}_i/\beta_i\right)$ is jointly concave in $(\widetilde{x}_i, \beta_i)$. The maximization problem

$$ERC_i(p) = \max_{\substack{\beta_i \in [0,1] \\ \widetilde{x}_i \in \beta_i \mathcal{X}_i}} \beta_i w_i - \langle p, \widetilde{x}_i\rangle$$

is convex, and its set of optimal solutions consists of proper participation levels β_i and real consumption bundles $\widetilde{x}_i$. Moreover, $ERC_i(p)$ is convex in p as the maximum of linear functions. The convex subdifferential of the excessive revenue ERC_i gives the opposite demand D_i of the i-th consumer, i.e.

$$\partial ERC_i(p) = -D_i(p). \tag{3.4}$$

The latter follows also from [**24**, Theorem 2.4.18].

REMARK 3.1 (Homogeneous case). For the homogeneous case we can give yet another explanation why marginal producers and consumers still stay at the market. Let us assume the homogeneity of the k-th producer's cost function $c_k(\cdot)$, and the homogeneity of the fixed cost $\kappa_k(\cdot)$, i.e.

$$\kappa_k(\alpha\mathcal{Y}_k) = \alpha\kappa_k(\mathcal{Y}_k), \quad \alpha \in [0,1].$$

Then,

$$ERP_k(p) = \max_{\alpha_k \in [0,1],\, \widetilde{y}_k \in \alpha_k\mathcal{Y}_k} \langle p, \widetilde{y}_k\rangle - c_k(\widetilde{y}_k) - \kappa_k(\alpha_k\mathcal{Y}_k).$$

For a marginal producer with $ERP_k(p) = 0$, this means that his activity, even within the maximal technological set $\mathcal{Y}_k$ does not generate any profit. The situation is not changing if the production activities (the set $\mathcal{Y}_k$) will be proportionally reduced by a factor $\alpha_k \in [0,1]$. Thus, it is natural to admit that in this marginal situation the producer can work with a reduced technological set $\alpha_k\mathcal{Y}_k$ by producing $\widetilde{y}_k \in \alpha_k\mathcal{Y}_k$. By doing so, he cannot cover the share $(1-\alpha_k)\kappa_k$ of the fixed cost. However, his unused capacities amounting to $(1-\alpha_k)\mathcal{Y}_k$ can be eventually exploited at other markets.

Now, we assume the homogeneity of the i-th consumer's utility function $u_i(\cdot)$, and that $X_i = \mathbb{R}^n_+$. Then,

$$ERC_i(p) = \max_{\substack{\beta_i \in [0,1],\, \widetilde{x}_i \geq 0 \\ u_i(\widetilde{x}_i) \geq \beta_i u_i}} \beta_i w_i - \langle p, \widetilde{x}_i\rangle .$$

If the revenue of a consumer is zero, then again, there is no special reason to allocate all the wealth w_i to this expensive market. The consumer can admit to spend here only a part of it, namely $\beta_i w_i$ with some $\beta_i \in [0,1]$, which is sufficient to guarantee the share $\beta_i u_i$ of his desirable utility level. Note that this does not change the zero level of the excessive revenue. The remaining part $(1-\beta_i)w_i$ of the wealth can be used then at other markets.

Overall, we define the *total excessive revenue* as the sum of excessive revenues of all agents:

$$TER(p) \overset{\text{def}}{=} \sum_{k=1}^{K} ERP_k(p) + \sum_{i=1}^{I} ERC_i(p). \tag{3.5}$$

Note that function $TER(\cdot)$ is convex since it is a sum of convex functions. Moreover, its convex subdifferential represents the excess supply operator due to (3.2) and (3.4).

THEOREM 3.2 (Excess supply and TER). *For $p \in \mathbb{R}^n_+$ it holds:*

$$\partial TER(p) = \sum_{k=1}^{K} S_k(p) - \sum_{i=1}^{I} D_i(p).$$

PROOF. We apply [**19**, Theorem 23.8] on the subdifferential of the sum of convex functions in order to obtain

$$\partial TER(p) = \sum_{k=1}^{K} \partial ERP_k(p) - \sum_{i=1}^{I} \partial ERC_i(p).$$

Together with (3.2) and (3.4) the assertion follows. □

Theorem 3.2 allows us to characterize equilibrium prices as minimizers of TER.

THEOREM 3.3 (Characterization of equilibrium prices). *$p \in \mathbb{R}^n_+$ is a system of equilibrium prices if and only if it solves the following convex minimization problem:*

$$(\mathbf{P}) \qquad \min_{p \in \mathbb{R}^n_+} TER(p).$$

PROOF. **1.** Assume that $p^* \in \mathbb{R}^n$ is an equilibrium prices. Then, in view of Definition 2.1, there exists an equilibrium market flow

$$\widetilde{F}^* = \left((\widetilde{y}_k^*)_{k=1}^K, (\widetilde{x}_i^*)_{i=1}^I \right) \in \prod_{k=1}^{K} S_k(p^*) \times \prod_{i=1}^{I} D_i(p^*),$$

satisfying the market clearing condition

$$p^* \geq 0, \quad \sum_{k=1}^{K} \widetilde{y}_k^* - \sum_{i=1}^{I} \widetilde{x}_i^* \geq 0, \quad \left\langle p^*, \sum_{k=1}^{K} \widetilde{y}_k^* - \sum_{i=1}^{I} \widetilde{x}_i^* \right\rangle = 0.$$

Denote $\zeta^* = \sum_{k=1}^{K} \widetilde{y}_k - \sum_{i=1}^{I} \widetilde{x}_i$. In view of Theorem 3.2, $\zeta^* \in \partial TER(p^*)$. Since TER is convex in p, for all $p \in \mathbb{R}^n_+$ we have:

$$TER(p) - TER(p^*) \geq \langle \zeta^*, p - p^* \rangle = \langle \zeta^*, p \rangle \geq 0.$$

Thus, p^* minimizes the total excessive revenue.

2. Assume that $p^* \in \mathbb{R}^n_+$ is optimal for the minimization problem (**P**). Then there exists $\zeta^* \in \partial TER(p^*)$ such that

$$\langle \zeta^*, p - p^* \rangle \geq 0, \quad \text{for all } p \in \mathbb{R}^n_+.$$

Considering $p = 0$ and $p = 2p^*$, we conclude that $\langle \zeta^*, p^* \rangle = 0$. Consequently, $\zeta^* \in \mathbb{R}^n_+$. Again due to Theorem 3.2, there exists a real market flow

$$\widetilde{F}^* = \left((\widetilde{y}_k^*)_{k=1}^K, (\widetilde{x}_i^*)_{i=1}^I \right) \in \prod_{k=1}^{K} S_k(p^*) \times \prod_{i=1}^{I} D_i(p^*),$$

such that

$$\zeta^* = \sum_{k=1}^{K} \widetilde{y}_k^* - \sum_{i=1}^{I} \widetilde{x}_i^*.$$

Hence, $\widetilde{F}^*$ satisfies the market clearing condition, meaning that it is actually an equilibrium market flow. In view of Definition 2.1, p^* is an equilibrium price. □

3.2. Existence. Theorem 3.3 says that equilibrium prices correspond to optimal solutions for the minimization problem:

$$(\mathbf{P}) \qquad \min_{p \in \mathbb{R}^n_+} TER(p).$$

This is the key to provide existence results for equilibrium prices. We denote by P^* the set of equilibrium prices. Let us introduce productive markets, at which the set of equilibrium prices P^* turns out to be nonempty and bounded.

DEFINITION 3.4 (Productive market). A market is called *productive* if there exist subsets of producers $\mathcal{K} \subset \{1, \dots, K\}$ and consumers $\mathcal{L} \subset \{1, \dots, L\}$, such that the corresponding production and consumption flows

$$\left(\{\bar{y}_k\}_{k \in \mathcal{K}}, \{\bar{x}_i\}_{i \in \mathcal{L}}\right) \in \prod_{k \in \mathcal{K}} \mathcal{Y}_k \times \prod_{i \in \mathcal{L}} \mathcal{X}_i$$

establish positive balances for goods:

$$\sum_{k \in \mathcal{K}} \bar{y}_k \;>\; \sum_{i \in \mathcal{L}} \bar{x}_i. \tag{3.6}$$

THEOREM 3.5 (Existence and boundedness of equilibrium prices). *At the productive markets, the set of equilibrium prices P^* is nonempty and bounded.*

PROOF. Due to Theorem 3.3, equilibrium prices in P^* form the set of optimal solutions of the minimization problem ($\mathbf{P}$). We show that the latter set is bounded. For that, it is sufficient to prove that the level sets of function $TER(\cdot)$ are bounded. Denote $\bar{\xi} = \sum\limits_{k \in \mathcal{K}} \bar{y}_k - \sum\limits_{i \in \mathcal{L}} \bar{x}_i$. For all $p \in \mathbb{R}^n_+$ we have

$$\begin{aligned}
TER(p) &= \sum_{k=1}^{K} [\pi(p) - \kappa_k]_+ + \sum_{i=1}^{L} [w_i - e_i(p)]_+ \\
&\geq \sum_{k \in \mathcal{K}} [\pi(p) - \kappa_k]_+ + \sum_{i \in \mathcal{L}} [w_i - e_i(p)]_+ \\
&\geq \sum_{k \in \mathcal{K}} \pi(p) - \kappa_k + \sum_{i \in \mathcal{L}} w_i - e_i(p) \\
&\geq \sum_{k \in \mathcal{K}} (\langle p, \bar{y}_k \rangle - c_k(\bar{y}_k) - \kappa_k) + \sum_{i \in \mathcal{L}} (w_i - \langle p, \bar{x}_i \rangle) \\
&= -\sum_{k \in \mathcal{K}} (\kappa_k + c_k(\bar{y}_k)) + \sum_{i \in \mathcal{L}} w_i + \langle \bar{\xi}, p \rangle .
\end{aligned}$$

Since $\bar{\xi} > 0$, the intersection of the level sets of function TER with $\mathbb{R}^n_+$ is bounded. □

As a direct consequence of Theorem 3.3, we state the following result.

THEOREM 3.6 (Convexity of equilibrium prices). *The set of equilibrium prices P^* is convex.*

Further, we formulate additional assumptions in order to guarantee that our market indeed works, i.e. the equilibrium prices do not vanish. Due to Theorem 3.3, we need to ensure that the optimal solution p^* of the minimization problem

$(\mathbf{P})$ is not at the origin. For that, we introduce the following condition rejecting the *Zero-Cost Production (ZCP)*:

$$\text{If } \alpha_k \kappa_k + \alpha_k c_k \left(\widetilde{y}_k / \alpha_k \right) = 0 \text{ with } \widetilde{y}_k \in \alpha_k \mathcal{Y}_k \text{ and } \alpha_k \in [0,1], \text{ then } \widetilde{y}_k = 0. \tag{3.7}$$

This condition is automatically satisfied for $\kappa_k > 0$. If $\kappa_k = 0$, then (3.7) implies that for the k-th producer there is no nonzero production plan with zero production cost. Recall that

$$ERP_k(p) = \max_{\substack{\alpha_k \in [0,1] \\ \widetilde{y}_k \in \alpha_k \mathcal{Y}_k}} \langle p, \widetilde{y}_k \rangle - \alpha_k c_k \left(\widetilde{y}_k / \alpha_k \right) - \alpha_k \kappa_k, \tag{3.8}$$

Therefore, condition (3.7) implies that $\partial ERP_k(0) = \{0\}$. Note that $\widetilde{y}_k = 0$ if $\alpha_k = 0$ in (3.8), hence, the term $\alpha_k c_k \left(\widetilde{y}_k / \alpha_k \right)$ is set to vanish in this case.

Assume now that the wealth w_i of i-th consumer is positive. Since

$$ERC_i(p) = \max_{\substack{\beta_i \in [0,1] \\ \widetilde{x}_i \in \beta_i \mathcal{X}_i}} [\beta_i w_i - \langle p, x_i \rangle ,$$

we conclude that $\partial ERC_i(0) = -\mathcal{X}_i$. Thus, we have proved the following statement.

LEMMA 3.7. *Let all producers satisfy ZCP-condition, and the wealths of all consumers be positive. Then,*

$$\partial TER(0) \quad = \quad - \sum_{i=1}^{L} \mathcal{X}_i. \tag{3.9}$$

COROLLARY 3.8 (Nonzero equilibrium prices). Existence of a consumer with nonzero life standard is sufficient for having $p^* \neq 0$.

PROOF. Indeed, assume that $p^* = 0$. In view of the first-order optimality conditions for $(\mathbf{P})$, there exists $\xi^* \in \partial TER(0)$ such that

$$\langle \xi^*, p \rangle \quad \geq \quad 0 \quad \forall p \geq 0.$$

Hence, $\xi^* = - \sum_{i=1}^{L} x_i^* \geq 0$ for some $x_i^* \in \mathcal{X}_i$. Therefore, all $x_i^* = 0$, implying zero life standards for all consumers. □

It is interesting that the last statement is formulated only in terms of consumption standards. This confirms the primary role of demand in generating supply.

3.3. Efficiency. Let us present the first welfare theorem for equilibrium market flow. We are going to prove that any equilibrium market flow is efficient in the sense of Pareto optimality. This means that no consumers or producers can improve his gain (excessive revenue) without worsening the gain of some others. Let us start from the definition of feasible market flows.

We recall that for a given vector of prices $p \in \mathbb{R}^n_+$ and a *tentative market flow*

$$F = \left(\{y_k\}_{k=1}^K, \{x_i\}_{i=1}^L \right) \in \prod_{k=1}^{K} \mathcal{Y}_k \times \prod_{i=1}^{L} \mathcal{X}_i,$$

the system of participation levels $\gamma = \left(\{\alpha_k\}_{k=1}^K, \{\beta_i\}_{i=1}^L \right) \in [0,1]^{K+L}$ is called *proper* (with respect to π and F) if it satisfies the following conditions:

$$\alpha_k = \begin{cases} 1, & \text{if } \langle p, y_k \rangle - c_k(y_k) > \kappa_k, \\ 0, & \text{if } \langle p, y_k \rangle - c_k(y_k) < \kappa_k, \end{cases}$$

$$\beta_i = \begin{cases} 1, & \text{if } \langle p, x_i \rangle < w_i, \\ 0, & \text{if } \langle p, x_i \rangle > w_i. \end{cases}$$

Such a triple (p, F, γ) defines a *real market flow*

$$\widetilde{F} = \left(\{\widetilde{y}_k = \alpha_k y_k\}_{k=1}^K, \{\widetilde{x}_i = \beta_i x_i\}_{i=1}^L \right).$$

DEFINITION 3.9 (Feasible market flow). The real market flow

$$\widetilde{F} = \left(\{\widetilde{y}_k\}_{k=1}^K, \{\widetilde{x}_i\}_{i=1}^L \right),$$

defined by the triple (p, F, γ), is called *feasible* if it satisfies the market clearing condition:

$$p \geq 0, \quad \sum_{k=1}^K \widetilde{y}_k - \sum_{i=1}^I \widetilde{x}_i \geq 0, \quad \left\langle p, \sum_{k=1}^K \widetilde{y}_k - \sum_{i=1}^I \widetilde{x}_i \right\rangle = 0.$$

Note that an equilibrium market flow is in particular feasible.

DEFINITION 3.10 (Pareto optimal market flow). A feasible market flow $\widetilde{F}$, defined by the triple

$$\left(p, F = \left(\{y_k\}_{k=1}^K, \{x_i\}_{i=1}^L \right), \gamma \right),$$

is called *Pareto optimal* if there is no feasible market flow $\widetilde{F}'$ defined by another triple

$$\left(p', F' = \left(\{y'_k\}_{k=1}^K, \{x'_i\}_{i=1}^L \right), \gamma' \right)$$

such that all inequalities

$$\begin{aligned} (\langle p', y'_k \rangle - c_k(y'_k) - \kappa_k)_+ &\geq (\langle p, y_k \rangle - c_k(y_k) - \kappa_k)_+, k = 1 \dots K, \\ (w_i - \langle p', x'_i \rangle)_+ &\geq (w_i - \langle p, x_i \rangle)_+, i = 1 \dots L, \end{aligned} \tag{3.10}$$

are satisfied, and at least one of them is strict.

Note that we define Pareto optimality with respect to excessive revenues. In our model they play a role of objective functions of the agents.

THEOREM 3.11 (Efficiency of equilibrium market flows). *Any equilibrium market flow is Pareto optimal.*

PROOF. Using notation of Definition 3.10, let $\widetilde{F}^*$ be the equilibrium market flow defined by the triple (p^*, F^*, γ^*). Assume that the inequalities (3.10) are all valid for some feasible market flow $\widetilde{F}'$ defined by the triple (p', F', γ'). And let at least one of these inequalities be strict. For $p \in \mathbb{R}^n_+$ and $F \in \Omega \stackrel{\text{def}}{=} \prod_{k=1}^K \mathcal{Y}_k \times \prod_{i=1}^L \mathcal{X}_i$, define the function

$$\varphi(p, F) = \sum_{k=1}^K (\langle p, y_k \rangle - c_k(y_k) - \kappa_k)_+ + \sum_{i=1}^L (w_i - \langle p, x_i \rangle)_+.$$

In view of our assumption, $\varphi(p', F') > \varphi(p^*, F^*)$. Since p^* is an equilibrium price, in view of Theorem 3.3 and definitions (3.1), (3.3) we have:

$$\varphi(p^*, F^*) = \min_{p\geq 0} \max_{F\in\Omega} \varphi(p, F) = \max_{F\in\Omega} \min_{p\geq 0} \varphi(p, F) \geq \min_{p\geq 0} \varphi(p, F').$$

It remains to note that the market clearance condition for the flow $\widetilde{F}'$ is exactly the necessary and sufficient characterization of point p' as the optimal solution to the latter minimization problem. Therefore, $\varphi(p^*, F^*) \geq \varphi(p', F')$, a contradiction. □

In view of Theorem 3.3, equilibrium prices minimize the total excessive revenue. Let us prove a very intuitive result that its optimal value is equal to the difference of the real consumers' wealths and the real producers' costs.

THEOREM 3.12 (Total excessive revenue of the market). *Let p^* be an equilibrium price, and*

$$\widetilde{F}^* = \left((\widetilde{y}_k^* = \alpha_k^* y_k^*)_{k=1}^K, (\widetilde{x}_i^* = \beta_i^* x_i^*)_{i=1}^I \right)$$

be an equilibrium market flow defined by the triple (p^*, F^*, γ^*). Then,

$$TER(p^*) = \sum_{i=1}^I \beta_i^* w_i - \sum_{k=1}^K \alpha_k^* (c_k(y_k^*) + \kappa_k) \geq 0.$$

PROOF. It holds:

$$\begin{aligned} TER(p^*) &= \sum_{k=1}^K \left(\langle p^*, y_k^* \rangle - c_k(u_k^*) - \kappa_k \right)_+ + \sum_{i=1}^I \left(w_i - \langle p^*, x_i^* \rangle \right)_+ \\ &= \sum_{k=1}^K \alpha_k^* \left(\langle p^*, y_k^* \rangle - c_k(u_k^*) - \kappa_k \right) + \sum_{i=1}^I \beta_i^* \left(w_i - \langle p^*, x_i^* \rangle \right) \\ &= \left\langle p^*, \sum_{k=1}^K \alpha_k^* y_k^* - \sum_{i=1}^I \beta_i^* x_i^* \right\rangle - \sum_{k=1}^K \alpha_k^* (c_k(y_k^*) + \kappa_k) + \sum_{i=1}^I \beta_i^* w_i^*. \end{aligned}$$

In view of the market clearance condition, we have

$$\left\langle p^*, \sum_{k=1}^K \alpha_k^* y_k^* - \sum_{i=1}^I \beta_i^* x_i^* \right\rangle = 0.$$

This gives us the desired expression for optimal value of TER. It is nonnegative since all terms in its definition (3.5) are nonnegative. □

Note that the nonnegative value

$$TER(p^*) = \sum_{i=1}^I \beta_i^* w_i - \sum_{k=1}^K \alpha_k^* (c_k(y_k^*) + \kappa_k) \tag{3.11}$$

represents the total rate of accumulation of the capital within the market. In general, equilibrium prices, market flows, and participation levels are not unique. Nevertheless, all of them ensure the same value of $TER^* \stackrel{\text{def}}{=} TER(p^*)$. We call it the *total excessive revenue of the market*.

4. Decentralization of prices

Theorem 3.3 reveals the origin of equilibrium prices at the market. Namely, in order to reach an equilibrium price one needs to solve the minimization problem:

$$(\mathbf{P}) \qquad \min_{p\in\mathbb{R}^n_+} TER(p).$$

Our goal is to explain how agents can efficiently tackle this nonsmooth convex minimization problem by successively updating prices. It is crucial for our approach that the updates of prices correspond to subgradient-type dynamics for solving (**P**). Due to Theorem 3.2, the subgradients $z(p) \in \partial TER(p)$ represent the *excess supply*, i.e.

$$z(p) = \sum_{k=1}^{K} \widetilde{y}_k - \sum_{i=1}^{I} \widetilde{x}_i, \text{ where } \widetilde{y}_k \in S_k(p), \widetilde{x}_i \in D_i(p). \tag{4.1}$$

This gives rise to various price dynamics. E.g., the change of prices can be taken proportional to the current excess demand:

$$\frac{dp}{dt} \sim -z(p).$$

Alternatively, the change of prices can be taken proportional to the excess demand measured in current prices:

$$\frac{dp}{dt} \sim -p \circ z(p).$$

However, as it can be seen from (4.1), the subgradients of TER are not known to individual agents. Indeed, $z(p)$ represents the *aggregate* excess supply. For getting access to its value, one would assume the existence of a manager who collects the information about agents' production and consumption bundles, and aggregates them over the whole market. Here, the full information about production and consumption over the market must be available to him. Besides, the prices need to be updated by the manager, thus, leading to price regulation. Clearly, these assumptions can be justified only within a centrally planned economy. Aiming to avoid this restriction, we decentralize prices.

The decentralization of prices can be implemented by the introduction of various price designs. In this paper, we focus just on the *trade design*:

k-th producer settles and updates his individual prices p_k,
and consumers buy at the lowest purchase price $\min\limits_{k=1,\dots,K} p_k$.

Note that for vectors $p_1, \dots, p_K \in \mathbb{R}^n$, we denote by $\min\limits_{k=1,\dots,K} p_k \in \mathbb{R}^n$ the vector with coordinates

$$\left(\min_{k=1,\dots,K} p_k\right)^{(j)} = \min_{k=1,\dots,K} p_k^{(j)}, \quad j = 1, \dots, n.$$

Now, the total excessive revenue depends on the producers' prices $(p_k)_{k=1}^K$ as follows:

$$TER(p_1, \dots, p_K) \stackrel{\text{def}}{=} \sum_{k=1}^{K} ERP_k(p_k) + \sum_{i=1}^{I} ERC_i\left(\min_{k=1,\dots,K} p_k\right) =$$

$$\sum_{k=1}^{K} \max_{y_k\in\mathcal{Y}_k} \left(\langle p, y_k\rangle - c_k(y_k)\right)_+ + \sum_{i=1}^{I} \max_{x_i\in\mathcal{X}_i} \left(w_i - \left\langle \min_{k=1,\dots,K} p_k, x_i\right\rangle\right)_+ . \tag{4.2}$$

The decentralization of prices makes the corresponding subdifferential information about excess demands available to producers. In fact, note that the total excessive revenue TER from (4.2) is convex in the variables $(p_k)_{k=1}^K$. Let us obtain an expression for its subgradients $z_k(p_1,\ldots,p_K) \in \partial_{p_k} TER(p_1,\ldots,p_K)$ w.r.t. p_k:

$$z_k(p_1,\ldots,p_K) = \widetilde{y}_k - \sum_{i=1}^{I} \mu_{ik} \circ \widetilde{x}_i, \quad k = 1,\ldots,K. \tag{4.3}$$

Here, $\widetilde{y}_k \in S_k(p_k)$ is the supply of k-th producer w.r.t. his individual price p_k, and $\widetilde{x}_i \in D_i\left(\min\limits_{k=1,\ldots,K} p_k\right)$ is the demand of i-th consumer w.r.t. the lowest purchase price $\min\limits_{k=1,\ldots,K} p_k$. Further, $\mu_{ik}^{(j)}$ denotes the share of i-th consumer's demand from k-th producer for good j. Indeed, the shares $\mu_{ik}^{(j)}$ for good j sum up to 1 over all producers $k = 1,\ldots,K$. Moreover, the share $\mu_{ik}^{(j)}$ vanishes if the k-th producer's price $p_k^{(j)}$ exceeds the lowest purchase price $\min\limits_{k=1,\ldots,K} p_k^{(j)}$ for good j. Thus, we write

$$(\mu_{ik})_{k=1}^K \in M(p_1,\ldots,p_K),$$

where

$$M(p_1,\ldots,p_K) \stackrel{\text{def}}{=} \left\{ (\mu_{ik})_{k=1}^K \in [0,1]^{n\times K} \;\middle|\; \begin{array}{c} \sum\limits_{k=1}^{K} \mu_{ik}^{(j)} = 1, \\ \mu_{ik}^{(j)} = 0 \text{ if } p_k^{(j)} \neq \min\limits_{k=1,\ldots,K} p_k^{(j)} \\ j = 1,\ldots,n, k = 1,\ldots,K \end{array} \right\}.$$

Note that the set of admissible market shares $M(p_1,\ldots,p_K)$ is the same for all consumers $i = 1,\ldots,I$. We claim that the subdifferential information in (4.3) is known to k-th producer. First, note that $\widetilde{y}_k$ is his real production. Despite to the fact that the shares μ_{ik} and the demands $\widetilde{x}_i$ cannot be estimated by k-th producer, their aggregated product $\sum_{i=1}^I \mu_{ik} \circ \widetilde{x}_i$ is perfectly available to him. Indeed, $\sum_{i=1}^I \mu_{ik} \circ \widetilde{x}_i$ forms the bundle of goods demanded by all consumers from k-th producer. Altogether, the subgradients $z_k(p_1,\ldots,p_K)$ represent the *individual excess* of k-th producer's supply over all consumers' demands. Overall, we obtain:

THEOREM 4.1 (Producers' excess supply and TER).

$$\partial_{p_k} TER(p_1,\ldots,p_K) = S_k(p_k) - \sum_{i=1}^{I} \mu_{ik} \circ D_i\left(\min_{k=1,\ldots,K} p_k\right), \quad k = 1,\ldots,K,$$

with demand shares $(\mu_{ik})_{k=1}^K \in M(p_1,\ldots,p_K)$.

Due to Theorem 4.1, the subdifferential of $TER(p_1,\ldots,p_k)$ is completely available to k-th producer. This fact suggests to adjust prices by solving the minimization problem

$$\textbf{(PD)} \qquad \min_{p_1,\ldots,p_K \in \mathbb{R}_+^n} TER(p_1,\ldots,p_K).$$

Note that the minimization problem (**PD**) is stated w.r.t. the decentralized producers' prices $(p_k)_{k=1}^K$, while previously in (**P**) one minimizes over the common prices p. We relate the minimization problems (**P**) and (**PD**). For that, let us call function

$f(x)$, $x \in \mathbb{R}^n$, nondecreasing (nonicreasing) in x if $f(x) \geq f(y)$ ($f(x) \leq f(y)$) for any $x \geq y$.

LEMMA 4.2 (Decentralization). *1) Let function of $K+1$ vector variables*

$$F(p_0, p_1, \ldots, p_K), \quad p_k \in \mathbb{R}^n, k = 0, \ldots, K,$$

be a1) nonincreasing in p_0, and b1) nondecreasing in all other variables p_k, $k = 1, \ldots, K$. Then,

$$\min_{p_1,\ldots,p_K \in \mathbb{R}^n_+} F\left(\min_{k=1,\ldots,K} p_k, p_1, \ldots, p_K\right) = \min_{p \in \mathbb{R}^n_+} F(p, \ldots, p).$$

2) Let function of $I+1$ vector variables

$$G(p_0, p_1, \ldots, p_I), \quad p_i \in \mathbb{R}^n, i = 0, \ldots, I,$$

be a2) nondecreasing in p_0, and b2) nonincreasing in all other variables p_i, $i = 1, \ldots, I$. Then,

$$\min_{p_1,\ldots,p_I \in \mathbb{R}^n_+} G\left(\max_{i=1,\ldots,I} p_i, p_1, \ldots, p_I\right) = \min_{p \in \mathbb{R}^n_+} G(p, \ldots, p).$$

PROOF. 1) Indeed,

$$\min_{p_1,\ldots,p_K \in \mathbb{R}^n_+} F\left(\min_{k=1,\ldots,K} p_k, p_1, \ldots, p_K\right) \overset{\text{a1)}}{=}$$
$$\min_{p, p_1,\ldots,p_K \in \mathbb{R}^n_+} \left\{ F\left(p, p_1, \ldots, p_K\right) \mid p_k \geq p, k = 1, \ldots, K \right\} \overset{\text{b1)}}{=}$$
$$\min_{p \in \mathbb{R}^n_+} F(p, \ldots, p).$$

2) The proof is analogous to 1). □

THEOREM 4.3 (Problems (**PD**) and (**P**)). *It holds:*

$$\min_{p_1,\ldots,p_K \in \mathbb{R}^n_+} TER(p_1, \ldots, p_K) = \min_{p \in \mathbb{R}^n_+} TER(p). \tag{4.4}$$

Moreover,

(i) *if $(p_k)_{k=1}^K$ solves (**PD**), then $\min_{k=1,\ldots,K} p_k$ solves (**P**),*

(ii) *if p solves (**P**), then $(p, \ldots, p)$ solves (**PD**).*

PROOF. We set

$$F(p_0, p_1, \ldots, p_K) \overset{\text{def}}{=} \sum_{k=1}^K ERP_k\left(p_k\right) + \sum_{i=1}^I ERC_i\left(p_0\right).$$

Note that F is nonincreasing in p_0, and nondecreasing in p_k, $k = 1, \ldots, K$. Applying Lemma 4.3 and in view of

$$F\left(\min_{k=1,\ldots,K} p_k, p_1, \ldots, p_K\right) = TER(p_1, \ldots, p_K), \quad F(p, \ldots, p) = TER(p),$$

(4.4) holds. Moreover, (4.4) implies (i) and (ii). □

Theorem 4.3 says that if $(p_k)_{k=1}^K$ is a solution of (**PD**) then the lowest purchase price $\min_{k=1,\ldots,K} p_k$ solves (**P**). Thus, $\min_{k=1,\ldots,K} p_k$ provides an equilibrium price due to Theorem 3.3. We exploit this fact to construct decentralized subgradient-type dynamics for producers' individual prices p_k.

5. Price dynamics

One of the most important questions within the scope of equilibrium theory is the description of price dynamics, which should result in equilibrium prices. Usually, for that we need to relate the variation of prices $p(t) \geq 0$ in time $t \in \mathbb{R}$ with the excess supply $z(t)$, which is the difference between the aggregate supply and demand, computed for the current price, i.e.

$$z(t) \stackrel{\text{def}}{=} z(p(t)) = \sum_{k=1}^{K} \widetilde{y}_k - \sum_{i=1}^{I} \widetilde{x}_i,$$

where $\widetilde{y}_k \in S_k(p(t)), \widetilde{x}_i \in D_i(p(t))$.

One suggestion is the classical tâtonnement process going back to Walras (e.g., [**13**]). It consists in postulating the following price dynamics

$$\frac{dp}{dt} = -z(t). \tag{5.1}$$

Note that $z(t) \in \partial TER(p(t))$ due to Theorem 3.2. That's why we refer to (5.1) as a subgradient-type dynamics. This very intuitive equation (5.1) has however two shortcomings. Firstly, it is not clear why the entries of the trajectory $p(t)$ never become negative. Secondly, in order to perform the price update in (5.1) the excess supply $z(t)$ need to be aggregated over the whole market. It assumes the existence of an auctioneer who actually computes $z(t)$ and then updates prices. Clearly, these assumptions can be justified only within a centrally planned economy.

To overcome these difficulties, we suggest a decentralized price dynamics. For that, the k-th producer relates the variation of his individual prices $p_k(t) \geq 0$ in time $t \in \mathbb{R}$ with his individual excess supply $z_k(t)$, $k = 1, \ldots, K$. Latter, $z_k(t)$ is the difference between his individual supply and all consumers' demands from him, computed for the current system of prices, i.e.

$$z_k(t) \stackrel{\text{def}}{=} z_k(p_1(t), \ldots, p_K(t)) = \widetilde{y}_k - \sum_{i=1}^{I} \mu_{ik} \circ \widetilde{x}_i.$$

Here, $\widetilde{y}_k \in S_k(p_k(t))$ is the supply of k-th producer w.r.t. his individual price $p_k(t)$, and $\widetilde{x}_i \in D_i\left(\min\limits_{k=1,\ldots,K} p_k(t)\right)$ is the demand of i-th consumer w.r.t. the lowest purchase price $\min\limits_{k=1,\ldots,K} p_k(t)$. The demand shares for k-th producer are $(\mu_{ik})_{k=1}^K \in M\left(p_1(t), \ldots, p_K(t)\right)$, $i = 1, \ldots, I$.

We study the following decentralized price dynamics:

$$\frac{dp_k}{dt} = -p_k(t) \circ z_k(t), \quad k = 1, \ldots, K. \tag{5.2}$$

Note that the individual excess supply $z_k(t)$ is available to the k-th producer, hence, he can successively perform the update of his individual prices p_k in (5.2). The equation (5.2) states that the k-th producer changes his individual prices p_k proportionally to his individual excess demand $z_k(t)$ measured in current prices. Due to Theorem 4.1, $z_k(t) \in \partial_{p_k} TER(p_1(t), \ldots, p_K(t))$. That's why (5.2) as also a *subgradient-type dynamics*. Additionally, we point out that the right-hand side of (5.2) is perfectly observable. From the practical point of view, the producer needs to compare his real production with the consumers' demands just at his place. For the latter it is necessary to solve a profit maximization problem for the producer and expenditure minimization problems for all consumers. As mentioned in Section 2, we may assume that a profit maximization problem can be efficiently tackled by the producer. For that, he employs internal capacities to analyse his production process or he alternatively asks for external consulting solutions. The assumption on the solvability of consumers' expenditure problems is more restrictive. Indeed, consumers usually do not enjoy any computational support in order to adjust their consumption bundles towards optimality. Instead of that, their adjusting process rely rather on buying, consuming and updating tastes. A variant of such an algorithm to successively solving the expenditure minimization problem is discussed in Section 7. Further development of consumption algorithms is a matter of future research.

In what follows, we prove that the subgradient-type dynamics (5.2) can have only a single limiting point. Moreover, this limiting point is formed by an equilibrium price.

THEOREM 5.1 (Convergence of price dynamics). *At productive markets, if $p_k(0)$ is positive, then the price trajectories $p_k(t)$ of (5.2) remain positive for all $t \geq 0$. Moreover, $p_k(t)$ converges to the same equilibrium price p^* for all $k = 1, \ldots, K$.*

PROOF. Consider the entropy function $\eta(\tau) = \tau \ln \tau$, $\tau \geq 0$. We can define the Bregman distance between two reals $\tau_1 > 0$ and $\tau_2 \geq 0$ as follows:

$$\begin{aligned} \rho(\tau_1, \tau_2) &= \eta(\tau_2) - \eta(\tau_1) - \eta'(\tau_1)(\tau_2 - \tau_1) \\ &= \tau_2 \ln \tfrac{\tau_2}{\tau_1} - \tau_2 + \tau_1 \;\geq\; 0. \end{aligned}$$

Note that $\rho(\tau_1, 0) = \tau_1$. Therefore, for any sequence of positive numbers $\{\tau_k\}$ and any value $\tau^* \geq 0$, we have $\lim\limits_{k\to\infty} \rho(\tau_k, \tau^*) = 0$ if and only if $\lim\limits_{k\to\infty} \tau_k = \tau^*$. It is important that

$$\rho_1'(\tau_1, \tau_2) \;\stackrel{\text{def}}{=}\; \tfrac{\partial}{\partial \tau_1} \rho(\tau_1, \tau_2) = 1 - \tfrac{\tau_2}{\tau_1}.$$

Recall that P^* denotes the set of equilibrium prices. Let us fix an arbitrary $p^* \in P^*$. Consider the function

$$V_{p^*}(t) = \sum_{k=1}^{K} \sum_{j=1}^{n} \rho\left(p_k^{(j)}(t), p^{*(j)}\right).$$

Since function $TER(\cdot, \ldots, \cdot)$ is convex and $z_k(t) \in \partial_{p_k} TER(p_1(t), \ldots, p_K(t))$, we have

(5.3)

$$\begin{aligned} \tfrac{d}{dt} V_{p^*}(t) &= \sum_{k=1}^{K} \sum_{j=1}^{n} \rho_1'\left(p_k^{(j)}(t), p^{*(j)}\right) \tfrac{dp_k^{(j)}}{dt} \\ &\stackrel{(5.2)}{=} -\sum_{k=1}^{K} \sum_{j=1}^{n} \left(1 - \tfrac{p^{*(i)}}{p_k^{(i)}(t)}\right) z_k^{(j)}(t) p_k^{(j)}(t) = -\sum_{k=1}^{K} \langle z_k(t), p_k(t) - p^* \rangle \\ &\leq -\left(TER(p_1(t), \ldots, p_K(t)) - TER(p^*, \ldots, p^*)\right). \end{aligned}$$

Due to Theorem 3.3, p^* solves the minimization problem (**P**). In view of Theorem 4.3, $(p^*, \ldots, p^*)$ solves the minimization problem (**PD**), i.e.

$$TER(p_1(t), \ldots, p_K(t)) \geq TER(p^*, \ldots, p^*).$$

Together with (5.3), it follows that the function $V_{p^*}(t)$ decreases monotonically. Since the market is productive, the set of equilibrium prices P^* is bounded in view of Theorem 3.5. This implies that the trajectory $(p_k(t))_{k=1}^K$ converges to the set

$$\{(p^*, \ldots, p^*) \,|\, p^* \in P^*\}.$$

Denote now by $(p^*, \ldots, p^*)$ with $p^* \in P$ a limiting point of any convergent sequence $(p_k(t_l))_{k=1}^K$, $t_l \to \infty$. Since the function $V_{p^*}(t)$ decreases monotonically, we conclude that $\lim\limits_{t\to\infty} p_k(t) = p^*$. □

Theorem 5.1 says that the decentralized price dynamics (5.2) leads to a unique equilibrium price. In fact, all trajectories $p_k(t)$ of producers' individual prices converge to the same equilibrium price p^*. Hence, the lowest purchase price $\min\limits_{k=1,\ldots,K} p_k(t)$ also converges to p^*.

REMARK 5.2 (Weber-Fechner law and price dynamics). The k-th producer's price dynamics (5.2) can be written in the following form:

$$\frac{d}{dt}\left(\ln p_k^{(j)}(t)\right) = -z_k^{(j)}(t), \quad j = 1, \ldots, n. \tag{5.4}$$

Hence, we obtain the following discrete-time dynamics for $t \in \mathbb{N}$:

$$\ln\left(\frac{p_k^{(j)}(t+1)}{p_k^{(j)}(t)}\right) = -z_k^{(j)}(t), \quad j = 1, \ldots, n. \tag{5.5}$$

Surprisingly enough, the latter logarithmic dependence may be interpreted as the *Weber-Fechner law* from psychophysics (e.g., [**4**]). The Weber-Fechner law describes the relationship between the physical magnitudes of stimuli and their perceived intensity. It states that the subjective sensation is proportional to the logarithm of the stimulus intensity. In our setting, the k-th producer's individual excess demand $-z_k^{(j)}(t)$ for good j is the perception resulting from the trade. The expression $\frac{p_k^{(j)}(t+1)}{p_k^{(j)}(t)}$ represents the relation of the previous price stimulus to the next one. This link to psychophysics can open a door for behavioral interpretations of price dynamics.

6. Extensions: markets of labor and resources

The excessive revenue model described in Section 2 can be easily extended in order to capture various additional features of the market. In this section we illustrate this by introducing in our model the markets of labor and resources.

For producers, the production cost includes now the cost of the labor, which is used for computing the salaries of workers. Assume that there exist m different professional skills of the workers. The corresponding vector of their unitary labor prices (e.g., on hourly basis) is $q \in \mathbb{R}^m_+$. For simplicity, we assume that all producers at the market apply the same labor prices. However, these prices are not fixed. They are finally determined by the clearance condition for the labor market. The technological labor matrix of k-th producer $L_k \in \mathbb{R}^{m\times n}$ describes the necessary amounts of labor in this particular production environment. Namely, the column $L_k e_j \in \mathbb{R}^m$ represents the amounts of working hours of all professions, which are necessary for producing single units of j-th product. Thus, the total remuneration paid by the k-th producer for the production bundle $y_k \in \mathcal{Y}_k$ is $\langle q, L_k y_k\rangle$.

The necessity of buying some row materials outside the local market leads to the market of resources. There are l different resources with corresponding vector of prices $r \in \mathbb{R}^l_+$. The technological matrix $T_k \in \mathbb{R}^{l\times n}$ describes the resources required by the production process of k-th producer. Its column $T_k e_j \in \mathbb{R}^l$ represents the volumes of corresponding resources needed for producing one unit of good j by k-th producer. The cost of resources is then given by $\langle r, T_k y_k \rangle$. We assume that the total amounts of available resources are bounded by some upper limits stored in vector $b \in \mathbb{R}^l_+$. Thus, the monetary value of available resources is $\langle r, b \rangle$.

Hence, for the market with labor, we need to extend our system of prices:

$$(p, q, r) \quad \in \quad \mathbb{R}^{n+m+l}.$$

All definitions are modified now in a natural way:

$$\begin{aligned} \pi_k(p,q,r) &= \max_{y_k \in \mathcal{Y}_k} \left[\langle p, y_k \rangle - c_k(y_k) - \langle q, L_k y_k \rangle - \langle r, T_k y_k \rangle \right], \\ \mathcal{Y}^*_k(p,q,r) &= \arg\max_{y_k \in \mathcal{Y}_k} \left[\langle p, y_k \rangle - c_k(y_k) - \langle q, L_k y_k \rangle - \langle r, T_k y_k \rangle \right], \\ ERP_k(p,q,r) &= (\, \pi_k(p,q,r) - \kappa_k \,)_+ . \end{aligned} \tag{6.1}$$

We assume that each consumer can increase his regular income by accepting a job. For that, i-th consumer has to decide on distribution $t_i \in \mathbb{R}^m_+$ of his total working time $\theta_i \in \mathbb{R}_+$ among m different jobs. Thus, the feasible set for time distribution of i-thth consumer looks as follows

$$\mathcal{T}_i \quad \stackrel{\text{def}}{=} \quad \left\{ t_i \in \mathbb{R}^m_+ : \sum_{j=1}^m t_i^{(j)} \le \theta_i \right\}.$$

Note that i-th consumer may have professional training for several jobs. Denoting by $s_i^{(j)}$ his productivity in performing job j, $j = 1, \dots, m$, (it can happen that some of these values are zeros), we introduce diagonal matrix $S_i = \mathrm{diag}\left(s_i^{(1)}, \dots, s_i^{(m)} \right) \in \mathbb{R}^{m\times m}$. Now, for a feasible time distribution $t_i \in \mathcal{T}_i$, the salary of this consumer can be computed as $\langle q, S_i t_i \rangle$.

In order to compute the optimal time distribution, i-th consumer maximizes his salary. Thus, he solves the maximization problem

$$\omega_i(q) \stackrel{\text{def}}{=} \max_{t_i \in \mathcal{T}_i} \langle q, S_i t_i \rangle, \quad \mathcal{T}_i^*(q) \stackrel{\text{def}}{=} \arg\max_{t_i \in \mathcal{T}_i} \langle q, S_i t_i \rangle.$$

The wage $\omega_i(q)$ modifies the excessive revenue of i-th consumer:

$$\begin{aligned} ERC_i(p,q) &= w_i + \omega_i(q) - e_i(p) \\ &= w_i + \max_{t_i \in \mathcal{T}_i} \langle q, S_i t_i \rangle - \min_{x_i \in \mathcal{X}_i} \langle p, x_i \rangle. \end{aligned}$$

Similar to the market of goods, we can give now an adapted definition of proper participation levels and real market flows. In the new definition, the time distributions play the same role as the production bundles.

DEFINITION 6.1 (Participation levels and real market flows). For a given system of prices and salaries $(p,q,r) \in \mathbb{R}^{n+m+l}_+$ and a system of *tentative market flows*

$$F = \left(\{y_k\}_{k=1}^K, \{x_i\}_{i=1}^I, \{t_i\}_{i=1}^I \right) \in \prod_{k=1}^K \mathcal{Y}_k \times \prod_{i=1}^I \mathcal{X}_i \times \prod_{i=1}^I \mathcal{T}_i,$$

the system of participation levels $\gamma = \left(\{\alpha_k\}_{k=1}^K, \{\beta_i\}_{i=1}^I\right) \in [0,1]^{K+I}$ is called *proper* (with respect to (p,q,r) and F) if it satisfies the following conditions:

$$\alpha_k = \begin{cases} 1, & \text{if } \langle p, y_k\rangle - c_k(y_k) - \langle q, L_k y_k\rangle - \langle r, T_k y_k\rangle > \kappa_k, \\ 0, & \text{if } \langle p, y_k\rangle - c_k(y_k) - \langle q, L_k y_k\rangle - \langle r, T_k y_k\rangle < \kappa_k, \end{cases}$$

$$\beta_i = \begin{cases} 1, & \text{if } \langle p, x_i\rangle < \tau_i + \langle q, S_i t_i\rangle, \\ 0, & \text{if } \langle p, x_i\rangle > \tau_i + \langle q, S_i t_i\rangle. \end{cases}$$

Such a triple $((p,q,r), F, \gamma)$ defines a *real market flow*

$$\widetilde{F} = \left(\{\widetilde{y}_k = \alpha_k y_k\}_{k=1}^K, \{\widetilde{x}_i = \beta_i x_i\}_{i=1}^I, \left\{\widetilde{t}_i = \beta_i t_i\right\}_{i=1}^I\right).$$

Now we can define the equilibrium market flows. Let

$$\widetilde{F} = \left(\{\widetilde{y}_k\}_{k=1}^K, \{\widetilde{x}_i\}_{i=1}^I, \left\{\widetilde{t}_i\right\}_{i=1}^I\right)$$

be a real market flow defined by some triple $((p,q,r), F_0, \gamma)$. We say that $\widetilde{F}$ clears the markets of goods, labor, and resources if

$$\text{(C1)} \qquad p \geq 0, \quad \sum_{k=1}^K \widetilde{y}_k - \sum_{i=1}^I \widetilde{x}_i \geq 0, \quad \left\langle p, \sum_{k=1}^K \widetilde{y}_k - \sum_{i=1}^I \widetilde{x}_i \right\rangle = 0,$$

$$\text{(C2)} \qquad q \geq 0, \quad \sum_{i=1}^I S_i\widetilde{x}_i - \sum_{k=1}^K L_k\widetilde{y}_k \geq 0, \quad \left\langle q, \sum_{i=1}^I S_i\widetilde{x}_i - \sum_{k=1}^K L_k\widetilde{y}_k \right\rangle = 0,$$

$$\text{(C3)} \qquad r \geq 0, \quad b - \sum_{k=1}^K T_k\widetilde{y}_k \geq 0, \quad \left\langle r, b - \sum_{k=1}^K T_k\widetilde{y}_k \right\rangle = 0,$$

As previously, the *market clearing condition* (C1) states that the real consumption never exceed the real production, and the markets of goods with positive prices are perfectly cleared. (C2) means that the amount of labor needed for fulfilling the real production does not exceed its available amount. Moreover, the labor markets for jobs with positives salaries clear. Finally, (C3) says that the resources used for production never exceed their limiting bounds. Moreover, the market of resources with positive prices clears.

DEFINITION 6.2 (Equilibrium market flow). We say that $(p^*, q^*, r^*) \in \mathbb{R}_+^{n+m+l}$ is the *system of equilibrium prices* if there exists a tentative market flow

$$F^* \in \prod_{k=1}^K \mathcal{Y}_k^*(p^*, q^*, r^*) \times \prod_{i=1}^I \mathcal{X}_i^*(p^*) \times \prod_{i=1}^I \mathcal{T}_i^*(q^*)$$

and a proper system of participation levels γ such that the corresponding real market flow $\widetilde{F}^*$ satisfies the balancing conditions (C1), (C2), and (C3).

In this case $\widetilde{F}^*$ is called the *equilibrium market flow.*

At equilibrium, all producers and consumers tend to maximize their excessive revenues. Again, we define the total excessive revenue of the market by summing up the excessive revenues of all agents:

$$TER(p,q,r) := \sum_{k=1}^K ERP_k(p,q,r) + \sum_{i=1}^I ERC_i(p,q) + \langle r, b\rangle.$$

At the same time, we can prove the statement similar to Theorem 3.3.

THEOREM 6.3 (Characterization of equilibrium prices). *$(p^*, q^*, r^*) \in \mathbb{R}^{n+m+l}_+$ is a system of equilibrium prices if and only if it solves the following convex minimization problem:*

$$\min_{(p,q,r)\in\mathbb{R}^{n+m+l}_+} TER(p,q,r).$$

The proof of this theorem is almost identical to the proof of Theorem 3.3.

In this section we have shown how we can enrich our model by additional elements, representing some economical relations between different agents. This can be done for many different types of goods trading at the market. The required structural changes are always the same. Namely, some agents introduce a new good with unknown price. Its production needs some expenditures, which reduce the excessive revenues of these agents. However, the new good can be sold at the market and the gain will increase the corresponding revenues. Similarly, we need to modify the excessive revenues of potential consumers of the new good. The equilibrium market price of the new good can be defined by minimizing the total excessive revenue of the market. Note that the whole structure of such market relations can be very complicated since the same agents can be producers of one good and consumers for another one. However, in many situations the characterization theorems can be justified by the same arguments as in Theorem 3.3.

7. Consumption strategies

In the previous sections we assumed that the consumer chooses his regular consumption pattern as a solution of the expenditure minimization problem:

$$\min_{x\in\mathbb{R}^n_+} \{\langle p, x\rangle : \; u(x) \geq u\},$$

where p is the price, $u(x)$ denotes the consumer's utility function, and u is his desirable utility level. In the real life, dimension of this problem is usually very big. It seems impossible that a normal consumer can approach somehow its solution without being involved in very heavy computations. Fortunately, this impression is wrong. We shall show how consumer can tackle the expenditure minimization problem by simple and natural consumption strategies for buying the daily goods. These consumption strategies can be seen as a numerical method for minimizing some special disutility function. This function is derived from the natural consumption cycle:

$$\begin{array}{ccc} \boxed{\begin{array}{c}\textit{Deciding on the budget}\\ \textit{for single shopping.}\end{array}} & \Rightarrow & \boxed{\begin{array}{c}\textit{Buying goods with the}\\ \textit{best quality/price ratio.}\end{array}} \\ \Uparrow & & \Downarrow \\ \boxed{\begin{array}{c}\textit{Observation of the}\\ \textit{results.}\end{array}} & \Leftarrow & \boxed{\begin{array}{c}\textit{Consumption of the}\\ \textit{goods bought.}\end{array}} \end{array} \tag{7.1}$$

We argue that the implementation of this cycle by a customer needs only regular updates of some subconscious estimates (individual prices for qualities), which allow to fight for a reasonable price of the goods available on the market. The corresponding average consumption pattern solves the expenditure minimization problem.

In what follows, we assume the Leontief utility function

$$u(x) \stackrel{\mathrm{def}}{=} \min_{i=1,\dots,d} \frac{(Qx)^{(i)}}{\sigma^{(i)}}, \quad \text{and } u \stackrel{\mathrm{def}}{=} 1. \tag{7.2}$$

Here, consumers judge the goods in accordance to the presence in them of d different qualities. The personal estimates of the volumes of these qualities are stored in the matrix $Q \in \mathbb{R}^{d\times n}$. The value $(Qe_j)^{(i)}$ is the consumer's estimate for the amount of quality i in one unit of the good j. Further, customer has some standards for regular consumptions of certain amounts of qualities during the standard interval of time. We store these standards in the vector $\sigma \in \mathbb{R}^d_+$. In view of (7.2), the consumer faces the optimization problem

$$\min_{x\in\mathbb{R}^n_+} \{\langle p, x\rangle : \; Qx \geq \sigma\}. \tag{7.3}$$

He tries to minimize the expenditure $\langle p, x\rangle$ while guaranteeing the standard σ on qualities. Let us look at the dual form of problem (7.3).

$$\begin{aligned} \min_{x\in\mathbb{R}^n_+} \{\langle p, x\rangle : \; Qx \geq \sigma\} &= \min_{x\in\mathbb{R}^n_+} \max_{\lambda\in\mathbb{R}^d_+} \{\langle p, x\rangle + \langle \lambda, \sigma - Qx\rangle\} \\ &= \max_{\lambda\in\mathbb{R}^d_+} \min_{x\in\mathbb{R}^n_+} \left\{\langle \lambda, \sigma\rangle + \langle p - Q^T\lambda, x\rangle\right\} \\ &= \max_{\lambda\in\mathbb{R}^d_+} \left\{\langle \sigma, \lambda\rangle : \; Q^T\lambda \leq p\right\}. \end{aligned}$$

Thus, the problem dual to (7.3) is as follows:

$$\max_{\lambda\in\mathbb{R}^d_+} \left\{\langle \sigma, \lambda\rangle : \; \langle q_j, \lambda\rangle \leq p^{(j)}, \; j = 1, \dots, n\right\}, \tag{7.4}$$

where $q_j = Qe_j$. In this problem, the elements $\lambda^{(i)}$ can be interpreted as personal prices for corresponding qualities, $i = 1, \dots, d$. Thus, the value $\langle q_j, \lambda\rangle$ is the personal estimate of the quality of j-th good. Let us introduce now the function

$$\psi(\lambda) = \max_{1\leq j\leq n} \tfrac{\langle q_j, \lambda\rangle}{p^{(j)}}. \tag{7.5}$$

This function selects the goods with the best *quality/price ratio.* Of course, the results of selection depend on our current prices for qualities.

Let us rewrite the dual problem (7.4) using the function ψ.

$$\max_{\lambda\in\mathbb{R}^d_+} \{\langle \sigma, \lambda\rangle : \; \psi(\lambda) \leq 1\} = \max_{\lambda\in\mathbb{R}^d_+} \tfrac{\langle \sigma, \lambda\rangle}{\psi(\lambda)} = \left[\min_{\lambda\in\mathbb{R}^d_+} \{\psi(\lambda) : \; \langle \sigma, \lambda\rangle = 1\} \right]^{-1}.$$

We are going to show that our consumer is able to approach the optimal solution of problem

$$\min_{\lambda} \{\psi(\lambda) : \; \lambda \in \Delta_d(\sigma)\}, \quad \Delta_d(\sigma) \stackrel{\mathrm{def}}{=} \left\{\lambda \in \mathbb{R}^d_+ : \; \langle \sigma, \lambda\rangle = 1\right\}, \tag{7.6}$$

by implementing the standard consumption cycle (7.1).

For simplicity, let us assume that for each shopping we have the same budget $h > 0$. During our consumption history, we update the vector of accumulated qualities $s_k \in \mathbb{R}^d$, and the bundle of cumulative consumption $\hat{x}_k \in \mathbb{R}^n$.

Clearly, $s_0 = 0$ and $\hat{x}_0 = 0$. Before starting the k-th shopping, we have already in mind some prices for qualities λ_k. It is convenient to normalize them by inclusion

$\lambda_k \in \Delta_d(\sigma)$, which means that the total budget for our life standards is equal to one.

During the k-th shopping, we subconsciously determine the set of goods with the best quality/price ratio:

$$J_*(\lambda_k) = \left\{ j : \tfrac{\langle q_j, \lambda_k \rangle}{p^{(j)}} = \psi(\lambda_k) \right\}.$$

If this set contains more than one element, we need to define the sharing vector $\nu_k \in \Delta_n(e)$, such that $\nu_k^{(j)} = 0$ for all $j \notin J_*(\lambda_k)$. Then, the budget for buying j-th good for $j \in J_*(\lambda_k)$ becomes $h\nu_k^{(j)}$, and we buy $\frac{h\nu_k^{(j)}}{p^{(j)}}$ quantity of this good. Thus, the vector of accumulated qualities is updated as follows:

$$s_{k+1} \;=\; s_k + hg_k, \quad g_k = \sum_{j=1}^{n} \tfrac{\nu_k^{(j)}}{p^{(j)}} q_j. \tag{7.7}$$

It is important, that $g_k \in \partial\psi(\lambda_k)$. Note that $\langle g_k, \lambda_k \rangle = \psi(\lambda_k)$.

The bundle of accumulated consumption is updated as

$$\hat{x}_{k+1}^{(j)} \;=\; \hat{x}_k^{(j)} + h\tfrac{\nu_k^{(j)}}{p^{(j)}}, \quad j = 1, \dots, n. \tag{7.8}$$

Note that these updates maintain the following relations:

$$\hat{x}_k \;\geq\; 0, \quad s_k \;=\; Q\hat{x}_k, \quad k \geq 0. \tag{7.9}$$

Let us look now at the *gap bound*

$$\delta_k \;=\; h \max_{\lambda \in \Delta_d(\sigma)} \sum_{l=0}^{k} \langle g_l, \lambda_l - \lambda \rangle. \tag{7.10}$$

Note that

$$\begin{aligned}
\delta_k &= h \sum_{l=0}^{k} \psi(\lambda_l) - \min_{\lambda \in \Delta_d(\sigma)} h \sum_{l=0}^{k} \langle g_l, \lambda \rangle \;=\; h \sum_{l=0}^{k} \psi(\lambda_l) - \min_{\lambda \in \Delta_d(\sigma)} \langle s_{k+1}, \lambda \rangle \\
&= h \sum_{l=0}^{k} \psi(\lambda_l) - \min_{1 \leq i \leq d} \tfrac{s_{k+1}^{(i)}}{\sigma^{(i)}}.
\end{aligned}$$

Let us writer down a problem dual to (7.6)

$$\begin{aligned}
\min_{\lambda} \{\psi(\lambda) : \lambda \in \Delta_d(\sigma)\} &= \min_{\lambda \in \mathbb{R}^d_+} \max_{\substack{\mu \in \mathbb{R}, \\ w \in \Delta_n(e)}} \left\{ \sum_{j=1}^{n} w^{(j)} \tfrac{\langle q_j, \lambda \rangle}{p^{(j)}} + \mu(1 - \langle \sigma, \lambda \rangle) \right\} \\
&= \max_{\substack{\mu \in \mathbb{R}, \\ w \in \Delta_n(e)}} \min_{\lambda \in \mathbb{R}^d_+} \left\{ \langle \sum_{j=1}^{n} w^{(j)} \tfrac{q_j}{p^{(j)}} - \mu\sigma, \lambda \rangle + \mu \right\} \\
(\tilde{x}^{(j)} \stackrel{\text{def}}{=} w^{(j)}/p^{(j)}) &= \max_{\mu \in \mathbb{R}, \tilde{x}} \{\mu : Q\tilde{x} \geq \mu\sigma, \ \tilde{x} \in \Delta_n(p)\} \\
&= \max_{s, \tilde{x}} \left\{ \min_{1 \leq i \leq d} \tfrac{s^{(i)}}{\sigma^{(i)}} : s = Q\tilde{x}, \ \tilde{x} \in \Delta_n(p) \right\}
\end{aligned}$$

Thus, the problem dual to (7.6) has the following form

$$\max_{s, \tilde{x}} \left\{ \min_{1 \leq i \leq d} \tfrac{s^{(i)}}{\sigma^{(i)}} : s = Q\tilde{x}, \ \tilde{v} \in \mathbb{R}^n_+, \ \langle p, \tilde{x} \rangle = 1 \right\}. \tag{7.11}$$

It is interesting that this problem belongs to the class of models suggested by Lancaster [**12**] with the utility function from (7.2), i.e.

$$u(\tilde{x}) = \min_{i=1,\dots,d} \frac{(Q\tilde{x})^{(i)}}{\sigma^{(i)}}.$$

However, in our framework this problem can be used only for finding the *structure* of optimal consumption. The consumption *volumes* must be found from problem (7.3).

Denote $S_k = (k+1)h$, the total budget for the first $k+1$ shoppings. Then by induction it is easy to see that $\langle p, \hat{x}_{k+1}\rangle(7.8)S_k$. Defining now

$$\tilde{\psi}_{k+1} \;=\; \tfrac{1}{k+1}\sum_{l=0}^{k}\psi(\lambda_l), \quad \tilde{s}_{k+1} \;=\; \tfrac{1}{S_k}s_{k+1}, \quad \tilde{x}_{k+1} \;=\; \tfrac{1}{S_k}\hat{x}_{k+1},$$

we can see that $\tilde{x}_{k+1} \in \Delta_n(p)$. Therefore

$$\tfrac{1}{S_k}\delta_k \;=\; \tilde{\psi}_{k+1} - u(\tilde{x}_{k+1}) \;\geq\; 0. \tag{7.12}$$

Thus, our goal is to present a behavioral strategy for updating the personal prices λ_k, which ensure $\frac{1}{S_k}\delta_k \to 0$ as $k \to \infty$.

For that, we need to introduce new variables $\xi^{(i)} = \frac{1}{\sigma^{(i)}}\lambda^{(i)}$, $i = 1,\dots,d$. If $\lambda \in \Delta_d(\sigma)$, then $\xi \in \Delta_d(e)$. Thus, the new objects have interpretation of probabilities, or *frequencies*.

The value $\xi_k^{(i)}$ has a sense of *relative importance* of quality i for the consumer after k iterations of the consumption process. We start from uniform distribution $\xi_0 = \frac{1}{d}e \in \Delta_d(e)$. For the next iterations, we apply the following rule:

$$\xi_{k+1}^{(i)} \;=\; \frac{\exp\left(-\frac{\bar{s}_{k+1}^{(i)}}{\mu_{k+1}\sigma^{(i)}}\right)}{\sum\limits_{m=1}^{d}\exp\left(-\frac{\bar{s}_{k+1}^{(m)}}{\mu_{k+1}\sigma^{(m)}}\right)}, \qquad i = 1,\dots,d. \tag{7.13}$$

This is a standard *Logit* model for detecting the maximal deficiency in the *average level* of consumption of qualities. It has a deviation parameter $\mu_k > 0$ and can be explained as follows.

Between two successive shoppings, we are trying to determine the most deficient quality in our consumption history. For that, we regularly perform subconscious inspections of the average level of their consumption. Since this a subconscious estimating process, its results are random. However, it is natural to assume that in our mind

> *the relative importance of each quality approaches the frequency of detecting its average consumption as the lowest one as compared to the standards of our life.*

These estimates, divided by the corresponding scale coefficients $\sigma^{(i)}$ become the prices of qualities for the next shopping.

All these computations are done, of course, subconsciously, without any visible efforts from the consumer. Similarly, we assume that our consumer can easily detect goods with the best quality-price ratio, which are understood as goods with *faire prices*.

It can be proved that with an appropriate strategy of tightening, (e.g. $\mu_k \approx \frac{1}{\sqrt{k+1}}$), the prices for qualities $\lambda_k^{(i)} = \xi_k^{(i)}/\sigma^{(i)}$ defined by (7.13) converge to the optimal solution of problem (7.6) (see Appendix). At the same time, the consumption bundles $\tilde{x}_k$ converge to the optimal solution of the dual problem in (7.11). Thus, our consumer is able to approach the optimal *structure* of the consumption. Playing now with the budget h, it is possible to ensure the necessary level σ for consumption of qualities.

Acknowledgement

The authors would like to thank the referee for his precise and constructive remarks.

Appendix. Primal-dual algorithm of optimal consumption

Let us present now a simple mathematical justification for numerical method for solving a scaled version of problem (7.6):

$$\min_{\xi \in \Delta_d(e)} \left\{ \phi(\xi) \stackrel{\mathrm{def}}{=} \psi(D^{-1}\xi) \right\},$$

where D is a diagonal matrix with vector σ at its diagonal. Note that $\nabla\phi(\xi) = D^{-1}\nabla\phi(D^{-1}\sigma)$. We denote $\lambda \equiv \lambda(\xi) = D^{-1}\xi$.

Let us introduce the entropy prox-function

$$\eta(\xi) \quad = \quad \sum_{i=1}^{d} \xi^{(i)} \ln \xi^{(i)}, \xi \in \Delta_d(e).$$

Note that this function is strongly convex on $\Delta_d(e)$ with respect to ℓ_1-norm with convexity parameter one.

Let us fix the step parameter $h > 0$. The primal-dual gradient method [**18**] looks as follows:

(7.14)

$$\xi_0 = \tfrac{1}{d}e, \quad \xi_{k+1} \quad = \quad \arg \min_{\xi \in \Delta_n(e)} \left\{ \sum_{l=0}^{k} \langle \nabla\phi(\xi_l), \xi - \xi_l \rangle + (k+1)\mu_k \eta(\xi) \right\}, \ k \geq 0,$$

where $\mu_k > 0$ are some scaling coefficients coefficients. Denote $\hat{s}_k = \sum\limits_{l=0}^{k-1} \nabla\phi(\xi_l)$ (thus, $\hat{s}_0 = 0$). Then, in accordance to the rule (7.14) we have

$$\xi_{k+1}^{(i)} \quad = \quad \frac{\exp\left(-\frac{\hat{s}_{k+1}^{(i)}}{(k+1)\mu_k} \right)}{\sum\limits_{m=1}^{d} \exp\left(-\frac{\hat{s}_{k+1}^{(m)}}{(k+1)\mu_k} \right)}, \quad i = 1, \ldots, d.$$

On the other hand,

$$\hat{s}_{k+1} \quad = \quad \sum_{l=0}^{k} D^{-1}\nabla\psi(\lambda_l),$$

where $\nabla\psi(\lambda_k) = \sum\limits_{j=1}^{n} \frac{\nu_k^{(j)}}{p^{(j)}} q_j \in \partial\psi(\lambda_k)$, and the sharing vectors $\nu_k \in \Delta_n(e)$ having elements $\nu_k^{(j)} = 0$ for $j \notin \left\{ j : \frac{\langle q_j, \lambda_k \rangle}{p^{(j)}} = \psi(\lambda_k) \right\}$. Thus, the process (7.14) constructs exactly the same sequence of the internal prices for qualities $\{\lambda_k\}$ as the behavioral strategy (7.7), (7.13). Moreover, it is easy to see that $hD\hat{s}_k = s_k$.

On the other hand, method (7.14) is a variant of the method of *dual averaging* (see (2.14) in [**18**]), with parameters

$$\lambda_k = 1, \quad \beta_{k+1} = (k+1)\mu_{k+1}, \quad k \geq 0.$$

In accordance, to Theorem 1 in [**18**], we have

$$\hat{\delta}_k \stackrel{\text{def}}{=} \max_{\xi \in \Delta_d} \sum_{l=0}^{k} \langle \nabla \phi(\xi_l), \xi_l - \xi \rangle \quad \leq \quad D(k+1)\mu_{k+1} + \tfrac{L^2}{2} \left(\tfrac{1}{\beta_0} + \sum_{l=1}^{k} \tfrac{1}{l\mu_l} \right),$$

where $L = \max\limits_{\xi \in \Delta_d(e)} \|\nabla \phi(\xi)\|_\infty = \max\limits_{i,j} \frac{1}{\sigma^{(i)}} Q^{(i,j)}$ and $D = \ln d$. Hence,

$$\tfrac{1}{k+1}\hat{\delta}_k \quad \leq \quad D\mu_{k+1} + \tfrac{L^2}{2(k+1)} \left(\tfrac{1}{\beta_0} + \sum_{l=1}^{k} \tfrac{1}{l\mu_l} \right).$$

Hence, if μ_k gradually goes to zero, then the right-hand side of this inequality vanishes. The best rate of convergence is achieved for $\mu_k \approx O\left(\frac{1}{\sqrt{k+1}}\right)$.

Finally, let us show that the above arguments ensure the convergence of the behavioral strategy (7.7), (7.13). Denote $\lambda_k = D^{-1}\xi_k$. Then $\nabla \phi(\xi_k) = D^{-1}\nabla\psi(\lambda_k)$, and we have

$$\begin{aligned} \sum_{l=0}^{k} \langle \nabla \phi(\xi_l), \xi_l - \xi \rangle \quad &= \quad \sum_{l=0}^{k} \langle D^{-1}\nabla\psi(\lambda_l), \xi_l - \xi \rangle \;=\; \sum_{l=0}^{k} \langle \nabla\psi(\lambda_l), D^{-1}\xi_l - D^{-1}\xi \rangle \\ &= \quad \sum_{l=0}^{k} \langle \nabla\psi(\lambda_l), \lambda_l - \lambda \rangle, \quad \xi \in \Delta_d(e), \quad \lambda = D^{-1}\xi \in \Delta_d(\sigma). \end{aligned}$$

Thus, $h\hat{\delta}_{k+1} = \delta_k$. It remains to use equation (7.12).

References

[1] L. Chen, Y. Ye, J. Zhang, *A Note on Equilibrium Pricing as Convex Optimization*, Internet and Network Economics, Lecture Notes in Computer Science, vol. 4858, 2007, pp. 7–16.

[2] Bruno Codenotti and Kasturi Varadarajan, *Efficient computation of equilibrium prices for markets with Leontief utilities*, Automata, languages and programming, Lecture Notes in Comput. Sci., vol. 3142, Springer, Berlin, 2004, pp. 371–382, DOI 10.1007/978-3-540-27836-8_33. MR2160947 (2006d:91130)

[3] Nikhil R. Devanur, Christos H. Papadimitriou, Amin Saberi, and Vijay V. Vazirani, *Market equilibrium via a primal-dual algorithm for a convex program*, J. ACM **55** (2008), no. 5, Art. 22, 18, DOI 10.1145/1411509.1411512. MR2456547 (2009k:91065)

[4] Jean-Claude Falmagne, *Elements of psychophysical theory*, Oxford Psychology Series, vol. 6, The Clarendon Press, Oxford University Press, New York, 1985. MR802072 (87b:92037)

[5] F. M. Fisher, *On Price Adjustment Without an Auctioneer*, The Review of Economic Studies **39** (1972), 1–15.

[6] David Gale, *The theory of linear economic models*, McGraw-Hill Book Co., Inc., New York-Toronto-London, 1960. MR0115801 (22 #6599)

[7] Douglas Gale, *Limit theorems for markets with sequential bargaining*, J. Econom. Theory **43** (1987), no. 1, 20–54, DOI 10.1016/0022-0531(87)90114-1. MR912000 (89c:90030)

[8] F. H. Hahn, T. Negishi, *A Theorem on Non-Tâtonnement Stability*, Econometrica **30** (1962), 463–469.

[9] A. Jofré, R. T. Rockafellar, and R. J.-B. Wets, *Convex analysis and financial equilibrium*, Math. Program. **148** (2014), no. 1-2, Ser. B, 223–239, DOI 10.1007/s10107-014-0747-3. MR3274851

[10] D. M. Kreps, *Microeconomic Foundations I:Choice and Competitive Markets*, Princeton University Press, Princeton, 2012.

[11] V. Krishna, H. Sonnenschein, *Duality in Consumer Theory*, in Chipman J., McFadden D., Richter M.: Prefereences, Uncertainty and Optimality, Westview Press, 1990, pp. 44–55.

[12] K. J. Lancaster, *Introduction to Modern Economics*, Rand McNally & Company, Chicago, 1972.

[13] A. Mas-Colell, M. D. Whinston, J. R. Green, *Microeconomic Theory*, Oxford University Press, New York, 1995.

[14] P. Milgrom, *Putting Auction Theory to Work: The Simultaneous Ascending Auction*, Journal of Political Economy **108** (2000), 245–272.

[15] Dale T. Mortensen and Randall Wright, *Competitive pricing and efficiency in search equilibrium*, Internat. Econom. Rev. **43** (2002), no. 1, 1–20, DOI 10.1111/1468-2354.t01-1-00001. MR1893436 (2002m:91027)

[16] S. Mullainathan, R. H. Thaler, *Behavioral Economics*, in International Encyclopedia of Social and Behavioral Sciences, Smelser, N. J., Baltes, P. B. (Editors-in-Chief), Elsevier, 2001.

[17] Yurii Nesterov, *Introductory lectures on convex optimization*, Applied Optimization, vol. 87, Kluwer Academic Publishers, Boston, MA, 2004. A basic course. MR2142598 (2005k:90001)

[18] Yurii Nesterov, *Primal-dual subgradient methods for convex problems*, Math. Program. **120** (2009), no. 1, Ser. B, 221–259, DOI 10.1007/s10107-007-0149-x. MR2496434 (2010e:90092)

[19] R. Tyrrell Rockafellar, *Convex analysis*, Princeton Mathematical Series, No. 28, Princeton University Press, Princeton, N.J., 1970. MR0274683 (43 #445)

[20] J. B. Rosen, *Existence and uniqueness of equilibrium points for concave n-person games*, Econometrica **33** (1965), 520–534. MR0194210 (33 #2423)

[21] A. Rubinstein, *Lecture Notes in Microeconomic Theory: The Economic Agent*, Princeton University Press, Princeton, 2012.

[22] Rubinstein, A., *Equilibrium in the jungle,* The Economic journal **117** (2007), 883–896.

[23] Ariel Rubinstein and Asher Wolinsky, *Equilibrium in a market with sequential bargaining*, Econometrica **53** (1985), no. 5, 1133–1150, DOI 10.2307/1911015. MR809907

[24] C. Zălinescu, *Convex analysis in general vector spaces*, World Scientific Publishing Co., Inc., River Edge, NJ, 2002. MR1921556 (2003k:49003)

Center for Operations Research and Econometrics (CORE), Catholic University of Louvain (UCL), 34 voie du Roman Pays, 1348 Louvain-la-Neuve, Belgium
E-mail address: `yurii.nesterov@uclouvain.be`

Center for Operations Research and Econometrics (CORE), Catholic University of Louvain (UCL), 34 voie du Roman Pays, 1348 Louvain-la-Neuve, Belgium
E-mail address: `vladimir.shikhman@uclouvain.be`

Contemporary Mathematics
Volume **659**, 2016
http://dx.doi.org/10.1090/conm/659/13161

On differential linear-quadratic games with mixed constraints

D. A. Schiro and J. S. Pang

ABSTRACT. This paper addresses solution existence and numerical algorithms for *symmetric* and *asymmetric* open-loop differential linear-quadratic Nash games. A sufficient condition is provided under which a symmetric game is equivalent to a certain concatenated linear-quadratic optimal control problem. This equivalent formulation facilitates the application of a time-stepping algorithm whose convergence to a continuous-time trajectory was established under certain conditions in a recent paper; the identified continuous-time trajectory represents a differential Nash equilibrium. While there is no such equivalence for an asymmetric game, a convergent distributed algorithm is described for computing a differential Nash equilibrium under a matrix-theoretic condition.

1. Introduction

In the mathematical programming literature, there has been extensive research on the "static" non-cooperative Nash equilibrium problem in recent years; two comprehensive surveys that contain the basic theory, iterative algorithms, and diverse applications in engineering and economics can be found in [**4**, **5**]. In a nutshell, such a problem involves a finite number of players each solving a finite-dimensional constrained optimization problem in which an objective function dependent on all players' decision variables is optimized subject to constraints that may be of the private, coupled, or shared type. While private constraints do not involve the decision variables of competing players, both coupled and shared constraints involve multiple players' decision variables and therefore lead to "generalized" Nash equilibrium problems which will not be dealt with in this paper. A Nash equilibrium is a set of decisions for all players in which no player can unilaterally alter their decision to increase their objective function value. In contrast, the continuous-time non-cooperative Nash game has at best been minimally studied from a constrained optimization perspective, especially in terms of the development of effective iterative algorithms for approximating a solution trajectory of the game and the convergence analysis of such algorithms.

The starting point of this work is the recent paper [**7**] that analyzes the convergence of a time-stepping method for solving a linear-quadratic (LQ) optimal control problem with mixed state-control constraints. Such a LQ optimal control

---

2010 *Mathematics Subject Classification.* Primary 49N70, 49M25, 49K15.

The work was based on research partially supported by the U.S. National Science Foundation grant CMMI 0969600.

problem is the building block of an open-loop differential linear-quadratic Nash game in which each of a finite number of selfish players solves a LQ optimal control problem with an objective function depending on rivals' variables and constrained by private, mixed state-control constraints. The main contribution of this work is to provide a constructive analysis of this differential LQ game by considering two cases:

• The **Symmetric** case. We provide a "symmetry" condition under which a differential LQ game with mixed state-control constraints is equivalent to a "concatenated" LQ optimal control problem; the implication of this equivalence is that all such differential LQ games can be solved by the convergent time-stepping algorithm described in [**7**].

• The **Asymmetric** case. We provide a "spectral radius" condition under which the differential LQ game can be solved by a distributed algorithm of the Jacobi or Gauss-Seidel type. Such a distributed algorithm has been used extensively for solving static Nash games arising from signal processing problems for communication; see e.g. [**10, 12–14, 16**]. A general convergence theory of such an algorithm for solving static Nash games is developed in [**5**], which is extended herein to a differential LQ game under the specified setting.

The remainder of the paper is organized in five sections. Section 2 formulates the LQ optimal control problem being solved by each player, develops a variational principle regarding solution optimality, and summarizes the assumptions and existence theorem presented in [**7**]. Section 3 proves the equivalence of a symmetric differential LQ game with a concatenated LQ optimal control problem, and Section 4 proves the convergence of a Jacobi-type iterative scheme for solving an asymmetric differential LQ game. Illustrative examples of both symmetric and asymmetric differential LQ games are presented in § 5. The paper is concluded in Section 6.

2. The Differential LQ Game

We consider the linear-quadratic $\mathcal{F}$-player noncooperative game on the time interval $[0, \mathcal{T}]$, where $\mathcal{T} < \infty$ is the finite horizon and $\mathcal{F}$ is the number of players of the game. Each player $i \in \{1, \cdots, \mathcal{F}\}$ chooses an absolutely continuous state function $x_i : [0, \mathcal{T}] \to \mathbb{R}^{n_i}$ and a bounded measurable (thus integrable) control function $u_i : [0, \mathcal{T}] \to \mathbb{R}^{m_i}$ for some positive integers n_i and m_i to solve a LQ optimal control problem; these state and control variables are constrained by a player-specific linear inequality system. We write $\mathbf{x} \triangleq (x_i)_{i=1}^{\mathcal{F}}$ and $\mathbf{u} \triangleq (u_i)_{i=1}^{\mathcal{F}}$ for all players' state and control variables, respectively; we also write $x_{-i} \triangleq (x_j)_{j \neq i}$ and $u_{-i} \triangleq (u_j)_{j \neq i}$ for the state and control tuples of the rival players $j \neq i$ of player i. In the formulation (2.1) below, $\xi_i \in \mathbb{R}^{n_i}$ is a given initial state, $c_i \in \mathbb{R}^{n_i}$, $W_{ii'} \in \mathbb{R}^{n_i \times n_{i'}}$, $P_{ii'} \in \mathbb{R}^{n_i \times n_{i'}}$, $Q_{ii'} \in \mathbb{R}^{n_i \times m_{i'}}$, $R_{ii'} \in \mathbb{R}^{m_i \times n_{i'}}$, $S_{ii'} \in \mathbb{R}^{m_i \times m_{i'}}$, $A_i \in \mathbb{R}^{n_i \times n_i}$, $B_i \in \mathbb{R}^{n_i \times m_i}$, $r_i \in \mathbb{R}^{n_i}$, $f_i \in \mathbb{R}^{s_i}$, $C_i \in \mathbb{R}^{s_i \times n_i}$, and $D_i \in \mathbb{R}^{s_i \times m_i}$ for some nonnegative integer s_i, $p_i : [0, \mathcal{T}] \to \mathbb{R}^{n_i}$, and $q_i : [0, \mathcal{T}] \to \mathbb{R}^{m_i}$. The matrices W_{ii}, P_{ii}, and S_{ii} are symmetric for all players i. Taking (x_{-i}, u_{-i}) as constant, each player i solves the following continuous-time optimal control problem:

$$\underset{x_i,u_i}{\text{minimize}} \quad \theta_i(x_i,x_{-i},u_i,u_{-i}) \triangleq x_i(\mathcal{T})^T\left(c_i+\sum_{i'=1}^{\mathcal{F}} W_{ii'}x_{i'}(\mathcal{T})\right)+$$

$$\int_0^{\mathcal{T}}\begin{pmatrix} x_i(t)\\ u_i(t)\end{pmatrix}^T\left(\begin{pmatrix} p_i(t)\\ q_i(t)\end{pmatrix}+\sum_{i'=1}^{\mathcal{F}}\begin{bmatrix} P_{ii'} & Q_{ii'}\\ R_{ii'} & S_{ii'}\end{bmatrix}\begin{pmatrix} x_{i'}(t)\\ u_{i'}(t)\end{pmatrix}\right)dt \tag{2.1}$$

$$\begin{aligned}
\text{subject to}\quad & x_i(0) = \xi_i \text{ and for almost all } t\in[0,\mathcal{T}]:\\
& \dot{x}_i(t) = r_i + A_ix_i(t)+B_iu_i(t)\\
\text{and}\quad & f_i + C_ix_i(t)+D_iu_i(t) \geq 0.
\end{aligned}$$

A *feasible solution* of (2.1) is a pair (x_i,u_i) such that x_i is absolutely continuous, u_i is integrable, and the constraints of (2.1) are satisfied as stated. A noteworthy feature of the problem (2.1) is the mixed state-control constraint $f_i + C_ix_i(t) + D_iu_i(t) \geq 0$; throughout the paper, we permit the matrix C_i to equal zero (i.e., pure control constraints). However, a key condition to be imposed for the solvability of this problem yields a special relation among the matrices (A_i,B_i,C_i,D_i); see condition (E$_i$) in § 2.1. An aggregated pair $(\mathbf{x}^*,\mathbf{u}^*)$, where $\mathbf{x}^* \triangleq (x_i^*)_{i=1}^{\mathcal{F}}$ and $\mathbf{u}^* \triangleq (u_i^*)_{i=1}^{\mathcal{F}}$, is a *Nash equilibrium* (NE) of the above game if for each $i = 1,\cdots,\mathcal{F}$,

$$(x_i^*,u_i^*) \in \underset{(x_i,u_i)}{\text{argmin}} \quad \theta_i(x_i,x_{-i}^*,u_i,u_{-i}^*) \quad \text{subject to} \quad (x_i,u_i) \text{ feasible to (2.1).} \tag{2.2}$$

2.1. Properties of (2.1): A synopsis. For player i, suppose (2.2) holds for an arbitrary pair (x_{-i},u_{-i}). Let (x_i,u_i) be a feasible pair to (2.1) and let $(x_i^\tau,u_i^\tau) \triangleq (x_i^*,u_i^*)+\tau(x_i-x_i^*,u_i-u_i^*)$ for $\tau\in[0,1]$. Since (x_i^τ,u_i^τ) is also feasible to (2.1), we have

$$\begin{aligned}
0 &\leq \theta_i(x_i^\tau,x_{-i},u_i^\tau,u_{-i})-\theta_i(x_i^*,x_{-i},u_i^*,u_{-i})\\
&= \tau\left(x_i(\mathcal{T})-x_i^*(\mathcal{T})\right)\left(c_i+2W_{ii}x_i^*(\mathcal{T})+\sum_{i'\neq i}W_{ii'}x_{i'}(\mathcal{T})\right)+\\
&\quad \tau^2\left(x_i(\mathcal{T})-x_i^*(\mathcal{T})\right)^TW_{ii}\left(x_i(\mathcal{T})-x_i^*(\mathcal{T})\right)+\tau\int_0^{\mathcal{T}}\begin{pmatrix} x_i(t)-x_i^*(t)\\ u_i(t)-u_i^*(t)\end{pmatrix}^T\left(\begin{pmatrix} p_i(t)\\ q_i(t)\end{pmatrix}\right.\\
&\quad \left.+2\begin{bmatrix} P_{ii} & \frac{1}{2}\left(Q_{ii}+R_{ii}^T\right)\\ \frac{1}{2}\left(Q_{ii}^T+R_{ii}\right) & S_{ii}\end{bmatrix}\begin{pmatrix} x_i^*(t)\\ u_i^*(t)\end{pmatrix}+\sum_{i'\neq i}\begin{bmatrix} P_{ii'} & Q_{ii'}\\ R_{ii'} & S_{ii'}\end{bmatrix}\begin{pmatrix} x_{i'}(t)\\ u_{i'}(t)\end{pmatrix}\right)dt\\
&\quad +\tau^2\int_0^{\mathcal{T}}\begin{pmatrix} x_i(t)-x_i^*(t)\\ u_i(t)-u_i^*(t)\end{pmatrix}^T\begin{bmatrix} P_{ii} & \frac{1}{2}\left(Q_{ii}+R_{ii}^T\right)\\ \frac{1}{2}\left(Q_{ii}^T+R_{ii}\right) & S_{ii}\end{bmatrix}\begin{pmatrix} x_i(t)-x_i^*(t)\\ u_i(t)-u_i^*(t)\end{pmatrix}dt.
\end{aligned}$$

Consequently, a necessary condition for (2.2) to hold with an arbitrary pair (x_{-i}, u_{-i}) is that

$$(x_i(\mathcal{T}) - x_i^*(\mathcal{T}))^T \left(c_i + 2W_{ii}x_i^*(\mathcal{T}) + \sum_{i' \neq i} W_{ii'}x_{i'}(\mathcal{T}) \right)$$

$$+\int_0^{\mathcal{T}} \begin{pmatrix} x_i(t) - x_i^*(t) \\ u_i(t) - u_i^*(t) \end{pmatrix}^T \left(\begin{pmatrix} p_i(t) \\ q_i(t) \end{pmatrix} + \sum_{i' \neq i} \begin{bmatrix} P_{ii'} & Q_{ii'} \\ R_{ii'} & S_{ii'} \end{bmatrix} \begin{pmatrix} x_{i'}(t) \\ u_{i'}(t) \end{pmatrix} \right. \tag{2.3}$$

$$\left. + 2 \begin{bmatrix} P_{ii} & \frac{1}{2}\left(Q_{ii} + R_{ii}^T\right) \\ \frac{1}{2}\left(Q_{ii}^T + R_{ii}\right) & S_{ii} \end{bmatrix} \begin{pmatrix} x_i^*(t) \\ u_i^*(t) \end{pmatrix} \right) dt \geq 0$$

for all pairs (x_i, u_i) feasible to (2.1), which is a variational principle for optimality to this optimal control problem.

For notational convenience, let

$$\Xi_{ii'} \triangleq \begin{cases} 2 \begin{bmatrix} P_{ii} & \frac{1}{2}\left(Q_{ii} + R_{ii}^T\right) \\ \frac{1}{2}\left(Q_{ii}^T + R_{ii}\right) & S_{ii} \end{bmatrix} & \text{if } i = i' \\ \begin{bmatrix} P_{ii'} & Q_{ii'} \\ R_{ii'} & S_{ii'} \end{bmatrix} & \text{if } i \neq i'. \end{cases}$$

In general, $\Xi_{ii'} \neq \Xi_{i'i}^T$ for $i \neq i'$, reflecting the asymmetric impact of the strategy of player i' on player i's objective function and vice versa. The case where symmetry between these matrices holds is addressed in § 3.

In [**7**], a set of postulates was introduced under which a LQ optimal control problem such as (2.1) is shown to have a solution; such a solution is obtained as the limit of a sequence of piecewise trajectories formed by interpolating the discrete-time iterates computed from a time-stepping procedure. Except for postulate (E_i), the other postulates provided in the reference are fairly mild.

In what follows, we summarize the results in [**7**] for the problem (2.1) by starting with the assumptions made throughout the paper:

($\mathbf{A}_i$) the matrices W_{ii} and Ξ_{ii} are symmetric positive semidefinite;
($\mathbf{B}_i$) the functions $p_i(t)$ and $q_i(t)$ are Lipschitz continuous on $[0, \mathcal{T}]$;
($\mathbf{C}_i$) a continuously differentiable function $\widehat{x}_i^{\text{fs}}$ with $\widehat{x}_i^{\text{fs}}(0) = \xi_i$ and a continuous function $\widehat{u}_i^{\text{fs}}$ exist such that for all $t \in [0, \mathcal{T}]$: $d\widehat{x}_i^{\text{fs}}(t)/dt = r_i + A_i\widehat{x}_i^{\text{fs}}(t) + B_i\widehat{u}_i^{\text{fs}}(t)$ and $\widehat{u}_i^{\text{fs}}(t) \in U_i(\widehat{x}_i^{\text{fs}}(t))$, where

$$U_i(x_i) \triangleq \{ u \in \mathbb{R}^{m_i} \mid f_i + C_i x_i + D_i u_i \geq 0 \};$$

($\mathbf{D}_i$) $[S_{ii}u_i = 0,\ D_i u_i \geq 0]$ implies $u_i = 0$ (a primal condition);
($\mathbf{E}_i$) $[D_i^T \mu_i = 0,\ \mu_i \geq 0]$ implies $(C_i A_i^j B_i)^T \mu_i = 0$ for all integers $j = 0, \cdots, n_i - 1$, or equivalently, for all nonnegative integers j (a dual condition).

To derive the necessary and sufficient optimality conditions for the LQ optimal control problem (2.1), we start with the definition of the Hamiltonian function [t is removed for notational convenience] $H_i(x, u, \lambda_i)$:

$$\begin{pmatrix} x_i \\ u_i \end{pmatrix}^T \left(\begin{pmatrix} p_i \\ q_i \end{pmatrix} + \sum_{i'=1}^{\mathcal{F}} \begin{bmatrix} P_{ii'} & Q_{ii'} \\ R_{ii'} & S_{ii'} \end{bmatrix} \begin{pmatrix} x_{i'} \\ u_{i'} \end{pmatrix} \right) + \lambda_i^T \left(r_i + A_i x_i + B_i u_i \right),$$

where λ_i is the vector of costate (also called adjoint) variables of the ordinary differential equation $\dot{x}_i = r_i + A_i x_i + B_i u$. The Lagrangian function for the optimal control problem is therefore

$$L_i(x, u, \lambda_i, \mu_i) \triangleq H_i(x, u, \lambda_i) - \mu_i^T \left(f_i + C_i x_i + D_i u_i \right),$$

where μ_i is the vector of Lagrange multipliers for $f_i + C_i x_i + D_i u_i \geq 0$. The (boundary-value) differential affine variational inequality (DAVI) associated with the LQ problem (2.1) is

$$\begin{pmatrix} \dot{\lambda}_i(t) \\ \dot{x}_i(t) \end{pmatrix} = \begin{pmatrix} -p_i(t) - \sum_{i' \neq i} \left(P_{ii'} x_{i'}(t) + Q_{ii'} u_{i'}(t) \right) \\ r_i \end{pmatrix} +$$

$$\begin{bmatrix} -A_i^T & -2\,P_{ii} \\ 0 & A_i \end{bmatrix} \begin{pmatrix} \lambda_i(t) \\ x_i(t) \end{pmatrix} + \begin{bmatrix} -\left[Q_{ii} + R_{ii}^T \right] \\ B_i \end{bmatrix} u_i(t) + \begin{bmatrix} C_i^T \\ 0 \end{bmatrix} \mu_i(t)$$

$$\text{(2.4)} \quad \left\{ \begin{array}{l} 0 = q_i(t) + \sum_{i' \neq i} \left(R_{ii'} x_{i'}(t) + S_{ii'} u_{i'}(t) \right) + \left[Q_{ii}^T + R_{ii} \right] x_i(t) \\ \qquad\qquad + 2\,S_{ii} u_i(t) + B_i^T \lambda_i(t) - D_i^T \mu_i(t) \\ 0 \leq \mu_i(t) \perp f_i + C_i x_i(t) + D_i u_i(t) \geq 0 \end{array} \right\}$$

$$\left[\begin{array}{c} \text{the two bracketed conditions imply that} \\ u_i(t) \in \underset{u_i \in U_i(x_i(t))}{\text{argmin}} \; H_i(x(t), u_i, u_{-i}(t), \lambda_i(t)) \end{array} \right]$$

$$x_i(0) = \xi_i \quad \text{and} \quad \lambda_i(\mathcal{T}) = c_i + 2\,W_{ii} x_i(\mathcal{T}) + \sum_{i' \neq i} W_{ii'} x_{i'}(\mathcal{T}).$$

As a parameterized DAVI with (x_i, λ_i) as the pair of differential variables, (u_i, μ_i) as the pair of algebraic variables, and (x_{-i}, u_{-i}) as parameters, the tuple $(x_i, u_i, \lambda_i, \mu_i)$ is a weak solution of (2.4) in the sense of Carathéodory if (i) (x_i, λ_i) is absolutely continuous and (u_i, μ_i) is square-integrable on $[0, \mathcal{T}]$, (ii) the differential equation and the two algebraic conditions hold for almost all $t \in (0, \mathcal{T})$, and (iii) the initial and boundary conditions are satisfied.

With detailed proofs given in [**7**], Theorem 2.1 below summarizes the main properties of the optimal control problem (2.1), the DAVI (2.4), and their relationship. The theorem starts with the postulates (A_i–E_i), under which part (I) asserts the existence of a weak solution of the DAVI (2.4) that is proved by a constructive numerical method. Part (II) of Theorem 2.1 states that any weak solution of the DAVI (2.4) is an optimal solution of (2.1), thus proving sufficiency of the Pontryagin optimality principle [**1, 2, 6, 11, 15**]. Part (III) asserts several properties of an optimal solution of (2.1); this part is analogous to well-known results for a finite-dimensional convex QP. From these properties, part (IV) proves the reverse implication between solutions of optimal control problem (2.1) and DAVI (2.4), namely that any optimal solution of (2.1) must be a weak solution of the DAVI (2.4). An alternative characterization of optimality in terms of the variational inequality (2.3) is stated in part (V). Since this part was not explicitly proved in [**7**], we give a proof below. Finally, part (VI) asserts solution uniqueness from part (I) under the positive definiteness of the matrix S_{ii}.

THEOREM 2.1. *Under conditions* (A_i–E_i), *the following statements* (I–VI) *hold for arbitrary integrable pair* (x_{-i}, u_{-i}).

(I: Solvability of the DAVI) *The DAVI* (2.4) *has a weak solution* $(x_i^*, u_i^*, \lambda_i^*, \mu_i^*)$, *provided that the pair* (x_{-i}, u_{-i}) *is Lipschitz continuous.*
(II: Sufficiency of Pontryagin) *If* $(x_i^*, u_i^*, \lambda_i^*, \mu_i^*)$ *is any weak solution of* (2.4), *then the pair* (x_i^*, u_i^*) *is an optimal solution of the problem* (2.1).
(III: Gradient characterization of optimal solutions) *If* $(\widehat{x}_i, \widehat{u}_i)$ *and* $(\widetilde{x}_i, \widetilde{u}_i)$ *are any two optimal solutions of* (2.1), *the following three properties hold:*

(a) *for almost all* $t \in [0, \mathcal{T}]$,

$$\begin{bmatrix} P_{ii} & \frac{1}{2}(Q_{ii} + R_{ii}^T) \\ \frac{1}{2}(Q_{ii}^T + R_{ii}) & S_{ii} \end{bmatrix} \begin{pmatrix} \widehat{x}_i(t) - \widetilde{x}_i(t) \\ \widehat{u}_i(t) - \widetilde{u}_i(t) \end{pmatrix} = 0;$$

(b) $W_{ii}\widehat{x}_i(\mathcal{T}) = W_{ii}\widetilde{x}_i(\mathcal{T})$;

(c) $0 = (\widehat{x}_i(\mathcal{T}) - \widetilde{x}_i(\mathcal{T}))^T \left(c_i + \sum_{i' \neq i} W_{ii'} x_{i'}(\mathcal{T}) \right) +$

$$\int_0^T \begin{pmatrix} \widehat{x}_i(t) - \widetilde{x}_i(t) \\ \widehat{u}_i(t) - \widetilde{u}_i(t) \end{pmatrix}^T \left(\begin{pmatrix} p_i(t) \\ q_i(t) \end{pmatrix} + \sum_{i' \neq i} \begin{bmatrix} P_{ii'} & Q_{ii'} \\ R_{ii'} & S_{ii'} \end{bmatrix} \begin{pmatrix} x_{i'}(t) \\ u_{i'}(t) \end{pmatrix} \right) dt;$$

moreover, given any optimal solution $(\widehat{x}_i, \widehat{u}_i)$ *of* (2.1), *a feasible tuple* $(\widetilde{x}_i, \widetilde{u}_i)$ *of* (2.1) *is optimal if and only if conditions* (a), (b), *and* (c) *hold.*

(IV: Necessity of Pontryagin) *Let* $(x_i^*, u_i^*, \lambda_i^*, \mu_i^*)$ *be the tuple obtained from part* (I). *A feasible tuple* $(\widetilde{x}_i, \widetilde{u}_i)$ *of* (2.1) *is optimal if and only if* $(\widetilde{x}_i, \widetilde{u}_i, \lambda_i^*, \mu_i^*)$ *is a weak solution of* (2.4).

(V: Variational characterization of optimality) *The optimal solution* (x_i^*, u_i^*) *is characterized by the variational principle* (2.3) *holding for all pairs* (x_i, u_i) *feasible to* (2.1).

(VI: Uniqueness under positive definiteness) *If* S_{ii} *is positive definite and the pair* (x_{-i}, u_{-i}) *is Lipschitz continuous, then* (2.1) *has a unique optimal solution* $(\widehat{x}_i, \widehat{u}_i)$ *such that* $\widehat{x}_i$ *is continuously differentiable and* $\widehat{u}_i$ *is Lipschitz continuous on* $[0, \mathcal{T}]$; *moreover, for any optimal* $\widehat{\lambda}_i$ *and all* $t \in [0, \mathcal{T}]$,
$\widehat{u}_i(t) \in \underset{u_i \in U_i(\widehat{x}_i(t))}{\text{argmin}} \; H_i(\widehat{x}_i(t), x_{-i}(t), u_i, u_{-i}(t), \widehat{\lambda}_i(t))$.

PROOF. The derivation of (2.3) has proved the necessity. The sufficiency follows easily from the positive semidefinitess assumption (A_i). Incidentally, the validity of this part requires only condition (A_i) and not the others. □

3. The Symmetric Case

Theorem 2.1 pertains only to the individual players' optimal control problems and says nothing about the game that is the equilibration of these problems. In what follows, under an additional symmetry condition, we show that this inequality constrained differential LQ game is equivalent to a single concatenated optimal control problem that is under the scope of applicability of Theorem 2.1. For this purpose, we define aggregate vector $\mathbf{c} \triangleq (c_i)_{i=1}^{\mathcal{F}}$ with similar definitions holding for p and q and aggregate matrix $\mathbf{W} \triangleq \left[[W_{ii'}]_{i,i'=1}^{\mathcal{F}} + \text{diag}(W_{ii})_{i=1}^{\mathcal{F}} \right]$ with similar definitions for

$$\begin{bmatrix} \mathbf{P} & \mathbf{Q} \\ \mathbf{R} & \mathbf{S} \end{bmatrix} \triangleq \begin{bmatrix} [P_{ii'}]_{i,i'=1}^{\mathcal{F}} + \operatorname{diag}(P_{ii})_{i=1}^{\mathcal{F}} & [Q_{ii'}]_{i,i'=1}^{\mathcal{F}} + \operatorname{diag}(Q_{ii})_{i=1}^{\mathcal{F}} \\ [R_{ii'}]_{i,i'=1}^{\mathcal{F}} + \operatorname{diag}(R_{ii})_{i=1}^{\mathcal{F}} & [S_{ii'}]_{i,i'=1}^{\mathcal{F}} + \operatorname{diag}(S_{ii})_{i=1}^{\mathcal{F}} \end{bmatrix}.$$

We also define $\boldsymbol{\Xi} \triangleq [\Xi_{ii'}]_{i,i'=1}^{\mathcal{F}}$. Using this notation, define the aggregated LQ optimal control problem in the variables $(\mathbf{x}, \mathbf{u})$:

$$\begin{aligned} &\underset{\mathbf{x},\mathbf{u}}{\text{minimize}} \quad \mathbf{x}(\mathcal{T})^T \left(\mathbf{c} + \tfrac{1}{2}\mathbf{W}\mathbf{x}(\mathcal{T}) \right) + \\ &\int_0^{\mathcal{T}} \begin{pmatrix} \mathbf{x}(t) \\ \mathbf{u}(t) \end{pmatrix}^T \left(\begin{pmatrix} \mathbf{p}(t) \\ \mathbf{q}(t) \end{pmatrix} + \frac{1}{2} \begin{bmatrix} \mathbf{P} & \mathbf{Q} \\ \mathbf{R} & \mathbf{S} \end{bmatrix} \begin{pmatrix} \mathbf{x}(t) \\ \mathbf{u}(t) \end{pmatrix} \right) dt \\ &\text{subject to} \quad \text{for all } i \in \{1, \cdots, \mathcal{F}\}: \\ &\qquad \dot{x}_i(t) = r_i + A_i x_i(t) + B_i u_i(t) \quad \text{for almost all } t \in [0, \mathcal{T}] \\ &\qquad f_i + C_i x_i(t) + D_i u_i(t) \geq 0 \quad \text{for almost all } t \in [0, \mathcal{T}] \\ &\text{and} \qquad x_i(0) = \xi_i. \end{aligned} \tag{3.1}$$

Note that the matrix $\begin{bmatrix} \mathbf{P} & \mathbf{Q} \\ \mathbf{R} & \mathbf{S} \end{bmatrix}$ in the objective function is a principal rearrangement of $\boldsymbol{\Xi}$, thereby implying that properties of $\boldsymbol{\Xi}$ such as symmetry and positive semidefiniteness will hold for $\begin{bmatrix} \mathbf{P} & \mathbf{Q} \\ \mathbf{R} & \mathbf{S} \end{bmatrix}$. Lastly, we aggregate the assumptions (A_i) through (E_i) to obtain a set of conditions pertaining to the above single LQ optimal control problem.

(A) the matrices $\mathbf{W}$ and $\boldsymbol{\Xi}$ are symmetric positive semidefinite;
(B) condition (B_i) holds for all $i \in \{1, \cdots, \mathcal{F}\}$;
(C) condition (C_i) holds for all $i \in \{1, \cdots, \mathcal{F}\}$;
(D) $[\mathbf{Su} = 0$ and $D_i u_i \geq 0$ for all $i \in \{1, \cdots, \mathcal{F}\}]$ implies

$[u_i = 0$ for all $i \in \{1, \cdots, \mathcal{F}\}]$;

(E) condition (E_i) holds for all $i \in \{1, \cdots, \mathcal{F}\}$.

Under the above conditions, we have the following result for the "symmetric" differential LQ game.

THEOREM 3.1. *Under condition* (A), *the following statements hold:*

(I: Equivalence) *A pair* $(\mathbf{x}^*, \mathbf{u}^*)$ *is a NE if and only if* $(\mathbf{x}^*, \mathbf{u}^*)$ *is an optimal solution of* (3.1);
(II: Existence) *Under the additional assumptions* (B–E), *a NE exists such that* $\mathbf{x}^*$ *is absolutely continuous and* $\mathbf{u}^*$ *is square-integrable on* $[0, \mathcal{T}]$;
(III: Uniqueness) *If* $\mathbf{S}$ *is positive definite in addition to* (A–E), *then* $(\mathbf{x}^*, \mathbf{u}^*)$ *is the unique NE such that* $\mathbf{x}^*$ *is continuously differentiable and* $\mathbf{u}^*$ *is Lipschitz continuous on* $[0, \mathcal{T}]$.

PROOF. (I) This follows from the following facts under assumption (A): (i) the pair $(\mathbf{x}^*, \mathbf{u}^*)$ is a NE if and only if the following variational property, which is an aggregation of (2.3),

$$\sum_{i=1}^{\mathcal{F}} (x_i(\mathcal{T}) - x_i^*(\mathcal{T}))^T \left(c_i + 2W_{ii}x_i^*(\mathcal{T}) + \sum_{i' \neq i} W_{ii'}x_{i'}^*(\mathcal{T}) \right) +$$

$$\sum_{i=1}^{\mathcal{F}} \int_0^{\mathcal{T}} \begin{pmatrix} x_i(t) - x_i^*(t) \\ u_i(t) - u_i^*(t) \end{pmatrix}^T \left(\begin{pmatrix} p_i(t) \\ q_i(t) \end{pmatrix} + \sum_{i' \neq i} \begin{bmatrix} P_{ii'} & Q_{ii'} \\ R_{ii'} & S_{ii'} \end{bmatrix} \begin{pmatrix} x_{i'}^*(t) \\ u_{i'}^*(t) \end{pmatrix} \right. \tag{3.2}$$

$$\left. + 2 \begin{bmatrix} P_{ii} & \frac{1}{2}\left(Q_{ii} + R_{ii}^T\right) \\ \frac{1}{2}\left(Q_{ii}^T + R_{ii}\right) & S_{ii} \end{bmatrix} \begin{pmatrix} x_i^*(t) \\ u_i^*(t) \end{pmatrix} \right) dt \geq 0$$

holds for all pairs $(\mathbf{x}, \mathbf{u})$ feasible to (3.1); and (ii) the variational condition (3.2) is necessary and sufficient for the pair $(\mathbf{x}^*, \mathbf{u}^*)$ to be optimal for the LQ optimal control problem (3.1). Assertions (II) and (III) follow from the theory in [**7**] applied to (3.1) and the equivalence in (I). □

A consequence of the equivalence between the differential LQ game and the single LQ optimal control problem is that we can apply the numerical time-stepping method presented in [**7**] to solve the former game problem. This method consists of solving convex quadratic subprograms obtained by discretizing (3.1) in time; the convergence of the method provides a constructive proof for the existence result in part (I) of Theorem 3.1. Details are omitted and can be found in the cited reference.

4. The Asymmetric Case

We next consider the asymmetric case where the matrix Ξ is not necessarily symmetric. We replace the symmetry assumption by a spectral condition on a certain condensed matrix obtained from Ξ. Under this matrix-theoretic spectral condition, we show that a distributed algorithm will converge to the unique NE of the differential LQ game. We simplify the analysis by making assumption (W) (see below) that removes the dependence of the rivals' terminal states in each player's objective function, i.e., the sum $\sum_{i' \neq i} W_{ii'}x_{i'}(\mathcal{T})$ is removed. The analysis can be extended to handle the case where some $W_{ii'}$ are nonzero; in this case, the corresponding diagonal block W_{ii} is required to be positive definite with its smallest eigenvalue "dominating" the sum of the nonzero off-diagonal blocks, roughly speaking.

The treatment in this section is an extension of that for the static Nash game whose details can be found in [**5**]. Unlike the algorithm for the latter game that solves finite-dimensional optimization problems, the algorithm below solves a sequence of LQ optimal control problems, which in turn need to be discretized for numerical implementation by, for example, the time-stepping method described in [**7**].

In addition to assumption (W), we need two more assumptions in this section, $(\widehat{\text{A}})$ and $(\widehat{\text{D}})$, which we state below. Assumption $(\widehat{\text{A}})$ assumes that the diagonal blocks Ξ_{ii} are positive definite matrices, albeit not necessarily symmetric. This assumption implies (C_i) for all $i = 1, \cdots, \mathcal{F}$. Assumption $(\widehat{\text{D}})$ yields the uniform boundedness Lemma 4.1 through a proof similar to that of Gronwall's lemma.

($\widehat{\mathbf{A}}$) For all $i = 1, \cdots, \mathcal{F}$, the matrices Ξ_{ii} are positive definite with minimum eigenvalues $\sigma_i^{\Xi} > 0$; the matrices W_{ii} remain (symmetric) positive semidefinite;
(W) For all $i = 1, \cdots, \mathcal{F}$, the matrices $W_{ii'} = 0$ for all $i' \neq i$.
($\widehat{\mathbf{D}}$) For all $i = 1, \cdots, \mathcal{F}$, the following implication holds: $D_i u_i \geq 0 \Rightarrow u_i = 0$.

LEMMA 4.1. *Under assumption* ($\widehat{\text{D}}$), *there exists a constant* $\zeta > 0$ *such that for all* $i = 1, \cdots, \mathcal{F}$,

$$\sup_{t \in [0,\mathcal{T}]} \| x_i(t) \| \leq \zeta \quad \textit{and} \quad \operatorname*{ess\,sup}_{t \in [0,\mathcal{T}]} \| u_i(t) \| \leq \zeta$$

for any trajectory (x_i, u_i) *feasible to* (2.1).

PROOF. Given ($\widehat{\text{D}}$) and the constraint $f_i + C_i x_i(t) + D_i u_i(t) \geq 0$ holding for almost all $t \in [0, \mathcal{T}]$, there exist positive α_i and $\widehat{\alpha}_i$ such that for all feasible (x_i, u_i),

$$\|u_i(t)\| \leq \alpha_i \|f_i + C_i x_i(t)\| \leq \widehat{\alpha}_i (1 + \|x_i(t)\|) \tag{4.1}$$

holds when $u_i(t)$ is well-defined. Now consider that for all $t \in [0, \mathcal{T}]$ with $\dot{x}_i(t)$ well-defined,

$$\begin{aligned} \frac{d}{dt}\|x_i(t)\| \leq \|\dot{x}_i(t)\| &= \|r_i + A_i x_i(t) + B_i u_i(t)\| \\ &\leq \|r_i\| + \|A_i\| \|x_i(t)\| + \|B_i\| \widehat{\alpha}_i (1 + \|x_i(t)\|) \\ &\leq \beta_i (1 + \|x_i(t)\|) \end{aligned}$$

where $\beta_i > 0$. With manipulations similar to the proof of Gronwall's lemma,

$$-\beta_i \|x_i(t)\| + \frac{d}{dt}\|x_i(t)\| \leq \beta_i, \text{ implying}$$

$$\frac{d}{dt}\left(e^{-\beta_i t}\|x_i(t)\|\right) \leq \beta_i e^{-\beta_i t};$$

$$\text{thus,} \quad \|x_i(t)\| \leq e^{\beta_i t}\left(\|\xi_i\| + \int_0^t \beta_i e^{-\beta_i s} ds\right).$$

It follows that a constant $\zeta' > 0$ exists such that $\|x_i(t)\| \leq \zeta'$ for all $t \in [0, \mathcal{T}]$. The existence of the desired constant ζ follows readily from (4.1). □

Being a fixed-point iteration applied to the best-response map of the game, the algorithm for solving the asymmetric differential LQ Nash game generates a sequence of functional iterates as follows. Given a pair of state-control trajectories (x^ν, u^ν) at the beginning of iteration $\nu + 1$, where x^ν is continuously differentiable and u^ν is Lipschitz continuous, we compute the next pair of such trajectory $(x^{\nu+1}, u^{\nu+1})$ by solving $\mathcal{F}$ LQ optimal control problems (2.1), where for $i = 1, \cdots, \mathcal{F}$, the i-th such LQ problem solves for the pair $(x_i^{\nu+1}, u_i^{\nu+1})$ from (2.1) by fixing (x_j, u_j) at (x_j^ν, u_j^ν) for all $j \neq i$. This is the *Jacobi* version of the method. The convergence of this method in the differential context has not been analyzed before; such an analysis is the focus of the rest of the paper. The analysis of the *Gauss-Seidel* version is the same and omitted.

Remarks. The Lipschitz continuity of (x_j^ν, u_j^ν) for $j \neq i$ is needed so that condition (B_i) can be satisfied for the computation of the next pair $(x_i^{\nu+1}, u_i^{\nu+1})$. In turn, the positive definiteness of Ξ_{ii} ensures the preservation of the Lipschitz property of $(x_i^{\nu+1}, u_i^{\nu+1})$, thus the entire sequence of such iterates. Thus under assumptions

($\widehat{\text{A}}$), (B), (W), ($\widehat{\text{D}}$), and (E), the sequence of iterates $\{(x^\nu, u^\nu)\}$ is well defined with x^ν being continuously differentiable and u^ν being Lipschitz continuous on $[0, \mathcal{T}]$ for all ν. $\square$

From the variational inequality (2.3), we have

$$\left(x_i^\nu(\mathcal{T}) - x_i^{\nu+1}(\mathcal{T}) \right)^T \left(c_i + 2W_{ii} x_i^{\nu+1}(\mathcal{T}) \right) +$$

$$\int_0^{\mathcal{T}} \begin{pmatrix} x_i^\nu(t) - x_i^{\nu+1}(t) \\ u_i^\nu(t) - u_i^{\nu+1}(t) \end{pmatrix}^T \left(\begin{pmatrix} p_i(t) \\ q_i(t) \end{pmatrix} + \sum_{i' \neq i} \begin{bmatrix} P_{ii'} & Q_{ii'} \\ R_{ii'} & S_{ii'} \end{bmatrix} \begin{pmatrix} x_{i'}^\nu(t) \\ u_{i'}^\nu(t) \end{pmatrix} \right.$$

$$\left. + 2 \begin{bmatrix} P_{ii} & \frac{1}{2} \left(Q_{ii} + R_{ii}^T \right) \\ \frac{1}{2} \left(Q_{ii}^T + R_{ii} \right) & S_{ii} \end{bmatrix} \begin{pmatrix} x_i^{\nu+1}(t) \\ u_i^{\nu+1}(t) \end{pmatrix} \right) dt \geq 0$$

and

$$\left(x_i^{\nu+1}(\mathcal{T}) - x_i^\nu(\mathcal{T}) \right)^T \left(c_i + 2W_{ii} x_i^\nu(\mathcal{T}) \right) +$$

$$\int_0^{\mathcal{T}} \begin{pmatrix} x_i^{\nu+1}(t) - x_i^\nu(t) \\ u_i^{\nu+1}(t) - u_i^\nu(t) \end{pmatrix}^T \left(\begin{pmatrix} p_i(t) \\ q_i(t) \end{pmatrix} + \sum_{i' \neq i} \begin{bmatrix} P_{ii'} & Q_{ii'} \\ R_{ii'} & S_{ii'} \end{bmatrix} \begin{pmatrix} x_{i'}^{\nu-1}(t) \\ u_{i'}^{\nu-1}(t) \end{pmatrix} \right.$$

$$\left. + 2 \begin{bmatrix} P_{ii} & \frac{1}{2} \left(Q_{ii} + R_{ii}^T \right) \\ \frac{1}{2} \left(Q_{ii}^T + R_{ii} \right) & S_{ii} \end{bmatrix} \begin{pmatrix} x_i^\nu(t) \\ u_i^\nu(t) \end{pmatrix} \right) dt \geq 0.$$

Adding, we deduce

$$\begin{aligned} 0 \leq\; & 2 \left(x_i^\nu(\mathcal{T}) - x_i^{\nu+1}(\mathcal{T}) \right)^T W_{ii} \left(x_i^{\nu+1}(\mathcal{T}) - x_i^\nu(\mathcal{T}) \right) + \\ & \int_0^{\mathcal{T}} \begin{pmatrix} x_i^\nu(t) - x_i^{\nu+1}(t) \\ u_i^\nu(t) - u_i^{\nu+1}(t) \end{pmatrix}^T \left(\sum_{i' \neq i} \begin{bmatrix} P_{ii'} & Q_{ii'} \\ R_{ii'} & S_{ii'} \end{bmatrix} \begin{pmatrix} x_{i'}^\nu(t) - x_{i'}^{\nu-1}(t) \\ u_{i'}^\nu(t) - u_{i'}^{\nu-1}(t) \end{pmatrix} \right. \\ & \left. + 2 \begin{bmatrix} P_{ii} & \frac{1}{2} \left(Q_{ii} + R_{ii}^T \right) \\ \frac{1}{2} \left(Q_{ii}^T + R_{ii} \right) & S_{ii} \end{bmatrix} \begin{pmatrix} x_i^{\nu+1}(t) - x_i^\nu(t) \\ u_i^{\nu+1}(t) - u_i^\nu(t) \end{pmatrix} \right) dt \\ \leq\; & \int_0^{\mathcal{T}} \left\{ -\sigma_i^\Xi \left\| \begin{pmatrix} x_i^\nu(t) - x_i^{\nu+1}(t) \\ u_i^\nu(t) - u_i^{\nu+1}(t) \end{pmatrix} \right\|^2 + \right. \\ & \left. \left\| \begin{matrix} x_i^\nu(t) - x_i^{\nu+1}(t) \\ u_i^\nu(t) - u_i^{\nu+1}(t) \end{matrix} \right\| \sum_{i' \neq i} \left\| \begin{bmatrix} P_{ii'} \; Q_{ii'} \\ R_{ii'} \; S_{ii'} \end{bmatrix} \right\| \left\| \begin{pmatrix} x_{i'}^\nu(t) - x_{i'}^{\nu-1}(t) \\ u_{i'}^\nu(t) - u_{i'}^{\nu-1}(t) \end{pmatrix} \right\| \right\} dt. \end{aligned}$$

By the Cauchy-Schwarz inequality of integrals, we deduce

$$\begin{aligned} & \sigma_i^\Xi \left(\int_0^{\mathcal{T}} \left\| \begin{pmatrix} x_i^\nu(t) - x_i^{\nu+1}(t) \\ u_i^\nu(t) - u_i^{\nu+1}(t) \end{pmatrix} \right\|^2 dt \right)^{1/2} \\ & \leq \sum_{i' \neq i} \left\| \begin{bmatrix} P_{ii'} & Q_{ii'} \\ R_{ii'} & S_{ii'} \end{bmatrix} \right\| \left(\int_0^{\mathcal{T}} \left\| \begin{pmatrix} x_{i'}^\nu(t) - x_{i'}^{\nu-1}(t) \\ u_{i'}^\nu(t) - u_{i'}^{\nu-1}(t) \end{pmatrix} \right\|^2 dt \right)^{1/2}. \end{aligned} \tag{4.2}$$

Defining the difference

$$e_i^\nu \triangleq \sqrt{\sigma_i^\Xi} \sqrt{\int_0^{\mathcal{T}} \left\| \begin{pmatrix} x_i^\nu(t) - x_i^{\nu+1}(t) \\ u_i^\nu(t) - u_i^{\nu+1}(t) \end{pmatrix} \right\|^2 dt},$$

the vector $\mathbf{e}^\nu \triangleq (e_i^\nu)_{i=1}^{\mathcal{F}}$, and the matrix $\mathbf{\Gamma} \triangleq [\Gamma_{ii'}]_{i,i'=1}^{\mathcal{F}}$, where

$$\Gamma_{ii'} \triangleq \begin{cases} 0 & \text{if } i = i' \\ \dfrac{1}{\sqrt{\sigma_i^\Xi \, \sigma_{i'}^\Xi}} \, \| \, \Xi_{ii'} \, \| & \text{if } i \neq i', \end{cases}$$

we deduce by aggregating the inequalities (4.2) for $i = 1, \cdots, \mathcal{F}$, the following vector inequality:

$$\mathbf{e}^\nu \leq \mathbf{\Gamma}\, \mathbf{e}^{\nu-1}. \tag{4.3}$$

Thus, if the spectral radius $\rho(\mathbf{\Gamma}) < 1$, then the sequence $\{\mathbf{e}^\nu\}$ contracts and converges to zero. In particular, the sequence $\{(x_i^\nu(t), u_i^\nu(t))\}$ converges strongly to a square-integrable, thus integrable, limit $(x_i^\infty(t), u_i^\infty(t))$ in the space $L^2[\,0, \mathcal{T}\,]$, i.e.,

$$\lim_{\nu \to \infty} \int_0^{\mathcal{T}} \left\| \begin{pmatrix} x_i^\nu(t) - x_i^\infty(t) \\ u_i^\nu(t) - u_i^\infty(t) \end{pmatrix} \right\| dt \; = \; 0.$$

Thus, for an infinite index set κ, the subsequence of functions $\{(x_i^\nu(t), u_i^\nu(t))\}_{\nu \in \kappa}$ converges pointwise to $(x_i^\infty(t), u_i^\infty(t))$ for almost all $t \in [\,0, \mathcal{T}\,]$ by the Riesz-Fischer Theorem. Hence $f_i + C_i x_i^\infty(t) + D_i u_i^\infty(t) \geq 0$ for almost all t. Moreover, we have

$$\lim_{\nu \to \infty} \int_0^t \left\| \begin{pmatrix} x_i^\nu(\tau) - x_i^\infty(\tau) \\ u_i^\nu(\tau) - u_i^\infty(\tau) \end{pmatrix} \right\| d\tau \; = \; 0 \quad \text{for all } t \in (\,0, \mathcal{T}\,].$$

Hence, for all $t \in (\,0, \mathcal{T}\,]$, we have

$$\lim_{\nu \to \infty} \int_0^t \left(r_i + A_i x_i^\nu(\tau) + B_i u_i^\nu(\tau) \right) d\tau \; = \; \int_0^t \left(r_i + A_i x_i^\infty(\tau) + B_i u_i^\infty(\tau) \right) d\tau.$$

At times t where $\{x_i^\nu(t)\}_{\nu \in \kappa}$ converges to $x_i^\infty(t)$, we have

$$\begin{aligned} x_i^\infty(t) \; = \; \lim_{\nu(\in \kappa) \to \infty} x_i^\nu(t) \; &= \; \xi_i + \lim_{\nu(\in \kappa) \to \infty} \int_0^t \left(r_i + A_i x_i^\nu(\tau) + B_i u_i^\nu(\tau) \right) d\tau \\ &= \; \xi_i + \int_0^t \left(r_i + A_i x_i^\infty(\tau) + B_i u_i^\infty(\tau) \right) d\tau. \end{aligned}$$

From Lemma 4.1, the sequence $\{(x_i^\nu(t), u_i^\nu(t))\}_{\nu \in \kappa}$ is uniformly bounded for $t \in [\,0, \mathcal{T}\,]$; hence the limit pair $(x^\infty(t), u^\infty(t))$ is bounded for almost all $t \in [\,0, \mathcal{T}\,]$. We have, for any two times $t < s$ in $[\,0, \mathcal{T}\,]$,

$$\begin{aligned} & x_i^\nu(t) - x_i^\nu(s) \\ & \quad = \int_t^s \left(r_i + A_i x_i^\nu(\tau) + B_i u_i^\nu(\tau) \right) d\tau \\ & \quad = \int_t^s \left(r_i + A_i x_i^\infty(\tau) + B_i u_i^\infty(\tau) \right) d\tau \\ & \qquad + \int_t^s \left(A_i \left(x_i^\nu(\tau) - x_i^\infty(\tau) \right) + B_i \left(u_i^\nu(\tau) - u_i^\infty(\tau) \right) \right) d\tau. \end{aligned}$$

Hence the family $\{x_i^\nu(t)\}_{\nu\in\kappa}$ is equicontinuous; by the well-known Arzelà-Ascoli theorem, this sequence has a subsequence that converges uniformly to a continuous function on $t \in [0, \mathcal{T}]$. Without loss of generality, we may assume that $\{x_i^\nu(t)\}_{\nu\in\kappa}$ converges uniformly to $x_i^\infty(t)$ which is continuous. It follows that for all $t \in [0, \mathcal{T}]$,

$$x_i^\infty(t) = \xi_i + \int_0^t (r_i + A_i x_i^\infty(\tau) + B_i u_i^\infty(\tau))\, d\tau.$$

Thus the pair (x_i^∞, u_i^∞) is feasible to (2.1). Moreover, since

$$\lim_{\nu\to\infty} \left(x_i^\nu(\mathcal{T}) - x_i^{\nu+1}(\mathcal{T}) \right)^T W_{ii} \left(x_i^{\nu+1}(\mathcal{T}) - x_i^\nu(\mathcal{T}) \right) = 0$$

and W_{ii} is symmetric positive semidefinite, it follows that

$$\lim_{\nu(\in\kappa)\to\infty} W_{ii} \left(x_i^{\nu+1}(\mathcal{T}) - x_i^\infty(\mathcal{T}) \right) = 0$$

and

$$\lim_{\nu(\in\kappa)\to\infty} \left(x_i^{\nu+1}(\mathcal{T}) - x_i^\infty(\mathcal{T}) \right)^T W_{ii} \left(x_i^{\nu+1}(\mathcal{T}) - x_i^\infty(\mathcal{T}) \right) = 0.$$

To show that the tuple $(x_i^\infty, u_i^\infty)_{i=1}^{\mathcal{F}}$ is a NE of the game, it suffice to show that

$$\begin{aligned}
&\left(x_i(\mathcal{T}) - x_i^\infty(\mathcal{T}) \right)^T \left(c_i + 2W_{ii} x_i^\infty(\mathcal{T}) \right) + \\
&\int_0^{\mathcal{T}} \begin{pmatrix} x_i(t) - x_i^\infty(t) \\ u_i(t) - u_i^\infty(t) \end{pmatrix}^T \left(\begin{pmatrix} p_i(t) \\ q_i(t) \end{pmatrix} + \sum_{i'\neq i} \begin{bmatrix} P_{ii'} & Q_{ii'} \\ R_{ii'} & S_{ii'} \end{bmatrix} \begin{pmatrix} x_{i'}^\infty(t) \\ u_{i'}^\infty(t) \end{pmatrix} \right. \\
&\left. + 2 \begin{bmatrix} P_{ii} & \frac{1}{2}(Q_{ii} + R_{ii}^T) \\ \frac{1}{2}(Q_{ii}^T + R_{ii}) & S_{ii} \end{bmatrix} \begin{pmatrix} x_i^\infty(t) \\ u_i^\infty(t) \end{pmatrix} \right) dt \geq 0
\end{aligned}$$

for every feasible pair (x_i, u_i) to (2.1). We have, for all ν,

$$\begin{aligned}
&\left(x_i(\mathcal{T}) - x_i^{\nu+1}(\mathcal{T}) \right)^T \left(c_i + 2W_{ii} x_i^{\nu+1}(\mathcal{T}) \right) + \\
&\int_0^{\mathcal{T}} \begin{pmatrix} x_i(t) - x_i^{\nu+1}(t) \\ u_i(t) - u_i^{\nu+1}(t) \end{pmatrix}^T \left(\begin{pmatrix} p_i(t) \\ q_i(t) \end{pmatrix} + \sum_{i'\neq i} \begin{bmatrix} P_{ii'} & Q_{ii'} \\ R_{ii'} & S_{ii'} \end{bmatrix} \begin{pmatrix} x_{i'}^\nu(t) \\ u_{i'}^\nu(t) \end{pmatrix} \right. \\
&\left. + 2 \begin{bmatrix} P_{ii} & \frac{1}{2}(Q_{ii} + R_{ii}^T) \\ \frac{1}{2}(Q_{ii}^T + R_{ii}) & S_{ii} \end{bmatrix} \begin{pmatrix} x_i^{\nu+1}(t) \\ u_i^{\nu+1}(t) \end{pmatrix} \right) dt \geq 0.
\end{aligned}$$

We can write

$$\begin{aligned}
&\left(x_i(\mathcal{T}) - x_i^{\nu+1}(\mathcal{T}) \right)^T \left(c_i + 2W_{ii} x_i^{\nu+1}(\mathcal{T}) \right) \\
&\quad = \left(x_i^\infty(\mathcal{T}) - x_i^{\nu+1}(\mathcal{T}) \right)^T \left(c_i + 2W_{ii} x_i^{\nu+1}(\mathcal{T}) \right) \\
&\qquad + \left(x_i(\mathcal{T}) - x_i^\infty(\mathcal{T}) \right)^T \left(c_i + 2W_{ii} x_i^{\nu+1}(\mathcal{T}) \right) \\
&\quad = 2 \left(x_i^\infty(\mathcal{T}) - x_i^{\nu+1}(\mathcal{T}) \right)^T W_{ii} \left(x_i^{\nu+1}(\mathcal{T}) - x_i^\infty(\mathcal{T}) \right) + \\
&\left(x_i(\mathcal{T}) - x_i^\infty(\mathcal{T}) \right)^T \left(c_i + 2W_{ii} x_i^{\nu+1}(\mathcal{T}) \right) + \left(x_i^\infty(\mathcal{T}) - x_i^{\nu+1}(\mathcal{T}) \right)^T \left(c_i + 2W_{ii} x_i^\infty(\mathcal{T}) \right).
\end{aligned}$$

Hence,

$$\lim_{\nu(\in\kappa)\to\infty} \left(x_i(\mathcal{T}) - x_i^{\nu+1}(\mathcal{T}) \right)^T \left(c_i + 2W_{ii}x_i^{\nu+1}(\mathcal{T}) \right) = \left(x_i(\mathcal{T}) - x_i^{\infty}(\mathcal{T}) \right)^T \left(c_i + 2W_{ii}x_i^{\infty}(\mathcal{T}) \right).$$

Similarly, we can establish that

$$\lim_{\nu\to\infty} \int_0^{\mathcal{T}} \begin{pmatrix} x_i(t) - x_i^{\nu+1}(t) \\ u_i(t) - u_i^{\nu+1}(t) \end{pmatrix}^T \left(\begin{pmatrix} p_i(t) \\ q_i(t) \end{pmatrix} + 2 \begin{bmatrix} P_{ii} & \frac{1}{2}\left(Q_{ii} + R_{ii}^T\right) \\ \frac{1}{2}\left(Q_{ii}^T + R_{ii}\right) & S_{ii} \end{bmatrix} \begin{pmatrix} x_i^{\nu+1}(t) \\ u_i^{\nu+1}(t) \end{pmatrix} \right) dt$$

$$= \int_0^{\mathcal{T}} \begin{pmatrix} x_i(t) - x_i^{\infty}(t) \\ u_i(t) - u_i^{\infty}(t) \end{pmatrix}^T \left(\begin{pmatrix} p_i(t) \\ q_i(t) \end{pmatrix} + 2 \begin{bmatrix} P_{ii} & \frac{1}{2}\left(Q_{ii} + R_{ii}^T\right) \\ \frac{1}{2}\left(Q_{ii}^T + R_{ii}\right) & S_{ii} \end{bmatrix} \begin{pmatrix} x_i^{\infty}(t) \\ u_i^{\infty}(t) \end{pmatrix} \right) dt.$$

Finally, we have

$$\int_0^{\mathcal{T}} \begin{pmatrix} x_i(t) - x_i^{\nu+1}(t) \\ u_i(t) - u_i^{\nu+1}(t) \end{pmatrix}^T \begin{bmatrix} P_{ii'} & Q_{ii'} \\ R_{ii'} & S_{ii'} \end{bmatrix} \begin{pmatrix} x_{i'}^{\nu}(t) \\ u_{i'}^{\nu}(t) \end{pmatrix} dt$$

$$= \int_0^{\mathcal{T}} \begin{pmatrix} x_i(t) - x_i^{\infty}(t) \\ u_i(t) - u_i^{\infty}(t) \end{pmatrix}^T \begin{bmatrix} P_{ii'} & Q_{ii'} \\ R_{ii'} & S_{ii'} \end{bmatrix} \begin{pmatrix} x_{i'}^{\nu}(t) - x_{i'}^{\infty}(t) \\ u_{i'}^{\nu}(t) - u_{i'}^{\infty}(t) \end{pmatrix} dt$$

$$+ \int_0^{\mathcal{T}} \begin{pmatrix} x_i(t) - x_i^{\infty}(t) \\ u_i(t) - u_i^{\infty}(t) \end{pmatrix}^T \begin{bmatrix} P_{ii'} & Q_{ii'} \\ R_{ii'} & S_{ii'} \end{bmatrix} \begin{pmatrix} x_{i'}^{\infty}(t) \\ u_{i'}^{\infty}(t) \end{pmatrix} dt$$

$$+ \int_0^{\mathcal{T}} \begin{pmatrix} x_i^{\infty}(t) - x_i^{\nu+1}(t) \\ u_i^{\infty}(t) - u_i^{\nu+1}(t) \end{pmatrix}^T \begin{bmatrix} P_{ii'} & Q_{ii'} \\ R_{ii'} & S_{ii'} \end{bmatrix} \begin{pmatrix} x_{i'}^{\nu}(t) \\ u_{i'}^{\nu}(t) \end{pmatrix} dt,$$

from which we deduce

$$\lim_{\nu\to\infty} \int_0^{\mathcal{T}} \begin{pmatrix} x_i(t) - x_i^{\nu+1}(t) \\ u_i(t) - u_i^{\nu+1}(t) \end{pmatrix}^T \begin{bmatrix} P_{ii'} & Q_{ii'} \\ R_{ii'} & S_{ii'} \end{bmatrix} \begin{pmatrix} x_{i'}^{\nu}(t) \\ u_{i'}^{\nu}(t) \end{pmatrix} dt$$

$$= \int_0^{\mathcal{T}} \begin{pmatrix} x_i(t) - x_i^{\infty}(t) \\ u_i(t) - u_i^{\infty}(t) \end{pmatrix}^T \begin{bmatrix} P_{ii'} & Q_{ii'} \\ R_{ii'} & S_{ii'} \end{bmatrix} \begin{pmatrix} x_{i'}^{\infty}(t) \\ u_{i'}^{\infty}(t) \end{pmatrix} dt.$$

Summarizing the above analysis, we obtain the following convergence, existence, and uniqueness result for the asymmetric differential LQ Nash game that is the counterpart of Theorem 3.1. Incidentally, the condition $\rho(\mathbf{\Gamma}) < 1$ is a familiar condition for the contraction of fixed-point iterations for computing static Nash equilibria; see e.g. [**5**, Section 6].

THEOREM 4.2. *Under assumptions* $(\widehat{\text{A}})$, (B), (W), $(\widehat{\text{D}})$, (E), *the following statements hold for the sequence* $\{(x_i^\nu, u_i^\nu)\}$ *generated by the Jacobi iterative algorithm.*

(I: Well-definedness) *The sequence* $\{(x_i^\nu, u_i^\nu)\}$ *is well-defined with* x_i^ν *being continuously differentiable and* u_i^ν *Lipschitz continuous on* $[0, \mathcal{T}]$ *for all* ν.
(II: Contraction and strong convergence) *If* $\rho(\boldsymbol{\Gamma}) < 1$, *the sequence* $\{(x_i^\nu, u_i^\nu)_{i=1}^{\mathcal{F}}\}$ *converges strongly to a pair* $(x_i^\infty, u_i^\infty)_{i=1}^{\mathcal{F}}$ *that is the unique NE of the differential LQ game.*

PROOF. It remains to show that the differential game has a unique NE. Let $(\widehat{x}, \widehat{u})$ and $(\widetilde{x}, \widetilde{u})$ be two Nash equilibria. Defining the error vector $\mathbf{e} \triangleq (e_i)_{i=1}^{\mathcal{F}}$, where

$$e_i \triangleq \sqrt{\sigma_i^{\Xi}} \sqrt{\int_0^{\mathcal{T}} \left\| \begin{pmatrix} \widehat{x}_i(t) - \widetilde{x}_i(t) \\ \widehat{u}_i(t) - \widetilde{u}_i(t) \end{pmatrix} \right\|^2 dt},$$

we can deduce, similar to (4.3), $\mathbf{e} \leq \boldsymbol{\Gamma}\mathbf{e}$. Since $\rho(\boldsymbol{\Gamma}) < 1$, $\mathbf{e} = 0$ and the NE must be identical. □

5. Example Problems

Illustrating the abstract framing of the symmetric and asymmetric problems in the previous two sections, we present two concrete examples of how such problems may arise in applied game theory. The first example model is an adaptation of the well-known Nash-Cournot equilibrium problem while the second is a conjectured supply function equilibrium problem. Although these types of problems are typically studied in a static setting, the differential formulations presented herein represent natural problem extensions for which solution existence can be established from the previous results.

5.1. A Nash-Cournot game. In non-cooperative microeconomic equilibrium problems, $\mathcal{F}$ players produce and sell a commodity to maximize profit subject to constraints such as manufacturing capacity and available budget. For the Nash-Cournot version of this problem, each player believes that their output affects the commodity price which is represented as a function of total output. For a two-player, two-node problem with a linear pricing function and quadratic cost functions, let player 1's optimal control problem be

$$\underset{\mathbf{g}_1,\mathbf{s}_1,\mathbf{r}_1}{\text{minimize}}$$

$$\int_0^{\mathcal{T}} \begin{pmatrix} g_{11}(t) \\ g_{12}(t) \\ s_{11}(t) \\ s_{12}(t) \\ r_{11}(t) \\ r_{12}(t) \end{pmatrix}^T \left(\begin{pmatrix} a_{11} - w(t) \\ a_{12} \\ -P_1^0 + w(t) \\ -P_2^0 \\ 0 \\ 0 \end{pmatrix} + \begin{bmatrix} b_{11} & & & & & \\ & b_{12} & & & & \\ & & \frac{P_1^0}{Q_1^0} & & & \\ & & & \frac{P_2^0}{Q_2^0} & & \\ & & & & 0 & \\ & & & & & 0 \end{bmatrix} \begin{pmatrix} g_{11}(t) \\ g_{12}(t) \\ s_{11}(t) \\ s_{12}(t) \\ r_{11}(t) \\ r_{12}(t) \end{pmatrix} \right.$$

$$\left. + \begin{bmatrix} 0 & & & & & \\ & 0 & & & & \\ & & \frac{P_1^0}{Q_1^0} & & & \\ & & & \frac{P_2^0}{Q_2^0} & & \\ & & & & 0 & \\ & & & & & 0 \end{bmatrix} \begin{pmatrix} g_{21}(t) \\ g_{22}(t) \\ s_{21}(t) \\ s_{22}(t) \\ r_{21}(t) \\ r_{22}(t) \end{pmatrix} \right) dt$$

$$= \int_0^{\mathcal{T}} \left\{ \sum_{j=1}^{2} \left(a_{1j} g_{1j}(t) + b_{1j} g_{1j}(t)^2 - \left[P_j^0 - \frac{P_j^0}{Q_j^0} (s_{1j}(t) + s_{2j}(t)) \right] s_{1j}(t) \right) \right.$$

$$\left. + (s_{11}(t) - g_{11}(t)) w(t) \right\} dt$$

subject to $g_{11}(0) = g_{11}^0$, $g_{12}(0) = g_{12}^0$, and for almost all $t \in [0, \mathcal{T}]$:

$$\left. \begin{array}{ll} \dot{g}_{11}(t) = r_{11}(t) & \\ \dot{g}_{12}(t) = r_{12}(t) & \\ -\underline{r}_{1j} + r_{1j}(t) \geq 0 & \text{for } j = 1,2 \\ \overline{r}_{1j} - r_{1j}(t) \geq 0 & \text{for } j = 1,2 \end{array} \right\} \Rightarrow \begin{array}{l} \underline{r}_{11} \leq \dot{g}_{11}(t) \leq \overline{r}_{11} \\ \underline{r}_{12} \leq \dot{g}_{12}(t) \leq \overline{r}_{12} \end{array}$$

$$g_{11}(t) + g_{12}(t) - s_{11}(t) - s_{12}(t) \geq 0 \quad \text{for } j = 1,2$$

and $$- g_{11}(t) - g_{12}(t) + s_{11}(t) + s_{12}(t) \geq 0 \quad \text{for } j = 1,2,$$

where the state variables are $\{g_{ij}(t)\}_{i,j=1}^2$, representing player i's production at node j at time t, and the control variables are $\{s_{ij}(t), r_{ij}(t)\}_{i,j=1}^2$, representing player i's sales and ramp rate (instantaneous change in production) at node j at time t, respectively. The term $a_{ij}g_{ij}(t) + b_{ij}g_{ij}(t)^2$ is the quadratic production cost function, $P_j^0 - \frac{P_j^0}{Q_j^0}\left(\sum_{i=1}^{\mathcal{F}} s_{ij}(t)\right)$ is the linear nodal pricing equation at time t with intercept P_j^0 and slope $\dfrac{P_j^0}{Q_j^0}$ where P_j^0 and Q_j^0 are positive, and $(s_{ij}(t) - g_{ij}(t))w(t)$ is the transportation cost with $w(t)$ being the marginal directional shipment cost at time t. We assume that $w(t)$ is a given Lipschitz continuous function on $t \in [0, \mathcal{T}]$. The first group of constraints describes generation ramp rates, namely that the rate of generation change for player i at node j is bounded by $\underline{r}_{ij}$ and $\overline{r}_{ij}$. The last two constraints equate total generation with total sales. Player 2's objective function is easily shown to be identical to that given above except with 1 and 2 interchanged in player index i. Therefore, it is apparent that $\boldsymbol{\Xi} \triangleq \begin{bmatrix} \Xi_{11} & \Xi_{12} \\ \Xi_{21} & \Xi_{22} \end{bmatrix}$ is the symmetric matrix

$$
\left[\begin{array}{cccccccccccc}
2b_{11} & & & & & & & & & & & \\
& 2b_{12} & & & & & & & & & & \\
& & 2\dfrac{P_1^0}{Q_1^0} & & & & & & \dfrac{P_1^0}{Q_1^0} & & & \\
& & & 2\dfrac{P_2^0}{Q_2^0} & & & & & & \dfrac{P_2^0}{Q_2^0} & & \\
& & & & 0 & & & & & & & \\
& & & & & 0 & & & & & & \\
& & & & & & 2b_{21} & & & & & \\
& & & & & & & 2b_{22} & & & & \\
& & \dfrac{P_1^0}{Q_1^0} & & & & & & 2\dfrac{P_1^0}{Q_1^0} & & & \\
& & & \dfrac{P_2^0}{Q_2^0} & & & & & & 2\dfrac{P_2^0}{Q_2^0} & & \\
& & & & & & & & & & 0 & \\
& & & & & & & & & & & 0
\end{array}\right].
$$

Furthermore, $\boldsymbol{\Xi}$ must be positive semidefinite by row diagonal dominance, thereby fulfilling condition (A). Condition (B) is obvious from the Lipschitz continuity of $w(t)$. Condition (C) follows from setting $r_{ij}(t) = 0$ and $s_{ij}(t) = g_{ij}(t)$ for $i, j = 1, 2$ and $t \in [0, \mathcal{T}]$ to obtain a feasible trajectory with $g_{ij}(t) = g_{ij}^0$ for $t \in [0, \mathcal{T}]$. Since

$$
\mathbf{S} \triangleq \begin{bmatrix} 2S_{11} & S_{12} \\ S_{21} & 2S_{22} \end{bmatrix} =
\left[\begin{array}{cccccccc}
2\dfrac{P_1^0}{Q_1^0} & & & & \dfrac{P_1^0}{Q_1^0} & & & \\
& 2\dfrac{P_2^0}{Q_2^0} & & & & \dfrac{P_2^0}{Q_2^0} & & \\
& & 0 & & & & & \\
& & & 0 & & & & \\
\dfrac{P_1^0}{Q_1^0} & & & & 2\dfrac{P_1^0}{Q_1^0} & & & \\
& \dfrac{P_2^0}{Q_2^0} & & & & 2\dfrac{P_2^0}{Q_2^0} & & \\
& & & & & & 0 & \\
& & & & & & & 0
\end{array}\right]
$$

and

$$D_i \triangleq \begin{bmatrix} & & 1 & \\ & & & 1 \\ & & -1 & \\ & & & -1 \\ -1 & -1 & & \\ 1 & 1 & & \end{bmatrix}$$

for players $i = 1, 2$, $\mathbf{Su} = 0$ implies that $s_{11} = s_{12} = s_{21} = s_{22} = 0$ and $D_i u_i \geq 0$ for $i = 1, 2$ gives $r_{11} = r_{12} = r_{21} = r_{22} = 0$. Hence, (D) holds. It can be seen that $\left[D_i^T \mu_i = 0,\ \mu_i \geq 0\right]$ can only hold when $\mu_{i1} = \mu_{i3}$, $\mu_{i2} = \mu_{i4}$, and $\mu_{i5} = \mu_{i6}$. Therefore, $\left(C_i A_i^0 B_i\right)^T \mu_i$ equals

$$\left[\begin{bmatrix} 0 & 0 \\ 0 & 0 \\ 0 & 0 \\ 0 & 0 \\ 0 & 0 \\ 1 & 1 \\ -1 & -1 \end{bmatrix} I \begin{bmatrix} 0 & 0 & 1 & 0 \\ 0 & 0 & 0 & 1 \end{bmatrix}\right]^T \begin{pmatrix} \mu_{i1} \\ \mu_{i2} \\ \mu_{i3} \\ \mu_{i4} \\ \mu_{i5} \\ \mu_{i6} \end{pmatrix} = \begin{pmatrix} 0 \\ 0 \\ \mu_{i5} - \mu_{i6} \\ \mu_{i5} - \mu_{i6} \end{pmatrix} = \begin{pmatrix} 0 \\ 0 \\ 0 \\ 0 \end{pmatrix}.$$

Since A_i is a zero matrix, A_i^j for every positive integer j is 0. Hence, (E) holds. We have established that this differential Nash-Cournot equilibrium problem can be formulated as an equivalent symmetric optimal control problem satisfying conditions (A–E). As a consequence, Theorem 2.1 guarantees the existence of an equilibrium solution and [**7**] provides an iterative method for finding an equilibrium.

5.2. A conjectured supply function game. We now turn our attention to a conjectured supply function (CSF) problem [**3**, **8**, **9**, **17**] to demonstrate the existence of games that take an asymmetric form. In the Nash-Cournot problem, symmetry arises from the assumptions that each player uses the same commodity pricing function and that no player anticipates competitor production/sales changes with respect to price. In a conjectured supply function equilibrium problem, players instead use a function to predict how total competitor production will change based on price. For this example, we will simplify the model to include only one node so that generation and sales quantities are equivalent and transmission is not needed.

For player i, let the function $\sigma_i(G_{-i}(t), p_i(t), t)$ represent the relationship between price and total competitor production in time t. For our linear-quadratic problem, we will define

$$\sigma_i(G_{-i}(t), p_i(t), t) \triangleq G_{-i}(t) + \beta_i(G_{-i}(t), p_i^*(t), t)(p_i(t) - p_i^*(t)),$$

where $G_{-i}(t)$ is the total amount of competitor generation expected at the specified equilibrium price $p_i^*(t)$ at time t. Notice that players may expect different equilibrium price trajectories here; this setting generalizes the case in which players use the same equilibrium price trajectory where $p_i^*(t) = p^*(t)$ for $i = 1, 2$. It follows that, depending on the specification of $\beta_i(G_{-i}(t), p_i^*(t), t)$, the conjectured total production from other players will rise or fall if the realized price $p_i(t)$ does not equal the equilibrium price $p_i^*(t)$. Upon substitution into the production-pricing relationship

$$g_i(t) + \sigma_i(G_{-i}(t), p_i(t), t) \;=\; Q^0 - \frac{Q^0}{P^0}\, p_i(t),$$

invertibility of $\dfrac{Q^0}{P^0} + \beta_i(G_{-i}(t), p_i^*(t), t)$ provides an explicit equation for player i's conjectured price $p_i(t)$. This invertibility will hold in realistic market settings since $\beta_i(G_{-i}(t), p_i^*(t), t)$ should be nonnegative so total competitor production levels are believed to change in the same direction as price differences (i.e., higher prices than expected at equilibrium should not decrease conjectured production). In the special case assumed here where $\beta_i(G_{-i}(t), p_i^*(t), t) \triangleq B_{-i}$ for some positive constant B_{-i}, we obtain

$$p_i(t) \;=\; \frac{Q^0 - G_i(t) + B_{-i}p_i^*(t)}{\dfrac{Q^0}{P^0} + B_{-i}}.$$

Using this conjectured price, we can formulate player 1's optimal control problem as a cost minimization problem in which the conjectured supply function price is used for determining revenue and costs include a quadratic production cost and a quadratic ramp rate cost:

$$\operatorname*{minimize}_{g_1,\, r_1} \int_0^{\mathcal{T}} \begin{pmatrix} g_1(t) \\ r_1(t) \end{pmatrix}^T \left(\begin{pmatrix} a_{11} - \dfrac{Q^0 + B_{-1}p_1^*(t)}{\dfrac{Q^0}{P^0} + B_{-1}} \\ 0 \end{pmatrix} \right.$$

$$\left. + \begin{bmatrix} b_{11} + \dfrac{1}{\dfrac{Q^0}{P^0} + B_{-1}} & 0 \\ 0 & a_{12} \end{bmatrix} \begin{pmatrix} g_1(t) \\ r_1(t) \end{pmatrix} + \begin{bmatrix} \dfrac{1}{\dfrac{Q^0}{P^0} + B_{-1}} & 0 \\ 0 & 0 \end{bmatrix} \begin{pmatrix} g_2(t) \\ r_2(t) \end{pmatrix} \right) dt$$

subject to $g_1(0) \;=\; g_1^0$ and for almost all $t \in [\,0, \mathcal{T}\,]$:

$$\left.\begin{array}{l} \dot{g}_1(t) \;=\; r_1(t) \\ -\underline{r}_1 + r_1(t) \;\geq\; 0 \\ \overline{r}_1 - r_1(t) \;\geq\; 0 \end{array}\right\} \Rightarrow \underline{r}_1 \;\leq\; \dot{g}_1(t) \;\leq\; \overline{r}_1$$

Similarly, player 2's optimal control problem just interchanges 1 and 2 for the player index. If the player supply conjectures are not identical (i.e., $B_{-1} \neq B_{-2}$),

$$\Xi_{12} \;=\; \begin{bmatrix} \dfrac{1}{\dfrac{Q^0}{P^0} + B_{-1}} & 0 \\ 0 & 0 \end{bmatrix} \;\neq\; \begin{bmatrix} \dfrac{1}{\dfrac{Q^0}{P^0} + B_{-2}} & 0 \\ 0 & 0 \end{bmatrix} \;=\; \Xi_{21}^T.$$

It follows that a conjectured supply function game in which players have different conjectures is not a symmetric game. With $p_1^*(t)$ and $p_2^*(t)$ being Lipschitz continuous functions, it is simple to see that ($\widehat{\text{A}}$), (B), (W), ($\widehat{\text{D}}$), and (E) hold.

To prove $\rho(\mathbf{\Gamma}) < 1$, we can use the fact that $\rho(\mathbf{\Gamma}) \leq \|\mathbf{\Gamma}^k\|^{\frac{1}{k}}$ for all natural numbers k. Examining $k = 1$ with the Euclidean norm, $\|\mathbf{\Gamma}\|$ is the largest eigenvalue of

$$(\mathbf{\Gamma}^T\mathbf{\Gamma})^{\frac{1}{2}} = \left(\begin{bmatrix} 0 & \frac{1}{\sqrt{\sigma_2^{\Xi}\sigma_1^{\Xi}}}\|\Xi_{21}\| \\ \frac{1}{\sqrt{\sigma_1^{\Xi}\sigma_2^{\Xi}}}\|\Xi_{12}\| & 0 \end{bmatrix} \begin{bmatrix} 0 & \frac{1}{\sqrt{\sigma_1^{\Xi}\sigma_2^{\Xi}}}\|\Xi_{12}\| \\ \frac{1}{\sqrt{\sigma_2^{\Xi}\sigma_1^{\Xi}}}\|\Xi_{21}\| & 0 \end{bmatrix} \right)^{\frac{1}{2}}$$

$$= \frac{1}{\sqrt{\sigma_1^{\Xi}\sigma_2^{\Xi}}} \begin{bmatrix} \|\Xi_{12}\| & 0 \\ 0 & \|\Xi_{21}\| \end{bmatrix}$$

where, as defined in ($\widehat{\text{A}}$), σ_i^{Ξ} is the minimum eigenvalue of Ξ_{ii}. For this problem,

$$\sigma_1^{\Xi} \triangleq \min\left(b_{11} + \frac{1}{\frac{Q_0}{P_0} + B_{-1}},\ a_{12} \right) \quad \text{and} \quad \sigma_2^{\Xi} \triangleq \min\left(b_{21} + \frac{1}{\frac{Q_0}{P_0} + B_{-2}},\ a_{22} \right).$$

Hence, if

$$\begin{array}{l} \|\,\Xi_{12}\,\| \\ \quad \| \\ \dfrac{1}{\frac{Q^0}{P^0} + B_{-1}} < \sqrt{\min\left(b_{11} + \dfrac{1}{\frac{Q_0}{P_0} + B_{-1}},\ a_{12} \right) \min\left(b_{21} + \dfrac{1}{\frac{Q_0}{P_0} + B_{-2}},\ a_{22} \right)} \\ \qquad\qquad\qquad\qquad\qquad\qquad \| \\ \qquad\qquad\qquad\qquad\qquad \sqrt{\sigma_1^{\Xi}\sigma_2^{\Xi}} \\ \qquad\qquad\qquad\qquad\qquad\qquad \| \\ \dfrac{1}{\frac{Q^0}{P^0} + B_{-2}} < \sqrt{\min\left(b_{11} + \dfrac{1}{\frac{Q_0}{P_0} + B_{-1}},\ a_{12} \right) \min\left(b_{21} + \dfrac{1}{\frac{Q_0}{P_0} + B_{-2}},\ a_{22} \right)} \\ \quad \| \\ \|\,\Xi_{21}\,\|, \end{array}$$

then $\rho(\mathbf{\Gamma}) < 1$. The above condition can clearly be satisfied for a wide variety of parameter values. We have thus proven that Theorem 4.2 holds for the above CSF problem specification and the presented Jacobi iterative algorithm will converge to the unique differential Nash equilibrium.

6. Conclusion

In this paper, solution existence of open-loop differential LQ Nash games was studied via constructive approaches. Under certain assumptions, the equivalence of a *symmetric* differential LQ game to a single concatenated linear-quadratic optimal control problem was established. Using this equivalence, the convergent numerical method developed in [**7**] for solving LQ optimal control problems can be leveraged to identify differential Nash equilibria. For *asymmetric* differential LQ games, a Jacobi-type iterative solution scheme was proven to converge under certain conditions to a unique differential Nash equilibrium. Examples of problems fulfilling the stated assumptions were presented. It is hoped that the results derived herein will pave the way for the practical application of differential LQ Nash games with mixed state-control constraints.

References

[1] A. E. Bryson and Y. C. Ho, *Applied Optimal Control: Optimization, Estimation, and Control*, John Wiley & Sons, Hoboken, NJ, 1979.

[2] Francis Clarke and M. R. de Pinho, *Optimal control problems with mixed constraints*, SIAM J. Control Optim. **48** (2010), no. 7, 4500–4524, DOI 10.1137/090757642. MR2683896 (2011i:49039)

[3] C. J. Day, B. F. Hobbs, and J. S. Pang, *Oligopolistic competition in power networks: A conjectured supply function approach*, IEEE Trans. Power Syst., 17 (2002), pp. 597–607.

[4] Francisco Facchinei and Christian Kanzow, *Generalized Nash equilibrium problems*, 4OR **5** (2007), no. 3, 173–210, DOI 10.1007/s10288-007-0054-4. MR2353415 (2008h:91011)

[5] Francisco Facchinei and Jong-Shi Pang, *Nash equilibria: the variational approach*, Convex optimization in signal processing and communications, Cambridge Univ. Press, Cambridge, 2010, pp. 443–493. MR2767574

[6] L. Göllmann, D. Kern, and H. Maurer, *Optimal control problems with delays in state and control variables subject to mixed control-state constraints*, Optimal Control Appl. Methods **30** (2009), no. 4, 341–365, DOI 10.1002/oca.843. MR2553382 (2010h:49043)

[7] Lanshan Han, M. Kanat Camlibel, Jong-Shi Pang, and W. P. Maurice H. Heemels, *A unified numerical scheme for linear-quadratic optimal control problems with joint control and state constraints*, Optim. Methods Softw. **27** (2012), no. 4-5, 761–799, DOI 10.1080/10556788.2011.593624. MR2946056

[8] Benjamin F. Hobbs and Jong-Shi Pang, *Spatial oligopolistic equilibria with arbitrage, shared resources, and price function conjectures*, Math. Program. **101** (2004), no. 1, Ser. B, 57–94, DOI 10.1007/s10107-004-0537-4. MR2085259 (2005c:91010)

[9] B. F. Hobbs and F. A. M. Rijkers, *Strategic generation with conjectured transmission price responses in a mixed transmission pricing system–Part I: Formulation*, IEEE Trans. Power Syst., 19 (2004), pp. 707–717.

[10] Z. Q. Luo and J. S. Pang, *Analysis of iterative waterfilling algorithm for multiuser power control in digital subscriber lines*, EURASIP J. on Applied Signal Processing, Article ID 24012 (2006), pp. 1–10.

[11] Karol Makowski and Lucien W. Neustadt, *Optimal control problems with mixed control-phase variable equality and inequality constraints*, SIAM J. Control **12** (1974), 184–228. Collection of articles dedicated to the memory of Lucien W. Neustadt. MR0380566 (52 #1466)

[12] J. S. Pang and G. Scutari, *Joint sensing and power allocation in nonconvex cognitive radio games: Quasi-Nash equilibria*, IEEE Trans. Signal Process., revision under review.

[13] Jong-Shi Pang and Gesualdo Scutari, *Nonconvex games with side constraints*, SIAM J. Optim. **21** (2011), no. 4, 1491–1522, DOI 10.1137/100811787. MR2869506

[14] Jong-Shi Pang, Gesualdo Scutari, Daniel P. Palomar, and Francisco Facchinei, *Design of cognitive radio systems under temperature-interference constraints: a variational inequality approach*, IEEE Trans. Signal Process. **58** (2010), no. 6, 3251–3271, DOI 10.1109/TSP.2010.2043138. MR2756856

[15] M. d. R. de Pinho, R. B. Vinter, and H. Zheng, *A maximum principle for optimal control problems with mixed constraints*, IMA J. Math. Control Inform. **18** (2001), no. 2, 189–205, DOI 10.1093/imamci/18.2.189. MR1837350 (2002c:49037)

[16] G. Scutari, J. S. Pang, D. P. Palomar, and F. Facchinei, *Convex optimization, game theory, and variational inequality in multiuser communication systems*, IEEE Signal Process. Mag., 27 (2010), pp. 35–49.

[17] Y. Song, Y. Ni, F. Wen, and F. F. Wu, *Analysis of strategic interations among generation companies using conjectured supply function equilibrium model*, IEEE Power Engineering Society General Meeting, 2003, pp. 849–853.

Department of Industrial and Enterprise Systems Engineering, University of Illinois at Urbana-Champaign, Urbana, Illinois, USA.

Current address: ISO New England, Holyoke, Massachusetts 01040

E-mail address: dschiro@iso-ne.com

Department of Industrial and Enterprise Systems Engineering, University of Illinois at Urbana-Champaign, Urbana, Illinois, USA

Current address: Department of Industrial and Systems Engineering, University of Southern California, Los Angeles, California 90089

E-mail address: jongship@usc.edu

Contemporary Mathematics
Volume **659**, 2016
http://dx.doi.org/10.1090/conm/659/13157

Nonlinear input-output equilibrium

Roman A. Polyak

Abstract. We introduce and study Nonlinear Input-Output Equilibrium (NIOE). The main difference between NIOE and the classical Wassily Leontief's input-output (IO) model is: both the production cost and the consumption are not fixed and are not given a priori.

Instead, the production cost is an operator, which maps the output into the cost per unit and the consumption is an operator, which maps the price for goods into the consumption.

The NIOE finds such output and such prices for goods that the following two important goals can be reached.

First, at the NIOE the production cost is in agreement with the output and the consumption is in agreement with the prices for goods.

Second, at the NIOE the total production cost reaches its minimum, while the total consumption reaches its maximum.

Finding NIOE is equivalent to solving a variational inequality (VI) on a simple feasible set $\Omega = \mathbb{R}^n_+ \times \mathbb{R}^n_+$.

Therefore for finding NIOE we use two first-order methods: Pseudo-Gradient Projection (PGP) and Extra Pseudo-Gradient (EPG), for which the projection on Ω is the main operation per step. Numerically both methods require two matrix by vector multiplications or $O(n^2)$ arithmetic operations per step.

It allows, along with proving convergence and global Q-linear convergence rate, also establishing complexity bounds for both methods under various assumptions on the input data.

1. Introduction

The input-output (IO) model has been introduced before the second World War (WW2). The main purpose was better understanding the interdependence of the production sectors of an economy.
Since WW2 the IO model has been widely used for analysis of economic activities, planning production, economic prognoses, international trade just to mention a few. The applications of IO range from a branch of a single economy to the World economy.

The main contributor to IO theory and, in particular, to the practical aspects of IO model was Wassily W. Leontief (1906-1999)(see [**17**]-[**19**]).

He received the Nobel Price in Economics in 1973 "for the development of the input-output method and for its application to important economic problems."

The input-output model assumes that an economy produces n products, which are also consumed in the production process. The elements $a_{ij} \geq 0$ $1 \leq i \leq n$; $1 \leq j \leq n$ of the consumption matrix $A = \|a_{ij}\|$ show the amount of product $1 \leq i \leq n$ consumed for production of one unit of product $1 \leq j \leq n$.

Let $x = (x_1, \ldots, x_n)$ be the production vector, i.e. x_j defines how many units of product $1 \leq j \leq n$ we are planning to produce. Then Ax is the total consumption needed to produce the vector $x \in \mathbb{R}^n_{++}$. Therefore the components of the vector $y = x - Ax = (I - A)x$ shows how much of each product is left after the production needs were covered. Vector y can be used for consumption, investments, trades just to mention a few possibilities.

Let $\lambda = (\lambda_1, \ldots, \lambda_n)$ be the price vector, i.e. λ_j defines the price of one unit of product $1 \leq j \leq n$, then $q = (I - A)\lambda$ is the profit vector, i.e. q_j defines the profit out of one unit of product $1 \leq j \leq n$ under the price vector $\lambda \in \mathbb{R}^n_{++}$.

The IO model solves two basic problems.

(1) For a fixed and given a priori consumption vector $c = (c_1, \ldots, c_n)^T \in \mathbb{R}^n_+$ the IO finds the production vector $x_c = (x_{c,1}, \ldots, x_{c,n})^T \in \mathbb{R}^n_+$, which guarantees the required consumption. Vector x_c is the solution of the following linear system

$$(I - A)x = c, \;\; x \in \mathbb{R}^n_+ \tag{1.1}$$

(2) For a fixed and given a priori profit vector $q = (q_1, \ldots, q_n)^T \in \mathbb{R}^n_+$ the IO finds such price vector $\lambda_q = (\lambda_{q,1}, .., \lambda_{q,n})^T \in \mathbb{R}^n_+$, which guarantees the given profit. Vector λ_q is the solution of the following system

$$q = (I - A)^T\lambda, \;\; \lambda \in \mathbb{R}^n_+. \tag{1.2}$$

For a **productive** economy both systems (1.1) and (1.2) have unique solutions and both vectors x_c and λ_q are positive (see, for example, [**2**], [**8**]). The productivity assumption will be kept throughout the paper.

Another possible use of IO model consists of finding "optimal" production vector $x^* \in \mathbb{R}^n_+$ and "optimal" price vector $\lambda^* \in \mathbb{R}^n_+$ by solving the dual Linear Programming (LP) pair.

Let $p = (p_1, \ldots, p_n)^T$ be fixed and a priori given production cost vector, and $c = (c_1, \ldots, c_n)^T \in \mathbb{R}^n_+$ be fixed and a priori given consumption vector. Finding "optimal" production output x^* and "optimal" consumption prices λ^* leads to the following dual pair of LP problems

$$x^* \in \operatorname{Argmin}\{(p, x) | (I - A)x \geq c, x \in \mathbb{R}^n\} \tag{1.3}$$

and

$$\lambda^* \in \operatorname{Argmax}\{(c, \lambda) | (I - A)^T\lambda \leq p, \lambda \in \mathbb{R}^n_+\} \tag{1.4}$$

For a productive economy both the primal feasible solution

$$x_c = (I - A)^{-1}c$$

and the dual feasible solution

$$\lambda_p = ((I - A)^T)^{-1}p$$

are positive vectors (see, for example, [**2**], [**16**]).

So x_c and λ_p are primal and dual feasible solution, for which the complementarity conditions are satisfied. Therefore $x^* = x_c$ and $\lambda^* = \lambda_p$.

In other words, the dual LP pair (1.3) and (1.4), unfortunately, does not produce results different from IO model.
The purpose of this paper is to introduce an study the Nonlinear Input-Output Equilibrium (NIOE), which extends the abilities of the classical IO model in a few directions.

The fixed cost vector $p = (p_1, \dots, p_n)^T \in \mathbb{R}^n_+$ we replace by a cost operator $p : \mathbb{R}^n_+ \to \mathbb{R}^n_+$, which maps the production output vector $x = (x_1, \dots, x_n)^T \in \mathbb{R}^n_+$ into cost per unit vector $p(x) = (p_1(x), \dots, p_n(x))^T \in \mathbb{R}^n_+$.
Similarly, the fixed consumption vector $c = (c_1, \dots, c_n)^T \in \mathbb{R}^n_+$ we replace by consumption operator $c : \mathbb{R}^n_+ \to \mathbb{R}^n_+$,which maps the prices for goods vector $\lambda = (\lambda_1, \dots, \lambda_n)^T \in \mathbb{R}^n_+$ into consumption vector $c(\lambda) = (c_1(\lambda), \dots, c_n(\lambda))^T \in \mathbb{R}^n_+$.

We will call a pair $y^* = (x^*; \lambda^*) \in \Omega = \mathbb{R}^n_+ \times \mathbb{R}^n_+$ the Nonlinear Input-Output Equilibrium (NIOE) if

$$x^* \in \text{Argmin}\{(p(x^*), x) | (I - A)x \geq c(\lambda^*), x \in \mathbb{R}^n_+\} \tag{1.5}$$

and

$$\lambda^* \in \text{Argmax}\{(c(\lambda^*), \lambda) | (I - A)^T \lambda \leq p(x^*), \lambda \in \mathbb{R}^n_+\} \tag{1.6}$$

The primal and dual LP (1.3) and (1.4), which one obtains in case of $p(x) \equiv p$ and $c(\lambda) \equiv c$, can be viewed as linear input-output equilibrium, which is identical to IO model.
Finding NIOE $y^* = (x^*; \lambda^*)$ is equivalent to solving a concave two person game, which is, in turn, equivalent to solving a particular variational inequality (VI) on $\Omega = \mathbb{R}^n_+ \times \mathbb{R}^n_+$.

Using the equivalence we found conditions on the production and consumption operators, under which the NIOE exists and it is unique.

For solving VI we use two projection type methods. Projection methods for convex optimization were introduced in the 60s (see [**11**], [**20**]) and used for solving VI in [**4**]. They got mainly theoretical value, because even in case of linear constraints projection methods require solving quadratic programming problem at each step. Projection on Ω, however, is a very simple operation, therefore for finding $y^* = (x^*; \lambda^*) \in \Omega$ we consider two first order projection type methods: Pseudo-Gradient Projection (PGP) and Extra Pseudo-Gradient (EPG), for which projection on Ω is the main operation per step. Numerically it requires two matrix by vector multiplications, which is at most $O(n^2)$ arithmetic operations per step.

It allows, along with proving convergence and establishing global Q-linear convergence rate, also estimate complexity bounds for both methods under various assumptions on the input date.

Both methods decompose the NIOE problem in the primal and in the dual spaces allowing computing the production and the price vectors simultaneously. Both PGP and EPG can be viewed as pricing mechanisms for establishing NIOE.

The main distinction of the EPG is its ability to handle NIOE problems when both the production and the consumption operators are not strongly monotone, but just monotone.

This is one of the finest property of the EPG method introduced by G.Korpelevich in the 70s (see [**15**]). It drawn much attention lately (see, for example, [**1**], [**5**]-[**7**],[**12**]-[**14**], [**21**]).

In case of NIOE the application of EPG leads to two stage algorithm. At the first stage EPG predicts the production vector $\hat{x} \in \mathbb{R}^n_+$ and the price vector $\hat{\lambda} \in \mathbb{R}^n_+$.

At the second stage EPG corrects them dependent on the production cost per unit vector $p(\hat{x})$ and consumption vector $c(\hat{\lambda})$.

The paper is organized as follows. In the following section we recall some basic facts about IO model.

In section 3 we show the equivalence of NIOE to two person concave game and, eventually, to a particular VI and prove existence and uniqueness of NIOE.

In section 4 we consider the PGP method and show its global Q-linear convergence rate under local strong monotonicity and local Lipschitz condition of both p and c operators.

In section 5 we prove convergence of the EPG for finding NIOE under minimum assumption on the input data.

In section 6 under local strong monotnicity and Lipschitz condition of both operators p and c the global Q-linear convergence rate was proven and complexity bound for EPG method was established. In section 7 we estimate the Lipschitz constant of the VI operator.

We conclude the paper with some remarks on complexity and numerical aspects of PGP and EPG methods.

2. Preliminaries

We consider an economy, which produces n products. The products are partially consumed in the production process.

The economy is called **productive** if for any consumption vector $c \in \mathbb{R}^n_+$ the following system

$$x - Ax = c,\ x \geq 0 \tag{2.1}$$

has a solution.
In such case the matrix A is called productive matrix. It is rather remarkable that if the system (2.1) has a solution for only one vector $c \in \mathbb{R}^n_{++}$ it has a solution for any given vector $c \in \mathbb{R}^n_+$.

It is obvious that for any consumption vector $c \in \mathbb{R}^n_+$ the system (2.1) has a positive solution if the inverse matrix $(I - A)^{-1}$ exists and it is positive.

To address the issue of matrix $B = (I - A)^{-1}$ positivity we have to recall the notion of indecomposability of a non-negative matrix A.

Let S be a subset of a set of indices $N = \{1, \dots, n\}$, i.e. $S \subset N$ and $S^{'} = N \setminus S$. The sets of indices S is called isolated set if $a_{ij} = 0$ for all $i \in S$ and $j \in S'$. It means that the products, which belong to set S are not used for production of any product from the set $S^{'}$.

In other words if the matrix is decomposable then there is a subset of indices that by simultaneous renumeration of rows and columns we can find $S = \{1, \dots, k\}$ and $S^{'} = \{k+1, \dots, n\}$ such that

$$A = \begin{array}{c} \\ S \\ S' \end{array} \begin{array}{c} \begin{array}{cc} S & S' \end{array} \\ \left[\begin{array}{c|c} A_{11} & 0 \\ \hline A_{21} & A_{22} \end{array}\right] \end{array} \tag{2.2}$$

where $A_{11} : \mathbb{R}^k \to \mathbb{R}^k$, $A_{22} : \mathbb{R}^{n-k} \to \mathbb{R}^{n-k}$. The matrix A is indecomposable if representation (2.2) is impossible by any simultaneous rows and columns renumeration.

For indecomposable matrix production of any product $1 \leq j \leq n$ require product $1 \leq i \leq n$ directly ($a_{ij} > 0$) or indirectly, i.e. for any i and j such that $a_{ij} = 0$ there is a set of indices $i = i_1, \dots, i_{m-1}, i_m = j$, that $a_{i_s,i_{s+1}} > 0, s = 1, 2, \dots, m-1$.

The following theorem (see, for example, [**2**], [**8**], [**10**]) plays an important role on the input-output model.

THEOREM 1 (Frobenius-Perron). *Any non-negative indecomposable matrix A has a real positive dominant eigenvalue λ_A, i.e. for any eigenvalue λ of matrix A, the following inequality $|\lambda| \leq \lambda_A$ holds and the correspondent to λ_A eigenvector $x_A \in \mathbb{R}^n_{++}$.*

We are ready to formulate (see, for example, [**2**], [**8**], [**9**], [**16**], [**19**]) the necessary and sufficient condition for Leontief's model to be productive .

THEOREM 2 (Leontief). *For non-negative indecomposable consumption matrix A the input-output model (2.1) is productive iff $\lambda_A < 1$.*

It follows from $\lambda_A < 1$ and Theorem 1 that for all eigenvalues λ of A we have $|\lambda| < 1$, therefore the matrix $B = \|b_{ij}\|$ has the following representation

$$B = (I - A)^{-1} = I + A + A^2 + A^3 + \dots$$

For non-negative indecomposable matrix A we have

$$B - A = I + A^2 + A^3 + \dots >> 0,$$

which means that total consumption b_{ij} of product $1 \leq i \leq n$ require to produce one item of product $1 \leq j \leq n$ is always greater than direct consumption a_{ij}.

There is another sufficient condition for Leontief's model to be productive. (see, for example, [**9**], [**16**], [**17**], [**19**])

Let $\lambda = (\lambda_1, \dots, \lambda_n) \in \mathbb{R}^n_{++}$ be the price vector, components of which are prices for one item of correspondent product. If for all $1 \leq i \leq n$ products but one we have balance, i.e. there is $1 \leq j \leq n$ that

$$\lambda_j > (\lambda^T A)_j \text{ and } \lambda_i = (\lambda^T A)_i, i \neq j,$$

then the economy is productive, i.e. for any given $c \in \mathbb{R}^n_+$ the system (1.1) has a positive solution and for any given $q \in \mathbb{R}^n_+$ the system (1.2) has a positive solution as well.

In what follows we will consider a non-negative indecomposable matrices A with $\lambda_A < 1$. Therefore the inverse matrix

$$B = (I - A)^{-1} \tag{2.3}$$

exists and it is positive (see, for example, [**16**]).

As we mentioned already using the dual LP pair (1.3)-(1.4) to find the optimal production and optimal prices does not produce results different from IO.

In order to find optimal production consistent with the production cost and optimal consumption consistent with the consumption prices we need a new tool, which we consider in the following section.

3. Nonlinear Input-Output Equilibrium

It follows from (1.5) that the production vector $x^* \in \mathbb{R}^n_+$ minimizes the total production cost and at the same time is such that the consumption $c(\lambda^*) \in \mathbb{R}^n_+$ defined by the price vector $\lambda^* \in \mathbb{R}^n_+$ is satisfied.

It follows from (1.6) that the price vector λ^* for goods maximizes the total consumption and at the same time guarantees that the price of each unit of goods does not exceeds the production cost given by vector $p(x^*)$.

We would like to emphasize that due to (1.5) the production cost $p(x^*)$ is in agreement with the optimal production vector x^* and due to (1.6) the consumption $c(\lambda^*)$ is in agreement with the optimal price vector λ^*.

Therefore our main concern is the existence of NIOE $y^* = (x^*, \lambda^*)$.

We assume at this point that both p and c are strongly monotone operators, i.e. there is $\alpha > 0$ and $\beta > 0$ that

$$(p(x_1) - p(x_2), x_1 - x_2) \geq \alpha \|x_1 - x_2\|^2, \ \forall x_1, x_2 \in \mathbb{R}^n_+ \tag{3.1}$$

and

$$(c(\lambda_1) - c(\lambda_2), \lambda_1 - \lambda_2) \leq -\beta \|\lambda_1 - \lambda_2\|^2, \ \forall \lambda_1, \lambda_2 \in \mathbb{R}^n_+ \tag{3.2}$$

Assumption (3.1) implies that the production increase of any goods when the production of the rest are fixed leads to the cost per item increase. Moreover the lower bound of the margin is $\alpha > 0$.

Assumption (3.2) implies that the price increase for an item of any product, when the prices for the rest are fixed, leads to consumption decrease of such product. Moreover, the margine has a negative upper bound $-\beta < 0$.

In order to prove NIOE existence we first show that finding NIOE y^* is equivalent to solving a particular variational inequality (VI) on $\Omega = \mathbb{R}^n_+ \times \mathbb{R}^n_+$.

THEOREM 3. *For $y^* = (x^*; \lambda^*) \in \Omega$ to be a solution of (1.5) and (1.6) it is necessary and sufficient for y^* to be a solution of the following VI*

$$(g(y^*), y - y^*) \leq 0, \forall y \in \Omega, \tag{3.3}$$

where the operator $g : \Omega \to \mathbb{R}^{2n}$ is given by the following formula

$$g(y) = ((I - A)^T \lambda - p(x); c(\lambda) - (I - A)x)) \tag{3.4}$$

PROOF. If $y^* = (x^*, \lambda^*)$ solves (1.5) and (1.6), then y^* is a saddle point of the Lagrangian

$$L(y^*; X, \Lambda) = (p(x^*), X) - (\Lambda, (I - A)X - c(\lambda^*)), \tag{3.5}$$

i.e.

$$y^* \in \operatorname*{Argmin}_{X \in \mathbb{R}^n_+} \max_{\Lambda \in \mathbb{R}^n_+} L(y^*; X, \Lambda)$$

In other words,

$$\begin{aligned} x^* &\in \operatorname{Argmin}\{L(y^*; X, \lambda^*) | X \in \mathbb{R}^n_+\} = \\ &= \operatorname{Argmin}\{(p(x^*) - (I - A)^T \lambda^*, X) | X \in \mathbb{R}^n_+\} \\ &= \operatorname{Argmax}\{((I - A)^T \lambda^* - p(x^*), X) | X \in \mathbb{R}^n_+\}. \end{aligned} \tag{3.6}$$

Therefore

$$(I - A)^T \lambda^* - p(x^*) \leq 0. \tag{3.7}$$

On the other hand

$$\lambda^* \in \text{Argmax}\{L(y^*; x^*, \Lambda) | \Lambda \in \mathbb{R}^n_+\}$$

$$= \text{Argmax}\{(c(\lambda^*) - (I - A)x^*, \Lambda) | \Lambda \in \mathbb{R}^n_+\}. \tag{3.8}$$

Therefore

$$c(\lambda^*) - (I - A)x^* \leq 0. \tag{3.9}$$

Keeping in mind the complementarity condition $(g(y^*), y^*) = 0$ for the dual LP pair (1.5)-(1.6) we conclude that (3.3) holds for any $y = (x, \lambda) \in \Omega$, i.e. $y^* = (x^*; \lambda^*)$ solves VI (3.3).

Now let assume that $\bar{y} \in \Omega$ is the solution of VI (3.3), then

$$(g(\bar{y}), y) \leq (g(\bar{y}), \bar{y}), \ \forall y \in \Omega. \tag{3.10}$$

It means that $g(\bar{y}) \leq 0$, otherwise the left hand side can be made as large as one wants by taking the correspondent component of vector y large enough. Therefore we have

$$(I - A)^T \bar{\lambda} - p(\bar{x}) \leq 0, \ \ \bar{x} \geq 0 \tag{3.11}$$

and

$$c(\bar{\lambda}) - (I - A)\bar{x} \leq 0, \ \ \bar{\lambda} \geq 0. \tag{3.12}$$

So $\bar{x}$ is a feasible solution for LP

$$\min\{(p(\bar{x}), x) | (I - A)x \geq c(\bar{x}), x \in \mathbb{R}^n_+\} \tag{3.13}$$

and $\bar{\lambda}$ is a feasible solution for the following dual LP

$$\max\{(c(\bar{x}), \lambda) | (I - A)^T \lambda \leq p(\bar{x}) | \lambda \in \mathbb{R}^n_+\}. \tag{3.14}$$

From (3.10) for $y = 0 \in \mathbb{R}^{2n}$ we have

$$(g(\bar{y}), \bar{y}) \geq 0,$$

which together with (3.11)-(3.12) leads to

$$(g(\bar{y}), \bar{y}) = 0. \tag{3.15}$$

Therefore $(\bar{x}; \bar{\lambda})$ is primal and dual feasible solution for the dual LP pair (3.13) - (3.14) and the complementarity condition (3.15) is satisfied.

Therefore $\bar{x}$ solves (3.13) and $\bar{\lambda}$ solves (3.14), hence $\bar{y} = y^*$ and the proof is completed □

Finding a saddle point of the Lagrangian (3.5), on the other hand, is equivalent to finding an equilibrium of two person concave game with payoff function

$$\varphi_1(y; X, \lambda) = -L(y; X, \lambda) = ((I - A)^T \lambda - p(x), X) + (\lambda, c(\lambda))$$

and strategy $X \in \mathbb{R}^n_+$ for the first player and payoff function

$$\varphi_2(y; x, \Lambda) = L(y; x, \Lambda) = (c(\lambda) - (I - A)x, \Lambda) + (p(x), x)$$

and strategy $\Lambda \in \mathbb{R}^n_+$ for the second player.

Let's consider the normalized payoff function $\Phi : \Omega \times \Omega \to R$, which is given by the following formula

$$\Phi(y; Y) = \varphi_1(y; X, \lambda) + \varphi_2(y; x, \Lambda),$$

then finding $y^* = (x^*; \lambda^*) \in \Omega$ is equivalent to finding a fixed point of the following map

$$y \to \omega(y) = \text{Argmax}\{\Phi(y; Y)|Y \in \Omega\}. \tag{3.16}$$

The normalized payoff function $\Phi(y; Y)$ is linear in $Y \in \Omega$ for any given $y \in \Omega$, but Ω is unbounded, which makes impossible the use of Kakutani's fixed point Theorem to prove the existence of $y^* \in \Omega : y^* \in \omega(y^*)$.

The operator $g : \Omega \to \mathbb{R}^{2n}$ will be called pseudo-gradient because of the following formula

$$g(y) = \nabla_Y \Phi(y; Y) = ((I - A)^T \lambda - p(x); c(\lambda) - (I - A)x) \tag{3.17}$$

We will prove the existence of the NIOE (1.5)-(1.6) later.

First, let us prove the following Lemma.

LEMMA 1. *If operators p and c are strongly monotone, i.e. (3.1)-(3.2) are satisfied, then the operator $g : \mathbb{R}^{2n}_+ \to \mathbb{R}^{2n}$, given by (3.17), is strongly monotone as well, i.e. there is $\gamma > 0$ that*

$$(g(y_1) - g(y_2), y_1 - y_2) \leq -\gamma \|y_1 - y_2\|^2 \quad \forall y_1, y_2 \in \Omega. \tag{3.18}$$

PROOF. Let consider $y_1 = (x_1; \lambda_1)$ and $y_2 = (x_2; \lambda_2) \in \Omega$, then

$$\begin{aligned}(g(y_1) - g(y_2), y_1 - y_2) &= ((I - A)^T \lambda_1 - p(x_1) - (I - A)^T \lambda_2 + p(x_2), x_1 - x_2) + \\ &+ (c(\lambda_1) - (I - A)x_1 - c(\lambda_2) + (I - A)x_2, \lambda_1 - \lambda_2) = \\ &((I - A)^T (\lambda_1 - \lambda_2), x_1 - x_2) - (p(x_1) - p(x_2), x_1 - x_2) + \\ &+ (c(\lambda_1) - c(\lambda_2), \lambda_1 - \lambda_2) - ((I - A)(x_1 - x_2), \lambda_1 - \lambda_2) \\ &\leq -\alpha \|x_1 - x_2\|^2 - \beta \|\lambda_1 - \lambda_2\|^2.\end{aligned}$$

Therefore for $\gamma = \min\{\alpha, \beta\}$ we have (3.18).

We are ready to prove the existence of NIOE $y^* = (x^*; \lambda^*)$ in (1.5)-(1.6).

THEOREM 4. *If p and c are continuous operators and the strong monotonicity assumption (3.1)-(3.2) hold, then the NIOE $y^* = (x^*, \lambda^*)$ defined by (1.5)-(1.6) exists and it is unique.*

PROOF. Let consider $y_0 \in \mathbb{R}^{2n}_{++}$ such that $\|y_0\| \leq 1$ and large enough number $M > 0$. We replace Ω by $\Omega_M = \{y \in \Omega : \|y - y_0\| \leq M\}$. Now the set Ω_M is bounded and convex and so is the set

$$\omega_M(y) = \text{Argmax}\{\Phi(y, Y)|Y \in \Omega_M\}, \tag{3.19}$$

because for any given $y \in \Omega$ the function $\Phi(y, Y)$ is linear in Y and Ω_M is convex and bounded.

Moreover, due to continuity p and c the map $y \to \omega_M(y)$ is upper semi-continuous.

Therefore Kakutani's Theorem can be applied, i.e. there is $y^*_M \in \Omega_M$ such that $y^*_M \in \omega_M(y^*_M)$ (see, for example, [**16**]). Now we will show that the constraint $\|y - y_0\| \leq M$ in (3.19) is irrelevant. Let us consider (3.18) with $y_1 = y_0$ and $y_2 = y^*_M$. We obtain

$$\begin{aligned}\gamma \|y_0 - y^*_M\|^2 &\leq (g(y^*_M) - g(y_0), y_0 - y^*_M) = \\ &= (g(y^*_M), y_0 - y^*_M) + (g(y_0), y^*_M - y_0)).\end{aligned} \tag{3.20}$$

From $y_M^* \in \omega_M(y_M^*)$ and (3.17) follows that y_M^* is solution of VI

$$(g(y), Y - y) \leq 0, \ \forall Y \in \Omega_M$$

i.e.

$$(g(y_M^*), y_0 - y_M^*) \leq 0.$$

Therefore from (3.20) we have

$$\gamma \|y_0 - y_M^*\|^2 \leq (g(y_0), y_M^* - y_0).$$

Using Couchy-Scwarz inequality we obtain $\|y_0 - y_M^*\| \leq \gamma^{-1}\|g(y_0)\|$, therefore

$$\|y_M^*\| \leq \|y_0\| + \|y_0 - y_M^*\| \leq 1 + \gamma^{-1}\|g(y_0)\|.$$

So, for M large enough the constraint $\|y_0 - y\| \leq M$ can not be active, therefore it is irrelevant in (3.19). Hence, $y^* = y_M^*$ solves the VI (3.3), which is equivalent to finding NIOE from (1.5)-(1.6).

The uniqueness follows from (3.18). In fact, assuming that there is two vectors y^* and $\bar{y}$, which solve VI (3.3) we obtain

$$(g(y^*), \bar{y} - y^*) \leq 0 \text{ and } (g(\bar{y}), y^* - \bar{y}) \leq 0$$

or

$$(g(y^*) - g(\bar{y}), y^* - \bar{y}) \geq 0,$$

which contradicts (3.18) with $y_1 = y^*$ and $y_2 = \bar{y}$, i.e. y^* is unique. □

The strong monotonicity assumption (3.1) and (3.2) are just sufficient for existence of NIOE. In the rest of the paper we assume the existence NIOE.

In what is following we replace the global strong monotonicity assumptions (3.1)-(3.2) for both operators p and c by less restrictive assumptions of local strong monotonicity only at the NIOE $y^* = (x^*; \lambda^*)$.

We assume the existence of $\alpha > 0$ and $\beta > 0$ such that

$$(3.21) \qquad (p(x) - p(x^*), x - x^*) \geq \alpha\|x - x^*\|^2, \ \forall x \in \mathbb{R}^n_+$$

$$(3.22) \qquad (c(\lambda) - c(x^*), \lambda - \lambda^*) \leq -\beta\|\lambda - \lambda^*\|^2, \forall \lambda \in \mathbb{R}^n_+$$

holds.

In the next section we also replace the global Lipschitz continuity of p and c by corresponding assumptions at the NIOE $y^* = (x^*; \lambda^*)$.

$$(3.23) \qquad \|p(x) - p(x^*)\| \leq L_p\|x - x^*\|, \ \forall x \in \mathbb{R}^n_+$$

and

$$(3.24) \qquad \|c(\lambda) - c(\lambda^*)\| \leq L_c\|\lambda - \lambda^*\|, \forall \lambda \in \mathbb{R}^n_+.$$

We will say that both production p and consumption c operators are well defined if (3.21)-(3.24) hold.

The assumption (3.21) means that the production cost operator p is sensitive to the production change only at the equilibrium. The assumption (3.22) means that the consumption operator c is sensitive to the prices change only at the equilibrium. Lipschitz condition (3.23)-(3.24) means that the production and consumption is under control in the neigbourhood of the NIOE $y^* = (x^*; \lambda^*)$.

4. Pseudo-Gradient Projection Method

The VI (3.3) has a simple feasible set $\Omega = \mathbb{R}^n_+ \times \mathbb{R}^n_+$, projection on which is a very simple operation. Therefore we will concentrate on two methods, for which projection on Ω is the main operation per step.

For a vector $u \in \mathbb{R}^q$ the projection on R^q_+ is given by formula

$$v = P_{R^q_+}(u) = [u]_+ = ([u_1]_+, \dots, [u_q]_+)^T,$$

where for $1 \le i \le q$ we have

$$[u_i]_+ = \begin{cases} u_i, \; u_i \ge 0 \\ 0, \; u_i < 0 \end{cases}$$

The operator $P_\Omega : \mathbb{R}^n \times \mathbb{R}^n \to \mathbb{R}^n_+ \times \mathbb{R}^n_+$ we define by formula

$$P_\Omega(y) = ([x]_+, [\lambda]_+),$$

where $y = (x; \lambda), \; x \in \mathbb{R}^n, \lambda \in \mathbb{R}^n$.

The following two well known properties of the projection operator P_Ω will be used later.

First, the operator P_Ω is not expansive, i.e. for any $u_1 \in \mathbb{R}^{2n}, u_2 \in \mathbb{R}^{2n}$ we have

$$\|P_\Omega(u_1) - P_\Omega(u_2)\| \le \|u_1 - u_2\| \tag{4.1}$$

Second, the vector u^* solves the VI (3.3) iff for any $t > 0$ the vector u^* is a fixed point of the map $P_Q(I + tg) : \Omega \to \Omega$, i.e.

$$u^* = P_\Omega(u^* + tg(u^*)). \tag{4.2}$$

Let us recall that the VI operator $g : \Omega \to \mathbb{R}^{2n}$ is defined by the following formula

$$g(y) = ((I - A)^T\lambda - p(x); c(\lambda) - (I - A)x).$$

We are ready to describe the Pseudo-Gradient Projection (PGP) method for finding NIOE $y^* \in \Omega$.

Let $y^0 = (x^0; \lambda^0) \in \mathbb{R}^n_{++} \times \mathbb{R}^n_{++}$be the starting point and let the pair $y_s = (x_s; \lambda_s)$ has been found already.

The PGP method finds the next approximation by the following formula

$$y_{s+1} = P_\Omega(y_s + tg(y_s)). \tag{4.3}$$

We will specify the step length $t > 0$ later.

In other words, the PGP method simultaneously updates production vector x_s and the price vector λ_s by the following formulas

$$x_{j,s+1} = [x_{j,s} + t((I - A)^T\lambda_s - p(x_s))_j]_+, \quad 1 \le j \le n \tag{4.4}$$

$$\lambda_{i,s+1} = [\lambda_{i,s} + t(c(\lambda_s) - (I - A)x_s)_i]_+, \quad 1 \le i \le n. \tag{4.5}$$

Therefore PGP is primal-dual decomposition method, which allows computing the primal and dual vectors independently and simultaneously.

Formulas (4.4)-(4.5) can be viewed as a pricing mechanism for eatablishing NIOE.

From (4.4) follows: if the price $((I - A)^T\lambda_s)_j$ for one unit of product $1 \le j \le n$ is greater then the production cost $p_j(x_s)$ then the production of $x_{j,s}$ has to be increased.

On the other hand if production cost $p_j(x_s)$ is greater than the price $((I-A)^T\lambda_s)_j$ of one unit of product $1 \leq j \leq n$ than the production $x_{j,s}$ has to be reduced.

From (4.5) follows: if the consumption $c_i(\lambda_s)$ of product $1 \leq i \leq n$ is greater then the production $((I-A)x_s)_i$ then the price $\lambda_{i,s}$ has to be increased. If the consumption $c_i(\lambda_s)$ is less than the production $((I-A)x_s)$ then the price $\lambda_{i,s}$ has to be reduced.

The PGP method (4.5)-(4.6) can be viewed as a projected explicit Euler method for solving the following system of ordinary differential equations

$$\frac{dx}{dt} = (I-A)^T\lambda - p(x)$$

$$\frac{d\lambda}{dt} = c(\lambda) - (I-A)x$$

with $x(0) = x_0$ and $\lambda(0) = \lambda_0$.

To prove convergence of the PGP method (4.3) we need the following lemma similar to Lemma 1.

LEMMA 2. *If operators p and c are strongly monotone at x^* and λ^*, i.e. (3.21)-(3.22) holds, then the operator $g : \Omega \to \mathbb{R}^{2n}$ given by (3.4) is strongly monotone at y^* i.e. for $\gamma = \min\{\alpha, \beta\} > 0$ the following bound holds.*

$$(g(y) - g(y^*), y - y^*) \leq -\gamma\|y - y^*\|^2, \forall y \in \Omega. \tag{4.6}$$

The proof is similar to the proof of Lemma 1.

LEMMA 3. *If operators p and c satisfy local Lipschitz condition (3.23)-(3.24), then operator $g : \Omega \to \mathbb{R}^{2n}$ given by (3.4) satisfies local Lipschitz condition at y^*, i.e. there is $L > 0$ such that*

$$\|g(y) - g(y^*)\| \leq L\|y - y^*\|, \ \forall y \in \Omega. \tag{4.7}$$

For the proof of Lemma 3 and the upper bound for L see Appendix.

REMARK 1. *Let us assume that for a given $x \in \mathbb{R}^n_+$ computing $p(x)$ and for a given $\lambda \in \mathbb{R}^n_+$ computing $c(\lambda)$ does not require more then $O(n^2)$ operations. It is true, for example, if $c(\lambda) = \nabla(\frac{1}{2}\lambda^T C\lambda + d^T\lambda)$ and $p(x) = \nabla(\frac{1}{2}x^T Px + q^T x)$, where $C : \mathbb{R}^n \to \mathbb{R}^n$ is symmetric negative semidefinite matrix and $P : \mathbb{R}^n \to \mathbb{R}^n$ is symmetric positive semidefinite matrix. Then each PGP step does not require more then $O(n^2)$ operations, because the rest is matrix $(I-A)$ by vector multiplication twice.*

Let $\varkappa = \gamma L^{-1}$ be the condition number of the VI operator $g : \Omega \to \mathbb{R}^{2n}$. The following theorem establishes global Q-linear convergence rate and complexity of the PGP method (4.3). The proof is similar to the proof of Theorem 3 in [**21**]. We will sketch the proof for completeness.

THEOREM 5. *If operators p and c are well defined (i.e. (3.21)-(3.24) holds) then:*

(1) *for any $0 < t < 2\gamma L^{-2}$ the PGP method (4.3) globally converges to NIOE $y^* = (x^*; \lambda^*)$ with Q-linear rate and the ratio $0 < q(t) = (1 - 2t\gamma + t^2L^2)^{\frac{1}{2}} < 1$, i.e.*

$$\|y_{s+1} - y^*\| \leq q(t)\|y_s - y^*\| \tag{4.8}$$

(2) *for* $t = \gamma L^{-2} = \min\{q(t)|t > 0\}$ *the following bound holds*

$$\|y_{s+1} - y^*\| \leq (1 - \varkappa^2)^{\frac{1}{2}} \|y_s - y^*\| \tag{4.9}$$

(3) *the PGP complexity is given by the following bound*

$$comp(PGP) = O(n^2 \varkappa^{-2} \ln \varepsilon^{-1}), \tag{4.10}$$

where $\varepsilon > 0$ *is the given accuracy.*

PROOF. From (4.1) - (4.3) follows

$$\|y_{s+1} - y^*\|^2 = \|P_\Omega(y_s + tg(y_s)) - P_\Omega(y^* + tg(y^*))\|^2 \leq$$

$$\leq \|y_s + tg(y_s) - y^* - tg(y^*)\|^2 = \tag{4.11}$$

$$= \|y_s - y^*\|^2 + 2t(y_s - y^*, g(y_s) - g(y^*)) + t^2\|g(y_s) - g(y^*)\|^2$$

From Lemma 2, Lemma 3 and (4.11) follows

$$\|y_{s+1} - y^*\| \leq \|y_s - y^*\|(1 - 2t\gamma + t^2 L^2)^{\frac{1}{2}},$$

therefore for any $0 < t < 2\gamma L^{-2}$ we have $0 < q(t) = (1 - 2t\gamma + t^2 L^2)^{\frac{1}{2}} < 1$.

For $t = \gamma L^{-2} = \text{Argmin}\{q(t)|t > 0\}$ we have

$$q = q(\gamma L^{-2}) = (1 - (\gamma L^{-1})^2)^{\frac{1}{2}} = (1 - \varkappa^2)^{\frac{1}{2}}.$$

Let $0 < \varepsilon << 1$ be the required accuracy . From (4.8) follows that it takes $O((\ln q)^{-1} \ln \varepsilon)$ steps to find an ε- approximation for NIOE $y^* = (x^*, \lambda^*)$. Each step requires $O(n^2)$ operations, therefore the complexity bound for the PGP method (4.3) is

$$\mathcal{N} = O\left(n^2 \frac{\ln \varepsilon^{-1}}{\ln q^{-1}}\right).$$

From $(\ln q^{-1})^{-1} = (-\frac{1}{2}\ln(1 - \varkappa^2))^{-1}$ and $\ln(1 + x) \leq x, \ \ \forall x > -1$ follows $(\ln q^{-1})^{-1} \leq 2\varkappa^{-2}$, which leads to the bound (4.10). □

In case $\gamma = \min\{\alpha, \beta\} = 0$. Theorem 5 can't guarantee even convergence of the PGP method. We need another tool for solving VI (3.3).

In the following section we apply Extra Pseudo-Gradient (EPG) method for solving VI (3.3). The EPG method was first introduced by G. Korepelevich [**15**] in the 70s for finding saddle points. Since then the EPG became a popular tool for solving VI (see [**5**]-[**7**],[**12**]-[**14**], [**21**]).

We first assume that the operators p and c are monotone and satisfy Lipschitz condition, so is the operator g, i.e. for $\forall y_1, y_2 \in \Omega$ we have

$$(g(y_1) - g(y_2), y_1 - y_2) \leq 0 \tag{4.12}$$

and

$$\|g(y_1) - g(y_2)\| \leq L\|y_1 - y_2\|. \tag{4.13}$$

The monotonicity property (4.12) follows from monotonicity properties of operators p and c (see Lemma 1). The Lipschitz condition (4.13) follows from correspondent condition of c and p.

5. Extra Pseudo-Gradient method for finding NIOE

Application of G.Korpelevich extra-gradient method for solving VI (3.3) leads to the following method for finding NIOE.

Each step of the EPG method consists of two phases: in the prediction phase we predict the production vector $\hat{x} \in \mathbb{R}^n_+$ and the price vector $\hat{\lambda} \in \mathbb{R}^n_+$, in the corrector phase we correct the production and the price vectors by using the predicted cost per unit vector $p(\hat{x})$ and the predicted consumption vectors $c(\hat{\lambda})$.

Let $y_0 = (x_0, \lambda_0) \in \mathbb{R}^n_{++} \times \mathbb{R}^n_{++}$ be the starting point. We assume that the approximation $y_s = (x_s, \lambda_s)$ has been found already.

The predictor phase finds

$$\hat{y}_s = P_\Omega(y_s + tg(y_s)) = [y_s + tg(y_s)]_+. \tag{5.1}$$

The corrector phase finds the new approximation

$$y_{s+1} = P_\Omega(y_s + tg(\hat{y}_s)) = [y_s + tg(\hat{y}_s)]_+. \tag{5.2}$$

The step length $t > 0$ will be specified later.

In other words, the EPG method first predicts the production vector

$$\hat{x}_s = [x_s + t((I - A)^T\lambda_s - p(x_s))]_+ \tag{5.3}$$

and the price vector

$$\hat{\lambda}_s = [\lambda_s + t(c(\lambda_s) - (I - A)x_s)]_+. \tag{5.4}$$

Then EPG finds the new production vector

$$x_{s+1} = [x_s + t((I - A)^T\hat{\lambda}_s - p(\hat{x}_s))]_+ \tag{5.5}$$

and the new price vector

$$\lambda_{s+1} = [\lambda_s + t(c(\hat{\lambda}_s) - (I - A)\hat{x}_s)]_+. \tag{5.6}$$

The meaning of formulas (5.3)-(5.6) is similar to the meaning of formulas (4.4)-(4.5).

The EPG method (5.1)-(5.2), in fact, is a pricing mechanism for establishing NIOE. The following Theorem establishes the convergence of EPG.

THEOREM 6. *If p and c are monotone operators and Lipschitz condition (4.13) is satisfied, then for any $t \in (0, (\sqrt{2}L)^{-1})$ the EPG method (5.1)-(5.2) generates a converging sequence $\{y_s\}_{s=0}^{\infty}$ and $\lim_{s\to\infty} y_s = y^*$.*

(1) PROOF. It follows from (5.1)-(5.2), non-expansiveness of P_Q and Lipschitz condition (4.13) that

$$\begin{aligned}\|y_{s+1} - \hat{y}_s\| &= \|P_\Omega(y_s + tg(\hat{y}_s)) - P_\Omega(y_s + tg(y_s))\| \\ &\le t\|g(\hat{y}_s) - g(y_s)\| \le tL\|\hat{y}_s - y_s\|.\end{aligned} \tag{5.7}$$

From (5.1) follows

$$(tg(y_s) + (y_s - \hat{y}_s), y - \hat{y}_s) \le 0, \ \ \forall y \in \Omega,$$

therefore by taking $y = y_{s+1}$, we obtain

$$(tg(y_s) + (y_s - \hat{y}_s, y_{s+1} - \hat{y}_s) \le 0$$

or

$$\begin{aligned}&(y_s - \hat{y}_s, y_{s+1} - \hat{y}_s) + t(g(\hat{y}_s), y_{s+1} - \hat{y}_s)- \\ &-t(g(\hat{y}_s) - g(y_s), y_{s+1} - \hat{y}_s) \le 0.\end{aligned} \tag{5.8}$$

It follows from (5.7) and Lipschitz condition (4.13)

$$(g(\hat{y}_s)-g(y_s),y_{s+1}-\hat{y}_s)\le\|g(\hat{y}_s)-g(y_s)\|\|y_{s+1}-\hat{y}_s\|\le tL^2\|\hat{y}_s-y_s\|^2. \tag{5.9}$$

From (5.8) and (5.9) we obtain

$$(y_s-\hat{y}_s,y_{s+1}-\hat{y}_s)+t(g(\hat{y}_s),y_{s+1}-\hat{y}_s)-(tL)^2\|\hat{y}_s-y_s\|^2\le 0. \tag{5.10}$$

From (5.2) follows

$$(tg(\hat{y}_s)+y_s-y_{s+1},y-y_{s+1})\le 0,\ \ \forall y\in\Omega. \tag{5.11}$$

Therefore for $y=y^*$ we have

$$(y_s-y_{s+1}+tg(\hat{y}_s),y^*-y_{s+1})\le 0. \tag{5.12}$$

Also from $(g(y^*),y-y^*)\le 0,\ \ \forall y\in\Omega$ we obtain $(g(y^*),\hat{y}_s-y^*)\le 0$, so for $\forall t>0$ we have

$$t(-g(y^*),y^*-\hat{y}_s)\le 0. \tag{5.13}$$

By adding (5.10),(5.12) and (5.13) and using the monotonicity of g, i.e.

$$(g(\hat{y}_s)-g(y^*),y^*-\hat{y}_s)\ge 0$$

we obtain

$$2(y_s-y_{s+1},y^*-y_{s+1})+2(y_s-\hat{y}_s,y_{s+1}-\hat{y}_s)-2(tL)^2\|\hat{y}_s-y_s\|^2\le 0. \tag{5.14}$$

Using the following itentity

$$2(u-v,w-v)=\|u-v\|^2+\|v-w\|^2-\|u-w\|^2 \tag{5.15}$$

twice first with $u=y_s,v=y_{s+1}$ and $w=y^*$ and second with $u=y_s,v=\hat{y}_s$ and $w=y_{s+1}$ we obtain

$$2(y_s-y_{s+1},y^*-y_{s+1})=\|y_s-y_{s+1}\|^2+\|y_{s+1}-y^*\|^2-\|y_s-y^*\|^2 \tag{5.16}$$

and

$$2(y_s-\hat{y}_s,y_{s+1}-\hat{y}_s)=\|y_s-\hat{y}_s\|^2+\|\hat{y}_s-y_{s+1}\|^2-\|y_s-y_{s+1}\|^2. \tag{5.17}$$

From (5.14), (5.16) and (5.17) follows

$$\|y_{s+1}-y^*\|^2+(1-2(tL)^2)\|y_s-\hat{y}_s\|^2+\|\hat{y}_s-y_{s+1}\|^2\le\|y_s-y^*\|^2. \tag{5.18}$$

Summing up (5.18) from $s=0$ to $s=N$ we obtain

$$\|y_{N+1}-y^*\|^2+(1-2(tL)^2)\sum_{s=0}^{N}\|y_s-\hat{y}_s\|^2+ \tag{5.19}$$

$$+\sum_{s=0}^{N}\|\hat{y}_s-y_{s+1}\|^2\le\|y_0-y^*\|^2.$$

It follows from (5.19) that for $0<t\le\frac{1}{\sqrt{2}L}$ we have $\sum_{s=0}^{\infty}\|y_s-\hat{y}_s\|^2<\infty$ and $\sum_{s=0}^{\infty}\|\hat{y}_s-y_{s+1}\|^2<\infty$, which means that

$$(a)\ \lim_{s\to\infty}\|y_s-\hat{y}_s\|\to 0\ \text{ and } (b)\ \lim_{s\to\infty}\|\hat{y}_s-y_{s+1}\|\to 0. \tag{5.20}$$

Also from (5.18) follows that

$$\|y_{s+1}-y^*\|\le\|y_s-y^*\|\ \ \forall s\ge 0. \tag{5.21}$$

Thus $\{y_s\}_{s=0}^{\infty}$ is bounded sequence, therefore there is a converging sub-sequence $\{y_{s_k}\}_{s_k>1}^{\infty} : \lim_{k\to\infty} y_{s_k} = \bar{y}$. It follows from (5.20a) that $\lim_{k\to\infty} \hat{y}_{s_k} = \bar{y}$ and from (5.20b) follows $\lim_{k\to\infty} y_{s_k+1} = \bar{y}$. □

From the continuity of the operator g we have

$$\bar{y} = \lim_{k\to\infty} y_{s_k+1} = \lim_{k\to\infty} [y_{s_k} + tg(\hat{y}_{s_k})]_+ = [\bar{y} + tg(\bar{y})]_+.$$

From (4.2) we obtain that $\bar{y} = y^*$, which together (5.21) leads to $\lim_{s\to\infty} y_s = y^*$.

6. Convergence rate and complexity of the EPG method

In this section we establish the convergence rate and complexity of the EPG method under Lipschitz continuity and local strong monotonicity of both operators $p : \mathbb{R}^n_+ \to \mathbb{R}^n_+$ and $c : \mathbb{R}^n_+ \to \mathbb{R}^n_+$. The proof has some similarities with correspondent proof in [**21**], but the complexity bound is better due to the structure of matrix D. We have (7.1) without rescaling matrix D.

The EPG require two projection on Ω while PGP requires only one such projection. Keeping in mind that the most costly operation per projection is matrix by vector multiplication the EPG does not increase the complexity bound $O(n^2)$ per step.

On the other hand, EPG converges in case of just monotonicity of g, i.e. when $\gamma = 0$. Moreover EPG has much better then PGP complexity bound in case of local strong monotonicity, in particular, when the condition number $0 < \varkappa < 1$ is small.

To establish the convergence rate and complexity of EPG we will need two inequalities, which are following from the local strong monotonicity of g.

First, by adding (5.10) and (5.12) we obtain

$$(y_s - y_{s+1}, y^* - y_{s+1}) + t(g(\hat{y}_s), y^* - \hat{y}_s) + (y_s - \hat{y}_s, y_{s+1} - \hat{y}_s) - \\ -(tL)^2\|y_s - \hat{y}_s\|^2 \leq 0. \tag{6.1}$$

Second, it follows from (4.6) that

$$(g(y), y - y^*) - (g(y^*), y - y^*) \leq -\gamma\|y - y^*\|^2. \tag{6.2}$$

Keeping in mind $(g(y^*), y - y^*) \leq 0, \ \forall y \in \Omega$ from (6.2) we obtain

$$(g(y), y^* - y) \geq \gamma\|y - y^*\|^2, \ \forall y \in \Omega. \tag{6.3}$$

THEOREM 7. *If for both p and c operators the local strong monotonicity (3.21)-(3.22) holds and Lipschitz condition (4.13) satisfies, then*

(1) *there exists* $0 < q(t) < 1, \ \forall t \in (0, (\sqrt{2}L)^{-1})$ *that*

$$\|y_{s+1} - y^*\| \leq q(t)\|y_s - y^*\| \tag{6.4}$$

(2)

$$\|y_{s+1} - y^*\| \leq \sqrt{1 - 0.5\varkappa}\|y_s - y^*\|, \ \forall \varkappa \in [0, 0.5] \tag{6.5}$$

(3)

$$Comp(EPG) \leq O(n^2\varkappa^{-1}\ln\varepsilon^{-1}) \tag{6.6}$$

where $\varepsilon > 0$ *is the required accuracy.*

PROOF. (1) From (6.3) with $y = \hat{y}_s$ we have

$$(g(\hat{y}_s), y^* - \hat{y}_s) \geq \gamma \|\hat{y}_s - y^*\|^2.$$

Therefore (6.1) we can rewrite as follows.

$$\begin{aligned} 2(y_s - y_{s+1}, y^* - y_{s+1}) + 2(y_s - \hat{y}_s, y_{s+1} - \hat{y}_s) \\ +2\gamma t\|\hat{y}_s - y^*\|^2 - 2(tL)^2\|\hat{y}_s - y_s\|^2 \leq 0. \end{aligned} \tag{6.7}$$

Using identity (5.15), first, with $u = y_s, v = y_{s+1}, w = y^*$ and second with $u = y_s, v = y_{s+1}, w = \hat{y}_s$ from (6.7) we obtain

$$\begin{aligned} \|y_{s+1} - y^*\|^2 + \|\hat{y}_s - y_{s+1}\|^2 + (1 - 2(tL)^2)\|y_s - \hat{y}_s\|^2 \\ +2\gamma t\|\hat{y}_s - y^*\|^2 \leq \|y_s - y^*\|^2. \end{aligned} \tag{6.8}$$

Using

$$\|\hat{y}_s - y^*\|^2 = \|\hat{y}_s - y_s\|^2 + 2(\hat{y} - y_s, y_s - y^*) + \|y_s - y^*\|^2$$

from (6.8) we obtain

$$\|y_{s+1} - y^*\|^2 + \|\hat{y}_s - y_{s+1}\|^2 + (1 - 2(tL)^2)\|\hat{y}_s - y_s\|^2 +$$
$$+2\gamma t\|\hat{y}_s - y_s\|^2 + 4\gamma t(\hat{y}_s - y_s, y_s - y^*) + 2\gamma t\|y_s - y^*\|^2 \leq \|y_s - y^*\|^2$$

or

$$\begin{aligned} \|y_{s+1} - y^*\|^2 + \|\hat{y}_s - y_{s+1}\|^2 + (1 + 2\gamma t - 2(tL)^2)\|\hat{y}_s - y_s\|^2 + \\ +4\gamma t(\hat{y}_s - y_s, y_s - y^*) \leq (1 - 2\gamma t)\|y_s - y^*\|^2. \end{aligned} \tag{6.9}$$

Let $\mu(t) = 1 + 2\gamma t - 2(tL)^2$, then the third and the fourth term of the left hand side can be rewritten as follows

$$\|\sqrt{\mu(t)}(\hat{y} - y_s) + 2(y_s - y^*)\frac{\gamma t}{\sqrt{\mu(t)}}\|^2 - \frac{4(\gamma t)^2}{\mu(t)}\|y_s - y^*\|^2.$$

Therefore from (6.9) we obtain bound (6.4)

$$\|y_{s+1} - y^*\| \leq q(t)\|y_s - y^*\|$$

with $q(t) = (1 - 2\gamma t + 4(\gamma t)^2(\mu(t))^{-1})^{\frac{1}{2}}$. It is easy to see that for $t \in (0, (\sqrt{2}L)^{-1})$ we have $0 < q(t) < 1$.

(2) For $\varkappa = \gamma L^{-1}$ and $t = (2L)^{-1}$ we obtain

$$q((2L)^{-1}) = [(1 + \varkappa)(1 + 2\varkappa)^{-1}]^{\frac{1}{2}}.$$

For $0 \leq \varkappa \leq 0.5$ we have

$$q = q((2L)^{-1}) \leq \sqrt{1 - 0.5\varkappa},$$

therefore the bound (6.5) holds

(3) It follows from (6.5) that for a given accuracy $0 < \varepsilon << 1$ the EPG method requires

$$s = O\left(\frac{\ln \varepsilon^{-1}}{\ln q^{-1}}\right)$$

steps to get $y_s : \|y_s - y^*\| \leq \varepsilon$. Then we have

$$(\ln q^{-1})^{-1} = \left(-\frac{1}{2}\ln(1 - 0.5\varkappa))^{-1}\right).$$

Using $\ln(1 + x) \leq x, \ \forall x > -1$ we obtain $\ln(1 - 0.5\varkappa) \leq -0.5\varkappa$ or

$$-0.5\ln(1 - 0.5\varkappa) \geq 0.25\varkappa.$$

Therefore $(\ln q^{-1})^{-1} \leq 4\varkappa^{-1}$. Keeping in mind that each EPG step requires $O(n^2)$ operation we obtain bound (6.6) □

REMARK 2. *For small* $0 < \varkappa < 1$ *the EPG complesity bound* (6.6)*is much better than PGP bound* (4.10)*, therefore the necessity to project on* Ω *twice per step is easily compensated by faster convergence of the EPG.*

In case of $1 > \varkappa > 0.5$ *and large* n*, however, the PGP could be more efficient.*

7. Appendix

Let us estimate the Lipschitz constant $L > 0$ in (4.13), which plays an important role in both PGP and EPG methods.

First of all due to the productivity of the matrix A we have $\max_{1 \leq j \leq n} \sum_{i=1}^n a_{ij} \leq 1$ and $\max_{1 \leq i \leq n} \sum_{i=1}^n a_{ij} \leq 1$.

Therefore, for matrix

$$D = I - A = \begin{pmatrix} 1 - a_{11} & \cdots & -a_{1n} & \\ \vdots & \vdots & \ddots & \vdots \\ -a_{n1} & \cdots & 1 - a_{nn} & \end{pmatrix} = \begin{pmatrix} d_{11} & \cdots & d_{1n} & \\ \vdots & \vdots & \ddots & \vdots \\ d_{n1} & \cdots & d_{nn} & \end{pmatrix}$$

we obtain

$$\|D\|_I = \max_{1 \leq j \leq n} \sum_{i=1}^n |a_{ij}| \leq 2, \ \|D\|_{II} = \max_{1 \leq i \leq n} \sum_{j=1}^n |d_{ij}| \leq 2. \tag{7.1}$$

Let consider operators $p : \mathbb{R}^n_+ \to \mathbb{R}^n_+$ and $c : \mathbb{R}^n_+ \to \mathbb{R}^n_+$. We assume that for both operators Lipschitz condition is satisfied for each component of correspodent vector-function, i.e. for $x_1, x_2 \in \mathbb{R}^n_+$ we have

$$|p_j(x_1) - p_j(x_2)| \leq L_{p,j} \|x_1 - x_2\|, 1 \leq j \leq n \tag{7.2}$$

and for $\lambda_1, \lambda_2 \in \mathbb{R}^n_+$ we have

$$|c_i(\lambda_1) - c_i(\lambda_2)| \leq L_{c,i} \|\lambda_1 - \lambda_2\|, \ \ 1 \leq i \leq n. \tag{7.3}$$

It follows from (7.2) that for any $x_1, x_2 \in \mathbb{R}^n_+$

$$\|p(x_1) - p(x_2)\| = \sqrt{\sum_{j=1}^n (p_j(x_1) - p_j(x_2))^2} \leq \sqrt{\sum_{i=1}^n L^2_{p,j} \|x_1 - x_2\|^2} \leq$$

$$L_p \sqrt{n \|x_1 - x_2\|^2} = L_p \sqrt{n} \|x_1 - x_2\|, \tag{7.4}$$

where $L_p = \max_{1 \leq j \leq n} L_{p,j}$.

Similarly, we obtain

$$\|c(\lambda_1) - c(\lambda_2)\| = \sqrt{\sum (c_i(\lambda_1) - c_i(\lambda_2)^2} \leq$$

$$\sqrt{\sum_{i=1}^n L^2_{c,i} \|\lambda_1 - \lambda_2\|^2} \leq L_c \sqrt{n} \|\lambda_1 - \lambda_2\| \tag{7.5}$$

where $L_c = \max_{1 \leq i \leq n} L_{c,i}$.

We are ready to find the upper bound for L in (4.13). Using (3.4) for any pair $y_1, y_2 \in \Omega = \mathbb{R}^n_+ \times \mathbb{R}^n_+$ we have

$$\begin{aligned}\|g(y_1)-g(y_2)\| &\le \|(I-A)^T\lambda_1 - p(x_1) - (I-A)^T\lambda_2 + p(x_2)\| \\ &+\|c(\lambda_1)-(I-A)x_1 - c(\lambda_2) + (I-A)x_2\| \\ &\le \|p(x_1)-p(x_2)\| + \|(I-A)^T\|\|\lambda_1-\lambda_2\|+ \\ &+\|c(\lambda_1)-c(\lambda_2)\| + \|(I-A)^T\|\|x_1-x_2\| \le\end{aligned}$$

$$(L_p\sqrt{n}+\|I-A\|)\|x_1-x_2\| + (L_c\sqrt{n}+\|(I-A)^T\|)\|\lambda_1-\lambda_2\|. \tag{7.6}$$

For $D = I - A$ and $D^T = (I-A)^T$ we have

$$\|D\| = \sqrt{\lambda_{max}(D^TD)} \text{ and } \|D^T\| = \sqrt{\lambda_{max}(DD^T)}$$

Keeping in mind (7.1) we obtain (see [**10**])

$$\|D\| \le \sqrt{n}\|D\|_I = 2\sqrt{n}, \quad \|D^T\| \le \sqrt{n}\|D^T\|_{II} = 2\sqrt{n}. \tag{7.7}$$

Using (7.4),(7.5) and (7.8) from (7.6) we obtain

$$\begin{aligned}\|g(y_1)-g(y_2)\| &\le \sqrt{n}(L_p+2)\|x_1-x_2\|+ \\ +\sqrt{n}(L_c+2)\|\lambda_1-\lambda_2\| &\le \sqrt{2}\hat{L}(\sqrt{n}+2)\|y_1-y_2\|\end{aligned}$$

where $\hat{L} = \max\{L_p, L_c\}$. Therefore for the Lipschitz constant in (4.13) we have the following bound

$$L \le \sqrt{2}\hat{L}(\sqrt{n}+2) = O(\sqrt{n}). \tag{7.8}$$

8. Concluding remarks

The NIOE adds some new features to the classical IO model. In particular, it finds the output consistent with production cost and consumption consistent with prices for goods.

At the NIOE the output minimizes the total production cost while the price vector maximizes the total consumption.

It looks like solving problems (1.5)-(1.6) is much more difficult then solving IO systems (1.1)-(1.2). In reality both the PGP and EPG are very simple methods, which require neither solving subproblems nor systems of equations. At most each step requires few matrix by vector multiplications.

Therefore in same instances finding NIOE can be even less expensive than solving large systems of linear equations.

In particular, it follows from the bound (6.6) and (7.8) that EPG complexity is $O(n^{2.5}\ln\varepsilon^{-1})$ for any given fixed $\gamma > 0$.

Therefore for very large n the EPG method can be used when solving linear systems (2.1)-(2.2) is very difficult or even impossible.

Full primal- dual decomposition for both methods allows computing the primal and dual vectors in parallel.

Along with parallel technique for matrix by vector multiplication (see, for example, [**22**] and references therein) both PGP and EPG can be used for developing numerically efficient methods for large scale NIOE.

Three main issues are left for further research.

First, numerical realization and correspondent software based on both PGP and EPG methods need to be developed.

Second, numerical experiments with real data have to be performed.

Third, economic analyses of the numerical results and its comparison with results obtained by IO has to be conducted.

References

[1] Antipin A., The Gradient and Exstragradient Approaches in Bilinear Equilibrium Programming. A. *Dorodnizin Computing Center RAS (in Russian)* (2002)

[2] S. A. Ashmanov, *Vvedenie v matematicheskuyu ekonomiku* (Russian), "Nauka", Moscow, 1984. MR771664 (86j:90001)

[3] Alfred Auslender and Marc Teboulle, *Interior projection-like methods for monotone variational inequalities*, Math. Program. **104** (2005), no. 1, Ser. A, 39–68, DOI 10.1007/s10107-004-0568-x. MR2167425 (2006d:90232)

[4] Bakushinskij A. B., Polyak B. T., On the solution of variational inequalities. Sov. Math Doklady 14, 1705-1710 (1974)

[5] Yair Censor, Aviv Gibali, and Simeon Reich, *Extensions of Korpelevich's extragradient method for the variational inequality problem in Euclidean space*, Optimization **61** (2012), no. 9, 1119–1132, DOI 10.1080/02331934.2010.539689. MR2966121

[6] Y. Censor, A. Gibali, and S. Reich, *The subgradient extragradient method for solving variational inequalities in Hilbert space*, J. Optim. Theory Appl. **148** (2011), no. 2, 318–335, DOI 10.1007/s10957-010-9757-3. MR2780566 (2011k:49014)

[7] Yair Censor, Aviv Gibali, and Simeon Reich, *Strong convergence of subgradient extragradient methods for the variational inequality problem in Hilbert space*, Optim. Methods Softw. **26** (2011), no. 4-5, 827–845, DOI 10.1080/10556788.2010.551536. MR2837800 (2012g:49012)

[8] Robert Dorfman, Paul A. Samuelson, and Robert M. Solow, *Linear programming and economic analysis*, A Rand Corporation Research Study, McGraw-Hill Book Co., Inc., New York-Toronto-London, 1958. MR0128543 (23 #B1582)

[9] David Gale, *The theory of linear economic models*, McGraw-Hill Book Co., Inc., New York-Toronto-London, 1960. MR0115801 (22 #6599)

[10] Gantmacher F., Theory of Matrices, AMS (1959)

[11] A. A. Goldstein, *Convex programming in Hilbert space*, Bull. Amer. Math. Soc. **70** (1964), 709–710. MR0165982 (29 #3262)

[12] A. N. Iusem and B. F. Svaiter, *A variant of Korpelevich's method for variational inequalities with a new search strategy*, Optimization **42** (1997), no. 4, 309–321, DOI 10.1080/02331939708844365. MR1609571 (98m:49020)

[13] E. N. Khobotov, *A modification of the extragradient method for solving variational inequalities and some optimization problems* (Russian), Zh. Vychisl. Mat. i Mat. Fiz. **27** (1987), no. 10, 1462–1473, 1597. MR918546 (88m:49005)

[14] Igor Konnov, *Combined relaxation methods for variational inequalities*, Lecture Notes in Economics and Mathematical Systems, vol. 495, Springer-Verlag, Berlin, 2001. MR1795730 (2001i:49001)

[15] G. M. Korpelevič, *An extragradient method for finding saddle points and for other problems* (Russian), Èkonom. i Mat. Metody **12** (1976), no. 4, 747–756. MR0451121 (56 #9408)

[16] Kelvin Lancaster, *Mathematical economics*, The Macmillan Co., New York; Collier-Macmillan Ltd., London, 1968. MR0234708 (38 #3024)

[17] Leontief W. Quantitative Input and Output Relations in the Economic System of the United States, Review of Economic Statistics 18, pp 105-125 (1936)

[18] Leontief W.The structure of American Economy, 1919-1939 2d ed. Oxford University Pres NY (1951)

[19] Leontief W., Input - Output Economics, Oxford University Press (1966)

[20] Levitin E., Polyak B., Constrained minimization methods, Journal of Computational Math and Math Physics. Vol. 6,5 pp 787-823 (1966)

[21] Polyak R., Nonlinear Equilibrium for Resorce Allocation Problems. Contemporary Mathematics AMS (2014)

[22] Quinn M. Parallel Programming in C with MPI and Open MP, New York NY McGraw-Hill (2004)

Current address: Department of Mathematics, The Technion - Israel Institute of Technology, 32000 Haifa, Israel

E-mail address: `rpolyak@techunix.technion.ac.il`

Contemporary Mathematics
Volume **659**, 2016
http://dx.doi.org/10.1090/conm/659/13150

A convex analytic inequality revisited

Simeon Reich and Xianfu Wang

ABSTRACT. We first give a direct proof of the Reich-Simons convex analytic inequality and then use this inequality to derive a number of related results in convex analysis.

1. Introduction

Let $(H, \langle\cdot,\cdot\rangle)$ be a real Hilbert space with induced norm $\|\cdot\|$ and let the function $f : H \to (-\infty, +\infty]$ be convex. Recall that the *Fenchel conjugate* $f^* : H \to (-\infty, +\infty]$ of f is defined by $f^*(x^*) := \sup_H[x^* - f]$ for $x^* \in H$. In [**6**, Remark 7, page 2659] the authors write that it would be interesting to find a direct proof of the following result.

THEOREM 1.1 (Reich & Simons). *If the function $f : H \to (-\infty, +\infty]$ is convex and*

$$f(x) + \frac{\|x\|^2}{2} \geq 0 \quad \text{for all } x \in H, \tag{1.1}$$

then there exists a point $y \in H$ such that

$$f^*(y) + \frac{\|y\|^2}{2} \leq 0,$$

that is,

$$f(x) \geq \langle x, y\rangle + \frac{\|y\|^2}{2} \quad \text{for all } x \in H. \tag{1.2}$$

This result has turned out to be very useful in the study of the Kirszbraun-Valentine extension theorem and several aspects of monotone operator theory (see [**6**], [**9**] and [**3**]). It follows from Rockafellar's version of Fenchel's duality theorem [**8**, Theorem 1, pages 82–83]. In this connection, see also [**2**].

The purpose of this note is first to present a direct proof of Theorem 1.1 and then use this theorem to derive a number of related results in convex analysis the original proofs of which relied on the Fenchel-Rockafellar duality theorem. Our

---

2010 *Mathematics Subject Classification.* Primary 46C05, 46N10; Secondary 47H05, 90C25.

Key words and phrases. Convex function, Fitzpatrick function of order n, Hilbert space, maximally monotone operator, Minty's theorem, n-cyclically monotone operator.

S. Reich was partially supportd by the Israel Science Foundation (Grant 389/12), the Fund for the Promotion of Research at the Technion and by the Technion General Research Fund.

X. Wang was partially supported by the Natural Sciences and Engineering Research Council of Canada.

proof is based on the weak compactness of bounded, closed and convex subsets of H. The approach we have taken has the advantage that the arguments we employ are self-contained and elementary.

We write $f^+(x) := \max\{f(x), 0\} = f(x) \vee 0$ and denote the lower semicontinuous hull of f either by $\mathrm{cl}f$ or by $\underline{f}$. These two functions are also convex. When f is proper, the *subdifferential* of f is the set-valued operator

$$\partial f : H \rightrightarrows H : x \mapsto \{x^* \in H : \ (\forall z \in H)\ \langle z - x, x^*\rangle + f(x) \leq f(z)\}.$$

The remainder of the paper is organized as follows. Section 2 contains our direct proof of the Reich-Simons inequality (Theorem 1.1). In Section 3 we employ this inequality to extend n-cyclic monotone operators by using n-th order Fitzpatrick functions (Theorem 3.2) and to prove the Simons-Zălinescu inequality (Theorem 3.3).

2. The proof of Theorem 1.1

If $f \equiv +\infty$, the result obviously holds. Moreover, we can also assume that f is lower semicontinuous because

$$\begin{aligned} f(x) + \frac{1}{2}\|x\|^2 \geq 0 \quad &\Rightarrow \quad f(x) \geq -\frac{1}{2}\|x\|^2 \\ &\Rightarrow \quad \underline{f}(x) \geq -\frac{1}{2}\|x\|^2 \Rightarrow \underline{f}(x) + \frac{1}{2}\|x\|^2 \geq 0, \end{aligned}$$

and if Theorem 1.1 holds for the lower semicontinuous convex function $\underline{f}$, then there exists a point $y \in H$ such that

$$f(x) \geq \underline{f}(x) \geq \langle y, x\rangle + \frac{1}{2}\|y\|^2 \quad \text{for all } x \in H,$$

so that Theorem 1.1 is also valid for f. Therefore in the sequel we may assume that

$$f(x_0) < +\infty \quad \text{for some } x_0 \in H$$

and that f is lower semicontinuous on H.

Define $g : H \to (-\infty, +\infty]$ by

$$g(x) := f(x) + \frac{1}{2}\|x\|^2 \quad \text{for } x \in H.$$

We consider two cases.

Case 1: *f is bounded below by some number $M \in \mathbb{R}$.*

In this case we have $g(x) \geq M + 1/2\|x\|^2$, so that the function g is *coercive*, that is, $\lim_{\|x\| \to +\infty} g(x) = +\infty$. Each norm closed and convex set in H is weakly closed. Since g is convex and lower semicontinuous, it is weakly lower semicontinuous [**10**, Theorem 2.2.1, page 60]. As g is weakly lower semicontinuous and coercive, there exists a point $y \in H$ such that $g(y) = \min_{x \in H} g(x)$. In this connection, see also [**3**, Proposition 11.14(i), page 159]. That is, for all $x \in H$,

$$f(x) + \frac{1}{2}\|x\|^2 \geq f(y) + \frac{1}{2}\|y\|^2 \Rightarrow f(x) - f(y) \geq -\frac{1}{2}\|x\|^2 + \frac{1}{2}\|y\|^2,$$

from which it follows that

$$-y \in \partial f(y).$$

Indeed, for every $t > 0$ and $v \in H$, we have

$$f(y + tv) - f(y) \geq -\frac{1}{2}\|y + tv\|^2 + \frac{1}{2}\|y\|^2.$$

Dividing both sides by $t > 0$ and then letting $t \downarrow 0$, we get

$$f'(y; v) \geq \langle -y, v \rangle.$$

Since f is convex, $f'(y; v) \leq f(y+v) - f(y)$, so that $-y \in \partial f(y)$.

Therefore we have for all $x \in H$,

$$\begin{aligned} f(x) &\geq f(y) + \langle -y, x - y \rangle = f(y) + \|y\|^2 + \langle -y, x \rangle \\ &= \left(f(y) + \frac{1}{2}\|y\|^2 \right) + \frac{1}{2}\|y\|^2 + \langle -y, x \rangle \\ &\geq \frac{1}{2}\|y\|^2 + \langle -y, x \rangle = \langle -y, x \rangle + \frac{1}{2}\| - y\|^2 \quad \text{by using (1.1),} \end{aligned}$$

which is (1.2).

In order to proceed, we estimate y. If (1.2) holds, then using the Cauchy-Bunyakovsky-Schwarz inequality, we get

$$f(x) \geq \frac{1}{2}\|y\|^2 - \|x\|\|y\|, \text{ that is,}$$

$$\|y\|^2 - 2\|x\|\|y\| \leq 2f(x).$$

It follows that

$$2f(x) + \|x\|^2 \geq \|y\|^2 - 2\|y\|\|x\| + \|x\|^2 = (\|y\| - \|x\|)^2,$$

from which it follows that

$$\|y\| \leq \left(\|x\| + \sqrt{\|x\|^2 + 2f(x)} \right) \quad \forall x \in H. \tag{2.1}$$

In particular,

$$\|y\| \leq \left(\|x_0\| + \sqrt{\|x_0\|^2 + 2f(x_0)} \right) < +\infty.$$

Case 2: f is an arbitrary proper lower semicontinuous convex function.

Consider the functions $f_n : H \to (-\infty, +\infty]$ defined by

$$f_n(x) := \max\{f(x), -n\} \quad \text{for } n \in \{1, 2, 3, \ldots\}.$$

We have

$$f^+ \geq f_n \geq f, \tag{2.2}$$

$$\lim_{n\to\infty} f_n = f \quad \text{pointwise} \tag{2.3}$$

and

$$f_n \text{ is convex on } H. \tag{2.4}$$

If f satisfies (1.1), then $\forall x \in H$,

$$f_n(x) + \frac{1}{2}\|x\|^2 \geq 0,$$

because $f_n \geq f$ by (2.2). Since $f_n \geq -n$, we may apply Case 1 to each f_n to find $y_n \in H$ so that

$$f_n(x) \geq \langle y_n, x \rangle + \frac{1}{2}\|y_n\|^2 \quad \forall x \in H. \tag{2.5}$$

In view of (2.1) and (2.2), we have $\forall n \geq 1$,

$$\|y_n\| \leq \left(\|x_0\| + \sqrt{\|x_0\|^2 + 2f_n(x_0)} \right) \leq \left(\|x_0\| + \sqrt{\|x_0\|^2 + 2f^+(x_0)} \right). \tag{2.6}$$

By (2.6), the sequence $(y_n)_{n=1}^\infty$ is bounded. Since the unit ball of H is weakly compact, the sequence $(y_n)_{n=1}^\infty$ has a weakly convergent subsequence. After relabelling, we may assume that $y_n \rightharpoonup y \in H$ weakly. Using (2.5), (2.3) and the fact that $\|\cdot\|^2$ is weakly lower semicontinuous, we obtain $\forall x \in H$,

$$\begin{aligned} f(x) &= \lim_{n\to\infty} f_n(x) \geq \lim_{n\to\infty} \langle y_n, x\rangle + \liminf_{n\to\infty} \frac{1}{2}\|y_n\|^2 \\ &\geq \langle y, x\rangle + \frac{1}{2}\|y\|^2, \end{aligned}$$

as required. Of course, as in (2.1), we have the following estimate:

$$\|y\| \leq \inf_{x\in H} \left(\|x\| + \sqrt{\|x\|^2 + 2f(x)}\right) \quad \forall x \in H.$$

This completes the proof of Theorem 1.1. □

3. Applications

Theorem 1.1 allows us to derive a number of known results the original proofs of which rely on Rockafellar's version of the Fenchel duality theorem. We start off with some terminology. Let $A\colon H \rightrightarrows H$ be a set-valued operator with $\operatorname{gra} A \neq \varnothing$, and let $n \in \{2, 3, \ldots\}$. We say that A is *n-monotone* (or *n-cyclically monotone*) if

$$\left.\begin{array}{c} (a_1, a_1^*) \in \operatorname{gra} A, \\ \vdots \\ (a_n, a_n^*) \in \operatorname{gra} A \\ a_{n+1} := a_1 \end{array}\right\} \Rightarrow \sum_{i=1}^{n} \langle a_{i+1} - a_i, a_i^*\rangle \leq 0, \tag{3.1}$$

where $\operatorname{gra} A := \{(x, x^*) \in H \times H : x^* \in Ax\}$ denotes the *graph* of A. Note that 2-monotonicity reduces to

$$(\forall (x, x^*) \in \operatorname{gra} A)(\forall (y, y^*) \in \operatorname{gra} A) \quad \langle x - y, x^* - y^*\rangle \geq 0, \tag{3.2}$$

that is, to ordinary monotonicity. The operator A is said to be *maximally n-monotone* if A is n-monotone and no proper extension (in the sense of inclusion of graphs) of A is n-monotone. Thanks to Zorn's Lemma, every n-monotone operator admits a maximally n-monotone extension. If A is maximally 2-monotone, then A is called *maximally monotone*. The *Fitzpatrick function of order n* associated with A is defined for $(x, x^*) \in H \times H$ by

$$F_{A,n}(x, x^*) := \sup_{\substack{(a_1, a_1^*) \in \operatorname{gra} A, \\ \vdots \\ (a_{n-1}, a_{n-1}^*) \in \operatorname{gra} A}} \left(\sum_{i=1}^{n-2} \langle a_{i+1} - a_i, a_i^*\rangle + \langle x - a_{n-1}, a_{n-1}^*\rangle + \langle a_1, x^*\rangle\right).$$

The classical Fitzpatrick function [**5**] is $F_{A,2}$, that is,

$$F_{A,2}\colon H \times H \to [-\infty, +\infty]\colon (x, x^*) \mapsto \sup_{(a,a^*)\in \operatorname{gra} A} (\langle x, a^*\rangle + \langle a, x^*\rangle - \langle a, a^*\rangle).$$

Combining [**1**, Propositions 2.3 and 2.4], we now recall several basic properties of $F_{A,n}$.

FACT 3.1. *Let $A : H \rightrightarrows H$ with $\operatorname{gra} A \neq \varnothing$ and $n \in \{2, 3, \ldots\}$. Then*

(1) *$F_{A,n}$ is lower semicontinuous and convex.*
(2) *A is n-cyclically monotone if and only if $F_{A,n} = \langle \cdot, \cdot \rangle$ on $\operatorname{gra} A$.*

Consequently, if A is n-cyclically monotone with $\operatorname{gra} A \neq \varnothing$, then $F_{A,n}$ is a proper, lower semicontinuous and convex function.

Our first application of Theorem 1.1 is to provide a new proof of a result regarding extensions of n-cyclic monotone operators which is due to Bauschke and Wang [**4**]. We shall need the following elementary result. As usual, we denote the inner product $\langle \cdot, \cdot \rangle$ by p.

LEMMA 3.1. *Let $A : H \rightrightarrows H$ be a set-valued operator and let $a, w \in H$.*

(i) *Define $A - w$ by $\operatorname{gra}(A - w) = \operatorname{gra} A - (0, w)$. Then $\forall (x, x^*), (y^*, y) \in H \times H$,*

$$F_{A-w,n}(x, x^*) = F_{A,n}(x, x^* + w) - \langle x, w \rangle \quad \text{and}$$
$$F^*_{A-w,n}(y^*, y) = F^*_{A,n}(y^* + w, y) - \langle y, w \rangle.$$

Consequently,

$$\begin{aligned} F_{A-w,n} \geq p \quad &\Leftrightarrow \quad F_{A,n} \geq p, \\ F^*_{A-w,n} \geq p \quad &\Leftrightarrow \quad F^*_{A,n} \geq p. \end{aligned} \tag{3.3}$$

(ii) *Define A_a by $A_a(x) := A(x + a)$. Then $\forall (x, x^*), (y^*, y) \in H \times H$,*

$$\begin{aligned} F_{A_a,n}(x, x^*) &= F_{A,n}(x + a, x^*) - \langle a, x^* \rangle, \quad \text{and} \\ F^*_{A_a,n}(y^*, y) &= F^*_{A,n}(y^*, y + a) - \langle y^*, a \rangle. \end{aligned} \tag{3.4}$$

Consequently,

$$\begin{aligned} F_{A_a,n} \geq p \quad &\Leftrightarrow \quad F_{A,n} \geq p, \\ F^*_{A_a,n} \geq p \quad &\Leftrightarrow \quad F^*_{A,n} \geq p. \end{aligned} \tag{3.5}$$

PROOF. (i) Observe that $\forall (x, x^*) \in H \times H$,

$$F_{A-w,n}(x, x^*) =$$

$$\sup_{\substack{(a_2, a_2^*) \in \operatorname{gra} A, \\ \vdots \\ (a_{n-1}, a_{n-1}^*) \in \operatorname{gra} A}} \Big(\sum_{i=1}^{n-2} \langle a_{i+1} - a_i, a_i^* - w \rangle\Big) + \langle x - a_{n-1}, a_{n-1}^* - w \rangle + \langle a_1, x^* \rangle$$

$$= \sup_{\substack{(a_2, a_2^*) \in \operatorname{gra} A, \\ \vdots \\ (a_{n-1}, a_{n-1}^*) \in \operatorname{gra} A}} \Big(\sum_{i=1}^{n-2} \langle a_{i+1} - a_i, a_i^* \rangle\Big) + \langle x - a_{n-1}, a_{n-1}^* \rangle + \langle a_1, x^* \rangle + \langle x - a_1, -w \rangle$$

$$= \sup_{\substack{(a_2, a_2^*) \in \operatorname{gra} A, \\ \vdots \\ (a_{n-1}, a_{n-1}^*) \in \operatorname{gra} A}} \Big(\sum_{i=1}^{n-2} \langle a_{i+1} - a_i, a_i^* \rangle\Big) + \langle x - a_{n-1}, a_{n-1}^* \rangle + \langle a_1, x^* + w \rangle - \langle x, w \rangle$$

$$= F_{A,n}(x, x^* + w) - \langle x, w \rangle,$$

and $\forall(y^*,y)\in H\times H$,

$$\begin{aligned}F^*_{A-w,n}(y^*,y) &= \sup_{(x,x^*)\in H\times H}\left(\langle (y^*,y),(x,x^*)\rangle - F_{A-w,n}(x,x^*)\right)\\ &= \sup_{(x,x^*)\in H\times H}\left(\langle y^*,x\rangle + \langle y,x^*\rangle - F_{A,n}(x,x^*+w) + \langle x,w\rangle\right)\\ &= \sup_{(x,x^*)\in H\times H}\left(\langle y^*+w,x\rangle + \langle y,x^*+w\rangle - F_{A,n}(x,x^*+w) - \langle y,w\rangle\right)\\ &= F^*_{A,n}(y^*+w,y) - \langle y,w\rangle.\end{aligned}$$

We now see that for each $(x,x^*),(y^*,y)\in H\times H$,

$$F_{A-w,n}(x,x^*)\geq\langle x,x^*\rangle \Leftrightarrow F_{A,n}(x,x^*+w)\geq\langle x,x^*+w\rangle,$$

$$F^*_{A-w,n}(y^*,y)\geq\langle y^*,y\rangle \Leftrightarrow F^*_{A,n}(y^*+w,y)\geq\langle y^*+w,y\rangle.$$

Therefore (3.3) follows.

(ii) We apply similar arguments to those used in the proof of (i). Assertion (3.5) follows immediately from (3.4). □

THEOREM 3.2 (Bauschke & Wang). *Let H be a Hilbert space and let the operator $A : H \rightrightarrows H$ be n-cyclically monotone with* $\operatorname{gra} A \neq \varnothing$. *If*

$$F^*_{A,n}(y,x)\geq\langle y,x\rangle \quad \forall(y,x)\in H\times H, \tag{3.6}$$

then for every $w\in H$, there exists $x\in H$ such that $\operatorname{gra} A\bigcup\{(x,w-x)\}$ *is n-cyclically monotone.*

PROOF. By [**1**, Proposition 2.7], for each $w\in H$, we must show that $\exists x\in H$ such that

$$F_{A,n}(x,w-x)\leq\langle x,w-x\rangle. \tag{3.7}$$

We proceed by considering two cases.

Case 1: $w=0$.

In this case inequality (3.7) becomes

$$F_{A,n}(x,-x)\leq\langle x,-x\rangle \Leftrightarrow \frac{1}{2}F_{A,n}(x,-x)+\frac{1}{2}\|x\|^2\leq 0. \tag{3.8}$$

Set $f(x):=\frac{1}{2}F_{A,n}(x,-x)$ and rewrite (3.8) as

$$f(x)+\frac{1}{2}\|x\|^2\leq 0. \tag{3.9}$$

We shall show that

$$f^*(y)+\frac{1}{2}\|y\|^2\geq 0 \quad \forall y\in H. \tag{3.10}$$

To do this, we estimate f^*. Define

$$L:=\{(x,x^*)\in H\times H:\ x=-x^*\}$$

and

$$\iota_L(x,x^*):=\begin{cases}0 & \text{if } (x,x^*)\in L,\\ +\infty & \text{otherwise,}\end{cases}$$

so that

$$L^\perp:=\{(x^*,x)\in H\times H:\ x=x^*\}$$

and

$$\iota^*_L(x,x^*)=\iota_{L^\perp}(x,x^*). \tag{3.11}$$

Recall that if f and g are proper, convex and lower semicontinuous functions, and $\underline{f^*\Box g^*}$ is proper, then

$$(f+g)^* = \underline{f^*\Box g^*}. \tag{3.12}$$

Here $\Box$ denotes the *infimal-convolution.* See [**7**, Theorem 16.4, page 145] for the case of a finite dimensional space and [**10**, Corollary 2.35, page 78] for the case of an infinite dimensional space. By Fact 3.1, $F_{A,n}$ is proper, lower semicontinuous and convex on $H \times H$, so $F^*_{A,n}$ is proper, lower semicontinuous and convex by [**3**, Theorem 13.32]. In particular, $\operatorname{dom} F^*_{A,n} \neq \varnothing$. Since

$$\operatorname{dom} \underline{F^*_{A,n}\Box\iota^*_L} \supset \operatorname{dom} F^*_{A,n}\Box\iota^*_L = \operatorname{dom} F^*_{A,n} + L^\perp \neq \varnothing$$

and $\forall (y^*,y) \in H \times H$,

$$\begin{aligned}
\underline{F^*_{A,n}\Box\iota^*_L}(y^*,y) &= \operatorname{cl}\inf\{F^*_{A,n}(y^*-x^*,y-x) + \iota^*_L(x^*,x) : \ (x^*,x) \in H\times H\}\\
&\geq \operatorname{cl}\inf\{\langle y^*-x^*, y-x\rangle + \iota_{L^\perp}(x^*,x) : \ (x^*,x)\in H\times H\} \quad \text{(using (3.6) and (3.11))}\\
&= \operatorname{cl}\inf\{\langle y^*-x, y-x\rangle : x\in H\} = \operatorname{cl}\inf\{\langle y^*,y\rangle - \langle y^*+y,x\rangle + \langle x,x\rangle : \ x\in H\}\\
&= \frac{-\|y^*-y\|^2}{4},
\end{aligned}$$

that is,

$$\underline{F^*_{A,n}\Box\iota^*_L}(y^*,y) \geq \frac{-\|y^*-y\|^2}{4}, \tag{3.13}$$

so the lower semicontinuous convex function $\underline{F^*_{A,n}\Box\iota^*_L}$ is proper. We also have

$$\begin{aligned}
f^*(y) &= \sup_{x\in H}(\langle y,x\rangle - \frac{1}{2}F_{A,n}(x,-x))\\
&= \sup\{\langle y,x\rangle - (\frac{1}{2}F_{A,n}(x,x^*) + \iota_L(x,x^*)) : \ (x,x^*)\in H\times H\}\\
&= \sup\{\frac{1}{2}\langle (y,-y),(x,x^*)\rangle - \frac{1}{2}(F_{A,n}(x,x^*) + \iota_L(x,x^*)) : \ (x,x^*)\in H\times H\}\\
&= \frac{1}{2}(F_{A,n}+\iota_L)^*(y,-y) = \frac{1}{2}\operatorname{cl}(F^*_{A,n}\Box\iota^*_L)(y,-y) \quad \text{(using (3.12))}\\
&\geq -\frac{1}{2}\|y\|^2 \quad \text{(using (3.13) with } y^* = -y),
\end{aligned}$$

from which (3.10) follows. Now we apply Theorem 1.1 to inequality (3.10) to find $x \in H$ satisfying (3.9).

Case 2: $w \in H$ is arbitrary.

Define $\operatorname{gra} B := \operatorname{gra} A + (0,-w)$. By Lemma 3.1(i), inequality (3.6) holds for the n-cyclic monotone operator B. Therefore we may apply the above arguments to find a pair $(x,-x) \in H\times H$ such that $\{(x,-x)\}\cup \operatorname{gra} B$ is n-cyclically monotone. Then $\{(x,w-x)\}\cup\operatorname{gra} A$ is also n-cyclically monotone and the proof is complete. □

Our next application is a new proof of the following inequality of Simons and Zălinescu [**9**, Lemma 1.3] in the proof of which they used Fenchel's duality theorem. Define $\Delta : H\times H \to \mathbb{R}$ by

$$\Delta(y,y^*) := \frac{1}{2}\|y\|^2 + \langle y,y^*\rangle + \frac{1}{2}\|y^*\|^2 = \frac{1}{2}\|y+y^*\|^2 \geq 0.$$

THEOREM 3.3 (Simons & Zălinescu). *Let H be a Hilbert space, $k : H \times H \to (-\infty, +\infty]$ be proper and convex, $(w, w^*) \in H \times H$ and assume that*

$$(x, x^*) \in H \times H \Rightarrow k(x, x^*) - \langle x, x^* \rangle + \Delta(w - x, w^* - x^*) \geq 0. \tag{3.14}$$

Then $\exists(\hat{x}^*, \hat{x}) \in H \times H$ such that

$$k^*(\hat{x}^*, \hat{x}) - \langle \hat{x}, \hat{x}^* \rangle + \Delta(w - \hat{x}, w^* - \hat{x}^*) \leq 0. \tag{3.15}$$

PROOF. Note that

$$\begin{aligned} -\langle x, x^* \rangle + \Delta(w - x, w^* - x^*) = \\ \langle w, w^* \rangle - \langle (x, x^*), (w^*, w) \rangle + \tfrac{1}{2}\|(w, w^*) - (x, x^*)\|^2. \end{aligned} \tag{3.16}$$

Inequality (3.14) may be rewritten in the following way:

$$k(x, x^*) + \langle w, w^* \rangle - \langle (x, x^*), (w^*, w) \rangle + \frac{1}{2}\|(w, w^*) - (x, x^*)\|^2 \geq 0.$$

Using $(y, y^*) := (x - w, x^* - w^*)$, we obtain

$$[k(y + w, y^* + w^*) + \langle w, w^* \rangle - \langle (y + w, y^* + w^*), (w^*, w) \rangle] + \frac{1}{2}\|(y, y^*)\|^2 \geq 0.$$

Define $g : H \times H \to (-\infty, +\infty]$ by

$$g(y, y^*) := k(y + w, y^* + w^*) + \langle w, w^* \rangle - \langle (y + w, y^* + w^*), (w^*, w) \rangle.$$

Inequality (3.14) may be also rewritten as follows:

$$g(y, y^*) + \frac{1}{2}\|(y, y^*)\|^2 \geq 0 \qquad \forall (y, y^*) \in H \times H. \tag{3.17}$$

We now compute g^*:

$$\begin{aligned} &g^*(x^*, x) = \\ &\sup\left(\langle (x^*, x), (y, y^*) \rangle - k(y + w, y^* + w^*) - \langle w, w^* \rangle + \langle (y + w, y^* + w^*), (w^*, w) \rangle\right) \\ &= \sup\left(\langle (x^*, x), (y{+}w, y^*{+}w^*) \rangle + \langle (w^*, w), (y{+}w, y^*{+}w^*) \rangle - k(y{+}w, y^*{+}w^*)\right) \\ &- \langle w, w^* \rangle - \langle (x^*, x), (w, w^*) \rangle \\ &= \sup\left(\langle (x^* + w^*, x + w), (y + w, y^* + w^*) \rangle - k(y + w, y^* + w^*)\right) \\ &- \langle w, w^* \rangle - \langle (x^*, x), (w, w^*) \rangle \\ &= k^*(x^* + w^*, x + w) - \langle w, w^* \rangle - \langle (x^*, x), (w, w^*) \rangle. \end{aligned}$$

Applying Theorem 1.1 to (3.17), we find a pair $(x^*, x) \in H \times H$ such that

$$g^*(x^*, x) + \frac{1}{2}\|(x^*, x)\|^2 \leq 0,$$

that is,

$$k^*(x^* + w^*, x + w) - \langle w, w^* \rangle - \langle (x^*, x), (w, w^*) \rangle + \frac{1}{2}\|(x^*, x)\|^2 \leq 0.$$

Put $(\hat{x}^*, \hat{x}) := (x^* + w^*, x + w)$. Then

$$k^*(\hat{x}^*, \hat{x}) - \langle w, w^* \rangle - \langle (\hat{x}^* - w^*, \hat{x} - w), (w, w^*) \rangle + \frac{1}{2}\|(\hat{x}^* - w^*, \hat{x} - w)\|^2 \leq 0,$$

and this gives

$$k^*(\hat{x}^*, \hat{x}) + \langle w, w^* \rangle - \langle (\hat{x}^*, \hat{x}), (w, w^*) \rangle + \frac{1}{2}\|(\hat{x}^*, \hat{x}) - (w^*, w)\|^2 \leq 0.$$

In view of (3.16), this establishes inequality (3.15). □

REMARK 3.4. We note that Theorem 3.3 implies Minty's theorem (see, for example, [**3**, Theorem 21.1, page 311]): Let the operator $A : H \rightrightarrows H$ be monotone. Then A is maximally monotone if and only if $\operatorname{ran}(\mathrm{Id} + A) = H$.

Indeed, if A is maximally monotone, then (see [**5**])

$$F^*_{A,2}(x^*, x) \geq F_{A,2}(x, x^*) \geq \langle x^*, x\rangle \quad \forall (x, x^*) \in H \times H.$$

Letting $k := F^*_{A,2}$ and applying Theorem 3.3 to $(0, w^*)$, we find a pair (x, x^*) such that

$$F_{A,2}(x, x^*) - \langle x^*, x\rangle + \Delta(0 - x, w^* - x) \leq 0.$$

Since $F_{A,2}(x, x^*) \geq \langle x, x^*\rangle$ and $\Delta \geq 0$, we have

$$F_{A,2}(x, x^*) = (x, x^*) \Rightarrow (x, x^*) \in \operatorname{gra} A$$

and

$$\Delta(-x, w^* - x^*) \leq 0 \Rightarrow w^* - x^* = x \Rightarrow w^* = x + x^*,$$

so that $w^* \in (\mathrm{Id} + A)x$. Since the point $w^* \in H$ is arbitrary, we conclude that $\operatorname{ran}(\mathrm{Id} + A) = H$. The proof of the reverse implication is standard: Let $(w, w^*) \in H \times H$ be such that

$$\langle w - x, w^* - x^*\rangle \geq 0 \quad \forall (x, x^*) \in \operatorname{gra} A. \tag{3.18}$$

We claim that $(w, w^*) \in H \times H$. Since $\operatorname{ran}(\mathrm{Id} + A) = H$, we have $w^* + w = x + x^*$ for some $(x, x^*) \in \operatorname{gra} A$. Put $w^* - x^* = -(w - x)$ into (3.18) to obtain $0 \geq \langle w - x, -(w - x)\rangle \geq 0$. Thus, $w = x$ and $w^* = x^*$. This means that $(w, w^*) \in \operatorname{gra} A$. Thus the operator A is indeed maximally monotone, as asserted.

Acknowledgments

Both authors would like to thank Heinz Bauschke for his useful suggestions regarding this note, and an anonymous referee for several pertinent comments and helpful remarks.

References

[1] Sedi Bartz, Heinz H. Bauschke, Jonathan M. Borwein, Simeon Reich, and Xianfu Wang, *Fitzpatrick functions, cyclic monotonicity and Rockafellar's antiderivative*, Nonlinear Anal. **66** (2007), no. 5, 1198–1223, DOI 10.1016/j.na.2006.01.013. MR2286629 (2008a:47082)

[2] Heinz H. Bauschke, *Fenchel duality, Fitzpatrick functions and the extension of firmly nonexpansive mappings*, Proc. Amer. Math. Soc. **135** (2007), no. 1, 135–139 (electronic), DOI 10.1090/S0002-9939-06-08770-3. MR2280182 (2008a:49017)

[3] Heinz H. Bauschke and Patrick L. Combettes, *Convex analysis and monotone operator theory in Hilbert spaces*, CMS Books in Mathematics/Ouvrages de Mathématiques de la SMC, Springer, New York, 2011. With a foreword by Hédy Attouch. MR2798533 (2012h:49001)

[4] Heinz H. Bauschke and Xianfu Wang, *A convex-analytical approach to extension results for n-cyclically monotone operators*, Set-Valued Anal. **15** (2007), no. 3, 297–306, DOI 10.1007/s11228-006-0029-1. MR2328863 (2008e:47119)

[5] Simon Fitzpatrick, *Representing monotone operators by convex functions*, Workshop/Miniconference on Functional Analysis and Optimization (Canberra, 1988), Proc. Centre Math. Anal. Austral. Nat. Univ., vol. 20, Austral. Nat. Univ., Canberra, 1988, pp. 59–65. MR1009594 (90i:47054)

[6] Simeon Reich and Stephen Simons, *Fenchel duality, Fitzpatrick functions and the Kirszbraun-Valentine extension theorem*, Proc. Amer. Math. Soc. **133** (2005), no. 9, 2657–2660 (electronic), DOI 10.1090/S0002-9939-05-07983-9. MR2146211 (2006d:46025)

[7] R. Tyrrell Rockafellar, *Convex analysis*, Princeton Mathematical Series, No. 28, Princeton University Press, Princeton, N.J., 1970. MR0274683 (43 #445)

[8] R. Tyrrell Rockafellar, *Extension of Fenchel's duality theorem for convex functions*, Duke Math. J. **33** (1966), 81–89. MR0187062 (32 #4517)
[9] Stephen Simons and Constantin Zălinescu, *Fenchel duality, Fitzpatrick functions and maximal monotonicity*, J. Nonlinear Convex Anal. **6** (2005), no. 1, 1–22. MR2138099 (2005k:49102)
[10] Constantin Zălinescu, *Convex analysis in general vector spaces*, World Scientific Publishing Co., Inc., River Edge, NJ, 2002. MR1921556 (2003k:49003)

Department of Mathematics, The Technion – Israel Institute of Technology, 32000 Haifa, Israel
E-mail address: sreich@tx.technion.ac.il

Mathematics, The University of British Columbia, Kelowna, British Columbia V1V 1V7, Canada
E-mail address: shawn.wang@ubc.ca

Contemporary Mathematics
Volume **659**, 2016
http://dx.doi.org/10.1090/conm/659/13149

Asymptotic centers, inexact orbits, and fixed points

Simeon Reich and Alexander J. Zaslavski

ABSTRACT. We use asymptotic centers and their variants to establish fixed point theorems for nonexpansive mappings in uniformly convex Banach spaces, and for holomorphic mappings in the Hilbert ball and its powers.

1. Asymptotic Centers

Let (X, ρ) be a metric space. For each $x \in X$ and each $r > 0$, set

$$B(x, r) = \{y \in X : \ \rho(x, y) \leq r\}.$$

Recall that a subset of X is said to be *bounded* if it is contained in a ball $B(x, r)$ for some $x \in X$ and $r > 0$.

Let $C \subset X$ be nonempty. We assume that for each bounded sequence

$$\{u_n\}_{n=1}^{\infty} \subset C,$$

there exists a unique point $c \in C$ such that for each $y \in C \setminus \{c\}$,

$$\begin{aligned} &\limsup_{n\to\infty} \rho(u_n, c) \\ &= \lim_{m\to\infty} \sup\{\rho(u_k, c) : k \geq m \text{ is an integer}\} \\ &< \lim_{m\to\infty} \sup\{\rho(u_k, y) : k \geq m \text{ is an integer}\} \\ &= \limsup_{n\to\infty} \rho(u_n, y). \end{aligned} \tag{1.1}$$

This point is called the *asymptotic center* of $\{u_n\}_{n=1}^{\infty}$ with respect to C (see [**3**, page 206], [**13**, page 251] and [**4**, pages 18 and 115]). The usefulness of this concept is illustrated by the following simple observation (*cf.* [**4**, pages 24 and 120]).

THEOREM 1.1. *Let $f : C \to C$ satisfy*

$$\rho(f(x), f(y)) \leq \rho(x, y), \ x, y \in X. \tag{1.2}$$

If a sequence $\{u_n\}_{n=1}^{\infty} \subset C$ is bounded and satisfies

$$\lim_{n\to\infty} \rho(u_n, f(u_n)) = 0, \tag{1.3}$$

2010 *Mathematics Subject Classification.* Primary 32A10, 46G20, 46T25, 47H09, 47H10.

Key words and phrases. Asymptotic center, fixed point, Hilbert ball, holomorphic mapping, inexact orbit, iterate, Kobayashi metric, nonexpansive mapping.

then the asymptotic center c of $\{u_n\}_{n=1}^{\infty}$ with respect to C is a fixed point of f, that is, $f(c) = c$.

PROOF. Let c be the asymptotic center of $\{u_n\}_{n=1}^{\infty}$. By (1.3), c is also the asymptotic center of $\{f(u_n)\}_{n=1}^{\infty}$ and

$$\lim_{m\to\infty} \sup\{\rho(u_k, c) : k \geq m \text{ is an integer}\}$$

$$= \lim_{m\to\infty} \sup\{\rho(f(u_k), c) : k \geq m \text{ is an integer}\}. \tag{1.4}$$

In view of (1.2),

$$\lim_{m\to\infty} \sup\{\rho(f(u_k), f(c)) : k \geq m \text{ is an integer}\}$$

$$\leq \lim_{m\to\infty} \sup\{\rho(u_k, c) : k \geq m \text{ is an integer}\}.$$

Since c is the unique asymptotic center of $\{f(u_n)\}_{n=1}^{\infty}$, it follows that $c = f(c)$, as asserted. Theorem 1.1 is proved. □

2. Uniformly Convex Banach Spaces

Let C be a closed and convex set in a uniformly convex Banach space $(X, \|\cdot\|)$. Recall that X is called *uniformly convex* if its modulus of convexity

$$\delta(\epsilon) := \inf\{1 - \|x + y\|/2 : x, y \in X, \ \|x\|, \|y\| \leq 1, \ \|x - y\| \geq \epsilon\} > 0$$

for all $\epsilon \in (0, 2]$.

Let a sequence $\{u_n\}_{n=1}^{\infty} \subset C$ be bounded. For any integer $m \geq 1$, define

$$r_m(y) := \sup\{\|u_k - y\| : k \geq m \text{ is an integer}\}, \ y \in X. \tag{2.1}$$

It is known that for any integer $m \geq 1$, there exists a unique point $c_m \in C$ such that

$$r_m := r_m(c_m) = \inf\{r_m(y) : y \in C\}. \tag{2.2}$$

Clearly, $r_m \geq r_{m+1}$ and $r_m \geq 0$ for all natural numbers m. Therefore there exists

$$r = \lim_{m\to\infty} r_m = \inf\{r_m : m = 1, 2, \dots\}. \tag{2.3}$$

Clearly, if $r = 0$, then the sequence $\{u_n\}_{n=1}^{\infty}$ converges.

The following two results were obtained in [**3**].

THEOREM 2.1. *The sequence $\{c_m\}_{m=1}^{\infty}$ converges to a point $c \in C$, which is the asymptotic center of $\{u_n\}_{n=1}^{\infty}$ with respect to C.*

THEOREM 2.2. *Let $f : C \to C$ satisfy*

$$\|f(x) - f(y)\| \leq \|x - y\|$$

for all $x, y \in C$, $x_0 \in C$, $u_n = f^n(x_0)$ for all integers $n \geq 1$, and let the sequence $\{f^n(x_0)\}_{n=1}^{\infty}$ be bounded. Then the asymptotic center c of $\{f^n(x_0)\}_{n=1}^{\infty}$ with respect to C is a fixed point of f, that is, $f(c) = c$.

The next result follows from Theorem 1.1.

THEOREM 2.3. *Let* $f : C \to C$ *satisfy*

$$\|f(x) - f(y)\| \le \|x - y\|$$

for all $x, y \in C$ *and let* $\{u_n\}_{n=1}^{\infty} \subset C$ *be a bounded sequence such that*

$$\lim_{n\to\infty} \|u_n - f(u_n)\| = 0.$$

Then the asymptotic center c *of* $\{u_n\}_{n=1}^{\infty}$ *with respect to* C *is a fixed point of* f*, that is,* $f(c) = c$.

For each $z \in X$ and each $\alpha > 0$, set

$$B(z, \alpha) = \{y \in X : \|z - y\| \le \alpha\}.$$

The next theorem is the main result of this section. We give its proof in the next one.

THEOREM 2.4. *Assume that* $f : C \to C$ *satisfies*

$$\|f(x) - f(y)\| \le \|x - y\|$$

for all $x, y \in C$, $M > 0$*, and that for each* $\epsilon > 0$*, there exists a sequence* $\{u_n\}_{n=1}^{\infty} \subset C \cap B(0, M)$ *such that*

$$\|u_{n+1} - f(u_n)\| \le \epsilon,\ n = 1, 2, \ldots.$$

Then there exists a point $p \in C$ *such that* $f(p) = p$.

3. Proof of Theorem 2.4

We use all the notations and definitions introduced in Section 2.

Let $\eta \in (0, 1)$. Choose a number $\epsilon \in (0, 1)$ such that

$$\epsilon < 8^{-1}\eta^2 \min\{1, \delta(\eta(6M + 3)^{-1})\}. \tag{3.1}$$

There exists a sequence

$$\{u_i\}_{i=1}^{\infty} \subset C \cap B(0, M) \tag{3.2}$$

such that

$$\|u_{i+1} - f(u_i)\| \le \epsilon,\ \ i = 1, 2, \ldots. \tag{3.3}$$

Let $c_m \in C$, $m = 1, 2, \ldots$, satisfy (2.2). By Theorem 2.1, there exists

$$c = \lim_{m\to\infty} c_m. \tag{3.4}$$

In view of (2.2) and (3.2), for each integer $m \ge 1$,

$$r_m(c_m) \le r_m(u_m) = \sup\{\|u_k - u_m\| : k \ge m \text{ is an integer}\} \le 2M$$

and

$$\|c_m\| \le 3M. \tag{3.5}$$

It follows from (3.4) and (3.5) that

$$\|c\| \le 3M. \tag{3.6}$$

We claim that

$$\|f(c) - c\| \le \eta. \tag{3.7}$$

If $r = 0$, then the sequence $\{u_i\}_{i=1}^{\infty}$ converges,

$$c = \lim_{n\to\infty} u_n,$$

and in view of (3.1) and (3.3),

$$\|f(c) - c\| = \lim_{n\to\infty} \|u_{n+1} - f(u_n)\| \le \epsilon < \eta,$$

so in this case (3.7) is true.

Assume now that $r > 0$ and (3.7) does not hold. Then

$$\|f(c) - c\| > \eta. \tag{3.8}$$

By (3.4), there exists an integer $n_1 \ge 1$ such that for each integer $n \ge n_1$,

$$\|c_n - c\| \le \eta/3. \tag{3.9}$$

In view of (3.9), for each integer $n \ge n_1$,

$$\|f(c_n) - f(c)\| \le \eta/3. \tag{3.10}$$

It follows from (3.8), (3.9) and (3.10) that for each integer $n \ge n_1$,

$$\|c_n - f(c_n)\| \ge \|c - f(c)\| - \|c_n - c\| - \|f(c_n) - f(c)\| \ge \eta/3. \tag{3.11}$$

Let a pair of integers k and n satisfy $k \ge n \ge n_1 + 1$. In view of (2.2),

$$\|u_k - c_n\| \le r_n. \tag{3.12}$$

By (3.3) and (3.12),

$$\|u_k - f(c_n)\| \le \|u_k - f(u_k)\| + \|f(u_k) - f(c_n)\|$$

$$\le \epsilon + \|u_k - c_n\| \le r_n + \epsilon. \tag{3.13}$$

It follows from (3.1), (3.2), (3.11)–(3.13) and the uniform convexity of the space X that

$$(r_n + \epsilon)^{-1}\|u_k - (c_n + f(c_n))/2\| \le 1 - \delta(\eta(3(r_n+\epsilon))^{-1}) \le 1 - \delta(\eta(3(2M+1))^{-1})$$

and so,

$$\|u_k - (c_n + f(c_n))/2\| \le (r_n + \epsilon)(1 - \delta(\eta(6M+3)^{-1})). \tag{3.14}$$

On the other hand, in view of (3.11),

$$(c_n + f(c_n))/2 \ne c_n.$$

Therefore there exists an integer $k > n$ such that

$$r_n < \|u_k - (c_n + f(c_n))/2\|.$$

When combined with (3.14), this inequality implies that

$$r_n < (r_n + \epsilon)(1 - \delta(\eta(6M+3)^{-1})).$$

Together with (2.3) and (3.1), this implies that

$$r \le (r + \epsilon)(1 - \delta(\eta(6M+3)^{-1})),$$

$$r\delta(\eta(6M+3)^{-1}) \le \epsilon$$

and

$$r \le \epsilon(\delta(\eta(6M+3)^{-1}))^{-1} < 8^{-1}\eta^2. \tag{3.15}$$

By (3.15), there exists a natural number n_* such that

$$r_n < 8^{-1}\eta^2 \text{ for all integers } n \ge n_* \tag{3.16}$$

and

$$\|c_n - c\| \le 8^{-1}\eta^2 \text{ for all integers } n \ge n_*. \tag{3.17}$$

In view of (3.12) and (3.16), for each pair of integers $k \geq n \geq n_*$,

$$\|u_k - c_n\| < 8^{-1}\eta^2. \tag{3.18}$$

It follows from (3.17) and (3.18) that for each integer $k \geq n_*$,

$$\|u_k - c\| \leq \|u_k - c_k\| + \|c_k - c\| < \eta^2/4.$$

When combined with (3.3), this implies that for each integer $k \geq n_*$,

$$\begin{aligned}\|f(c) - c\| &\leq \|f(c) - f(u_k)\| + \|f(u_k) - u_{k+1}\| + \|u_{k+1} - c\| \\ &\leq \|c - u_k\| + \epsilon + \|u_{k+1} - c\| \\ &\leq \eta^2/4 + \eta^2/4 + \epsilon < (3/4)\eta^2.\end{aligned}$$

This, however, contradicts (3.8). The contradiction we have reached shows that (3.7) does indeed hold, as claimed.

Therefore for each $\eta > 0$, there exists

$$c \in B(0, 3M) \cap C$$

satisfying (3.7). By Theorem 2.3, there exists a point $p \in C$ such that $f(p) = p$. This completes the proof of Theorem 2.4. □

4. Inexact Orbits of Holomorphic Mappings

In this and the next three sections we are going to show that if holomorphic self-mappings of the Hilbert ball (and certain self-mappings of other domains) have an inexact orbit which lies strictly inside the ball (respectively, the domain), then they have a fixed point. Several related results regarding inexact orbits are also obtained. The proofs employ variants of the asympotic center concept.

Let $(X, \|\cdot\|)$ be a Banach space over the complex field $\mathbf{C}$. For each point $x \in X$ and each number $r > 0$, set

$$B(x, r) := \{y \in X : \|x - y\| \leq r\}.$$

Recall that an open and connected subset of X is called a *domain*. We say that a nonempty subset K of a domain D is *strictly inside* D [**2**, page 61] if there exists a number $r > 0$ such that

$$\bigcup\{B(z, r) : z \in K\} \subset D.$$

We are concerned here with (infinite dimensional) holomorphic fixed point theory, where this concept has already been shown to be useful. See, for example, [**2**], [**8**], [**7**], [**17**], [**18**] and references therein.

We begin by recalling the following known result (*cf.* [**5**, Theorem 11, page 1019]; see also [**4**, Theorem 23.1, page 120]).

THEOREM 4.1. *Let $\mathbb{B}$ be the open unit ball in a complex Hilbert space and let the mapping $h : \mathbb{B} \to \mathbb{B}$ be holomorphic. If for some point $x_0 \in \mathbb{B}$, the sequence of iterates $\{h^n(x_0)\}_{n=1}^{\infty}$ lies strictly inside $\mathbb{B}$, then h has a fixed point in $\mathbb{B}$.*

In view of the recent interest in the influence of computational errors on the performance of iterative methods (see, for instance, [**12**], [**16**] and references therein), our main purpose in the next three sections is to establish the following two results concerning *inexact orbits* of holomorphic self-mappings.

THEOREM 4.2. *Let $\mathbb{B}$ be the open unit ball in a complex Hilbert space and let the mapping $h : \mathbb{B} \to \mathbb{B}$ be holomorphic. Assume that a sequence $\{x_n\}_{n=1}^{\infty}$ lies strictly inside $\mathbb{B}$ and satisfies*

$$\lim_{n\to\infty} \|x_{n+1} - h(x_n)\| = 0.$$

Then there exists a point $y_0 \in \mathbb{B}$ such that the sequence $\{h^n(y_0)\}_{n=1}^{\infty}$ lies strictly inside $\mathbb{B}$ and the mapping h has a fixed point in $\mathbb{B}$.

It is clear that Theorem 4.2 follows from Theorem 4.1 and the following general result, which we prove in Section 6.

THEOREM 4.3. *Let D be a bounded and convex domain in a complex Banach space $(X, \|\cdot\|)$ and let the mapping $h : D \to D$ be holomorphic. Assume that a sequence $\{x_n\}_{n=1}^{\infty}$ lies strictly inside D and satisfies*

$$\lim_{n\to\infty} \|x_{n+1} - h(x_n)\| = 0.$$

Then there exists a point $y_0 \in D$ such that the sequence $\{h^n(y_0)\}_{n=1}^{\infty}$ lies strictly inside D.

5. The Kobayashi Metric

Throughout this section we assume that $(X, \|\cdot\|)$ is a general complex Banach space and that D is a bounded and convex domain in X. Let k_D be the Kobayashi metric [**9**, page 462] on D. Each holomorphic self-mapping of D is *nonexpansive* (that is, 1-Lipschitz) with respect to this metric (see, for example, [**4**, page 91]).

Set

$$\mathrm{diam}(D) := \sup\{\|x - y\| : x, y \in D\},$$

denote by ∂D the boundary of D and for all $x \in X$, put

$$\mathrm{dist}(x, \partial D) := \inf\{\|x - y\| : y \in \partial D\}.$$

In the next section we use the following well-known result (see [**6**, pages 353 and 355], [**4**, pages 89–90], [**10**, page 442] and [**15**, page 99]).

THEOREM 5.1. *For all $x, y \in D$,*

$$\mathrm{argtanh}(\|x - y\|/\mathrm{diam}(D)) \le k_D(x, y)$$

and if $\|x - y\| < \mathrm{dist}(x, \partial D)$, then

$$k_D(x, y) \le \mathrm{argtanh}(\|x - y\|/\mathrm{dist}(x, \partial D)).$$

At this point we also recall the following characterization of those subsets which lie strictly inside a bounded and convex domain [**6**, Proposition 23, page 381]. See also [**10**, page 442] and [**15**, page 101].

THEOREM 5.2. *Let D be a bounded and convex domain in a complex Banach space. A nonempty subset K of D lies strictly inside D if and only if it is k_D-bounded.*

We remark in passing that a recent result related to Theorem 5.2 can be found in [**1**, Theorem 2.4].

6. Proof of Theorem 4.3

There exist numbers

$$M_0 > \sup\{\|z_1 - z_2\| : z_1, z_2 \in D\} \tag{6.1}$$

and $r_* \in (0,1)$ such that

$$B(x_i, 2r_*) \subset D,\ i = 1, 2, \dots. \tag{6.2}$$

There also exists an integer $i_1 \geq 1$ such that for all integers $i \geq i_1$,

$$\|x_{i+1} - h(x_i)\| \leq r_*/4. \tag{6.3}$$

By (6.2), (6.3) and Theorem 5.1, for every integer $i \geq i_1$, there exists a real number

$$\xi \in [0, r_*^{-1}\|x_{i+1} - h(x_i)\|] \subset [0, 1/4]$$

such that

$$\begin{aligned} k_D(x_{i+1}, h(x_i)) &\leq \operatorname{argtanh}(\|x_{i+1} - h(x_i)\| r_*^{-1}) \\ &= \operatorname{argtanh}'(\xi)\|x_{i+1} - h(x_i)\| r_*^{-1} \\ &= (1 - \xi^2)^{-1} r_*^{-1}\|x_{i+1} - h(x_i)\| \\ &\leq 2r_*^{-1}\|x_{i+1} - h(x_i)\|. \end{aligned} \tag{6.4}$$

Hence

$$\lim_{n\to\infty} k_D(x_{n+1}, h(x_n)) = 0. \tag{6.5}$$

Now fix a point $x_0 \in D$, let $R := \limsup_{n\to\infty} k_D(x_0, x_n)$, and consider the set

$$C := \{y \in D : \limsup_{n\to\infty} k_D(y, x_n) \leq R\}. \tag{6.6}$$

It is clear that this set is a nonempty and k_D-bounded subset of D. If $z \in C$, then by (6.5),

$$\begin{aligned} \limsup_{n\to\infty} k_D(h(z), x_n) &= \limsup_{n\to\infty} k_D(h(z), x_{n+1}) = \limsup_{n\to\infty} k_D(h(z), h(x_n)) \\ &\leq \limsup_{n\to\infty} k_D(z, x_n) \leq R. \end{aligned}$$

Hence C is invariant under h and therefore for each $y_0 \in C$, the sequence

$$\{h^n(y_0)\}_{n=1}^{\infty}$$

lies strictly inside D by Theorem 5.2, as asserted. This completes the proof of Theorem 4.3. □

7. Additional Results

Let H be a complex Hilbert space and let $d \geq 2$ be a natural number. When the finite power H^d is equipped with the maximum norm, then its open unit ball is the finite power $\mathbb{B}^d$ of the Hilbert ball $\mathbb{B}$. It is known [**10**, Theorem 7.3, page 462] that an analog of Theorem 4.1 holds for such powers. Combining this fact with Theorem 4.3, we obtain the following extension of Theorem 4.2.

THEOREM 7.1. *Let $\mathbb{B}^d$ be a finite power of the Hilbert ball $\mathbb{B}$ and let the mapping $h : \mathbb{B}^d \to \mathbb{B}^d$ be holomorphic. Assume that a sequence $\{x_n\}_{n=1}^{\infty}$ lies strictly inside $\mathbb{B}^d$ and satisfies*

$$\lim_{n\to\infty} \|x_{n+1} - h(x_n)\| = 0.$$

Then there exists a point $y_0 \in \mathbb{B}^d$ such that the sequence $\{h^n(y_0)\}_{n=1}^{\infty}$ lies strictly inside $\mathbb{B}^d$ and the mapping h has a fixed point in $\mathbb{B}^d$.

We also note that since the proofs of Theorems 4.1–4.3 and 7.1 have a metric character, these theorems continue to hold for all those self-mappings which are nonexpansive (that is, 1-Lipschitz) with respect to the Kobayashi metric (even if they are not holomorphic).

However, the example on page 107 of [**11**] (see also [**10**, Example 11.5, page 482]) shows that analogs of Theorems 4.1, 4.3 and 7.1 do *not* hold for the open unit ball $\mathbb{B}^\infty$ of $\ell^\infty(H)$ even if $H = \mathbf{C}$. This example exhibits a fixed point free holomorphic self-mapping of $\mathbb{B}^\infty$ all the orbits of which are strictly inside $\mathbb{B}^\infty$.

On the other hand, since an analog of Theorem 4.1 *does* hold for nonexpansive (that is, 1-Lipschitz with respect to the metric induced by the norm) self-mappings of certain subsets of Banach spaces (*cf.*, for example, [**14**, Theorem 1, page 196] and [**4**, Theorem 5.2, page 24]), the following analog of Theorems 4.3 and 7.1 can be established by using the idea of the proof of Theorem 4.3.

THEOREM 7.2. *Let C be a boundedly weakly compact and convex subset of a (real or complex) Banach space $(X, \|\cdot\|)$. Suppose that each weakly compact and convex subset of C has the fixed point property for nonexpansive mappings, and let the mapping $T : C \to C$ be nonexpansive. Assume that a bounded sequence $\{x_n\}_{n=1}^{\infty}$ satisfies*

$$\lim_{n\to\infty} \|x_{n+1} - T(x_n)\| = 0.$$

Then there exists a point $y_0 \in C$ such that the sequence $\{T^n(y_0)\}_{n=1}^{\infty}$ is bounded and the mapping T has a fixed point in C.

Acknowledgments. The first author was partially supported by the Israel Science Foundation (Grant No. 389/12), by the Fund for the Promotion of Research at the Technion and by the Technion General Research Fund. Both authors are grateful to an anonymous referee for several helpful remarks and corrections.

References

[1] Monika Budzyńska, Tadeusz Kuczumow, and Simeon Reich, *The common fixed point set of commuting holomorphic mappings in Cartesian products of Banach spaces*, Fixed Point Theory **16** (2015), no. 1, 49–66. MR3330992

[2] Clifford J. Earle and Richard S. Hamilton, *A fixed point theorem for holomorphic mappings*, Global Analysis (Proc. Sympos. Pure Math., Vol. XVI, Berkeley, Calif., (1968), Amer. Math. Soc., Providence, R.I., 1970, pp. 61–65. MR0266009 (42 #918)

[3] Michael Edelstein, *The construction of an asymptotic center with a fixed-point property*, Bull. Amer. Math. Soc. **78** (1972), 206–208. MR0291917 (45 #1005)

[4] Kazimierz Goebel and Simeon Reich, *Uniform convexity, hyperbolic geometry, and nonexpansive mappings*, Monographs and Textbooks in Pure and Applied Mathematics, vol. 83, Marcel Dekker, Inc., New York, 1984. MR744194 (86d:58012)

[5] K. Goebel, T. Sekowski, and A. Stachura, *Uniform convexity of the hyperbolic metric and fixed points of holomorphic mappings in the Hilbert ball*, Nonlinear Anal. **4** (1980), no. 5, 1011–1021, DOI 10.1016/0362-546X(80)90012-7. MR586863 (82f:32039)

[6] Lawrence A. Harris, *Schwarz-Pick systems of pseudometrics for domains in normed linear spaces*, Advances in holomorphy (Proc. Sem. Univ. Fed. Rio de Janeiro, Rio de Janeiro, 1977), North-Holland Math. Stud., vol. 34, North-Holland, Amsterdam-New York, 1979, pp. 345–406. MR520667 (80j:32043)

[7] Lawrence A. Harris, *Fixed points of holomorphic mappings for domains in Banach spaces*, Abstr. Appl. Anal. **5** (2003), 261–274, DOI 10.1155/S1085337503205042. MR1981265 (2004e:46091)

[8] V. Khatskevich, S. Reich, and D. Shoĭkhet, *Fixed point theorems for holomorphic mappings and operator theory in indefinite metric spaces*, Integral Equations Operator Theory **22** (1995), no. 3, 305–316, DOI 10.1007/BF01378779. MR1337378 (97b:46062)

[9] Shoshichi Kobayashi, *Invariant distances on complex manifolds and holomorphic mappings*, J. Math. Soc. Japan **19** (1967), 460–480. MR0232411 (38 #736)

[10] Tadeusz Kuczumow, Simeon Reich, and David Shoikhet, *Fixed points of holomorphic mappings: a metric approach*, Handbook of metric fixed point theory, Kluwer Acad. Publ., Dordrecht, 2001, pp. 437–515. MR1904285 (2003f:58018)

[11] Tadeusz Kuczumow, Simeon Reich, and Adam Stachura, *Holomorphic retracts of the open unit ball in the l_∞-product of Hilbert spaces*, Recent advances on metric fixed point theory (Seville, 1995), Ciencias, vol. 48, Univ. Sevilla, Seville, 1996, pp. 99–110. MR1440221 (98b:46061)

[12] Evgeniy Pustylnik, Simeon Reich, and Alexander J. Zaslavski, *Inexact orbits of nonexpansive mappings*, Taiwanese J. Math. **12** (2008), no. 6, 1511–1523. MR2444869 (2009i:47116)

[13] Simeon Reich, *Remarks on fixed points. II* (English, with Italian summary), Atti Accad. Naz. Lincei Rend. Cl. Sci. Fis. Mat. Natur. (8) **53** (1972), 250–254 (1973). MR0435953 (55 #8904)

[14] Simeon Reich, *Fixed point iterations of nonexpansive mappings*, Pacific J. Math. **60** (1975), no. 2, 195–198. MR0428130 (55 #1159)

[15] Simeon Reich and David Shoikhet, *Nonlinear semigroups, fixed points, and geometry of domains in Banach spaces*, Imperial College Press, London, 2005. MR2022955 (2006g:47105)

[16] Simeon Reich and Alexander J. Zaslavski, *A note on inexact infinite products*, Commun. Appl. Anal. **16** (2012), no. 4, 655–663. MR3051845

[17] Simeon Reich and Alexander J. Zaslavski, *Three generic results in holomorphic fixed point theory*, Complex Anal. Oper. Theory **8** (2014), no. 1, 51–56, DOI 10.1007/s11785-012-0266-2. MR3147712

[18] Simeon Reich and Alexander J. Zaslavski, *Inexact orbits of holomorphic mappings in complex Banach spaces*, Rend. Circ. Mat. Palermo (2) **63** (2014), no. 3, 439–445, DOI 10.1007/s12215-014-0171-9. MR3298594

Department of Mathematics, The Technion – Israel Institute of Technology, 32000 Haifa, Israel

E-mail address: `sreich@tx.technion.ac.il`

Department of Mathematics, The Technion – Israel Institute of Technology, 32000 Haifa, Israel

E-mail address: `ajzasl@tx.technion.ac.il`

Contemporary Mathematics
Volume **659**, 2016
http://dx.doi.org/10.1090/conm/659/13155

A discrete approximation of the entropy functional for probability measures on the plane

Gershon Wolansky

ABSTRACT. In this paper we use a recent result of R.Pinchasi to obtain a Gamma approximation of the Entropy functional. Pinchasi's Theorem was motivated by a generalization of Thue's theorem regarding the densest packing of the plane by unit discs. His Theorem provides an alternative proof to a generalized Thue's result, dealing with packing of discs of varying radii of positive homogeneity in the plane. We show that this result provides also a possibility to obtain the Entropy functional for positive measures on the plane as a limit of discrete functions on finitely supported empirical measures.

1. Introduction

The optimal unit disc packing in the plane is a celebrated classical problem that goes back to the 1663 famous Kepler's problem about the densest unit sphere packing in $\mathbb{R}^3$. As for the planar version: In 1773 Lagrange proved that among all lattice packings, that is packing of unit discs where the sets of centers is a lattice, the densest one (the hexagonal lattice) has density $\frac{\pi}{2\sqrt{3}}$. The densest unit disc packing problem in the plane, without any additional assumption, was solved only in 1910 by Thue and ever since this result is known as Thue's theorem. Thue's proof was considered incomplete and a full complete proof of the theorem was given in 1943 by L. F. Tóth. Since then more proofs, each more elegant than the other, where presented, as this theorem and topic attracted quite some attention (see [**1**, **5**–**7**]).

Tóth extended his result to a collection of discs of different radii between two positive limits $R_1 \leq r \leq R_2$ (so called "positive homogeneity" [**10**], [**11**]). Recently, a Theorem of Pinchasi leads to another proof of this result. Surprisingly, this result of Pinchasi can be used for another application, namely the Gamma approximation of entropy like functionals in $\mathbb{R}^2$.

THEOREM 1.1. *(R. Pinchasi [**8**]) Let $\{D_i\} \subset \mathbb{R}^2$ a finite collection of disjoint discs in $\mathbb{R}^2$. Let $r_i > 0$ be the radius of D_i and x_i its center. For each pair D_i, D_j consider the line l_i^j perpendicular to the segment connecting x_i to x_j, and dividing this segment by ratio r_i/r_j. Given a third disk D_k in this collection, consider the quadrilateral $\Delta_{i,j,k}$ bounded by the lines l_i^j, l_i^k and the segments $[x_i, x_j], [x_i, x_k]$. Then $|\Delta_{i,j,k}| \geq 2\sqrt{3}|\Delta_{i,j,k} \cap D_i|/\pi$. (here $|\cdot|$ stands for the aria).*

See Figure 1 for an illustration.

2010 *Mathematics Subject Classification.* Primary 37M99 28D99, 35Q70.

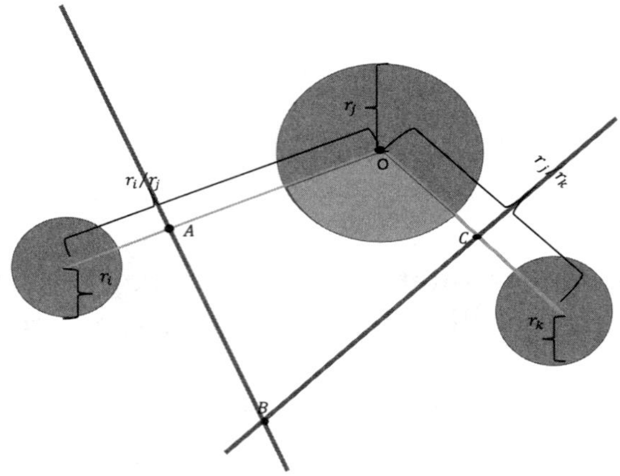

FIGURE 1. The quadrilateral $\Delta_{i,j,k}$ bounded by the lines l_i^j, l_i^k and the segments $[x_i, x_j], [x_i, x_k]$

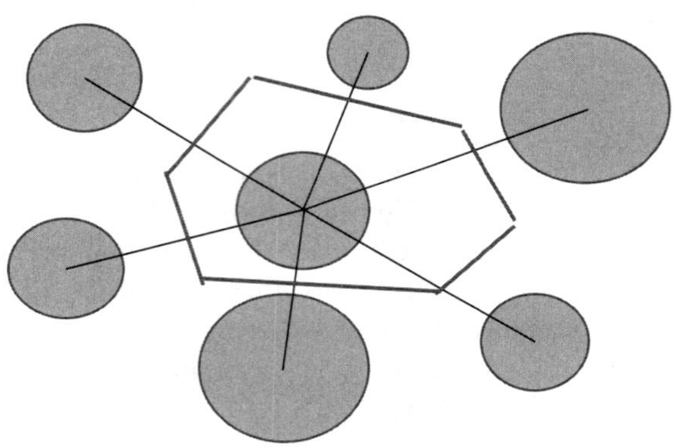

FIGURE 2. An illustration of Cell(D_i)

COROLLARY 1.2. Let H_i^j be the half plane bounded by the line l_i^j above and containing x_i. Let

$$Cell(D_i) := \cap_{j \neq i} H_i^j \ .$$

Then $|Cell(D_i)| \geq 2\sqrt{3} r_i^2$ (Figure 2) .

Note that $2\sqrt{3}r_i^2$ is just the area of the perfect hexagon inscribing D_i. Evidently, the number $2\sqrt{3}$ is sharp, and is obtained for hexagonal packing of identical discs (see Figure 3).

We now describe the notion of Gamma approximation of the entropy.

Consider the set $\mathcal{B}(\Omega)$ of Borel probability measures on a "nice", compact set $\Omega \subset \mathbb{R}^2$ (we may assume it is a disc, or square). The *entropy* of a measure $\mu \in \mathcal{B}(\Omega)$ is defined as the Lebesgue integral

$$E(\mu) := \int_\Omega \rho \ln \rho dx \ ,$$

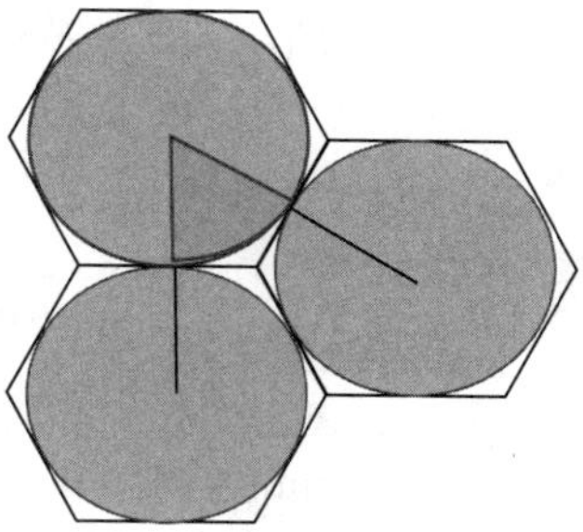

FIGURE 3. Optimal covering of the plan by regular hexagons

where $\mu := \rho dx$ if such a density exists, or $E(\mu) = +\infty$ if such a density does not exist.

A *gradient descent flow under Wasserstein metric* for a functional E in $\mathcal{B}(\Omega)$ (see [**3**]) is defined by a path $t \mapsto \rho(\cdot, t)$ where $\rho(\cdot, t)$ is a density of a probability measure $d\mu_{(t)} := \rho(x,t)dx$ in $\mathcal{B}(\Omega)$, defined by

$$\frac{\partial \rho}{\partial t} = \nabla \cdot \left(\rho \nabla \frac{\delta E(\mu_{(t)})}{\delta \mu_{(t)}} \right) \ , \ t > 0 \ , \ \lim_{t \to 0} \rho(\cdot, t) := \rho_0 \tag{1.1}$$

here $\delta E / \delta \mu$ is the variational derivative (in that case just $\ln(\rho) \mod \mathbb{R}$ if $E(\mu) < \infty$, reducing (1.1) to the heat equation)).

Our object is to find a proper approximation of the entropy on the class of $N-$*empirical measures*:

$$\mathcal{B}_N(\Omega) := \left\{ \mu_N = N^{-1} \sum_{i=1}^{N} \delta_{x_i} \ ; \quad x_i \in \Omega \ , \ x_i \neq x_j \text{ for } i \neq j, \ 1 \leq i, j \leq N \right\} \subset \mathcal{B}(\Omega)$$

where $N \in \mathbb{N}$. Let $\mathcal{B}_\infty(\Omega) := \cup_{N \in \mathbb{N}} \mathcal{B}_N(\Omega)$.

Since Ω is a compact set, $\mathcal{B}(\Omega)$ is compact with respect to the weak $(C_b^*(\Omega))$ topology, that is, for every sequence $\{\mu_j\} \in \mathcal{B}(\Omega)$ there exists a subsequence μ_{j_k} and a measure $\mu \in \mathcal{B}(\Omega)$ such that $\mu_{j_k} \rightharpoonup \mu$ as $k \to \infty$, that is:

$$\lim_{k \to \infty} \int_\Omega \phi d\mu_{j_k} = \int_\Omega \phi d\mu$$

for any bounded continuous ϕ on Ω. It is also evident that $\mathcal{B}_\infty(\Omega)$ is dense in $\mathcal{B}(\Omega)$ with respect to the weak topology.

Let $\Omega^\otimes := \cup_{N \in \mathbb{N}} \Omega^{\otimes N}$ where $X \in \Omega^{\otimes N}$ iff $X := (x_1, \ldots x_N)$ is an *unordered* sequence of N distinct points in Ω. Let $N(X) := N$ iff $X \in \Omega^{\otimes N}$. We first note that each $\mu \in \mathcal{B}_\infty(\Omega)$ can be identified with a point $X \in \Omega^{\otimes N}$. Thus, a measure $\mu \in \mathcal{B}_N(\Omega)$ can be identified with $X \in \Omega^{\otimes N}$ via $\mu := \delta_X \equiv N(X)^{-1} \sum_{i=1}^{N(X)} \delta_{x_i}$.

DEFINITION 1.3. A $\Gamma-$approximation of the entropy E is a continuous function $\mathcal{E} : \Omega^\otimes \to \mathbb{R} \cup \{\infty\}$ such that

i): For *any* sequence $\{X_N\}$ such that $X_N \in \Omega^{\otimes N}$ and $\delta_{X_N} \rightharpoonup \mu \in \mathcal{B}(\Omega)$,

$$\liminf_{N \to \infty} \mathcal{E}(X_N) \geq E(\mu) \ .$$

ii): For any $\mu \in \mathcal{B}(\Omega)$ such that $E(\mu) < \infty$ *there exists* a sequence $\{\tilde{X}_N\} \in \Omega^{\otimes}$ such that $\delta_{\tilde{X}_N} \rightharpoonup \mu$ and

$$\lim_{N\to\infty} \mathcal{E}(\tilde{X}_N) = E(\mu) .$$

Given $X = (x_1, \ldots x_N) \in \Omega^{\otimes N}$, let $r_i(X)$ be half the minimal distance of x_i to the rest of the points in X:

$$r_i(X) := \frac{1}{2} \min_{j\neq i} |x_i - x_j| .$$

We now pose the following result:

THEOREM 1.4.

$$\mathcal{E}(X) := -\frac{2}{N(X)} \sum_{i=1}^{N(X)} \ln(r_i(X)) - \ln\left(2\sqrt{3}N(X)\right)$$

is a Γ *approximation of the entropy.*

REMARK 1.5. The referee to this paper asked why not define the discrete entropy by counting the number of particles in fixed disks? The answer is that, in that case, the discrete entropy will not be continuous, as its value will jump as soon as a particle leaves or enters a prescribed disk.

1.1. Application. Given such a Gamma approximation of a functional E, it is natural to ask the following question: Can we approximate (1.1) by the following system of ODE for $X^N \in \Omega^{\otimes N}$:

$$\frac{1}{N}\dot{X^N} = -\nabla_X \mathcal{E}(X^N(t))$$

for some *"well prepared"* initial data $X_0^N \in \Omega^{\otimes N}$ satisfying

$$\lim_{N\to\infty} \delta_{X_0^N} = \rho_0(x)dx \ ? \tag{1.2}$$

It turns out that the Gamma approximation of E by $\mathcal{E}$ and well prepared initial data (1.2) are, indeed, necessary condition for such a convergence of gradient descent systems. In [**9**], S. Serfaty presents a formalism which provides also Sufficient conditions for this convergence. The applicability of these conditions in the present setting is the subject of another publication [**4**].

2. Sketch the proof of Theorem 1.4

Let $D_i(X)$ be the disc of radius $r_i(X)$ centered at x_i.

A *partition rule* W is defined as a mapping between $\Omega^{\otimes}$ to a partition of $\mathbb{R}^2$ into an essentially disjoint sets. For $X \in \Omega^{\oplus}$ the partition rule associate a collection of $N(X)$ measurable sets $W_i := W_i(X) \subset \mathbb{R}^2$ such that

a): $W_i(X) \supset D_i(X)$ for any $i \in \{1, \ldots N(X)\}$

b): $\cup_{i=1}^{N(X)} W_i(X) \supseteq \Omega$.

c): $|W_i(X) \cap W_j(X)| = 0$ for any $i \neq j \in \{1, \ldots N(X)\}$. Here $|\cdot|$ stands for the Lebesgue measure of a measurable set in $\mathbb{R}^2$.

d): If $\{X_N\}$ is a density set in Ω, i.e for any open set $U \subset \Omega$, $\#(U \cap X_N) > 0$ for all N large enough, then $\lim_{N\to\infty} \max_{i\in\{1,\ldots N\}}$ *diameter* $(W_i(X_N) \cap \Omega) = 0$.

Given a partition rule W we may associate with every $\delta_X \in \mathcal{B}_\infty(\Omega)$ another measure $\mu \in \mathcal{B}(\Omega)$ which admits a density

$$\rho_X^W(x) := N^{-1}(X) \sum_{j=1}^{N(X)} \frac{1_{W_j(X)\cap\Omega}(x)}{|W_j(X)\cap\Omega|} .$$

Here $1_A(x)$ is the characteristic function for a set A, i.e $1_A(x) = 1, 0$ if $x \in A, x \notin A$ respectively.

We now define the approximation entropy corresponding to the partition rule $\overrightarrow{W}$,

$$\mathcal{E}^W : \Omega^{\otimes} \to \mathbb{R} \cup \{\infty\}$$

as the entropy of $\rho_X^W dx$ associated with $\delta_X \in \mathcal{B}_\infty(\Omega)$:

$$\mathcal{E}^W(X) := E(\rho_X^W dx) = \sum_{i=1}^{N(X)} \ln\left(\frac{1}{|W_i(X)\cap\Omega|}\right) - \ln N(X) . \tag{2.1}$$

We show (cf. [**4**]):

PROPOSITION 2.1. *For any partition rule W verifying (a-d) above, $\mathcal{E}^W$ is a Γ−approximation of the entropy.*

It is not too difficult to find partition rules. Recall the definition of Voronoi tessellation corresponding to $X = (x_1, \ldots) \in \Omega^{\otimes}$:

$$V_i(X) := \left\{ y \in \mathbb{R}^2 \ ; \ \ |y - x_i| \le |y - x_j| \ \ \forall 1 \le j \le N(X) \ \right\} . \tag{2.2}$$

Indeed, one can easily show that $X \to \overrightarrow{V}$ is a partition rule.

We now define another partition rule: Let $cell(D_i(X))$ as defined in Corollary 1.2. We know that $cell(D_i(X))$ are pairwise disjoint, while $\cup_{i=1}^N cell(D_i(X))\cap\Omega \subset \Omega$ with a (possibly) strict inclusion. Thus, the partition X into $\{cell(D_i(X))\}$ is *not* a partition rule, since it may violate condition (b) above.

Let $Cell(X) := \cup_{i=1}^N cell(D_i(X))$. Define

$$W_i(X) := cell(D_i(X)) \cup (V_i(X) - Cell(X)) . \tag{2.3}$$

It is now easy to see that $W(X)$ so defined is a partition rule, verifying (a-d) above. In particular, $\mathcal{E}^W$ as defined in (2.1) under the partition rule (2.3) is a Γ−approximation of the entropy, via Proposition 2.1.

By Corollary 1.2 we find out that the area of $cell(D_i(X))$ is not smaller than $2\sqrt{3}r_i^2(X)$, i.e $|W_i(X)| \ge 2\sqrt{3}r_i^2(X)$. In addition we can show that for $X \in \Omega^{\otimes}$ for which δ_X approximate (in $C^*(\Omega)$) a measure $\mu \in \mathcal{B}(\Omega)$ satisfying $E(\mu) < \infty$, it follows that $W_i(X) \not\subset \Omega$ for only $o(N(X))$ of the points. So, we replace $\ln(|W_i(X)\cap\Omega|)$ in (2.1) by $\ln\left(2\sqrt{3}r_i^2(X)\right)$, taking advantage of the monotonicity of ln, and obtain that $\mathcal{E}$ defined in Theorem 1.4 is not smaller, asymptotically, than $\mathcal{E}^W$ with W given by (2.3). Thus $\mathcal{E}$ satisfies condition (i) of definition 1.3.

To verify condition (ii) we recall that an optimal ratio of $|W_i(X)|/(2\sqrt{3}r_i^2(X)) \gtrsim 1$ is obtained for hexagonal grids. Given $\mu \in \mathcal{B}(\Omega)$, we can approximate it (in the weak topology) by a sequentially constant density. Then we construct a hexagonal grid on every domain in Ω on which this density is a constant. For the details of the proof see [**4**].

References

[1] H. C. Chang, L. C. Wang, A Simple Proof of Thue's Theorem on Circle Packing, arXiv:1009.4322.

[2] Gerd Blind, *Über Unterdeckungen der Ebene durch Kreise* (German), J. Reine Angew. Math. **236** (1969), 145–173. MR0275291 (43 #1048)

[3] Luigi Ambrosio, Nicola Gigli, and Giuseppe Savaré, *Gradient flows in metric spaces and in the space of probability measures*, Lectures in Mathematics ETH Zürich, Birkhäuser Verlag, Basel, 2005. MR2129498 (2006k:49001)

[4] J. Carrillo, G. Wolansky, On Γ limit of generalized entropy on measure spaces and its applications *(In preparation)*

[5] H. Davenport, *Problems of packing and covering*, Univ. e Politec. Torino Rend. Sem. Mat. **24** (1964/1965), 41–48. MR0187150 (32 #4604)

[6] Wu-Yi Hsiang, *A simple proof of a theorem of Thue on the maximal density of circle packings in E^2*, Enseign. Math. (2) **38** (1992), no. 1-2, 125–131. MR1175523 (93k:52018)

[7] B. Segre and K. Mahler, *On the densest packing of circles*, Amer. Math. Monthly **51** (1944), 261–270. MR0010427 (6,16c)

[8] Pinchasi, R, and Wolansky, G. A generalization of Thue's theorem to packings of nonequal discs, and an application to a discrete approximation of entropy, http://arxiv.org/abs/1410.3240

[9] Sylvia Serfaty, *Gamma-convergence of gradient flows on Hilbert and metric spaces and applications*, Discrete Contin. Dyn. Syst. **31** (2011), no. 4, 1427–1451, DOI 10.3934/dcds.2011.31.1427. MR2836361 (2012h:49023)

[10] L. Fejes Tóth, *Minkowskian distribution of discs*, Proc. Amer. Math. Soc. **16** (1965), 999–1004. MR0180921 (31 #5151)

[11] L. Fejes Tóth, *Minkowskian circle-aggregates*, Math. Ann. **171** (1967), 97–103. MR0221386 (36 #4438)

Department of Mathematics, Technion, Haifa, Israel 32000

Contemporary Mathematics
Volume **659**, 2016
http://dx.doi.org/10.1090/conm/659/13151

Bolza variational problems with extended-valued integrands on large intervals

Alexander J. Zaslavski

ABSTRACT. We study the structure of approximate solutions of autonomous variational problems with a lower semicontinuous extended-valued integrand. In the first part of the paper we discuss our recent results obtained for Lagrange problems. In the second part we prove the extension of these results for Bolza problems. We show that the turnpike phenomenon holds for approximate solutions of variational problems on large intervals and study the structure of approximate solutions in regions close to the endpoints of the time intervals.

1. Introduction

In this paper we study the following variational problems:

$$(P_1) \qquad \int_0^T f(v(t), v'(t))dt \to \min,$$

$v : [0, T] \to R^n$ is an absolutely continuous (a. c.) function such that

$$v(0) = x, \ v(T) = y;$$

$$(P_2) \qquad \int_0^T f(v(t), v'(t))dt \to \min,$$

$v : [0, T] \to R^n$ is an a. c. function such that $v(0) = x$;

$$(P_3) \qquad \int_0^T f(v(t), v'(t))dt \to \min$$

$v : [0, T] \to R^n$ is an a. c. function;

$$(P_4) \qquad \int_0^T f(v(t), v'(t))dt + g(v(T)) \to \min,$$

$v : [0, T] \to R^n$ is an a. c. function such that $v(0) = x$;

$$(P_5) \qquad \int_0^T f(v(t), v'(t))dt + h(v(0), v(T)) \to \min,$$

$v : [0, T] \to R^n$ is an a. c. function,

where $x, y \in R^n$ and $T > 0$. Here R^n is the n-dimensional Euclidean space with the Euclidean norm $|\cdot|$, $f : R^n \times R^n \to R^1 \cup \{\infty\}$ is an extended-valued and

---

2010 *Mathematics Subject Classification.* Primary 49J99.

lower semicontinuous integrand and $g : R^n \to R^1$ and $h : R^n \times R^n \to R^1$ are lower semicontinuous functions which are bounded on bounded sets. We are interested in the structure of approximate solutions of these problems considered on large intervals.

It should be mentioned that the study of (approximate) solutions of variational and optimal control problems defined on infinite intervals and on sufficiently large intervals has recently been a rapidly growing area of research [**2–11, 16, 18, 21, 22, 24, 25, 27, 33, 34**]. This is due not only to theoretical achievements in this area, but also because of numerous applications to engineering [**1, 14, 27**], models of economic dynamics [**13, 17, 20, 23, 26, 27, 34**], the game theory [**12, 27, 32, 34**] and the theory of thermodynamical equilibrium for materials [**15, 19**].

In this paper we study the structure of solutions of autonomous variational problems with a lower semicontinuous integrand considered in [**28–31, 34, 35**]. In the first part of the paper we discuss our recent results obtained for Lagrange problems $(P_1) - (P_3)$. In the second part we prove the extension of these results for Bolza problems (P_4) and (P_5). Our results provide a full description of the structure of approximate solutions of variational problems on large intervals.

In [**28, 30, 31, 34**] we studied the turnpike phenomenon for approximate solutions of problems (P_1) and (P_2) which is independent of the length of the interval T, for all sufficiently large intervals. To have this phenomenon means that the approximate solutions are determined mainly by the integrand, and are essentially independent of the choice of time interval and data, except in regions close to the endpoints of the time interval. The term was first coined by Samuelson in 1948 (see [**26**]).

The problems (P_1) and (P_2) were studied in [**28, 31**] where it was established, under certain assumptions, that the turnpike phenomenon holds and that the turnpike $\bar{x}$ is a unique minimizer of the optimization problem $f(x, 0) \to \min$, $x \in R^n$.

The structure of approximate solutions of the problems (P_2) and (P_3), in regions close to the endpoints of the time intervals, was analyzed in [**35**]. It was shown there that in regions close to the right endpoint T of the time interval these approximate solutions are determined only by the integrand, and are essentially independent of the choice of interval and endpoint value x. For the problems (P_3), approximate solutions are determined only by the integrand function also in regions close to the left endpoint 0 of the time interval. The main goal of the present paper is to extend these results to approximate solutions of problems (P_4) and (P_5).

In this section we discuss the turnpike results for problems $(P_1) - (P_3)$.

We denote by $\mathrm{mes}(E)$ the Lebesgue measure of a Lebesgue measurable set $E \subset R^1$, denote by $|\cdot|$ the Euclidean norm of the n-dimensional Euclidean space R^n and by $\langle \cdot, \cdot \rangle$ the inner product of R^n. For each function $f : X \to R^1 \cup \{\infty\}$, where X is a nonempty, set

$$\mathrm{dom}(f) = \{x \in X : f(x) < \infty\}.$$

Let a be a real positive number, $\psi : [0, \infty) \to [0, \infty)$ be an increasing function such that

$$\lim_{t \to \infty} \psi(t) = \infty \tag{1.1}$$

and let $f : R^n \times R^n \to R^1 \cup \{\infty\}$ be a lower semicontinuous function such that the set

$$\mathrm{dom}(f) = \{(x,y) \in R^n \times R^n :\ f(x,y) < \infty\} \tag{1.2}$$

is nonempty, convex and closed and that

$$f(x,y) \geq \max\{\psi(|x|),\ \psi(|y|)|y|\} - a \text{ for each } x, y \in R^n. \tag{1.3}$$

A function v defined on an infinite subinterval of R^1 with values in R^n is called absolutely continuous (a. c.) if v is absolutely continuous on any finite subinterval of its domain.

For each $x, y \in R^n$ and each $T > 0$ define

$$\sigma(f,T,x) = \inf\{\int_0^T f(v(t),v'(t))dt :\ v : [0,T] \to R^n$$

$$\text{is an a. c. function satisfying } v(0) = x\}, \tag{1.4}$$

$$\sigma(f,T,x,y) = \inf\{\int_0^T f(v(t),v'(t))dt :\ v : [0,T] \to R^n$$

$$\text{is an a. c. function satisfying } v(0) = x,\ v(T) = y\}, \tag{1.5}$$

$$\sigma(f,T) = \inf\{\int_0^T f(v(t),v'(t))dt :\ v : [0,T] \to R^n \text{ is an a. c. function}\}, \tag{1.6}$$

$$\widehat{\sigma}(f,T,y) = \inf\{\int_0^T f(v(t),v'(t))dt :\ v : [0,T] \to R^n$$

$$\text{is an a. c. function satisfying } v(T) = y\}. \tag{1.7}$$

(Here we assume that infimum over an empty set is infinity.)

We suppose that there exists a point $\bar{x} \in R^n$ such that

$$f(\bar{x},0) \leq f(x,0) \text{ for each } x \in R^n \tag{1.8}$$

and that the following assumptions hold:

(A1) $(\bar{x},0)$ is an interior point of the set $\mathrm{dom}(f)$ and the function f is continuous at the point $(\bar{x},0)$;

(A2) for each $M > 0$ there exists $c_M > 0$ such that

$$\sigma(f,T,x) \geq Tf(\bar{x},0) - c_M$$

for each $x \in R^n$ satisfying $|x| \leq M$ and each real number $T > 0$;

(A3) for each $x \in R^n$ the function $f(x,\cdot) : R^n \to R^1 \cup \{\infty\}$ is convex.

Assumption (A2) implies that for each a.c. function $v : [0,\infty) \to R^n$ the function

$$T \to \int_0^T f(v(t),v'(t))dt - Tf(\bar{x},0),\ T \in (0,\infty)$$

is bounded from below.

It should be mentioned that inequality (1.8) and assumptions (A1)-(A3) are common in the literature and hold for many infinite horizon optimal control problems. In particular, we need inequality (1.8) and assumption (A2) in the cases when the problems (P_1) and (P_2) possess the turnpike property and the point $\bar{x}$ is its turnpike. Assumption (A2) means that the constant function $\bar{v}(t) = \bar{x}$, $t \in [0,\infty)$ is an approximate solution of the infinite horizon variational problem with the integrand f related to the problems (P_1) and (P_2).

We say that an a. c. function $v : [0, \infty) \to R^n$ is (f)-good [**27, 33, 34**] if

$$\sup\{|\int_0^T f(v(t), v'(t))dt - Tf(\bar{x}, 0)| : T \in (0, \infty)\} < \infty.$$

The following result obtained in [**28**] (see also Proposition 3.1 of [**34**]) is very useful in our study.

PROPOSITION 1.1. Let $v : [0, \infty) \to R^n$ be an a. c. function. Then either the function v is (f)-good or

$$\int_0^T f(v(t), v'(t))dt - Tf(\bar{x}, 0) \to \infty \text{ as } T \to \infty.$$

Moreover, if the function v is (f)-good, then $\sup\{|v(t)| : t \in [0, \infty)\} < \infty$.

For each pair of numbers $T_1 \in R^1$, $T_2 > T_1$ and each a. c. function $v : [T_1, T_2] \to R^n$ put

$$I^f(T_1, T_2, v) = \int_{T_1}^{T_2} f(v(t), v'(t))dt \tag{1.9}$$

and for any $T \in [T_1, T_2]$ set $I^f(T, T, v) = 0$.

Definition 1. For every positive real number M denote by $X_{M,f}$ the set of all points $x \in R^n$ such that $|x| \leq M$ and there exists an a. c. function $v : [0, \infty) \to R^n$ which satisfies

$$v(0) = x, \ I^f(0, T, v) - Tf(\bar{x}, 0) \leq M \text{ for each } T \in (0, \infty). \tag{1.10}$$

Evidently, $\cup\{X_{M,f} : M \in (0, \infty)\}$ is the collection of all $x \in X$ for which there exists an (f)-good function $v : [0, \infty) \to R^n$ satisfying $v(0) = x$.

We suppose that the following assumption holds:

(A4) (the asymptotic turnpike property) for each (f)-good function $v : [0, \infty) \to R^n$, $\lim_{t\to\infty} |v(t) - \bar{x}| = 0$.

Examples of integrands f which satisfy assumptions (A1)-(A4) are considered in [**28, 34**].

The following result obtained in [**28**] (see also Theorem 3.2 of [**34**]) shows that the turnpike phenomenon holds for approximate solutions of problem (P_2).

THEOREM 1.2. *Let ϵ, M be positive numbers. Then there exist an integer $L \geq 1$ and a real number $\delta > 0$ such that for each real number $T > 2L$ and each a. c. function $v : [0, T] \to R^n$ which satisfies*

$$v(0) \in X_{M,f} \text{ and } I^f(0, T, v) \leq \sigma(f, T, v(0)) + \delta$$

there exist a pair of numbers $\tau_1 \in [0, L]$ and $\tau_2 \in [T - L, T]$ such that

$$|v(t) - \bar{x}| \leq \epsilon \text{ for all } t \in [\tau_1, \tau_2]$$

and if $|v(0) - \bar{x}| \leq \delta$, then $\tau_1 = 0$.

Definition 2. Let M be a positive real number. Denote by $Y_{M,f}$ the collection of all $x \in R^n$ such that there are a real number $T \in (0, M]$ and an a. c. function $v : [0, T] \to R^n$ which satisfy $v(0) = \bar{x}$, $v(T) = x$ and $I^f(0, T, v) \leq M$.

The following result obtained in [**31**] (see also Theorem 3.5 of [**34**]) shows that the turnpike phenomenon holds for approximate solutions of problem (P_1).

THEOREM 1.3. *Let $\epsilon, M_0, M_1 > 0$. Then there exist numbers $L,\ \delta > 0$ such that for each number $T > 2L$, each point $z_0 \in X_{M_0,f}$ and each point $z_1 \in Y_{M_1,f}$, the value $\sigma(f,T,z_0,z_1)$ is finite and for each a. c. function $v : [0,T] \to R^n$ which satisfies*

$$v(0) = z_0,\ v(T) = z_1,\ I^f(0,T,v) \le \sigma(f,T,z_0,z_1) + \delta$$

there exists a pair of numbers $\tau_1 \in [0,L], \tau_2 \in [T-L,T]$ such that

$$|v(t) - \bar{x}| \le \epsilon,\ t \in [\tau_1, \tau_2].$$

Moreover if $|v(0) - \bar{x}| \le \delta$, then $\tau_1 = 0$ and if $|v(T) - \bar{x}| \le \delta$, then $\tau_2 = T$.

In the sequel we use a notion of an overtaking optimal function which plays an important role in the turnpike theory and the infinite horizon optimal control [**27**, **34**].

An a. c. function $v : [0,\infty) \to R^n$ is called (f)-overtaking optimal if for each a. c. function $u : [0,\infty) \to R^n$ satisfying $u(0) = v(0)$ the inequality

$$\limsup_{T\to\infty} [I^f(0,T,v) - I^f(0,T,u)] \le 0$$

holds.

The following existence result was obtained in [**28**] (see also Theorem 3.3 of [**34**].)

THEOREM 1.4. *Assume that $x \in R^n$ and that there exists an (f)-good function $v : [0,\infty) \to R^n$ satisfying $v(0) = x$. Then there exists an (f)-overtaking optimal function $u_* : [0,\infty) \to R^n$ such that $u_*(0) = x$.*

In view of assumption (A1) there exists a real number $\bar{r} \in (0,1)$ such that:

$$\Omega_0 := \{(x,y) \in R^n \times R^n :\ |x - \bar{x}| \le \bar{r} \text{ and } |y| \le \bar{r}\} \subset \text{dom}(f), \tag{1.11}$$

$$\Delta_0 := \sup\{|f(z_1,z_2)| :\ (z_1,z_2) \in \Omega_0\} < \infty. \tag{1.12}$$

Evidently, the value $\sigma(f,T,x,y)$ is finite for every real number $T \ge 1$ and every pair of points $x, y \in R^n$ satisfying $|x - \bar{x}|,\ |y - \bar{x}| \le \bar{r}/2$.

Definition 3. Let $M > 0$. Denote by $\bar{Y}_{M,f}$ the set of all points $x \in R^n$ such that $|x| \le M$ and for which there exist a number $T \in (0,M]$ and an a. c. function $v : [0,T] \to R^n$ such that $v(0) = x,\ v(T) = \bar{x}$ and $I^f(0,T,v) \le M$.

It is easy to see that the following result holds.

PROPOSITION 1.5. For each $M > 0$ there exists $M_0 > 0$ such that $\bar{Y}_{M,f} \subset X_{M_0,f}$.

In the sequel we use the following auxiliary result.

PROPOSITION 1.6. (Proposition 1.6 of [**35**]) For each $M > 0$ there exists $M_0 > 0$ such that $X_{M,f} \subset \bar{Y}_{M_0,f}$.

An a. c. function $v : [0,\infty) \to R^n$ is called (f)-minimal [**27**, **34**] if for each $T_1 \ge 0$, each $T_2 > T_1$ and each a. c. function $u : [T_1,T_2] \to R^n$ satisfying $u(T_i) = v(T_i)$, $i = 1,2$, we have

$$\int_{T_1}^{T_2} f(v(t),v'(t))dt \le \int_{T_1}^{T_2} f(u(t),u'(t))dt.$$

The next theorem obtained in [**30**] (see also Theorem 3.32 of [**34**]) shows that the optimality notions introduced above are equivalent.

THEOREM 1.7. *Assume that $x \in R^n$ and that there exists an (f)-good function $\tilde{v} : [0, \infty) \to R^n$ satisfying $\tilde{v}(0) = x$. Let $v : [0, \infty) \to R^n$ be an a. c. function such that $v(0) = x$. Then the following conditions are equivalent:*

(i) the function v is (f)-overtaking optimal; (ii) the function v is (f)-good and (f)-minimal; (iii) the function v is (f)-minimal and $\lim_{t\to\infty} v(t) = \bar{x}$; (iv) the function v is (f)-minimal and $\liminf_{t\to\infty} |v(t) - \bar{x}| = 0$.

The following two theorems obtained in [**30**] (see also Theorems 3.33 and 3.34 of [**34**]) describe the asymptotic behavior of overtaking optimal functions.

THEOREM 1.8. *Let ϵ be a positive number. Then there exists a positive number δ such that:*

(i) For each point $x \in R^n$ satisfying $|x - \bar{x}| \leq \delta$ there exists an (f)-overtaking optimal and (f)-good function $v : [0, \infty) \to R^n$ such that $v(0) = x$.

(ii) For each (f)-overtaking optimal function $v : [0, \infty) \to R^n$ satisfying $|v(0) - \bar{x}| \leq \delta$, the inequality $|v(t) - \bar{x}| \leq \epsilon$ holds for all numbers $t \in [0, \infty)$.

THEOREM 1.9. *Let $\epsilon, M > 0$. Then there exists $L > 0$ such that for each $x \in X_{M,f}$ and each (f)-overtaking optimal function $v : [0, \infty) \to R^n$ satisfying $v(0) = x$ the following inequality holds:*

$$|v(t) - \bar{x}| \leq \epsilon \text{ for all } t \in [L, \infty).$$

In [**35**] we prove the following turnpike result for approximate solutions of the problems of the type (P_3).

THEOREM 1.10. *Let $\epsilon > 0$. Then there exist numbers L, $\delta > 0$ such that for each number $T > 2L$ and each a. c. function $v : [0, T] \to R^n$ which satisfies*

$$I^f(0, T, v) \leq \sigma(f, T) + \delta$$

there exists a pair of numbers $\tau_1 \in [0, L], \tau_2 \in [T - L, T]$ such that

$$|v(t) - \bar{x}| \leq \epsilon, \ t \in [\tau_1, \tau_2].$$

Moreover if $|v(0) - \bar{x}| \leq \delta$, then $\tau_1 = 0$ and if $|v(T) - \bar{x}| \leq \delta$, then $\tau_2 = T$.

Our paper is organized as follows. Section 2 contains preliminaries. In Section 3 we discuss the results of [**35**] which describe the structure of approximate solutions of Lagrange problems on large intervals in the regions close to the endpoints. Several auxiliary results which are used in [**35**] and in the present paper are presented in Section 4. Our turnpike result for problem (P_4) (Theorem 5.2) is proved in Section 5 which also contains the statement of Theorem 5.3 describing the structure of approximate solutions of problems of the type (P_4) on large intervals in the regions close to the endpoints. An auxiliary result for Theorem 5.3 is proved in Section 6 while the proof of Theorem 5.3 is given in Section 7. Our turnpike result for problem (P_5) (Theorem 8.1) is proved in Section 8 which also contains the statement of Theorem 8.2 describing the structure of approximate solutions of problems of the type (P_5) on large intervals in the regions close to the endpoints. An auxiliary result for Theorem 8.2 is proved in Section 9. Theorem 8.2 is proved in Section 10.

2. Preliminaries

We use the notation, definitions and assumptions introduced in Section 1. We define a function $\pi^f(x)$, $x \in R^n$ which plays an important role in our study. For all $x \in R^n \setminus \cup\{X_{M,f} : M \in (0,\infty)\}$ (see Definition 1) set

$$\pi^f(x) = \infty.$$

Let

$$x \in \cup\{X_{M,f} : M \in (0,\infty)\}. \tag{2.1}$$

Denote by $\Lambda(f,x)$ the set of all (f)-overtaking optimal functions $v : [0,\infty) \to R^n$ satisfying $v(0) = x$. By (2.1) and Theorem 1.4, the set $\Lambda(f,x)$ is nonempty. In view of (2.1), any element of $\Lambda(f,x)$ is an (f)-good function. Define

$$\pi^f(x) = \liminf_{T\to\infty}[I^f(0,T,v) - Tf(\bar{x},0)], \tag{2.2}$$

where $v \in \Lambda(f,x)$. Clearly, $\pi^f(x)$ does not depend on the choice of v. In view of (A2) and (2.1), $\pi^f(x)$ is finite. Definition (2.2) and the definition of overtaking optimal functions imply the following result.

PROPOSITION 2.1. 1. Let $v : [0,\infty) \to R^n$ be an (f)-good function. Then

$$\pi^f(v(0)) \le \liminf_{T\to\infty}[I^f(0,T,v) - Tf(\bar{x},0)]$$

and for each $T \ge 0$ and each $S > T$,

$$\pi^f(v(T)) \le I^f(T,S,v) - (S-T)f(\bar{x},0) + \pi^f(v(S)). \tag{2.3}$$

2. Let $S > T \ge 0$ and $v : [0,S] \to R^n$ be an a. c. function such that $\pi^f(v(T)), \pi^f(v(S)) < \infty$. Then (2.3) holds.

The next result follows from definition (2.2).

PROPOSITION 2.2. Let $v : [0,\infty) \to R^n$ be an (f)-overtaking optimal and (f)-good function. Then for each $T \ge 0$ and each $S > T$,

$$\pi^f(v(T)) = I^f(T,S,v) - (S-T)f(\bar{x},0) + \pi^f(v(S)).$$

The following three results were proved in [**35**].

PROPOSITION 2.3. $\pi^f(\bar{x}) = 0$.

PROPOSITION 2.4. The function π^f is finite in a neighborhood of $\bar{x}$ and continuous at $\bar{x}$.

PROPOSITION 2.5. For each $M > 0$ the set $\{x \in R^n : \pi^f(x) \le M\}$ is bounded.

(Here we assume that an empty set is bounded.)

Set

$$\inf(\pi^f) = \inf\{\pi^f(z) : z \in R^n\}. \tag{2.4}$$

By (A2) and Proposition 2.5, $\inf(\pi^f)$ is finite. Set

$$X_f = \{x \in R^n : \pi^f(x) \le \inf(\pi^f) + 1\}. \tag{2.5}$$

The following two results were also proved in [**35**].

PROPOSITION 2.6. Assume that $x \in \cup\{X_{M,f} : M \in (0,\infty)\}$ and $v \in \Lambda(f,x)$. Then

$$\pi^f(x) = \lim_{T\to\infty}[I^f(0,T,v) - Tf(\bar{x},0)].$$

PROPOSITION 2.7. There exists $M > 0$ such that $X_f \subset X_{M,f}$.

Propositions 1.6 and 2.7 imply the following result.

PROPOSITION 2.8. There exists $L > 0$ such that $X_f \subset \bar{Y}_{L,f}$.

In [**35**] we proved the following proposition.

PROPOSITION 2.9. The function $\pi^f : R^n \to R^1 \cup \{\infty\}$ is lower semicontinuous.

Set

$$\mathcal{D}(f) = \{x \in R^n : \pi^f(x) = \inf(\pi^f)\}. \tag{2.6}$$

By Propositions 2.5 and 2.9, the set $\mathcal{D}(f)$ is nonempty, bounded and closed subset of R^n. The following proposition is proved in [**35**].

PROPOSITION 2.10. Let $v : [0, \infty) \to R^n$ be an (f)-good function such that for all $T > 0$,

$$I^f(0, T, v) - Tf(\bar{x}, 0) = \pi^f(v(0)) - \pi^f(v(T)). \tag{2.7}$$

Then v is an (f)-overtaking optimal function.

The next result easily follows from (2.5), (2.6), Proposition 2.7 and Theorem 1.9.

PROPOSITION 2.11. For each $\epsilon > 0$ there exists $T_\epsilon > 0$ such that for each $z \in \mathcal{D}(f)$ and each $v \in \Lambda(f, z)$ the inequality $|v(t) - \bar{x}| \le \epsilon$ holds for all $t \ge T_\epsilon$.

In order to study the structure of solutions of the problems (P_2) and (P_3) we introduce the following notation and definitions.

Define a function $\bar{f} : R^n \times R^n \to R^1 \cup \{\infty\}$ by

$$\bar{f}(x, y) = f(x, -y) \text{ for all } x, y \in R^n. \tag{2.8}$$

Clearly,

$$\mathrm{dom}(\bar{f}) = \{(x, y) \in R^n \times R^n : (x, -y) \in \mathrm{dom}(f)\}, \tag{2.9}$$

$\mathrm{dom}(\bar{f})$ is a nonempty closed convex set, $\bar{f}$ is a lower semicontinuous function satisfying

$$\bar{f}(x, y) \ge \max\{\psi(|x|),\ \psi(|y|)|y|\} - a \text{ for each } x, y \in R^n. \tag{2.10}$$

The notation introduced for the function f is also used for the function $\bar{f}$. Namely, for each pair of numbers $T_1 \in R^1$, $T_2 > T_1$ and each a. c. function $v : [T_1, T_2] \to R^n$ put

$$I^{\bar{f}}(T_1, T_2, v) = \int_{T_1}^{T_2} \bar{f}(v(t), v'(t))dt$$

and for each $x, y \in R^n$ and each $T > 0$ define

$$\sigma(\bar{f}, T, x) = \inf\{I^{\bar{f}}(0, T, v) : v : [0, T] \to R^n \text{ is an a. c. function satisfying } v(0) = x\},$$

$$\sigma(\bar{f}, T, x, y) = \inf\{I^{\bar{f}}(0, T, v) : v : [0, T] \to R^n \text{ is an a. c. function satisfying } v(0) = x,\ v(T) = y\},$$

$$\sigma(\bar{f}, T) = \inf\{I^{\bar{f}}(0, T, v) : v : [0, T] \to R^n \text{ is an a. c. function}\},$$

$$\widehat{\sigma}(\bar{f}, T, y) = \inf\{I^{\bar{f}}(0, T, v) : v : [0, T] \to R^n \text{ is an a. c. function satisfying } v(T) = y\}.$$

Let $v : [0, T] \to R^n$ be an a. c. function. Set

$$\bar{v}(t) = v(T - t),\ t \in [0, T].$$

It is not difficult to see that

$$\int_0^T \bar{f}(\bar{v}(t),\bar{v}'(t))dt = \int_0^T f(\bar{v}(t),-\bar{v}'(t))dt$$

$$= \int_0^T f(v(T-t),v'(T-t))dt = \int_0^T f(v(t),v'(t))dt. \tag{2.11}$$

Evidently, for all $x \in R^n$,

$$\bar{f}(\bar{x},0) = f(\bar{x},0) \le f(x,0) = \bar{f}(x,0), \tag{2.12}$$

$(\bar{x},0)$ is an interior point of the set $\mathrm{dom}(\bar{f})$ and the function $\bar{f}$ is continuous at the point $(\bar{x},0)$. Therefore (A1) holds for the function $\bar{f}$. It is clear that for each $x \in R^n$ the function $\bar{f}(x,0) : R^n \to R^1 \cup \{\infty\}$ is convex. Thus (A3) holds for $\bar{f}$. The next result easily follows from (2.11).

PROPOSITION 2.12. Let $T > 0$, $M \ge 0$ and $v_i : [0,T] \to R^n$, $i = 1,2$ be a. c. functions. Then

$$I^f(0,T,v_1) \ge I^f(0,T,v_2) - M$$

if and only if

$$I^{\bar{f}}(0,T,\bar{v}_1) \ge I^{\bar{f}}(0,T,\bar{v}_2) - M.$$

Proposition 2.12 implies the following result.

PROPOSITION 2.13. Let $T > 0$ and $v : [0,T] \to R^n$ be an a. c. function. Then the following assertions hold:

$$I^f(0,T,v) \le \sigma(f,T) + M \text{ if and only if}$$

$$I^{\bar{f}}(0,T,\bar{v}) \le \sigma(\bar{f},T) + M;$$

$$I^f(0,T,v) \le \sigma(f,T,v(0),v(T)) + M \text{ if and only if}$$

$$I^{\bar{f}}(0,T,\bar{v}) \le \sigma(\bar{f},T,\bar{v}(0),\bar{v}(T)) + M;$$

$$I^f(0,T,v) \le \widehat{\sigma}(f,T,v(T)) + M \text{ if and only if } I^{\bar{f}}(0,T,\bar{v}) \le \sigma(\bar{f},T,\bar{v}(0)) + M;$$

$$I^f(0,T,v) \le \sigma(f,T,v(0)) + M \text{ if and only if } I^{\bar{f}}(0,T,\bar{v}) \le \widehat{\sigma}(\bar{f},T,\bar{v}(T)) + M.$$

The next result is proved in [**35**].

PROPOSITION 2.14. 1. For each $M > 0$ there exists $c_M > 0$ such that $\sigma(\bar{f},T,x) \ge T\bar{f}(\bar{x},0) - c_M$ for each $x \in R^n$ satisfying $|x| \le M$ and each $T > 0$.
2. For each $(\bar{f})$-good function $v : [0,\infty) \to R^n$, $\lim_{t\to\infty} v(t) = \bar{x}$.

In view of Proposition 2.14 the function $\bar{f}$ satisfies assumptions (A2) and (A4). We have already mentioned that assumptions (A1) and (A3) hold for the function $\bar{f}$. Therefore all the results stated above for the function f are also true for the function $\bar{f}$.

3. Structure of solutions of Lagrange problems in the regions close to the endpoints

We use the notation, definitions and assumptions introduced in Sections 1 and 2. Recall that $f : R^n \times R^n \to R^1 \cup \{\infty\}$ is a lower semicontinuous function with a nonempty closed convex $\mathrm{dom}(f)$ satisfying (1.3), $\bar{x} \in R^n$ satisfies (1.8) and that (A1)-(A4) hold.

The following two theorems which describe the structure of approximate solutions of the problems (P_2) and (P_3) respectively, were obtained in [**35**].

THEOREM 3.1. *Let $L_0 > 0$, $\tau_0 > 0$ and $\epsilon > 0$. Then there exist $\delta > 0$ and $T_0 > \tau_0$ such that for each $T \geq T_0$ and each a. c. function $v : [0, T] \to R^n$ which satisfies*

$$v(0) \in \bar{Y}_{L_0, f},\ I^f(0, T, v) \leq \sigma(f, T, v(0)) + \delta$$

there exists an $(\bar{f})$-overtaking optimal function $v^* : [0, \infty) \to R^n$ such that $v^*(0) \in \mathcal{D}(\bar{f})$ and $|v(T - t) - v^*(t)| \leq \epsilon$ for all $t \in [0, \tau_0]$.

THEOREM 3.2. *Let $\tau_0 > 0$ and $\epsilon > 0$. Then there exist $\delta > 0$ and $T_0 > \tau_0$ such that for each $T \geq T_0$ and each a. c. function $v : [0, T] \to R^n$ which satisfies*

$$I^f(0, T, v) \leq \sigma(f, T) + \delta$$

there exist an (f)-overtaking optimal function $u^* : [0, \infty) \to R^n$ and an $(\bar{f})$-overtaking optimal function $v^* : [0, \infty) \to R^n$ such that $u^*(0) \in \mathcal{D}(f)$, $v^*(0) \in \mathcal{D}(\bar{f})$ and that for all $t \in [0, \tau_0]$,

$$|v(t) - u^*(t)| \leq \epsilon \text{ and } |v(T - t) - v^*(t)| \leq \epsilon.$$

4. Auxiliary results

This section contains several auxiliary results used in [**35**] and in the present paper.

LEMMA 4.1. *(Lemma 5.1 of [**35**]) Let $M, \epsilon > 0$. Then there exists $L_0 > 0$ such that for each number $T \geq L_0$, each a. c. function $v : [0, T] \to R^n$ satisfying*

$$I^f(0, T, v) \leq T f(\bar{x}, 0) + M$$

and each number $s \in [0, T - L_0]$ the inequality

$$\min\{|v(t) - \bar{x}| : t \in [s, s + L_0]\} \leq \epsilon$$

holds.

LEMMA 4.2. *([**28**], Lemma 3.11 of [**34**]) Let $\epsilon > 0$. Then there exists a number $\delta \in (0, \bar{r}/2)$ such that for each number $T \geq 2$ and each a. c. function $v : [0, T] \to R^n$ which satisfies*

$$|v(0) - \bar{x}|,\ |v(T) - \bar{x}| \leq \delta,$$

$$I^f(0, T, v) \leq \sigma(f, T, v(0), v(T)) + \delta$$

the inequality $|v(t) - \bar{x}| \leq \epsilon$ is true for all numbers $t \in [0, T]$.

LEMMA 4.3. *([**28**], Proposition 3.6 of [**34**]) Let $M_0, M_1 > 0$. Then there exists $M_2 > 0$ such that for each $T > 0$ and each a. c. function $v : [0, T] \to R^n$ which satisfies*

$$|v(0)| \leq M_0,\ I^f(0, T, v) \leq T f(\bar{x}, 0) + M_1$$

the following inequality holds:

$$|v(t)| \leq M_2 \text{ for all } t \in [0,T].$$

PROPOSITION 4.4. ([**28**], Proposition 3.7 of [**34**]) Let $T > 0$ and let $v_k : [0,T] \to R^n$, $k = 1, 2, \ldots$ be a sequence of a. c. functions such that the sequence $\{I^f(0,T,v_k)\}_{k=1}^{\infty}$ is bounded and that the sequence $\{v_k(0)\}_{k=1}^{\infty}$ is bounded. Then there exist a strictly increasing sequence of natural numbers $\{k_i\}_{i=1}^{\infty}$ and an a. c. function $v : [0,T] \to R^n$ such that

$$v_{k_i}(t) \to v(t) \text{ as } i \to \infty \text{ uniformly on } [0,T],$$

$$I^f(0,T,v) \leq \liminf_{i\to\infty} I^f(0,T,v_{k_i}).$$

LEMMA 4.5. *([**28**], Proposition 3.8 of [**34**]) Let $\epsilon > 0$. Then there exists $\delta > 0$ such that for each a.c. function $v : [0,1] \to R^n$ satisfying $|v(0) - \bar{x}|$, $|v(1) - \bar{x}| \leq \delta$,*

$$I^f(0,1,v) \geq f(\bar{x},0) - \epsilon.$$

The next result is proved in [**35**].

LEMMA 4.6. *Let $0 < L_0 < L_1$ and $M_0 > 0$. Then there exists $M_1 > 0$ such that for each $T \in [L_0, L_1]$ and each a. c. function $v : [0,T] \to R^n$ satisfying $I^f(0,T,v) \leq M_0$ the inequality $|v(t)| \leq M_1$ holds for all $t \in [0,T]$.*

5. The Bolza problem (P4)

We use the notation, definitions and assumptions introduced in Sections 1 and 2.

For each nonempty set Y and each function $h : Y \to R^1 \cup \{\infty\}$ put

$$\inf(h) = \inf\{h(y) : y \in Y\}.$$

Let $a_0 > 0$. Denote by $\mathfrak{A}(R^n)$ the set of all lower semicontinuous functions $h : R^n \to R^1$ which are bounded on bounded subsets of R^n and such that

$$h(z) \geq -a_0 \text{ for all } z \in R^n. \tag{5.1}$$

For simplicity, we set $\mathfrak{A} = \mathfrak{A}(R^n)$. The set $\mathfrak{A}$ is equipped with the uniformity determined by the base

$$E(N,\epsilon) = \{(h_1,h_2) \in \mathfrak{A} \times \mathfrak{A} :$$

$$|h_1(z) - h_2(z)| \leq \epsilon \text{ for all } z \in R^n \text{ satisfying } |z| \leq N\},$$

where $N, \epsilon > 0$. It is not difficult to see that the uniform space $\mathfrak{A}$ is metrizable and complete.

For each $x \in R^n$, each pair of numbers $T_1, T_2 \geq 0$ satisfying $T_1 < T_2$ and each $h \in \mathfrak{A}$ define

$$\sigma(f,h,T_1,T_2,x) = \inf\{\int_{T_1}^{T_2} f(v(t),v'(t))dt + h(v(T_2)) :$$

$$v : [T_1,T_2] \to R^n \text{ is an a. c. function satisfying } v(T_1) = x\}.$$

(Here we assume that the infimum over an empty set is ∞.)

By (A1), there exists $\bar{r} \in (0,1)$ such that

$$\{(z_1, z_2) \in R^n \times R^n : |z_1 - \bar{x}| \le \bar{r} \text{ and } |z_2| \le \bar{r}\} \subset \text{dom}(f), \tag{5.2}$$

$$|f(z_1, z_2) - f(\bar{x}, 0)| \le 1 \text{ for each } (z_1, z_2) \in R^n \times R^n$$

$$\text{satisfying } |z_1 - \bar{x}| \le \bar{r},\ |z_2| \le \bar{r}. \tag{5.3}$$

It is not difficult to see that the following result holds.

PROPOSITION 5.1. Let $z \in R^n$ satisfy $|z - \bar{x}| \le \bar{r}$,

$$u_1(t) = z + t(\bar{x} - z),\ u_2(t) = \bar{x} + t(z - \bar{x}),\ t \in [0,1].$$

Then

$$|I^f(0,1,u_i) - f(\bar{x}, 0)| \le 1,\ i = 1,2.$$

We prove the following result which show that the turnpike phenomenon holds for approximate solutions of problems (P_4). Its proof is based on the turnpike property for the corresponding Lagrange problem (Theorem 1.3).

THEOREM 5.2. *Let ϵ, M be positive numbers. Then there exist $L \ge 1$ and a real number $\delta > 0$ such that for each real number $T > 2L$, each $h \in \mathfrak{A}$ satisfying $|h(\bar{x})| \le M$ and each a. c. function $v : [0,T] \to R^n$ which satisfies*

$$v(0) \in X_{M,f} \text{ and } I^f(0,T,v) + h(v(T)) \le \sigma(f,h,0,T,v(0)) + \delta$$

there exist a pair of numbers $\tau_1 \in [0,L]$ and $\tau_2 \in [T-L,T]$ such that

$$|v(t) - \bar{x}| \le \epsilon \text{ for all } t \in [\tau_1, \tau_2]$$

and if $|v(0) - \bar{x}| \le \delta$, then $\tau_1 = 0$.

PROOF. By Proposition 1.6, there exists $M_0 > 0$ such that

$$X_{M,f} \subset \bar{Y}_{M_0,f}. \tag{5.4}$$

By Lemma 4.1, there exists $L_0 > 0$ such that the following property holds:

(i) for each number $T \ge L_0$, each a. c. function $u : [0,T] \to R^n$ satisfying

$$I^f(0,T,u) \le Tf(\bar{x},0) + 2a_0 + M_0 + M + 1 + M_0|f(\bar{x},0)|$$

and each number $s \in [0, T - L_0]$, the inequality

$$\min\{|u(t) - \bar{x}| : t \in [s, s + L_0]\} \le \bar{r}/4$$

holds.

By Theorem 1.3, there exist $L_1 > 0$ and a real number $\delta \in (0,1)$ such that the following property holds:

(ii) for each real number $T > 2L_1$, each point $z_0 \in X_{M,f}$ and each point $z_1 \in Y_{|f(\bar{x},0)|+1,f}$, the value $\sigma(f,T,z_0,z_1)$ is finite and for each a. c. function $u : [0,T] \to R^n$ which satisfies

$$u(0) = z_0,\ u(T) = z_1,$$

$$I^f(0,T,u) \le \sigma(f,T,z_0,z_1) + \delta$$

there exists a pair of numbers $\tau_1 \in [0, L_1], \tau_2 \in [T - L_1, T]$ such that

$$|u(t) - \bar{x}| \le \epsilon,\ t \in [\tau_1, \tau_2];$$

moreover if $|u(0) - \bar{x}| \le \delta$, then $\tau_1 = 0$.

Fix

$$L \ge 2M_0 + 2 + 2L_0 + 2L_1. \tag{5.5}$$

Assume that

$$T > 2L,\ h \in \mathfrak{A},\ |h(\bar{x})| \le M \tag{5.6}$$

and that an a. c. function $v : [0,T] \to R^n$ satisfies

$$v(0) \in X_{M,f}, \tag{5.7}$$

$$I^f(0,T,v) + h(v(T)) \le \sigma(f,h,0,T,v(0)) + \delta. \tag{5.8}$$

By (5.4) and (5.7),

$$|v(0)| \le M_0. \tag{5.9}$$

It follows from (5.4) and (5.7) that there are $\tau_0 \in (0, M_0]$ and an a. c. function $u : [0,\tau_0] \to R^n$ such that

$$u(0) = v(0),\ u(\tau_0) = \bar{x},\ I^f(0,\tau_0,u) \le M_0. \tag{5.10}$$

For all numbers $t > \tau_0$ set

$$u(t) = \bar{x}. \tag{5.11}$$

By (5.5), (5.6), (5.8), (5.10) and (5.11),

$$\begin{aligned} I^f(0,T,v) + h(v(T)) &\le \delta + I^f(0,T,u) + h(u(T)) \\ &\le \delta + M_0 + h(\bar{x}) + (T-\tau_0)f(\bar{x},0) \\ &\le Tf(\bar{x},0) + M_0 + 1 + M + M_0|f(\bar{x},0)|. \end{aligned}$$

Together with (5.1) this implies that

$$I^f(0,T,v) \le a_0 + Tf(\bar{x},0) + M_0 + 1 + M + M_0|f(\bar{x},0)| + a_0. \tag{5.12}$$

In view of (5.6), (5.12) and property (i) there exists

$$T_1 \in [T - L_0, T] \tag{5.13}$$

such that

$$|v(T_1) - \bar{x}| \le \bar{r}/4. \tag{5.14}$$

Proposition 5.1 and (5.14) imply that

$$v(T_1) \in Y_{|f(\bar{x},0)|+1,f}. \tag{5.15}$$

In view of (5.8) and (5.13),

$$I^f(0,T_1,v) \le \sigma(f,T_1,v(0),v(T_1)) + \delta. \tag{5.16}$$

It follows from (5.5), (5.6) and (5.13) that

$$T_1 > T - L_0 > 2L_1. \tag{5.17}$$

By (5.5), (5.7), (5.13), (5.15), (5.16), (5.17) and property (ii), there exist

$$\tau_1 \in [0, L_1],\ \tau_2 \in [T_1 - L_1, T_1] \subset [T - L, T]$$

such that

$$|v(t) - \bar{x}|,\ t \in [\tau_1, \tau_2],$$

and if $|v(0) - \bar{x}| \le \delta$, then $\tau_1 = 0$. Theorem 5.2 is proved. □

Let $g \in \mathfrak{A}$. By (5.1) and Propositions 2.5 and 2.9, $\pi^f + g : R^n \to R^1 \cup \{\infty\}$ is a lower semicontinuous and bounded from below function such that

$$\lim_{|z|\to\infty} (\pi^f + g)(z) = \infty$$

and

$$\mathcal{D}(f,g) := \{x \in R^n : (\pi^f + g)(x) = \inf(\pi^f + g)\} \neq \emptyset. \tag{5.18}$$

The following result is proved in Section 7. It describes the structure of solutions of Bolza problems of the type (P_4) in the regions close to the endpoints.

THEOREM 5.3. *Let $g \in \mathfrak{A}$ and $L_0 > 0$, $\tau_0 > 0$, $\epsilon > 0$. Then there exist $\delta > 0$, $T_0 > \tau_0$ and a neighborhood $\mathcal{U}$ of g in $\mathfrak{A}$ such that for each $T \geq T_0$, each $h \in \mathcal{U}$ and each a. c. function $v : [0,T] \to R^n$ which satisfies*

$$v(0) \in \bar{Y}_{L_0,f},\ I^f(0,T,v) + h(v(T)) \leq \sigma(f,h,0,T,v(0)) + \delta$$

there exists an $(\bar{f})$-overtaking optimal function $v^* : [0,\infty) \to R^n$ such that $v^*(0) \in \mathcal{D}(\bar{f},g)$ and $|v(T-t) - v^*(t)| \leq \epsilon$ for all $t \in [0,\tau_0]$.

The key ingredients of the proof of Theorem 5.3 are Lemma 6.1, which is stated and proved in the next section, and Theorem 5.2 (the turnpike result).

6. The basic lemma for Theorem 5.3

LEMMA 6.1. *Let $T_0 > 0$, $\epsilon \in (0,1)$ and $g \in \mathfrak{A}$. Then there exists $\delta \in (0,\epsilon)$ and a neighborhood $\mathcal{U}$ of g in $\mathfrak{A}$ such that for each $h \in \mathcal{U}$ and each a. c. function $v : [0,T_0] \to R^n$ which satisfies*

$$(\pi^f + h)(v(0)) \leq \inf(\pi^f + h) + \delta, \tag{6.1}$$

$$I^f(0,T_0,v) - T_0 f(\bar{x},0) - \pi^f(v(0)) + \pi^f(v(T_0)) \leq \delta \tag{6.2}$$

there exists an (f)-overtaking optimal function $u : [0,\infty) \to R^n$ which satisfies

$$(\pi^f + g)(u(0)) = \inf(\pi^f + g), \tag{6.3}$$

$$|u(t) - v(t)| \leq \epsilon \text{ for all } t \in [0,T_0]. \tag{6.4}$$

PROOF. Let d be the metric of the metrizable uniform space $\mathfrak{A}$. Assume that the lemma does not hold. Then there exist a sequence $\{\delta_k\}_{k=1}^\infty \subset (0,1]$, a sequence $\{h_k\}_{k=1}^\infty \subset \mathfrak{A}$ and a sequence of a. c. functions $v_k : [0,T_0] \to R^n$, $k = 1,2,\dots$ such that

$$\lim_{k\to\infty} \delta_k = 0, \tag{6.5}$$

$$d(h_k,g) \leq \delta_k,\ k = 1,2,\dots \tag{6.6}$$

and that for each integer $k \geq 1$ and each (f)-overtaking optimal function $u : [0,\infty) \to R^n$ satisfying

$$(\pi^f + g)(u(0)) = \inf(\pi^f + g), \tag{6.7}$$

we have

$$(\pi^f + h_k)(v_k(0)) \leq \inf(\pi^f + h_k) + \delta_k, \tag{6.8}$$

$$I^f(0,T_0,v_k) - T_0 f(\bar{x},0) - \pi^f(v_k(0)) + \pi^f(v_k(T_0)) \leq \delta_k, \tag{6.9}$$

$$\sup\{|u(t) - v_k(t)| : t \in [0, T_0]\} > \epsilon. \tag{6.10}$$

By (5.1), Propositions 2.3, 2.5 and 2.9 and the choice of the topology in $\mathfrak{A}$, the sequence $\{\inf(\pi^f + h_k)\}_{k=1}^{\infty}$ is bounded. Combined with (5.1) and (6.8) this implies that the sequence $\{\pi^f(v_k(0))\}_{k=1}^{\infty}$ is bounded. Together with Proposition 2.5 this implies that the sequence $\{v_k(0)\}_{k=1}^{\infty}$ is bounded. In view of (6.9), (A2) and the boundedness from below of the function π^f, for each integer $k \geq 1$, $\pi^f(v_k(T_0))$ is finite and the sequence $\{I^f(0, T_0, v_k)\}_{k=1}^{\infty}$ is bounded. By Proposition 4.4 and the boundedness of the sequences $\{v_k(0)\}_{k=1}^{\infty}$ and $\{I^f(0, T_0, v_k)\}_{k=1}^{\infty}$, extracting a subsequence and re-indexing if necessary, we may assume without loss of generality that there exists an a. c. function $v : [0, T_0] \to R^n$ such that

$$v_k(t) \to v(t) \text{ as } k \to \infty \text{ uniformly on } [0, T_0], \tag{6.11}$$

$$I^f(0, T_0, v) \leq \liminf_{k\to\infty} I^f(0, T_0, v_k). \tag{6.12}$$

Since the sequences $\{\inf(\pi^f + h_k)\}_{k=1}^{\infty}$ and $\{v_k(0)\}_{k=1}^{\infty}$ are bounded it follows from (6.5), (6.6) and (6.8) that

$$\lim_{k\to\infty} |(\pi^f + h_k)(v_k(0)) - (\pi^f + g)(v_k(0))| = 0. \tag{6.13}$$

By (6.5), (6.6), (6.9), (6.11), (6.13) and Proposition 2.9,

$$(\pi^f + g)(v(0)) \leq \liminf_{k\to\infty} (\pi^f + g)(v_k(0))$$

$$\leq \liminf_{k\to\infty} (\pi^f + h_k)(v_k(0)) \leq \liminf_{k\to\infty} \inf(\pi^f + h_k)$$

$$\leq \limsup_{k\to\infty} \inf(\pi^f + h_k) \leq \inf(\pi^f + g). \tag{6.14}$$

Relation (6.14) implies that

$$(\pi^f + g)(v(0)) = \inf(\pi^f + g). \tag{6.15}$$

It follows from (6.8), (6.14) and (6.15) that

$$(\pi^f + g)(v(0)) = \liminf_{k\to\infty} (\pi^f + g)(v_k(0))$$

$$= \limsup_{k\to\infty} \inf(\pi^f + h_k) = \limsup_{k\to\infty} (\pi^f + h_k)(v_k(0)). \tag{6.16}$$

In view of (6.13) and (6.16),

$$(\pi^f + g)(v(0)) = \lim_{k\to\infty} (\pi^f + g)(v_k(0)). \tag{6.17}$$

By (6.11),

$$\pi^f(v(0)) \leq \liminf_{k\to\infty} \pi^f(v_k(0)), \; g(v(0)) \leq \liminf_{k\to\infty} g(v_k(0)).$$

Together with (6.17) this implies that

$$\pi^f(v(0)) = \lim_{k\to\infty} \pi^f(v_k(0)), \; g(v(0)) = \lim_{k\to\infty} g(v_k(0)). \tag{6.18}$$

Relation (6.11) implies that

$$\pi^f(v(T_0)) \leq \liminf_{k\to\infty} \pi^f(v_k(T_0)). \tag{6.19}$$

It follows from (6.5), (6.9), (6.12), (6.16), (6.18) and (6.19) that

$$I^f(0, T_0, v) - T_0 f(\bar{x}, 0) - \pi^f(v(0)) + \pi^f(v(T_0))$$

$$\leq \liminf_{k\to\infty} I^f(0, T_0, v_k) - T_0 f(\bar{x}, 0) - \lim_{k\to\infty} \pi^f(v_k(0)) + \liminf_{k\to\infty} \pi^f(v_k(T_0))$$

$$\leq \liminf_{k\to\infty} [I^f(0, T_0, v_k) - T_0 f(\bar{x}, 0) - \pi^f(v_k(0)) + \pi^f(v_k(T_0))] \leq 0. \tag{6.20}$$

Relations (6.18) and (6.20) imply that $\pi^f(v(T_0))$ is finite. Together with (6.18) and Proposition 2.1 this implies that

$$I^f(0, T_0, v) - T_0 f(\bar{x}, 0) - \pi^f(v(0)) + \pi^f(v(T_0)) \geq 0.$$

By the inequality above and (6.20),

$$I^f(0, T_0, v) - T_0 f(\bar{x}, 0) - \pi^f(v(0)) + \pi^f(v(T_0)) = 0. \tag{6.21}$$

Since $\pi^f(v(T_0))$ is finite Theorem 1.4 implies that there is an (f)-overtaking optimal and (f)-good function $w : [0, \infty) \to R^n$ such that

$$w(0) = v(T_0).$$

For all $t > T_0$ set

$$v(t) = w(t - T_0). \tag{6.22}$$

Clearly, $v : [0, \infty) \to R^n$ is an (f)-good function. Since the function v is (f)-good and the function w is (f)-overtaking optimal it follows from Propositions 2.1 and 2.2 and (6.21) and (6.22) that for all $T > 0$,

$$I^f(0, T, v) - T f(\bar{x}, 0) - \pi^f(v(0)) + \pi^f(v(T)) = 0. \tag{6.23}$$

In view of (6.23) and Proposition 2.10, v is an (f)-overtaking optimal function satisfying (6.15). By (6.11), for all sufficiently large natural numbers k, $|v_k(t) - v(t)| \leq \epsilon/2$ for all $t \in [0, T_0]$. This contradicts (6.10). The contradiction we have reached proves Lemma 6.1. □

Since the function $\bar{f}$ satisfies the same assumptions as the function f Lemma 6.1 can be applied with the function $\bar{f}$.

7. Proof of Theorem 5.3

Recall (see (1.11)) that $\bar{r} \in (0, 1)$ and

$$\Omega_0 = \{(x, y) \in R^n \times R^n : |x - \bar{x}| \leq \bar{r} \text{ and } |y| \leq \bar{r}\} \subset \mathrm{dom}(f). \tag{7.1}$$

We may assume without loss of generality that

$$|f(x, y) - f(\bar{x}, 0)| \leq 1/8 \text{ for all } (x, y) \in \Omega_0. \tag{7.2}$$

By Lemma 6.1 applied to the function $\bar{f}$, there exist a neighborhood $\mathcal{U}_0$ of g in $\mathfrak{A}$ and

$$\delta_1 \in (0, \min\{\epsilon, \bar{r}/2\})$$

such that the following property holds:

(Pi) for each $h \in \mathcal{U}_0$ and each a. c. function $v : [0, \tau_0] \to R^n$ which satisfies

$$(\pi^{\bar{f}} + h)(v(0)) \leq \inf(\pi^{\bar{f}} + h) + \delta_1, \tag{7.3}$$

$$I^{\bar{f}}(0, \tau_0, v) - \tau_0 f(\bar{x}, 0) - \pi^{\bar{f}}(v(0)) + \pi^{\bar{f}}(v(\tau_0)) \leq \delta_1 \tag{7.4}$$

there exists an $(\bar{f})$-overtaking optimal function $u : [0, \infty) \to R^n$ such that

$$(\pi^{\bar{f}} + g)(u(0)) = \inf(\pi^{\bar{f}} + g), \tag{7.5}$$

$$|u(t) - v(t)| \le \epsilon \text{ for all } t \in [0, \tau_0]. \tag{7.6}$$

By Propositions 2.3 and 2.4, (A1) and Lemma 4.5, there exists $\delta_2 \in (0, \delta_1)$ such that for each $z \in R^n$ satisfying $|z - \bar{x}| \le 2\delta_2$,

$$|\pi^{\bar{f}}(z)| = |\pi^{\bar{f}}(z) - \pi^{\bar{f}}(\bar{x})| \le \delta_1/8, \tag{7.7}$$

for each $(x, y) \in R^n \times R^n$ satisfying $|x - \bar{x}| \le 4\delta_2$, $|y| \le 4\delta_2$,

$$|f(x, y) - f(\bar{x}, 0)| \le \delta_1/8 \tag{7.8}$$

and that the following property holds:

(Pii) for each a. c. function $v : [0, 1] \to R^n$ satisfying

$$|v(0) - \bar{x}|, \ |v(1) - \bar{x}| \le \delta_2$$

we have

$$I^f(0, 1, v) \ge f(\bar{x}, 0) - \delta_1/8.$$

By (5.1), Propositions 2.3 and 2.5 and the choice of the topology in $\mathfrak{A}$, there exists a neighborhood $\mathcal{U}_1$ of g in $\mathfrak{A}$ such that for each $h \in \mathcal{U}_1$,

$$|\inf(\pi^{\bar{f}} + g) - \inf(\pi^{\bar{f}} + h)| \le \delta_1/16, \ |g(u(0)) - h(u(0))| \le \delta_1/16. \tag{7.9}$$

By Theorem 5.2 and Proposition 1.5, there exist $L \ge 1$, $\delta_3 > 0$ and a neighborhood $\mathcal{U}_2$ of g in $\mathfrak{A}$ such that the following property holds:

(Piii) for each $T > 2L$, each $h \in \mathcal{U}_2$ and each a. c. function $v : [0, T] \to R^n$ which satisfies

$$v(0) \in \bar{Y}_{L_0, f},$$
$$I^f(0, T, v) + h(v(T)) \le \sigma(f, h, 0, T, v(0)) + \delta_3$$

we have

$$|v(t) - \bar{x}| \le \delta_2 \text{ for all } t \in [L, T - L]. \tag{7.10}$$

By Theorem 1.4, there exists an $(\bar{f})$-overtaking optimal and $(\bar{f})$-good function $u : [0, \infty) \to R^n$ such that

$$(\pi^{\bar{f}} + g)(u(0)) = \inf(\pi^{\bar{f}} + g). \tag{7.11}$$

In view of (A4),

$$\lim_{t \to \infty} u(t) = \bar{x}.$$

Therefore there exists $\tau_1 > 0$ such that

$$|u(t) - \bar{x}| \le \delta_2 \text{ for all } t \ge \tau_1. \tag{7.12}$$

There exists a neighborhood $\mathcal{U}$ of g in $\mathfrak{A}$ such that

$$\mathcal{U} \subset \cap_{i=0}^2 \mathcal{U}_i, \tag{7.13}$$

and

$$|h(u(0)) - g(u(0))| \le \delta_1/16 \text{ for all } h \in \mathcal{U}. \tag{7.14}$$

Choose $\delta > 0$ and $T_0 > 0$ such that

$$\delta < (16(L + \tau_1 + \tau_0 + 6))^{-1} \min\{\delta_1, \delta_2, \delta_3\}, \tag{7.15}$$

$$T_0 > 2L + 2\tau_0 + 2\tau_1 + 4. \tag{7.16}$$

Assume that $T \geq T_0$, $h \in \mathcal{U}$ and that an a. c. function $v : [0, T] \to R^n$ satisfies

$$v(0) \in \bar{Y}_{L_0,f} \text{ and } I^f(0, T, v) + h(v(T)) \leq \sigma(f, h, 0, T, v(0)) + \delta. \tag{7.17}$$

By (7.13), (7.15)-(7.17) and property(Piii), relation (7.10) holds. In view of (7.16),

$$[T - L - \tau_0 - \tau_1 - 4, T - L - \tau_0 - \tau_1] \subset [L, T - L - \tau_0 - \tau_1]. \tag{7.18}$$

Relations (7.10) and (7.18) imply that

$$|v(t) - \bar{x}| \leq \delta_2 \text{ for all } t \in [T - L - \tau_0 - \tau_1 - 4, T - L - \tau_0 - \tau_1]. \tag{7.19}$$

Define

$$\begin{aligned} &\tilde{v}(t) = v(t),\ t \in [0, T - L - \tau_0 - \tau_1 - 4], \\ &\tilde{v}(t) = u(T - t),\ t \in [T - L - \tau_0 - \tau_1 - 3, T], \\ &\tilde{v}(t + T - L - \tau_0 - \tau_1 - 4) = v(T - L - \tau_0 - \tau_1 - 4) \\ &+t[u(L + \tau_0 + \tau_1 + 3) - v(T - L - \tau_0 - \tau_1 - 4)],\ t \in (0, 1). \end{aligned} \tag{7.20}$$

Clearly, $\tilde{v} : [0, T] \to R^n$ is an a. c. function. In view of (7.12) and (7.20),

$$|\tilde{v}(T - L - \tau_0 - \tau_1 - 3) - \bar{x}| = |u(L + \tau_0 + \tau_1 + 3) - \bar{x}| \leq \delta_2. \tag{7.21}$$

It follows from (7.10), (7.19) and (7.21) that for all

$$t \in (T - L - \tau_0 - \tau_1 - 4, T - L - \tau_0 - \tau_1 - 3)$$

we have

$$|\tilde{v}'(t)| \leq |u(L + \tau_0 + \tau_1 + 3) - v(T - L - \tau_0 - \tau_1 - 4)| \leq 2\delta_2,$$

$$|\tilde{v}(t) - \bar{x}| \leq 3\delta_2. \tag{7.22}$$

By (7.8) and (7.22), for all

$$t \in (T - L - \tau_0 - \tau_1 - 4, T - L - \tau_0 - \tau_1 - 3)$$

we have

$$|f(\tilde{v}(t), \tilde{v}'(t)) - f(\bar{x}, 0)| \leq \delta_1/8. \tag{7.23}$$

It follows from (7.17), (7.20) and (2.11) that

$$\begin{aligned} \delta &\geq I^f(0, T, v) + h(v(T)) - (I^f(0, T, \tilde{v}) + h(\tilde{v}(T))) \\ &= I^f(T - L - \tau_0 - \tau_1 - 4, T, v) + h(v(T)) \\ &\quad -(I^f(T - L - \tau_0 - \tau_1 - 4, T, \tilde{v}) + h(\tilde{v}(T))). \end{aligned}$$

Together with (7.20) and (7.23) this implies that

$$\begin{aligned} &I^f(T - L - \tau_0 - \tau_1 - 4, T, v) + h(v(T)) \\ &\leq \delta + I^f(T - L - \tau_0 - \tau_1 - 4, T, \tilde{v}) + h(\tilde{v}(T)) \\ &\leq \delta + \delta_1/8 + f(\bar{x}, 0) + I^f(T - L - \tau_0 - \tau_1 - 3, T, \tilde{v}) + h(u(0)) \\ &\leq \delta + \delta_1/8 + f(\bar{x}, 0) + I^{\bar{f}}(0, L + \tau_0 + \tau_1 + 3, u) + h(u(0)). \end{aligned} \tag{7.24}$$

By (7.19) and property (Pii),

$$I^f(T - L - \tau_0 - \tau_1 - 4, T - L - \tau_0 - \tau_1 - 3, v) \geq f(\bar{x}, 0) - \delta_1/8.$$

Together with (7.24) this implies that

$$\begin{aligned} &I^f(T - L - \tau_0 - \tau_1 - 3, T, v) + h(v(T)) \\ &\leq \delta + \delta_1/4 + I^{\bar{f}}(0, L + \tau_0 + \tau_1 + 3, u) + h(u(0)). \end{aligned} \tag{7.25}$$

Set

$$y(t) = v(T - t),\ t \in [0, L + \tau_0 + \tau_1 + 3]. \tag{7.26}$$

Since the function u is $(\bar{f})$-good it follows from (7.25) and (7.26) that

$$\begin{aligned} &I^{\bar{f}}(0, L + \tau_0 + \tau_1 + 3, y) + h(y(0)) \\ &\le I^{\bar{f}}(0, L + \tau_0 + \tau_1 + 3, u) + h(u(0)) + \delta + \delta_1/4 < \infty. \end{aligned} \tag{7.27}$$

In view of (7.7), (7.19), (7.26) and (7.27), $\pi^{\bar{f}}(y(L+\tau_0+\tau_1+3))$, $\pi^{\bar{f}}(y(0))$ are finite. By (7.19), (7.26), (7.27), Propositions 2.1 and 2.2 and $(\bar{f})$-overtaking optimality of u,

$$\begin{aligned} &(\pi^{\bar{f}} + h)(y(0)) - \inf(\pi^{\bar{f}} + h) \\ &+I^{\bar{f}}(0, \tau_0, y) - \tau_0 f(\bar{x}, 0) - \pi^{\bar{f}}(y(0)) + \pi^{\bar{f}}(y(\tau_0)) \\ &\le (\pi^{\bar{f}} + h)(y(0)) - \inf(\pi^{\bar{f}} + h) \\ &+I^{\bar{f}}(0, L + \tau_0 + \tau_1 + 3, y) - (L + \tau_0 + \tau_1 + 3) f(\bar{x}, 0) \\ &-\pi^{\bar{f}}(y(0)) + \pi^{\bar{f}}(y(L + \tau_0 + \tau_1 + 3)) \\ &\le (\pi^{\bar{f}} + h)(y(0)) - \inf(\pi^{\bar{f}} + h) + \delta + \delta_1/4 + h(u(0)) - h(y(0)) \\ &+I^{\bar{f}}(0, L + \tau_0 + \tau_1 + 3, u) - (L + \tau_0 + \tau_1 + 3) f(\bar{x}, 0) \\ &-\pi^{\bar{f}}(y(0)) + \pi^{\bar{f}}(y(L + \tau_0 + \tau_1 + 3)) \\ &\le (\pi^{\bar{f}} + h)(y(0)) - \inf(\pi^{\bar{f}} + h) + \delta + \delta_1/4 + h(u(0)) - h(y(0)) \\ &+\pi^{\bar{f}}(u(0)) - \pi^{\bar{f}}(u(L + \tau_0 + \tau_1 + 3)) - \pi^{\bar{f}}(y(0)) - \pi^{\bar{f}}(y(L + \tau_0 + \tau_1 + 3)) \\ &\le h(u(0)) + \pi^{\bar{f}}(u(0)) - \inf(\pi^{\bar{f}} + h) + \delta + \delta_1/4 - \pi^{\bar{f}}(u(L + \tau_0 + \tau_1 + 3)) \\ &+\pi^{\bar{f}}(y(L + \tau_0 + \tau_1 + 3)). \end{aligned} \tag{7.28}$$

By (7.19) and (7.26),

$$|y(L + \tau_0 + \tau_1 + 3) - \bar{x}| \le \delta_2. \tag{7.29}$$

By (7.7), (7.12) and (7.29),

$$|\pi^{\bar{f}}(y(L + \tau_0 + \tau_1 + 3))|,\ |\pi^{\bar{f}}(u(L + \tau_0 + \tau_1 + 3))| \le \delta_1/8. \tag{7.30}$$

By (7.5), (7.9), (7.13) and Proposition 2.1,

$$\begin{aligned} &h(u(0)) + \pi^{\bar{f}}(u(0)) - \inf(\pi^{\bar{f}} + h) \\ &\le \delta_1/16 + g(u(0)) + \pi^{\bar{f}}(u(0)) - \inf(\pi^{\bar{f}} + g) + \delta_1/16 \le \delta_1/8. \end{aligned} \tag{7.31}$$

It follows from (7.28), (7.30) and (7.31) that

$$\begin{aligned} &(\pi^{\bar{f}} + h)(y(0)) - \inf(\pi^{\bar{f}} + h) \\ &+I^{\bar{f}}(0, \tau_0, y) - \tau_0 f(\bar{x}, 0) - \pi^{\bar{f}}(y(0)) + \pi^{\bar{f}}(y(\tau_0)) \\ &\le \delta_1/8 + \delta + \delta_1/4 + \delta_1/4 \le \delta_1. \end{aligned}$$

Combined with Proposition 2.1 the inequality above implies that

$$(\pi^{\bar{f}} + h)(y(0)) - \inf(\pi^{\bar{f}} + h) \le \delta_1, \tag{7.32}$$

$$I^{\bar{f}}(0, \tau_0, y) - \tau_0 f(\bar{x}, 0) - \pi^{\bar{f}}(y(0)) + \pi^{\bar{f}}(y(\tau_0)) \le \delta_1. \tag{7.33}$$

By (7.32), (7.33) and property (Pi) there exists an $(\bar{f})$-overtaking optimal function $w : [0, \infty) \to R^n$ such that

$$(\pi^{\bar{f}} + g)(w(0)) = \inf(\pi^{\bar{f}} + g)$$

and for all $t \in [0, \tau_0]$,

$$\epsilon \geq |y(t) - w(t)| = |v(T - t) - w(t)|.$$

Theorem 5.3 is proved. □

8. The Bolza problem (P_5)

We use the notation, definitions and assumptions introduced in Sections 1, 2 and 5.

Let $a_0 > 0$. Denote by $\mathfrak{A}(R^n \times R^n)$ the set of all lower semicontinuous functions $h : R^n \times R^n \to R^1$ which are bounded on bounded subsets of $R^n \times R^n$ and such that

$$h(z_1, z_2) \geq -a_0 \text{ for all } z_1, z_2 \in R^n. \tag{8.1}$$

For simplicity, we set $\mathfrak{A} = \mathfrak{A}(R^n \times R^n)$. The set $\mathfrak{A}$ is equipped with the uniformity determined by the base

$$E(N, \epsilon) = \{(h_1, h_2) \in \mathfrak{A} \times \mathfrak{A} :$$

$$|h_1(z) - h_2(z)| \leq \epsilon \text{ for all } z \in R^n \times R^n \text{ satisfying } |z| \leq N\},$$

where $N, \epsilon > 0$. It is not difficult to see that the uniform space $\mathfrak{A}$ is metrizable and complete.

For each pair of numbers $T_1, T_2 \geq 0$ satisfying $T_1 < T_2$ and each $h \in \mathfrak{A}$ define

$$\sigma(f, h, T_1, T_2) = \inf\{\int_{T_1}^{T_2} f(v(t), v'(t))dt + h(v(T_1), (v(T_2)) :$$

$$v : [T_1, T_2] \to R^n \text{ is an a. c. function}\}. \tag{8.2}$$

(Here we assume that the infimum over an empty set is ∞.)

By (A1), there exists $\bar{r} \in (0, 1)$ such that

$$\Omega_0 := \{(z_1, z_2) \in R^n \times R^n : \ |z_1 - \bar{x}| \leq \bar{r} \text{ and } |z_2| \leq \bar{r}\} \subset \operatorname{dom}(f), \tag{8.3}$$

$$|f(z_1, z_2) - f(\bar{x}, 0)| \leq 1 \text{ for each } (z_1, z_2) \in \Omega_0. \tag{8.4}$$

We prove the following result which show that the turnpike phenomenon holds for approximate solutions of problems (P_5). Its proof is based on the turnpike property for the corresponding Lagrange problem (Theorem 1.3).

THEOREM 8.1. *Let $\epsilon > 0$, $M > \bar{r}_v$. Then there exist $L \geq 1$ and $\delta > 0$ such that for each real number $T > 2L$, each $h \in \mathfrak{A}$ satisfying $|h(\bar{x}, \bar{x})| \leq M$ and each a. c. function $v : [0, T] \to R^n$ which satisfies*

$$I^f(0, T, v) + h(v(0), v(T)) \leq \sigma(f, h, 0, T) + \delta$$

there exist a pair of numbers $\tau_1 \in [0, L]$ and $\tau_2 \in [T - L, T]$ such that

$$|v(t) - \bar{x}| \leq \epsilon \text{ for all } t \in [\tau_1, \tau_2].$$

Moreover, if $|v(0) - \bar{x}| \leq \delta$, then $\tau_1 = 0$ and if $|v(T) - \bar{x}| \leq \delta$, then $\tau_2 = T$.

PROOF. By Lemma 4.1, there exists $L_0 > 0$ such that the following property holds:

(i) for each number $T \geq L_0$, each a. c. function $u : [0,T] \to R^n$ satisfying

$$I^f(0,T,u) \leq Tf(\bar{x},0) + 1 + a_0 + M$$

and each number $s \in [0, T - L_0]$ the inequality

$$\min\{|u(t) - \bar{x}| : t \in [s, s + L_0]\} \leq \bar{r}/4$$

holds.

By Theorem 1.3 and Proposition 1.5, there exist $L_1 > 0$ and $\delta \in (0, \bar{r}/8)$ such that the following property holds:

(ii) for each real number $T > 2L_1$, each point $z_0 \in \bar{Y}_{|f(\bar{x},0)|+1+M,f}$ and each point $z_1 \in Y_{|f(\bar{x},0)|+1+M,f}$, the value $\sigma(f,T,z_0,z_1)$ is finite and for each a.c. function $u : [0,T] \to R^n$ which satisfies

$$u(0) = z_0, \ u(T) = z_1,$$

$$I^f(0,T,u) \leq \sigma(f,T,z_0,z_1) + \delta$$

there exists a pair of numbers $\tau_1 \in [0, L_1], \tau_2 \in [T - L_1, T]$ such that

$$|u(t) - \bar{x}| \leq \epsilon, \ t \in [\tau_1, \tau_2];$$

moreover if $|u(0) - \bar{x}| \leq \delta$, then $\tau_1 = 0$ and if $|u(T) - \bar{x}| \leq \delta$, then $\tau_2 = T$.

Fix

$$L > 2M + 2 + 2L_0 + 2L_1. \tag{8.5}$$

Assume that

$$T > 2L, \ h \in \mathfrak{A}, \ |h(\bar{x},\bar{x})| \leq M \tag{8.6}$$

and that an a. c. function $v : [0,T] \to R^n$ satisfies

$$I^f(0,T,v) + h(v(0), v(T)) \leq \sigma(f,h,0,T) + \delta. \tag{8.7}$$

By (8.6) and (8.7),

$$I^f(0,T,v) + h(v(0), v(T)) \leq Tf(\bar{x},0) + M + 1.$$

Together with (8.1) this implies that

$$I^f(0,T,v) \leq Tf(\bar{x},0) + a_0 + M + 1. \tag{8.8}$$

In view of (8.5), (8.6), (8.8) and property (i), there exist

$$T_1 \in [0, L_0], \ T_2 \in [T - L_0, T] \tag{8.9}$$

such that

$$|v(T_i) - \bar{x}| \leq \bar{r}/4, \ i = 1, 2. \tag{8.10}$$

If $|v(0) - \bar{x}| \leq \delta$, then we may set $T_1 = 0$ and if $|v(T) - \bar{x}| \leq \delta$, then we may set $T_2 = T$. Proposition 5.1, (8.3), (8.4) and (8.10) imply that

$$v(T_i) \in Y_{|f(\bar{x},0)|+1,f} \cap \bar{Y}_{|f(\bar{x},0)|+M+1,f}, \ i = 1, 2. \tag{8.11}$$

In view of (8.7) and (8.9),

$$I^f(T_1, T_2, v) \leq \sigma(f, T_2 - T_1, v(T_1), v(T_2)) + \delta. \tag{8.12}$$

It follows from (8.5), (8.6), (8.9), (8.11), (8.12) and property (ii) that there exist

$$\tau_1 \in [T_1, T_1 + L_1] \subset [0, L], \ \tau_2 \in [T_2 - L_1, T_2] \subset [T - L, T]$$

such that

$$|v(t) - \bar{x}| \leq \epsilon, \ t \in [\tau_1, \tau_2],$$

and if $|v(0) - \bar{x}| \leq \delta$, then $\tau_1 = 0$ and if $|v(T) - \bar{x}| \leq \delta$, then $\tau_2 = T$. Theorem 8.1 is proved. □

Let $h \in \mathfrak{A}$. By (8.1) and Propositions 2.5 and 2.9, the function

$$\pi^f(z_1) + \pi^{\bar{f}}(z_2) + h(z_1, z_2), \ z_1, z_2 \in R^n$$

is lower semicontinuous and bounded from below,

$$\lim_{|(z_1,z_2)| \to \infty} (\pi^f(z_1) + \pi^{\bar{f}}(z_2) + h(z_1, z_2)) = \infty$$

and the set

$$\begin{aligned} \mathcal{D}(f,h) := \{(x,y) \in R^n \times R^n : \\ \pi^f(x) + \pi^{\bar{f}}(y) + h(x,y) \leq \pi^f(\xi) + \pi^{\bar{f}}(\eta) + h(\xi,\eta) \\ \text{for all } \xi, \eta \in R^n\} \end{aligned}$$

is nonempty and closed.

The following result is proved in Section 10. It describes the structure of solutions of Bolza problems of the type (P_5) in the regions close to the endpoints.

THEOREM 8.2. *Let $g \in \mathfrak{A}$, $\tau_0 > 0$ and $\epsilon > 0$. Then there exist $\delta > 0$, $T_0 > \tau_0$ and a neighborhood $\mathcal{U}$ of g in $\mathfrak{A}$ such that for each $T \geq T_0$, each $h \in \mathcal{U}$ and each a. c. function $v : [0, T] \to R^n$ which satisfies*

$$I^f(0, T, v) + h(v(0), v(T)) \leq \sigma(f, h, 0, T) + \delta$$

there exist an (f)-overtaking optimal function $v_1^* : [0, \infty) \to R^n$ and an $(\bar{f})$-overtaking optimal function $v_2^* : [0, \infty) \to R^n$ such that

$$\begin{aligned} \pi^f(v_1^*(0)) + \pi^{\bar{f}}(v_2^*(0)) + g(v_1^*(0), v_2^*(0)) \leq \pi^f(\xi_1) + \pi^{\bar{f}}(\xi_2) + g(\xi_1, \xi_2) \\ \text{for all } \xi_1, \xi_2 \in R^n \end{aligned}$$

and that for all $t \in [0, \tau_0]$,

$$|v(t) - v_1^*(t)| \leq \epsilon, \ |v(T - t) - v_2^*(t)| \leq \epsilon.$$

The key ingredients of the proof of Theorem 8.2 are Lemma 9.1, which is stated and proved in the next section, and Theorem 8.1 (the turnpike result).

9. The basic lemma for Theorem 8.2

For each $h \in \mathfrak{A}$ define

$$\phi_h(z_1, z_2) = \pi^f(z_1) + \pi^{\bar{f}}(z_2) + h(z_1, z_2), \ (z_1, z_2) \in R^n \times R^n. \tag{9.1}$$

LEMMA 9.1. *Let $T_0 \geq 1$, $\epsilon \in (0, 1)$ and $g \in \mathfrak{A}$. Then there exists $\delta \in (0, \epsilon)$ and a neighborhood $\mathcal{U}$ of g in $\mathfrak{A}$ such that for each $h \in \mathcal{U}$ and each pair of a. c. functions $v_i : [0, T_0] \to R^n$, $i = 1, 2$ which satisfies*

$$\phi_h(v_1(0), v_2(0)) \leq \phi_h(\xi, \eta) + \delta \text{ for all } \xi, \eta \in R^n, \tag{9.2}$$

$$I^f(0, T_0, v_1) - T_0 f(\bar{x}, 0) - \pi^f(v_1(0)) + \pi^f(v_1(T_0)) \leq \delta, \tag{9.3}$$

$$I^{\bar{f}}(0,T_0,v_2) - T_0 f(\bar{x},0) - \pi^{\bar{f}}(v_2(0)) + \pi^{\bar{f}}(v_2(T_0)) \le \delta \tag{9.4}$$

there exist an (f)-overtaking optimal function $u_1 : [0,\infty) \to R^n$ and an $(\bar{f})$-overtaking optimal function $u_2 : [0,\infty) \to R^n$ such that

$$(u_1(0), u_2(0)) \in \mathcal{D}(f,g)$$

and for all $t \in [0,T_0]$,

$$|u_1(t) - v_1(t)| \le \epsilon,\ |u_2(t) - v_2(t)| \le \epsilon.$$

PROOF. Let d be the metric of the metrizable uniform space $\mathfrak{A}$. Assume that the lemma does not hold. Then there exist a sequence $\{\delta_k\}_{k=1}^{\infty} \subset (0,1]$, a sequence $\{h_k\}_{k=1}^{\infty} \subset \mathfrak{A}$ and sequences of a. c. functions $v_{k,1} : [0,T_0] \to R^n$, $v_{k,2} : [0,T_0] \to R^n$, $k = 1,2,\dots$ such that

$$\lim_{k\to\infty} \delta_k = 0, \tag{9.5}$$

$$d(h_k, g) \le \delta_k,\ k = 1,2,\dots \tag{9.6}$$

and that for each integer $k \ge 1$ and each pair of an (f)-overtaking optimal function $u_1 : [0,\infty) \to R^n$ and an $(\bar{f})$-overtaking optimal function $u_2 : [0,\infty) \to R^n$ which satisfy

$$(u_1(0), u_2(0)) \in \mathcal{D}(f,g) \tag{9.7}$$

the following relations hold:

$$\phi_{h_k}((v_{k,1}(0), v_{k,2}(0)) \le \phi_{h_k}(\xi,\eta) + \delta_k \text{ for all } \xi,\eta \in R^n, \tag{9.8}$$

$$I^f(0,T_0,v_{k,1}) - T_0 f(\bar{x},0) - \pi^f(v_{k,1}(0)) + \pi^f(v_{k,1}(T_0)) \le \delta_k, \tag{9.9}$$

$$I^{\bar{f}}(0,T_0,v_{k,2}) - T_0 f(\bar{x},0) - \pi^{\bar{f}}(v_{k,2}(0)) + \pi^{\bar{f}}(v_{k,2}(T_0)) \le \delta_k, \tag{9.10}$$

$$\sup\{|u_1(t) - v_{k,1}(t)|,\ |u_2(t) - v_{k,2}(t)| : t \in [0,T_0]\} > \epsilon. \tag{9.11}$$

By (8.1),(9.1), (9.6) and Propositions 2.5 and 2.9, the sequence $\{\inf(\phi_{h_k})\}_{k=1}^{\infty}$ is bounded. Combined with (9.1), (9.8) and Propositions 2.5 and 2.9 this implies that the sequences

$$\{\pi^f(v_{k,1}(0))\}_{k=1}^{\infty}, \{\pi^{\bar{f}}(v_{k,2}(0))\}_{k=1}^{\infty} \text{ are bounded.} \tag{9.12}$$

Together with Proposition 2.5 this implies that the sequences

$$\{v_{k,1}(0)\}_{k=1}^{\infty},\ \{v_{k,2}(0)\}_{k=1}^{\infty} \text{ are bounded.} \tag{9.13}$$

In view of (9.9), (9.10) and (9.12), the sequences

$$\{I^f(0,T_0,v_{k,1})\}_{k=1}^{\infty},\ \{I^{\bar{f}}(0,T_0,v_{k,2})\}_{k=1}^{\infty} \text{ are bounded.} \tag{9.14}$$

By Proposition 4.4, (9.13) and (9.14), extracting a subsequence and re-indexing if necessary, we may assume without loss of generality that there exist a. c. functions $v_1 : [0,T_0] \to R^n$, $v_2 : [0,T_0] \to R^n$ such that for $i = 1,2$,

$$v_{k,i}(t) \to v_i(t) \text{ as } k \to \infty \text{ uniformly on } [0,T_0], \tag{9.15}$$

$$I^f(0,T_0,v_1) \le \lim_{k\to\infty} I^f(0,T_0,v_{k,1}),\ I^{\bar{f}}(0,T_0,v_2) \le \lim_{k\to\infty} I^{\bar{f}}(0,T_0,v_{k,2}). \tag{9.16}$$

By (9.5), (9.6), (9.12) and (9.13),

$$\lim_{k\to\infty} (\pi^f(v_{k,1}(0)) + \pi^{\bar{f}}(v_{k,2}(0)) + h_k(v_{k,1}(0), v_{k,2}(0))$$

$$-(\pi^f(v_{k,1}(0)) + \pi^{\bar{f}}(v_{k,2}(0)) - g(v_{k,1}(0), v_{k,2}(0))) = 0. \tag{9.17}$$

It follows from (9.1), (9.5), (9.6), (9.8), (9.12), (9.15) and (9.17) that

$$\begin{aligned}
&\pi^f(v_1(0)) + \pi^{\bar{f}}(v_2(0)) + g(v_1(0), v_2(0)) \\
&\le \liminf_{k\to\infty}[\pi^f(v_{k,1}(0)) + \pi^{\bar{f}}(v_{k,2}(0)) + g(v_{k,1}(0), v_{k,2}(0))] \\
&= \liminf_{k\to\infty}[\pi^f(v_{k,1}(0)) + \pi^{\bar{f}}(v_{k,2}(0)) + h_k(v_{k,1}(0), v_{k,2}(0))] \\
&\le \liminf_{k\to\infty} \inf(\phi_{h_k}) \le \limsup_{k\to\infty} \inf(\phi_{h_k}) \le \inf(\phi_g).
\end{aligned} \tag{9.18}$$

Relations (9.1) and (9.18) imply that

$$\pi^f(v_1(0)) + \pi^{\bar{f}}(v_2(0)) + g(v_1(0), v_2(0)) = \phi_g(v_1(0), v_2(0)) = \inf(\phi_g). \tag{9.19}$$

By (9.1), (9.5), (9.6), (9.8), (9.12), (9.13), (9.18) and (9.19),

$$\begin{aligned}
\pi^f(v_1(0)) + \pi^{\bar{f}}(v_2(0)) + g(v_1(0), v_2(0)) &= \liminf_{k\to\infty} \phi_g(v_{k,1}(0), v_{k,2}(0)) \\
&= \liminf_{k\to\infty} \inf(\phi_{h_k}) = \limsup_{k\to\infty} \inf(\phi_{h_k}) \\
&= \limsup_{k\to\infty} \phi_{h_k}(v_{k,1}(0), v_{k,2}(0)) = \limsup_{k\to\infty} \phi_g(v_{k,1}(0), v_{k,2}(0)).
\end{aligned} \tag{9.20}$$

In view of (9.1), (9.19) and (9.20),

$$\begin{aligned}
&\pi^f(v_1(0)) + \pi^{\bar{f}}(v_2(0)) + g(v_1(0), v_2(0)) \\
&= \phi_g(v_1(0), v_2(0)) = \lim_{k\to\infty} \phi_g(v_{k,1}(0), v_{k,2}(0)) \\
&= \lim_{k\to\infty} [\pi^f(v_{k,1}(0)) + \pi^{\bar{f}}(v_{k,2}(0)) + g(v_{k,1}(0), v_{k,2}(0))].
\end{aligned} \tag{9.21}$$

By (9.15),

$$\pi^f(v_1(0)) \le \liminf_{k\to\infty} \pi^f(v_{k,1}(0)),$$

$$\pi^{\bar{f}}(v_2(0)) \le \liminf_{k\to\infty} \pi^{\bar{f}}(v_{k,2}(0)), \tag{9.22}$$

$$g(v_1(0), v_2(0)) \le \liminf_{k\to\infty} g(v_{k,1}(0), v_{k,2}(0)). \tag{9.23}$$

Relations (9.21)-(9.23) imply that

$$\pi^f(v_1(0)) = \lim_{k\to\infty} \pi^f(v_{k,1}(0)),\ \pi^{\bar{f}}(v_2(0)) = \lim_{k\to\infty} \pi^{\bar{f}}(v_{k,2}(0)), \tag{9.24}$$

$$g(v_1(0), v_2(0)) = \lim_{k\to\infty} g(v_{k,1}(0), v_{k,2}(0)). \tag{9.25}$$

It view of (9.15),

$$\pi^f(v_1(T_0)) \le \liminf_{k\to\infty} \pi^f(v_{k,1}(T_0)),$$

$$\pi^{\bar{f}}(v_2(T_0)) \le \liminf_{k\to\infty} \pi^{\bar{f}}(v_{k,2}(T_0)). \tag{9.26}$$

It follows from (9.5), (9.9), (9.12), (9.14), (9.16), (9.24) and (9.26) that

$$\begin{aligned}
&I^f(0, T_0, v_1) - T_0 f(\bar{x}, 0) - \pi^f(v_1(0)) + \pi^f(v_1(T_0)) \\
&\le \liminf_{k\to\infty} I^f(0, T_0, v_{k,1}) - T_0 f(\bar{x}, 0) - \lim_{k\to\infty} \pi^f(v_{k,1}(0)) + \liminf_{k\to\infty} \pi^f(v_{k,1}(T_0))
\end{aligned}$$

$$\leq \liminf_{k\to\infty}[I^f(0,T_0,v_{k,1}) - T_0 f(\bar{x},0) - \pi^f(v_{k,1}(0)) + \pi^f(v_{k,1}(T_0))] \leq 0. \tag{9.27}$$

It follows from (9.5), (9.10), (9.12), (9.14), (9.16), (9.24) and (9.26) that

$$I^{\bar{f}}(0,T_0,v_2) - T_0 f(\bar{x},0) - \pi^{\bar{f}}(v_2(0)) + \pi^{\bar{f}}(v_2(T_0))$$

$$\leq \liminf_{k\to\infty} I^{\bar{f}}(0,T_0,v_{k,2}) - T_0 f(\bar{x},0) - \lim_{k\to\infty} \pi^{\bar{f}}(v_{k,2}(0)) + \liminf_{k\to\infty} \pi^{\bar{f}}(v_{k,2}(T_0))$$

$$\leq \liminf_{k\to\infty}[I^{\bar{f}}(0,T_0,v_{k,2}) - T_0 f(\bar{x},0) - \pi^{\bar{f}}(v_{k,2}(0)) + \pi^{\bar{f}}(v_{k,2}(T_0))] \leq 0. \tag{9.28}$$

Relations (9.12), (9.14), (9.16), (9.24), (9.27) and (9.28) imply that

$$\pi^f(v_1(T_0)),\ \pi^{\bar{f}}(v_2(T_0)) \text{ are finite.} \tag{9.29}$$

By (9.12), (9.27), (9.28), (9.29) and Proposition 2.1,

$$I^f(0,T_0,v_1) - T_0 f(\bar{x},0) - \pi^f(v_1(0)) + \pi^f(v_1(T_0)) \geq 0,$$

$$I^{\bar{f}}(0,T_0,v_2) - T_0 f(\bar{x},0) - \pi^{\bar{f}}(v_2(0)) + \pi^{\bar{f}}(v_2(T_0)) \geq 0$$

and

$$I^f(0,T_0,v_1) - T_0 f(\bar{x},0) - \pi^f(v_1(0)) + \pi^f(v_1(T_0)) = 0, \tag{9.30}$$

$$I^{\bar{f}}(0,T_0,v_2) - T_0 f(\bar{x},0) - \pi^{\bar{f}}(v_2(0)) + \pi^{\bar{f}}(v_2(T_0)) = 0. \tag{9.31}$$

In view of (9.29) and Theorem 1.4 there are an (f)-overtaking optimal and (f)-good function $w_1 : [0,\infty) \to R^n$ such that

$$w_1(0) = v_1(T_0) \tag{9.32}$$

and an $(\bar{f})$-overtaking optimal and $(\bar{f})$-good function $w_2 : [0,\infty) \to R^n$ such that

$$w_2(0) = v_2(T_0). \tag{9.33}$$

For all $t > T_0$ and $i = 1,2$ extend

$$v_i(t) = w_i(t - T_0). \tag{9.34}$$

Clearly, $v_1, v_2 : [0,\infty) \to R^n$ are a. c. functions, v_1 is an (f)-good function and v_2 is an $(\bar{f})$-good function. By Propositions 2.1 and 2.2, (9.12), (9.24), (9.29), (9.30) and (9.31), for all $T > 0$,

$$I^f(0,T,v_1) - T f(\bar{x},0) - \pi^f(v_1(0)) + \pi^f(v_1(T)) = 0,$$

$$I^{\bar{f}}(0,T,v_2) - T f(\bar{x},0) - \pi^{\bar{f}}(v_2(0)) + \pi^f(v_2(T)) = 0.$$

It follows from the equalities above and Proposition 2.10 that v_1 is an (f)-overtaking optimal function and v_2 is an $(\bar{f})$-overtaking optimal function satisfying (9.19). By (9.15), for all sufficiently large natural numbers k and $i = 1,2$, $|v_{k,i}(t) - v_i(t)| \leq \epsilon/2$ for all $t \in [0,T_0]$. This contradicts (9.11). The contradiction we have reached proves Lemma 9.1. □

10. Proof of Theorem 8.2

We use the function ϕ_h defined by (9.1). Recall (see (1.11)) that $\bar{r} \in (0,1)$ and

$$\Omega_0 = \{(x,y) \in R^n \times R^n : |x - \bar{x}| \leq \bar{r} \text{ and } |y| \leq \bar{r}\} \subset \operatorname{dom}(f). \tag{10.1}$$

We may assume without loss of generality that

$$|f(x,y) - f(\bar{x},0)| \leq 1/8 \text{ for all } (x,y) \in \Omega_0. \tag{10.2}$$

By Lemma 9.1, there exist a neighborhood $\mathcal{U}_0$ of g in $\mathfrak{A}$ and

$$\delta_1 \in (0, \min\{\epsilon, \bar{r}/4\})$$

such that the following property holds:

(Pi) for each $h \in \mathcal{U}_0$ and each pair of a. c. functions $v_i : [0,\tau_0] \to R^n$, $i = 1,2$ which satisfies

$$\pi^f(v_1(0)) + \pi^{\bar{f}}(v_2(0)) + h(v_1(0), v_2(0)) \leq \inf(\phi_h) + 4\delta_1, \tag{10.3}$$

$$I^f(0,\tau_0,v_1) - \tau_0 f(\bar{x},0) - \pi^f(v_1(0)) + \pi^f(v_1(\tau_0)) \leq 4\delta_1 \tag{10.4}$$

$$I^{\bar{f}}(0,\tau_0,v_2) - \tau_0 f(\bar{x},0) - \pi^{\bar{f}}(v_2(0)) + \pi^{\bar{f}}(v_2(\tau_0)) \leq 4\delta_1 \tag{10.5}$$

there exist an (f)-overtaking optimal function $u_1 : [0,\infty) \to R^n$ and an $(\bar{f})$-overtaking optimal function $u_2 : [0,\infty) \to R^n$ such that

$$(u_1(0), u_2(0)) \in \mathcal{D}(f,g)$$

and that for all $t \in [0,\tau_0]$ and $i = 1,2$,

$$|u_i(t) - v_i(t)| \leq \epsilon.$$

By Propositions 2.3 and 2.4, (A1) and Lemma 4.5, there exists $\delta_2 \in (0,\delta_1)$ such that for each $z \in R^n$ satisfying $|z - \bar{x}| \leq 2\delta_2$,

$$|\pi^{\bar{f}}(z)| = |\pi^{\bar{f}}(z) - \pi^{\bar{f}}(\bar{x})| \leq \delta_1/8, \tag{10.6}$$

$$|\pi^f(z)| = |\pi^f(z) - \pi^f(\bar{x})| \leq \delta_1/8, \tag{10.7}$$

for each $(x,y) \in R^n \times R^n$ satisfying $|x - \bar{x}| \leq 4\delta_2$, $|y| \leq 4\delta_2$,

$$|f(x,y) - f(\bar{x},0)| \leq \delta_1/8 \tag{10.8}$$

and that the following property holds:

(Pii) for each a. c. function $v : [0,1] \to R^n$ satisfying $|v(0) - \bar{x}|, |v(1) - \bar{x}| \leq \delta_2$,

$$I^f(0,1,v) \geq f(\bar{x},0) - \delta_1/8.$$

By (8.1), (9.1), Propositions 2.3 and 2.5 and the choice of the topology in $\mathfrak{A}$ there exists a neighborhood $\mathcal{U}_1$ of g in $\mathfrak{A}$ such that for each $h \in \mathcal{U}_1$,

$$|\inf(\phi_h) - \inf(\phi_g)| \leq \delta_1/16. \tag{10.9}$$

By Theorem 8.1, there exist $L \geq 1$, $\delta_3 > 0$ and a neighborhood $\mathcal{U}_2$ of g in $\mathfrak{A}$ such that the following property holds:

(Piii) for each $T > 2L$, each $h \in \mathcal{U}_2$ and each a. c. function $u : [0,T] \to R^n$ which satisfies

$$I^f(0,T,u) + h(u(0), u(T)) \leq \sigma(f,h,0,T) + \delta_3$$

we have

$$|u(t) - \bar{x}| \leq \delta_2 \text{ for all } t \in [L, T-L]. \tag{10.10}$$

By Theorem 1.4, there exist an (f)-overtaking optimal function $u_1 : [0,\infty) \to R^n$ and an $(\bar{f})$-overtaking optimal function $u_2 : [0,\infty) \to R^n$ such that

$$\pi^f(u_1(0)) + \pi^{\bar{f}}(u_2(0)) + g(u_1(0), u_2(0)) = \inf(\phi_g). \tag{10.11}$$

In view of (A4),

$$\lim_{t\to\infty} u_i(t) = \bar{x}, \ i = 1, 2. \tag{10.12}$$

Relation (10.12) implies that there exists $\tau_1 > 0$ such that for $i = 1, 2$,

$$|u_i(t) - \bar{x}| \le \delta_2 \text{ for all } t \ge \tau_1. \tag{10.13}$$

There exists a neighborhood $\mathcal{U}$ of g in $\mathfrak{A}$ such that

$$\mathcal{U} \subset \cap_{i=0}^2 \mathcal{U}_i, \tag{10.14}$$

and

$$|h(u_1(0), u_2(0)) - g(u_1(0), u_2(0))| \le \delta_1/8 \text{ for all } h \in \mathcal{U}. \tag{10.15}$$

Choose $\delta > 0$ and $T_0 > 0$ such that

$$\delta < (64(L + \tau_1 + \tau_0 + 8))^{-1} \min\{\delta_1, \delta_2, \delta_3\}, \tag{10.16}$$

$$T_0 > 4L + 4\tau_0 + 4\tau_1 + 8. \tag{10.17}$$

Assume that

$$T \ge T_0, \ h \in \mathcal{U} \tag{10.18}$$

and that an a. c. function $v : [0, T] \to R^n$ satisfies

$$I^f(0, T, v) + h(v(0), v(T)) \le \sigma(f, h, 0, T) + \delta. \tag{10.19}$$

By (10.8), (10.14), (10.16), (10.17), (10.19) and property (Piii), relation (10.10) holds.

Define

$$\begin{gathered}\tilde{v}(t) = u_1(t), \ t \in [0, L + \tau_0 + \tau_1 + 3],\\ \tilde{v}(t) = v(t), \ t \in [L + \tau_0 + \tau_1 + 4, T - L - \tau_0 - \tau_1 - 4],\\ \tilde{v}(t) = u_2(T - t), \ t \in [T - L - \tau_0 - \tau_1 - 3, T],\\ \tilde{v}(t + L + \tau_0 + \tau_1 + 3) = u_1(L + \tau_0 + \tau_1 + 3)\\ +t[v(L + \tau_0 + \tau_1 + 4) - u_1(L + \tau_0 + \tau_1 + 3)], \ t \in (0, 1),\\ \tilde{v}(t + T - L - \tau_0 - \tau_1 - 4) = v(T - L - \tau_0 - \tau_1 - 4)\\ +t[u_2(L + \tau_0 + \tau_1 + 3) - v(T - L - \tau_0 - \tau_1 - 4)], \ t \in (0, 1).\end{gathered} \tag{10.20}$$

Clearly, $\tilde{v} : [0, T] \to R^n$ is an a. c. function. In view of (10.10), (10.13), (10.17), (10.18) and (10.20),

$$|\tilde{v}(L + \tau_0 + \tau_1 + 3) - \bar{x}| = |u_1(L + \tau_0 + \tau_1 + 3) - \bar{x}| \le \delta_2, \tag{10.21}$$

$$|\tilde{v}(L + \tau_0 + \tau_1 + 4) - \bar{x}| = |v(L + \tau_0 + \tau_1 + 4) - \bar{x}| \le \delta_2, \tag{10.22}$$

$$|\tilde{v}(T - L - \tau_0 - \tau_1 - 4) - \bar{x}| = |v(T - L - \tau_0 - \tau_1 - 4) - \bar{x}| \le \delta_2. \tag{10.23}$$

$$|\tilde{v}(T - L - \tau_0 - \tau_1 - 3) - \bar{x}| = |u_2(L + \tau_0 + \tau_1 + 3) - \bar{x}| \le \delta_2. \tag{10.24}$$

It follows from (10.20), (10.21) and (10.22) that for all $t \in (L + \tau_0 + \tau_1 + 3, L + \tau_0 + \tau_1 + 4)$,

$$|\tilde{v}'(t)| = |v(L + \tau_0 + \tau_1 + 4) - u_1(L + \tau_0 + \tau_1 + 3)| \le 2\delta_2. \tag{10.25}$$

Relations (10.20), (10.23) and (10.24) imply that that for all $t \in (T - L - \tau_0 - \tau_1 - 4, T - L - \tau_0 - \tau_1 - 3)$,

$$|\tilde{v}'(t)| \le |u_2(L + \tau_0 + \tau_1 + 3) - v(T - L - \tau_0 - \tau_1 - 4)| \le 2\delta_2. \tag{10.26}$$

By (10.20), (10.21), (10.23), (10.25), (10.26) and the choice of δ_2 (see (10.8)), for all

$$t \in (L + \tau_0 + \tau_1 + 3, L + \tau_0 + \tau_1 + 4) \cup (T - L - \tau_0 - \tau_1 - 4, T - L - \tau_0 - \tau_1 - 3),$$

$$|f(\tilde{v}(t), \tilde{v}'(t)) - f(\bar{x}, 0)| \le \delta_1/8. \tag{10.27}$$

It follows from (10.19) and (10.20) that

$$\begin{aligned}
\delta \ge I^f(0, T, v) + h(v(0), v(T)) - (I^f(0, T, \tilde{v}) + h(\tilde{v}(0), \tilde{v}(T)))\\
= h(v(0), v(T)) - h(u_1(0), u_2(0))\\
+I^f(0, L + \tau_0 + \tau_1 + 4, v) + I^f(T - L - \tau_0 - \tau_1 - 4, T, v)\\
-I^f(0, L + \tau_0 + \tau_1 + 4, \tilde{v}) - I^f(T - L - \tau_0 - \tau_1 - 4, T, \tilde{v})\\
= h(v(0), v(T)) - h(u_1(0), u_2(0))\\
+I^f(0, L + \tau_0 + \tau_1 + 3, v) + I^f(L + \tau_0 + \tau_1 + 3, L + \tau_0 + \tau_1 + 4, v)\\
+I^f(T - L - \tau_0 - \tau_1 - 4, T - L - \tau_0 - \tau_1 - 3, v) + I^f(T - L - \tau_0 - \tau_1 - 3, T, v)\\
-I^f(0, L + \tau_0 + \tau_1 + 3, \tilde{v}) - I^f(L + \tau_0 + \tau_1 + 3, L + \tau_0 + \tau_1 + 4, \tilde{v})\\
-I^f(T-L-\tau_0-\tau_1-4, T-L-\tau_0-\tau_1-3, \tilde{v}) - I^f(T-L-\tau_0-\tau_1-3, T, \tilde{v}).
\end{aligned} \tag{10.28}$$

By (10.8), (10.10), (10.17) and property (Pii),

$$I^f(L + \tau_0 + \tau_1 + 3, L + \tau_0 + \tau_1 + 4, v) \ge f(\bar{x}, 0) - \delta_1/8, \tag{10.29}$$

$$I^f(T - L - \tau_0 - \tau_1 - 4, T - L - \tau_0 - \tau_1 - 3, v) \ge f(\bar{x}, 0) - \delta_1/8. \tag{10.30}$$

In view of (10.27),

$$I^f(L + \tau_0 + \tau_1 + 3, L + \tau_0 + \tau_1 + 4, \tilde{v}) \le f(\bar{x}, 0) + \delta_1/8, \tag{10.31}$$

$$I^f(T - L - \tau_0 - \tau_1 - 4, T - L - \tau_0 - \tau_1 - 3, \tilde{v}) \le f(\bar{x}, 0) + \delta_1/8. \tag{10.32}$$

It follows from (10.20) and (10.28)-(10.32) that

$$\begin{aligned}
\delta \ge h(v(0), v(T)) - h(u_1(0), u_2(0))\\
+I^f(0, L + \tau_0 + \tau_1 + 3, v) + I^f(T - L - \tau_0 - \tau_1 - 3, T, v)\\
-I^f(0, L + \tau_0 + \tau_1 + 3, \tilde{v}) - I^f(T - L - \tau_0 - \tau_1 - 3, T, \tilde{v}) - \delta_1/2\\
= h(v(0), v(T)) - h(u_1(0), u_2(0))\\
+I^f(0, L + \tau_0 + \tau_1 + 3, v) + I^f(T - L - \tau_0 - \tau_1 - 3, T, v)\\
-I^f(0, L + \tau_0 + \tau_1 + 3, u_1) - I^{\bar{f}}(0, L + \tau_0 + \tau_1 + 3, u_2) - \delta_1/2.
\end{aligned}$$

This implies that

$$\begin{aligned}
h(v(0), v(T)) + I^f(0, L + \tau_0 + \tau_1 + 3, v) + I^f(T - L - \tau_0 - \tau_1 - 3, T, v)\\
\le \delta + \delta_1/2 + h(u_1(0), u_2(0))\\
+I^f(0, L + \tau_0 + \tau_1 + 3, u_1) + I^{\bar{f}}(0, L + \tau_0 + \tau_1 + 3, u_2).
\end{aligned} \tag{10.33}$$

Set

$$y(t) = v(T - t), \ t \in [0, T]. \tag{10.34}$$

Relations (10.33) and (10.34) imply that

$$h(v(0),y(0))+I^f(0,L+\tau_0+\tau_1+3,v)+I^{\bar{f}}(0,L+\tau_0+\tau_1+3,y)$$
$$\leq \delta+\delta_1/2+h(u_1(0),u_2(0))$$
$$+I^f(0,L+\tau_0+\tau_1+3,u_1)+I^{\bar{f}}(0,L+\tau_0+\tau_1+3,u_2). \tag{10.35}$$

By (10.10), (10.19), (10.34), (10.35), Propositions 2.1 and 2.2 and the overtaking optimality of u_1, u_2,

$$\pi^f(v(0))+\pi^{\bar{f}}(v(T))+h(v(0),v(T))$$
$$+I^f(0,\tau_0,v)-\tau_0 f(\bar{x},0)-\pi^f(v(0))+\pi^f(v(\tau_0))$$
$$+I^{\bar{f}}(0,\tau_0,y)-\tau_0 f(\bar{x},0)-\pi^{\bar{f}}(y(0))+\pi^{\bar{f}}(y(\tau_0))$$
$$\leq \pi^f(v(0))+\pi^{\bar{f}}(y(0))+h(v(0),y(0))$$
$$+I^f(0,L+\tau_0+\tau_1+3,v)-(L+\tau_0+\tau_1+3)f(\bar{x},0)$$
$$-\pi^f(v(0))+\pi^f(v(L+\tau_0+\tau_1+3))$$
$$+I^{\bar{f}}(0,L+\tau_0+\tau_1+3,y)-(L+\tau_0+\tau_1+3)f(\bar{x},0)$$
$$-\pi^{\bar{f}}(y(0))+\pi^{\bar{f}}(y(L+\tau_0+\tau_1+3))$$
$$\leq \pi^f(v(0))+\pi^{\bar{f}}(y(0))+h(u_1(0),u_2(0))$$
$$+I^f(0,L+\tau_0+\tau_1+3,u_1)+I^{\bar{f}}(0,L+\tau_0+\tau_1+3,u_2)+\delta+\delta_1/2$$
$$-2(L+\tau_0+\tau_1+3)f(\bar{x},0)-\pi^f(v(0))+\pi^f(v(L+\tau_0+\tau_1+3))$$
$$-\pi^{\bar{f}}(y(0))+\pi^{\bar{f}}(y(L+\tau_0+\tau_1+3))$$
$$\leq \pi^f(v(0))+\pi^{\bar{f}}(y(0))+h(v(0),y(0))+h(u_1(0),u_2(0))-h(v(0),y(0))$$
$$+\pi^f(u_1(0))-\pi^f(u_1(L+\tau_0+\tau_1+3))+\pi^{\bar{f}}(u_2(0))-\pi^{\bar{f}}(u_2(L+\tau_0+\tau_1+3))$$
$$-\pi^f(v(0))+\pi^f(v(L+\tau_0+\tau_1+3))$$
$$-\pi^{\bar{f}}(y(0))+\pi^{\bar{f}}(y(L+\tau_0+\tau_1+3))+\delta+\delta_1/2$$
$$\leq h(u_1(0),u_2(0))+\pi^f(u_1(0))+\pi^{\bar{f}}(u_2(0))$$
$$-\pi^f(u_1(L+\tau_0+\tau_1+3))-\pi^{\bar{f}}(u_2(L+\tau_0+\tau_1+3))$$
$$+\pi^f(v(L+\tau_0+\tau_1+3))-\pi^{\bar{f}}(y(L+\tau_0+\tau_1+3))+\delta+\delta_1/2. \tag{10.36}$$

In view of the choice of δ_2 (see (10.6), (10.7)), (10.10), (10.13) and (10.34),

$$|\pi^{\bar{f}}(\xi)|,\ |\pi^f(\xi)|\leq \delta_1/8,$$

(10.37)
$$\xi\in\{u_1(L+\tau_0+\tau_1+3),\ u_2(L+\tau_0+\tau_1+3),\ v(L+\tau_0+\tau_1+3),\ y(L+\tau_0+\tau_1+3)\}.$$

Relations (10.15) and (10.18) imply that

$$|h(u_1(0),u_2(0))-g(u_1(0),u_2(0))|\leq \delta_1/8. \tag{10.38}$$

It follows from (10.9), (10.11), (10.14), (10.18), (10.34), (10.36) and (10.38) that

$$\pi^f(v(0))+\pi^{\bar{f}}(y(0))+h(v(0),y(0))$$
$$+I^f(0,\tau_0,v)-\tau_0 f(\bar{x},0)-\pi^f(v(0))+\pi^f(v(\tau_0))$$
$$+I^{\bar{f}}(0,\tau_0,y)-\tau_0 f(\bar{x},0)-\pi^{\bar{f}}(y(0))+\pi^{\bar{f}}(y(\tau_0))$$
$$\leq \delta_1/8+\pi^f(u_1(0))+\pi^{\bar{f}}(u_2(0))+g(u_1(0),u_2(0))+\delta_1+\delta$$

$$\leq \delta_1/8 + \delta_1 + \delta + \inf(\phi_g) \leq 5\delta_1/4 + \delta + \inf(\phi_h). \tag{10.39}$$

By (10.39),

$$\pi^f(v(0)) + \pi^{\bar{f}}(y(0)) + h(v(0), y(0)) - \inf(\phi_h)$$
$$+I^f(0, \tau_0, v) - \tau_0 f(\bar{x}, 0) - \pi^f(v(0)) + \pi^f(v(\tau_0))$$
$$+I^{\bar{f}}(0, \tau_0, y) - \tau_0 f(\bar{x}, 0) - \pi^{\bar{f}}(y(0)) + \pi^{\bar{f}}(y(\tau_0)) \leq 2\delta_1.$$

Together with Proposition 2.1 this implies that

$$\pi^f(v(0)) + \pi^{\bar{f}}(y(0)) + h(v(0), y(0)) \leq \inf(\phi_h) + 2\delta_1, \tag{10.40}$$

$$I^f(0, \tau_0, v) - \tau_0 f(\bar{x}, 0) - \pi^f(v(0)) + \pi^f(v(\tau_0)) \leq 2\delta_1, \tag{10.41}$$

$$I^{\bar{f}}(0, \tau_0, y) - \tau_0 f(\bar{x}, 0) - \pi^{\bar{f}}(y(0)) + \pi^{\bar{f}}(y(\tau_0)) \leq 2\delta_1. \tag{10.42}$$

By (10.14), (10.18), (10.40)-(10.42) and property (Pi) there exist an (f)-overtaking optimal function $u_1^* : [0, \infty) \to R^n$ and an $(\bar{f})$-overtaking optimal function $u_2^* : [0, \infty) \to R^n$ such that

$$(u_1^*(0), u_2^*(0)) \in \mathcal{D}(f, g)$$

and for all $t \in [0, \tau_0]$,

$$|v(t) - u_1^*(t)| \leq \epsilon, \; |v(T - t) - u_2^*(t)| = |y(t) - u_2^*(t)| \leq \epsilon.$$

Theorem 8.2 is proved. □

References

[1] Brian D. O. Anderson and John B. Moore, *Linear optimal control*, Prentice-Hall, Inc., Englewood Cliffs, N.J., 1971. MR0335000 (48 #13318)

[2] Sergey M. Aseev and Vladimir M. Veliov, *Maximum principle for infinite-horizon optimal control problems with dominating discount*, Dyn. Contin. Discrete Impuls. Syst. Ser. B Appl. Algorithms **19** (2012), no. 1-2, 43–63. MR2918248 (2012m:49035)

[3] J. Baumeister, A. Leitão, and G. N. Silva, *On the value function for nonautonomous optimal control problems with infinite horizon*, Systems Control Lett. **56** (2007), no. 3, 188–196, DOI 10.1016/j.sysconle.2006.08.011. MR2296645 (2007k:49060)

[4] Joël Blot, *Infinite-horizon Pontryagin principles without invertibility*, J. Nonlinear Convex Anal. **10** (2009), no. 2, 177–189. MR2548290 (2010j:91150)

[5] J. Blot and P. Cartigny, *Optimality in infinite-horizon variational problems under sign conditions*, J. Optim. Theory Appl. **106** (2000), no. 2, 411–419, DOI 10.1023/A:1004611816252. MR1788932 (2001h:49004)

[6] Joël Blot and Naïla Hayek, *Infinite-horizon optimal control in the discrete-time framework*, Springer Briefs in Optimization, Springer, New York, 2014. MR3154452

[7] Ido Bright, *A reduction of topological infinite-horizon optimization to periodic optimization in a class of compact 2-manifolds*, J. Math. Anal. Appl. **394** (2012), no. 1, 84–101, DOI 10.1016/j.jmaa.2012.03.042. MR2926206

[8] Pierre Cartigny and Philippe Michel, *On a sufficient transversality condition for infinite horizon optimal control problems*, Automatica J. IFAC **39** (2003), no. 6, 1007–1010, DOI 10.1016/S0005-1098(03)00060-8. MR2139173

[9] V. Gaitsgory, S. Rossomakhine, and N. Thatcher, *Approximate solution of the HJB inequality related to the infinite horizon optimal control problem with discounting*, Dyn. Contin. Discrete Impuls. Syst. Ser. B Appl. Algorithms **19** (2012), no. 1-2, 65–92. MR2918249

[10] Naïla Hayek, *Infinite horizon multiobjective optimal control problems in the discrete time case*, Optimization **60** (2011), no. 4, 509–529, DOI 10.1080/02331930903480352. MR2801395 (2012e:49049)

[11] Héctor Jasso-Fuentes and Onésimo Hernández-Lerma, *Characterizations of overtaking optimality for controlled diffusion processes*, Appl. Math. Optim. **57** (2008), no. 3, 349–369, DOI 10.1007/s00245-007-9025-6. MR2407317 (2009d:49032)

[12] Vassili Kolokoltsov and Wei Yang, *Turnpike theorems for Markov games*, Dyn. Games Appl. **2** (2012), no. 3, 294–312, DOI 10.1007/s13235-012-0047-6. MR2967552

[13] A. Leizarowitz, *Infinite horizon autonomous systems with unbounded cost*, Appl. Math. Optim. **13** (1985), no. 1, 19–43, DOI 10.1007/BF01442197. MR778419 (86g:49002)

[14] Arie Leizarowitz, *Tracking nonperiodic trajectories with the overtaking criterion*, Appl. Math. Optim. **14** (1986), no. 2, 155–171, DOI 10.1007/BF01442233. MR863337 (87m:49032)

[15] Arie Leizarowitz and Victor J. Mizel, *One-dimensional infinite-horizon variational problems arising in continuum mechanics*, Arch. Rational Mech. Anal. **106** (1989), no. 2, 161–194, DOI 10.1007/BF00251430. MR980757 (90b:49007)

[16] Valeriya Lykina, Sabine Pickenhain, and Marcus Wagner, *Different interpretations of the improper integral objective in an infinite horizon control problem*, J. Math. Anal. Appl. **340** (2008), no. 1, 498–510, DOI 10.1016/j.jmaa.2007.08.008. MR2376171 (2008m:49009)

[17] V. L. Makarov and A. M. Rubinov, *Mathematical theory of economic dynamics and equilibria*, Springer-Verlag, New York-Heidelberg, 1977. Translated from the Russian by Mohamed El-Hodiri. MR0439072 (55 #11973)

[18] Agnieszka B. Malinowska, Natália Martins, and Delfim F. M. Torres, *Transversality conditions for infinite horizon variational problems on time scales*, Optim. Lett. **5** (2011), no. 1, 41–53, DOI 10.1007/s11590-010-0189-7. MR2747272

[19] Moshe Marcus and Alexander J. Zaslavski, *The structure of extremals of a class of second order variational problems* (English, with English and French summaries), Ann. Inst. H. Poincaré Anal. Non Linéaire **16** (1999), no. 5, 593–629, DOI 10.1016/S0294-1449(99)80029-8. MR1712568 (2000h:49008)

[20] Lionel W. McKenzie, *Turnpike theory*, Econometrica **44** (1976), no. 5, 841–865. MR0459543 (56 #17735)

[21] Boris S. Mordukhovich, *Optimal control and feedback design of state-constrained parabolic systems in uncertainty conditions*, Appl. Anal. **90** (2011), no. 6, 1075–1109, DOI 10.1080/00036811003735840. MR2801265 (2012g:49078)

[22] Boris S. Mordukhovich and Ilya Shvartsman, *Optimization and feedback control of constrained parabolic systems under uncertain perturbations*, Optimal control, stabilization and nonsmooth analysis, Lecture Notes in Control and Inform. Sci., vol. 301, Springer, Berlin, 2004, pp. 121–132, DOI 10.1007/978-3-540-39983-4_8. MR2079680 (2005d:49045)

[23] Eladio Ocaña Anaya, Pierre Cartigny, and Patrice Loisel, *Singular infinite horizon calculus of variations. Applications to fisheries management*, J. Nonlinear Convex Anal. **10** (2009), no. 2, 157–176. MR2548289 (2010g:49041)

[24] Sabine Pickenhain, Valeriya Lykina, and Marcus Wagner, *On the lower semicontinuity of functionals involving Lebesgue or improper Riemann integrals in infinite horizon optimal control problems*, Control Cybernet. **37** (2008), no. 2, 451–468. MR2472885 (2009j:49004)

[25] Alessio Porretta and Enrique Zuazua, *Long time versus steady state optimal control*, SIAM J. Control Optim. **51** (2013), no. 6, 4242–4273, DOI 10.1137/130907239. MR3124890

[26] P. A. Samuelson, *A catenary turnpike theorem involving consumption and the golden rule*, Amer. Econom. Rev. **55** (1965), 486–496.

[27] Alexander J. Zaslavski, *Turnpike properties in the calculus of variations and optimal control*, Nonconvex Optimization and its Applications, vol. 80, Springer, New York, 2006. MR2164615 (2006f:49002)

[28] Alexander J. Zaslavski, *A turnpike result for a class of problems of the calculus of variations with extended-valued integrands*, J. Convex Anal. **15** (2008), no. 4, 869–890. MR2489618 (2010a:49042)

[29] Alexander J. Zaslavski, *Structure of approximate solutions of variational problems with extended-valued convex integrands*, ESAIM Control Optim. Calc. Var. **15** (2009), no. 4, 872–894, DOI 10.1051/cocv:2008053. MR2567250 (2010m:49029)

[30] Alexander J. Zaslavski, *Optimal solutions for a class of infinite horizon variational problems with extended-valued integrands*, Optimization **59** (2010), no. 2, 181–197, DOI 10.1080/02331930701764114. MR2765456 (2011k:49062)

[31] Alexander J. Zaslavski, *Two turnpike results for a continuous-time optimal control system*, Complex analysis and dynamical systems IV. Part 1, Contemp. Math., vol. 553, Amer. Math. Soc., Providence, RI, 2011, pp. 305–317, DOI 10.1090/conm/553/10947. MR2884389

[32] Alexander J. Zaslavski, *The existence and structure of approximate solutions of dynamic discrete time zero-sum games*, J. Nonlinear Convex Anal. **12** (2011), no. 1, 49–68. MR2816408 (2012d:91038)

[33] Alexander J. Zaslavski, *Structure of solutions of variational problems*, Springer Briefs in Optimization, Springer, New York, 2013. MR3026270

[34] Alexander J. Zaslavski, *Turnpike phenomenon and infinite horizon optimal control*, Springer Optimization and Its Applications, vol. 99, Springer, Cham, 2014. MR3306947

[35] Alexander J. Zaslavski, *Structure of Solutions of Variational Problems with Extended-Valued Integrands in the Regions Close to the Endpoints*, Set-Valued Var. Anal. **23** (2015), no. 2, 249–275, DOI 10.1007/s11228-014-0294-3. MR3342733

Department of Mathematics, The Technion – Israel Institute of Technology, Technion City, Haifa 32000, Israel

E-mail address: ajzasl@techunix.technion.ac.il